VOLUME SIX HUNDRED AND FOURTEEN

METHODS IN ENZYMOLOGY

Biological NMR Part A

METHODS IN ENZYMOLOGY

VOLUME SIX HUNDRED AND FOURTEEN

METHODS IN ENZYMOLOGY

Biological NMR Part A

Edited by

A. JOSHUA WAND

Johnson Research Foundation and the Department of Biochemistry & Biophysics;
Graduate Group in Biochemistry & Molecular Biophysics, University of Pennsylvania, Perelman School of Medicine, Philadelphia, PA, United States

ELSEVIER

ACADEMIC PRESS
An imprint of Elsevier

Academic Press is an imprint of Elsevier
50 Hampshire Street, 5th Floor, Cambridge, MA 02139, United States
525 B Street, Suite 1650, San Diego, CA 92101, United States
The Boulevard, Langford Lane, Kidlington, Oxford OX5 1GB, United Kingdom
125 London Wall, London, EC2Y 5AS, United Kingdom

First edition 2019

ISBN: 978-0-12-813860-1
ISSN: 0076-6879

For information on all Academic Press publications
visit our website at https://www.elsevier.com/books-and-journals

Publisher: Fiona Pattison
Acquisition Editor: Ashlie Jackman
Editorial Project Manager: Shellie Bryant
Production Project Manager: Divya Krishna Kumar
Cover Designer: Greg Harris

Typeset by SPi Global, India

CONTENTS

CONTRIBUTORS

Frederic H-T. Allain
Department of Biology, ETH Zürich, Institute of Molecular Biology and Biophysics, Zürich, Switzerland

Haribabu Arthanari
Department of Biological Chemistry and Molecular Pharmacology; Dana Farber Cancer Institute, Harvard Medical School, Boston, MA, United States

Taysir K. Bader
University of Minnesota, Twin Cities, Minneapolis, MN, United States

Adam W. Barb
Roy J. Carver Department of Biochemistry, Biophysics and Molecular Biology, Iowa State University, Ames, IA, United States

Frank Bernhard
Institute of Biophysical Chemistry, Centre for Biomolecular Magnetic Resonance, J.W. Goethe-University, Frankfurt am Main, Germany

Kelly P. Brock
Department of Systems Biology, Harvard Medical School, Boston, MA, United States

Sebastien Campagne
Department of Biology, ETH Zürich, Institute of Molecular Biology and Biophysics, Zürich, Switzerland

Lindsay D. Clark
Department of Biophysics; Molecular Biophysics Graduate Program, The University of Texas Southwestern Medical Center, Dallas, TX, United States

Igor Dikiy
Structural Biology Initiative, CUNY Advanced Science Research Center, New York, NY, United States

Mark D. Distefano
University of Minnesota, Twin Cities, Minneapolis, MN, United States

Volker Dötsch
Institute of Biophysical Chemistry, Centre for Biomolecular Magnetic Resonance, J.W. Goethe-University, Frankfurt am Main, Germany

Daniel J. Falconer
Roy J. Carver Department of Biochemistry, Biophysics and Molecular Biology, Iowa State University, Ames, IA, United States

Andreas O. Frank
Structural and Biophysical Chemistry, Global Discovery Chemistry, Novartis Institutes for BioMedical Research, Emeryville, CA, United States

Alexandra Frommlet
Structural and Biophysical Chemistry, Global Discovery Chemistry, Novartis Institutes for BioMedical Research, Emeryville, CA, United States

Kevin H. Gardner
Structural Biology Initiative, CUNY Advanced Science Research Center; Department of Chemistry and Biochemistry, City College of New York; Biochemistry, Chemistry and Biology Ph.D. Programs, Graduate Center, City University of New York, New York, NY, United States

Starla D. Glover
Department of Biochemistry & Biophysics, University of Pennsylvania Perelman School of Medicine, Philadelphia, PA, United States

Erik Henrich
Institute of Biophysical Chemistry, Centre for Biomolecular Magnetic Resonance, J.W. Goethe-University, Frankfurt am Main, Germany

Yuanpeng Janet Huang
Center for Advanced Biotechnology and Medicine; Department of Molecular Biology and Biochemistry, Rutgers, The State University of New Jersey, Piscataway, NJ, United States

Sven G. Hyberts
Department of Biological Chemistry and Molecular Pharmacology, Harvard Medical School, Boston, MA, United States

Masayori Inouye
Center for Advanced Biotechnology and Medicine; Department of Biochemistry and Molecular Biology, Robert Wood Johnson Medical School, Rutgers, The State University of New Jersey, Piscataway, NJ, United States

Yojiro Ishida
CABM, Rutgers University; Center for Advanced Biotechnology and Medicine; Department of Biochemistry and Molecular Biology, Robert Wood Johnson Medical School, Rutgers, The State University of New Jersey, Piscataway, NJ, United States

Charalampos G. Kalodimos
Department of Structural Biology, St Jude Children's Research Hospital, Memphis, TN, United States

Miroslav Krepl
Institute of Biophysics of the Czech Academy of Sciences, Brno; Department of Physical Chemistry, Faculty of Science, Regional Centre of Advanced Technologies and Materials, Palacky University Olomouc, Olomouc, Czech Republic

Ganesan Senthil Kumar
Department of Chemistry and Biochemistry, University of Arizona, Tucson, AZ, United States

Aisha Laguerre
Institute of Biophysical Chemistry, Centre for Biomolecular Magnetic Resonance, J.W. Goethe-University, Frankfurt am Main, Germany

Andreas Lingel
Structural and Biophysical Chemistry, Global Discovery Chemistry, Novartis Institutes for BioMedical Research, Emeryville, CA, United States; Global Discovery Chemistry, Novartis Institutes for BioMedical Research, Novartis Campus, Basel, Switzerland

Frank Löhr
Institute of Biophysical Chemistry, Centre for Biomolecular Magnetic Resonance, J.W. Goethe-University, Frankfurt am Main, Germany

Debora S. Marks
Department of Systems Biology, Harvard Medical School, Boston, MA, United States

Julija Mezhyrova
Institute of Biophysical Chemistry, Centre for Biomolecular Magnetic Resonance, J.W. Goethe-University, Frankfurt am Main, Germany

Yoan R. Monneau
University Grenoble Alpes, CEA, CNRS, IBS, Grenoble, France

Gaetano T. Montelione
Center for Advanced Biotechnology and Medicine; Department of Molecular Biology and Biochemistry; Department of Biochemistry and Molecular Biology, Robert Wood Johnson Medical School, Rutgers, The State University of New Jersey, Piscataway, NJ, United States

Santrupti Nerli
Department of Chemistry and Biochemistry; Department of Computer Science, University of California Santa Cruz, Santa Cruz, CA, United States

Rebecca Page
Department of Chemistry and Biochemistry, University of Arizona, Tucson, AZ, United States

Wolfgang Peti
Department of Chemistry and Biochemistry, University of Arizona, Tucson, AZ, United States

Andrew Proudfoot
Structural and Biophysical Chemistry, Global Discovery Chemistry, Novartis Institutes for BioMedical Research, Emeryville, CA, United States

Todd M. Rappe
University of Minnesota, Twin Cities, Minneapolis, MN, United States

Scott Robson
Department of Biological Chemistry and Molecular Pharmacology, Harvard Medical School, Boston, MA, United States

Julien Roche
Roy J. Carver Department of Biochemistry, Biophysics and Molecular Biology, Iowa State University, Ames, IA, United States

Daniel M. Rosenbaum
Department of Biophysics; Molecular Biophysics Graduate Program, The University of Texas Southwestern Medical Center, Dallas, TX, United States

Paolo Rossi
Department of Structural Biology, St Jude Children's Research Hospital, Memphis, TN, United States

Christian Roumestand
Centre de Biochimie Structural CNRS Université de Montpellier UMR, Montpellier, France

Catherine A. Royer
Department of Biological Sciences, Rensselaer Polytechnic Institute, Troy, NY, United States

Chris Sander
Department of Cell Biology, Harvard Medical School and cBio Center, Dana-Farber Cancer Institute, Boston, MA, United States

Nikolaos G. Sgourakis
Department of Chemistry and Biochemistry, University of California Santa Cruz, Santa Cruz, CA, United States

Jiri Sponer
Institute of Biophysics of the Czech Academy of Sciences, Brno; Department of Physical Chemistry, Faculty of Science, Regional Centre of Advanced Technologies and Materials, Palacky University Olomouc, Olomouc, Czech Republic

Ganesh P. Subedi
Roy J. Carver Department of Biochemistry, Biophysics and Molecular Biology, Iowa State University, Ames, IA, United States

Gurla V.T. Swapna
Center for Advanced Biotechnology and Medicine; Department of Molecular Biology and Biochemistry, Rutgers, The State University of New Jersey, Piscataway, NJ, United States

Cecilia Tommos
Department of Biochemistry & Biophysics, University of Pennsylvania Perelman School of Medicine, Philadelphia, PA, United States

Gianlugi Veglia
University of Minnesota, Twin Cities, Minneapolis, MN, United States

Gerhard Wagner
Department of Biological Chemistry and Molecular Pharmacology, Harvard Medical School, Boston, MA, United States

Ulrich Weininger
Institute of Physics, Biophysics, Martin-Luther-University Halle-Wittenberg, Halle, Germany

Youlin Xia
Department of Structural Biology, St Jude Children's Research Hospital, Memphis, TN, United States

PREFACE

Nuclear magnetic resonance spectroscopy of biological molecules continues to be a central tool in biology, biochemistry, and biophysics. Since its first applications to problems in chemistry and biology, NMR spectroscopy has shown continual evolution and episodic advancements in its capabilities and applications. Much has revolved around optimization of samples to facilitate full exploitation of the physics of NMR in the context of macromolecules such as proteins. In this Part A of a two-part series on biological NMR spectroscopy, recent advancements in isotopic enrichment of proteins are presented and protocols described. Three chapters center on methyl-labeling strategies. A contribution by Gardner and colleagues describes recent advances in the production of methyl-labeled membrane proteins in yeast, while Lingel describes the strategies for methyl labeling of proteins recombinantly expressed in *E. coli*. A chapter by Weininger focuses on recent advances in optimal isotopic labeling of aromatic rings in proteins for detailed studies of dynamics by NMR. A chapter by Dötsch describes the production of membrane proteins in various membrane mimetics, while Kalodimos and colleagues provide strategies for using auxotrophs for the production of methyl labeling of large soluble proteins. Over the past decade the role of posttranslational modifications in the control of protein activity has become ever clearer and central to studies of eukaryotic biology. Three chapters of this Part A present strategies and protocols for the preparation of biophysical quantities of phosphorylated, lipidated, and glycosylated proteins by Peti, Distefano, Barb, and their colleagues, respectively. A chapter by Glover and Tommos examines an often-overlooked potential artifact arising from the commonly used purification by tags based on affinity for divalent cations such as nickel.

Continual improvements in analytical and computational strategies have accompanied these advances in sample preparation. In their chapter, Wagner and colleagues describe the recent advances in nonuniform data acquisition and processing. Kalodimos and colleagues present protocols for an elegant algorithm for obtaining methyl assignments. Allain and coworkers carry the protein–nucleic acid torch by showing how NMR-based structural restraints can be combined with molecular dynamics simulations to characterize protein–RNA complexes. Also highlighted in this volume are chapters illustrating the use of knowledge-based approaches in conjunction with more traditional structural restraints for the determination of structural models of proteins and their complexes. Nerli and Sgourakis present detailed protocols for the use of the CS-Rosetta structure determination software package, while Montelione and colleagues describe how knowledge of "evolutionary couplings" can be combined with sparse NMR data sampling to yield structural models. Finally, taking NMR deep into the territory of experimental biophysics, Royer and colleagues present the issues and strategies for employing high pressure in the context of solution NMR spectroscopy.

Part B in the series turns to more specialized topics. Sharp provides an overview of how direct connections between NMR observables and molecular dynamics simulations can be made. Chapters by Stetz and colleagues and Iwahara and colleagues center on recent developments in NMR relaxation studies of ps–ns motion of methyl-bearing and basic side chains in proteins, respectively. The former contribution also presents protocols for the use of the recently established NMR-based dynamical proxy for conformational entropy. Palmer and Koss contribute a comprehensive chapter on the use of chemical exchange phenomena manifested in NMR observables to characterize motions in the μs–ms time regime. There are three chapters describing various routes to characterization of protein hydration, which has resisted a comprehensive experimental description for many decades. Prosser and colleagues present an overview of the use of ^{19}F NMR to characterize the protein structural ensemble including its hydration. Jorge and coworkers utilize the advantageous physical and chemical properties of the nanoscale water core of reverse micelles to probe protein hydration dynamics using the classical NOE and ROE approach. As a companion chapter, Fuglestad and colleagues provide protocols and strategies for optimal encapsulation of single protein molecules within reverse micelles. Franck and Han introduce the Overhauser dynamic nuclear polarization (DNP) technique for the characterization of the hydration shell of macromolecules. To represent the dramatic advances of the general DNP approach to sensitivity enhancement in solid-state NMR, Frederick shows how DNP can be used to probe the structure of biomolecules in their native cellular environments. Amplifying the recent advances in solid-state NMR, Carlomagno and colleagues describe various approaches to the structural characterization of RNA. The volume then turns to small molecule characterization and drug discovery by NMR. Brüschweiler and coworkers show how a hybrid approach using NMR and mass spectrometry can enable resolution of the complex mixtures of metabolomics. Kim and Hilty further illustrate the power of DNP in the search for small molecule binders to proteins. Huang and Leung and Angulo and colleagues present the recent advances in utilizing water and small molecule magnetization transfer to proteins in drug discovery. Finally, Montelione and colleagues describe an intriguing strategy for identifying the ligand–receptor pairs that underlie complex eukaryotic biology and will be the ever-increasing focus of structural biology by NMR in the future.

On behalf of the contributing authors and myself I thank you for your interest in these volumes and hope that you find them useful.

A. Joshua Wand
Philadelphia, Pennsylvania

CHAPTER ONE

Selective Methyl Labeling of Proteins: Enabling Structural and Mechanistic Studies As Well As Drug Discovery Applications by Solution-State NMR

Andrew Proudfoot*, Andreas O. Frank*, Alexandra Frommlet*, Andreas Lingel*,†,1

*Structural and Biophysical Chemistry, Global Discovery Chemistry, Novartis Institutes for BioMedical Research, Emeryville, CA, United States
†Global Discovery Chemistry, Novartis Institutes for BioMedical Research, Novartis Campus, Basel, Switzerland
[1]Corresponding author: e-mail address: andreas.lingel@novartis.com

Contents

Abstract

Escherichia coli expression protocols for selective labeling of methyl groups in proteins have been essential in expanding the size range of targets that can be studied by biomolecular NMR. Based on the initial work achieving selective labeling of isoleucine, leucine, and valine residues, additional methods were developed over the past years

Methods in Enzymology, Volume 614
ISSN 0076-6879
https://doi.org/10.1016/bs.mie.2018.08.035

which enabled the individual and/or simultaneous combinatorial labeling of all methyl containing amino acids. Together with the introduction of new methyl-optimized NMR experiments, this now allows the detailed characterization of protein–ligand interactions as well as mechanistic and dynamic processes of protein–protein complexes up to 1 MDa in size. In this chapter, we provide a general introduction to selective labeling of proteins using *E. coli*-based expression systems, describe the considerations taken into account prior to the selective labeling of a protein, and include the protocols used to produce such proteins. An overview of applications using selectively labeled proteins with an emphasis on examples relevant to the drug discovery process is then presented.

1. INTRODUCTION TO SELECTIVE LABELING

Solution-state nuclear magnetic resonance (NMR) spectroscopy is routinely used for the elucidation of protein structures along with the study of dynamics and the analysis of the interactions of biomolecules with their respective ligands (Gardner & Kay, 1998; Garrett et al., 1997; Yamazaki, Lee, Arrowsmith, Muhandiram, & Kay, 1994). Using traditional protocols for expression of uniformly labeled protein in *Escherichia coli*, studies of small proteins up to ~25–30 kDa in molecular weight became fully enabled in the early 1990s. At the same time, it was realized in the biomolecular NMR field that the development of new methods was necessary to be able to study larger macromolecules, since many proteins of biological relevance exceed this size limit. One of the major breakthroughs to enable such work with larger proteins has been the development of more elaborate isotope labeling schemes, in particular perdeuteration and selective labeling of methyl group containing amino acid side chains (Gardner & Kay, 1997; Venters et al., 1995). To achieve this, the protein expressing *E. coli* cells were grown in D_2O-based media with selectively labeled, deuterated carbon sources and precursors instead of protonated growth medium. This results in the incorporation of deuterium into positions otherwise occupied by protons, while simultaneously introducing protonated, ^{13}C-labeled methyl groups as NMR-active labels at defined positions in a selective manner. These methyl groups now experience significantly slower relaxation rates due to the reduced number of relaxation pathways in their spatial vicinity as well as lack of one-bond $^{13}C—^{13}C$ coupling interactions (as present in uniformly carbon-labeled proteins). Together with the fact that methyl groups are generally good probes because of the presence of three magnetically equivalent protons and the fast rotation around the bond that connects the carbon atom of a methyl group with the neighboring atom, this results in high quality and well-resolved spectra with excellent signal to noise and sharp lines. Combining such labeling schemes with the use of improved hardware (i.e., cryoprobes and higher field spectrometers) as well as the development and optimization of new pulse programs, i.e., the methyl-TROSY (Tugarinov, Hwang, Ollerenshaw, & Kay, 2003; Yamazaki et al., 1994) has enabled detailed studies of the structure and function of proteins and complexes over a wide molecular weight range which extends up to 1 MDa (Jiang & Kalodimos, 2017; Rosenzweig & Kay, 2014; Sprangers & Kay, 2007).

The labeling schemes discussed in this chapter have been developed using the *E. coli*-based expression system, mainly due to the fact that *E. coli* can be conveniently manipulated, can be grown in minimal and D_2O-based media, and can metabolize a range of provided sugars or amino acid precursors. This allows easy utilization of the endogenous biosynthetic pathways to incorporate NMR-active nuclei at specific positions of individual amino acids. However, it is well known that many membrane and human proteins along with protein complexes cannot easily be expressed in *E. coli* and thus need to be produced using an eukaryotic expression system. Such methods for the production of labeled proteins in insect and mammalian cells have been developed but require the growth medium to be directly supplemented with labeled amino acids which significantly increase the cost of the experiment. When compared to *E. coli*-based expression systems, these methods benefit from a reduction in amino acid scrambling since the final amino acids are used instead of precursors. However, the level of isotope incorporation can sometimes be as low as 75%–80%, depending on the labeling scheme used (Skora, Shrestha, & Gossert, 2015), which is in contrast to *E. coli*-based methodologies where almost complete isotopic labeling can be consistently achieved. The methods involved with selective labeling of proteins in eukaryotic cells are beyond the scope of this chapter, however detailed protocols are available elsewhere (Skora et al., 2015).

The selective labeling of [^{1}H,^{13}C] methyl groups in an otherwise deuterated background was pioneered by Gardner, Goto, and Kay in the 1990s when they published work describing the selective incorporation of ^{1}H and ^{13}C isotopes into multiple methyl containing amino acids (Gardner & Kay, 1997; Goto, Gardner, Mueller, Willis, & Kay, 1999). By supplementing the *E. coli* growth medium with 2-keto-3-methyl-^{13}C-butyric-4-^{13}C,3-D acid ([3-^{2}H] α-ketoisovalerate) and [3,3-$^{2}H_2$] α-ketobutyrate, the authors were able to show the selective incorporation of labeled methyl groups at the δ- and γ-positions of leucine and valine residues along with the δ^1-position of isoleucine, respectively. Based on this work, protocols have been developed to selectively label all methyl containing amino acids, either by providing selectively labeled precursors or through the direct incorporation of selectively labeled amino acids (Fig. 1) (Ayala, Sounier, Usé, Gans, & Boisbouvier, 2009;

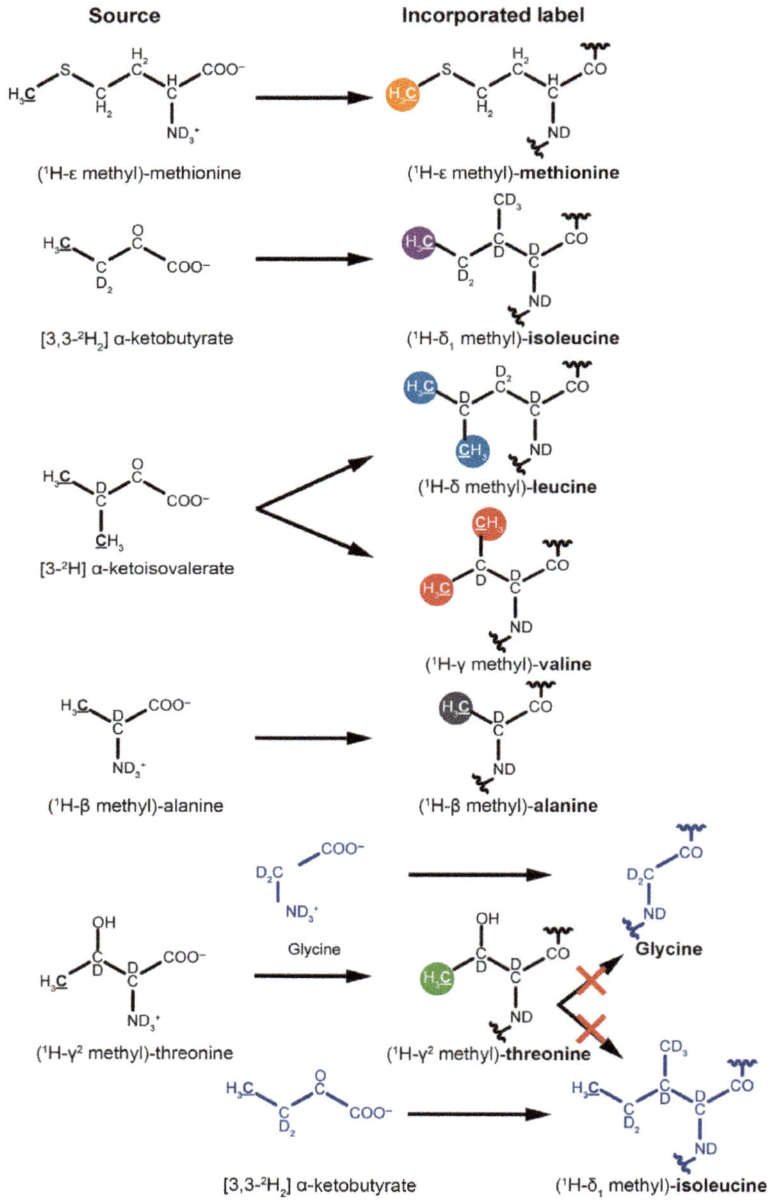

Fig. 1 Chemical structures of the precursors and amino acids used to supplement deuterated *E. coli* growth medium prior to induction (*left*) and the subsequent positions of the selective label in the respective amino acid (*colored spheres*) once incorporated into the protein (*right*). *Carbons that are underlined* correspond to those that are ^{13}C labeled. When selectively labeling threonine, deuterated glycine and [3,3-$^{2}H_2$] α-ketobutyrate must also be added to the growth medium at the same time to prevent scrambling of the selective label into glycine and isoleucine residues (*blue*).

Fischer et al., 2007; Gelis et al., 2007; Isaacson et al., 2007; Stoffregen, Schwer, Renschler, & Wiesner, 2012; Velyvis, Ruschak, & Kay, 2012). These methods have now also been expanded to enable the simultaneous selective labeling of multiple methyl-bearing amino acids, including MILV, ILVA, MILVT, and MILVAT-labeling schemes (Gelis et al., 2007; Godoy-Ruiz, Guo, & Tugarinov, 2010; Gross, Gelev, & Wagner, 2003; Lichtenecker, Ludwiczek, Schmid, & Konrat, 2004; Proudfoot, Frank, Ruggiu, Mamo, & Lingel, 2016; Saio, Guan, Rossi, Economou, & Kalodimos, 2014; Tugarinov & Kay, 2003a; Velyvis, Schachman, & Kay, 2009).

In addition to facilitating work with larger proteins, the selective methyl labeling of amino acid side chains has enabled a detailed characterization of the interactions made with their binding partners. This is achieved through the presence of a high density of probes around the interaction sites, in particular if multiple amino acids are labeled simultaneously (see below), and the favorable signal intensity compared to the well-established analysis of amide groups alone. This observation was described in a study by Stockman and Dalvit in 2002, which analyzed a sample of 191 nondegenerate X-ray protein–ligand costructures to probe the proximity of NMR-active nuclei to the bound ligand (Stockman & Dalvit, 2002). In the sample set, they found that 82% of the ligands have a heavy atom within 6 Å of a backbone ^{15}N atom, while 92% of the ligands have a heavy atom within the same radius of a methyl group. This observation is also exemplified in Fig. 2, which shows a costructure of the antibacterial target CoaD with compound **1** bound to the active site (Proudfoot, Bussiere, & Lingel, 2017). When analyzing a sphere of 6 Å radius around all heavy atoms of the bound ligand, only five amide groups are present (Fig. 2A), however

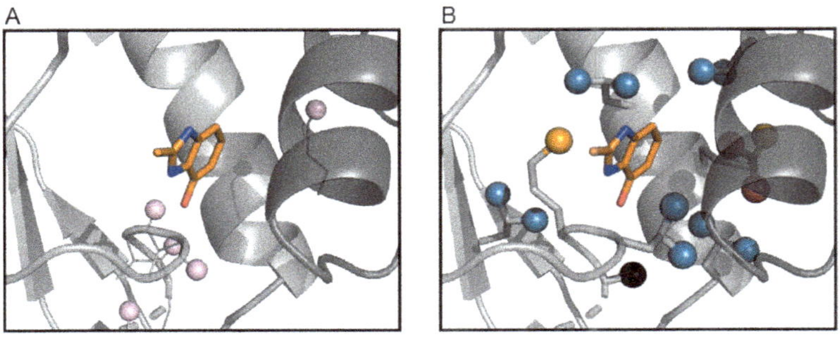

Fig. 2 Crystal structure of *E. coli* CoaD showing all (A) amide groups and (B) methyl groups within a 6 Å radius of any heavy atom of the bound compound **1** (PDB: 6B7A, methyl groups are colored according to Fig. 1).

when assessing the number of methyl groups within the same sphere, 15 groups are identified (Fig. 2B) that could be used as direct probes for ligand interaction studies.

2. LABELING SCHEMES

2.1 Commonly Used Selective Labeling Schemes

When selectively labeling a protein, the type of labeling scheme used will depend on the scientific questions that need to be answered, in addition to the types of experiments that will be performed with that protein. Working with protein that is labeled at a single amino acid is sometimes sufficient to answer the question posed, and while each individual methyl-bearing amino acid can be independently selectively labeled (except threonine), in reality work performed with methyl-labeled protein is generally done by labeling multiple residues simultaneously. In Fig. 3, we show examples of individual and combinatorial labeling schemes that are routinely used in our laboratory and described in the literature (Fischer et al., 2007; Goto et al., 1999; Proudfoot et al., 2016; Velyvis et al., 2012). As seen in the illustration of the location of amino acid methyl groups in the corresponding X-ray structure of our example protein CoaD (Fig. 3A, C, E, G), as the number of labels increases, the density of NMR-active probes evenly increases on both the surface and in the hydrophobic core of the protein.

The initial work performed by Gardner, Goto, and Kay pioneered the use of selectively labeled ILV protein. This continues to be a major labeling scheme used to this day since these residues are among the most abundant methyl containing amino acids and are generally found in ligand binding sites. Monitoring of these NMR signals can thus be used to characterize structure and function of a large portion of the protein, e.g., locating the binding site by assessing chemical shift perturbations (CSPs) upon addition of an interaction partner, or assessing dynamics by measuring relaxation dispersion. Extension of the selective labeling scheme to include threonine and methionine residues enables different areas and interactions to be probed due to the polar hydroxyl group and flexible aliphatic thioether moiety in their respective side chains, which can be beneficial for the characterization of protein–ligand interfaces and events at the protein surface.

In addition to the amino acid composition of the target, the location and enrichment of amino acids at functional sites of interest within the protein

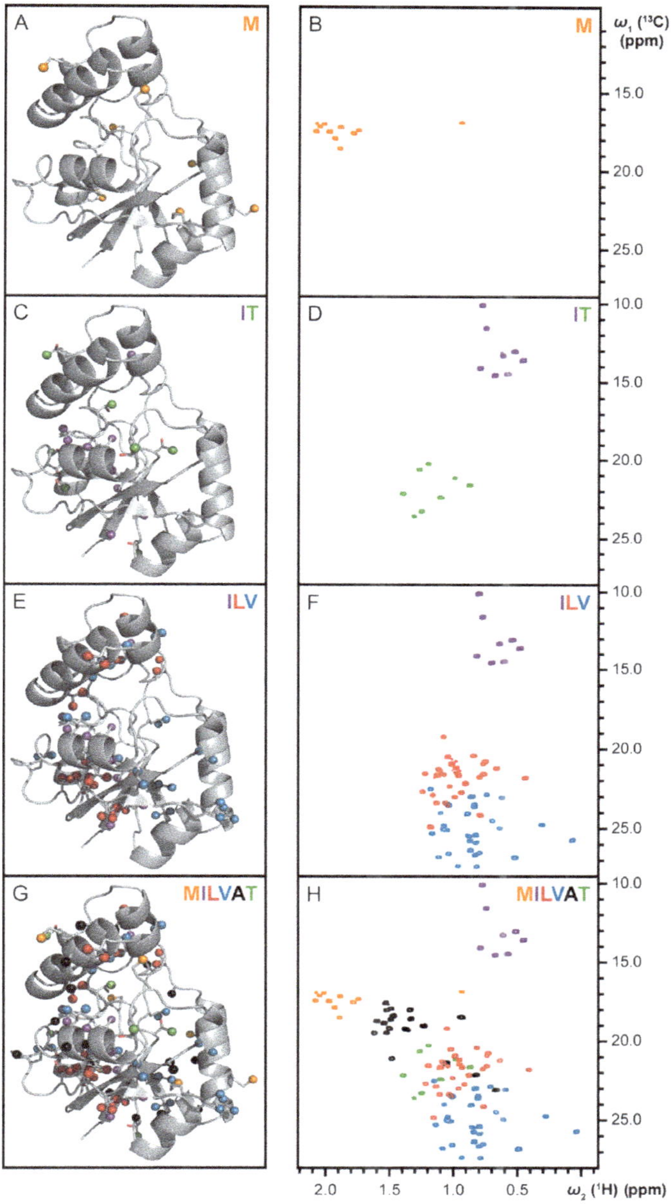

Fig. 3 (A, C, E, G) Structure of one *E. coli* CoaD subunit (PDB: 5JBN) with methyl groups highlighted corresponding to the 2D [^{13}C,^{1}H]-SOFAST HMQC spectra acquired on (B) M-, (D) IT-, (F) ILV-, and (H) MILVAT-labeled proteins.

are also important factors when considering different labeling schemes. It is therefore highly valuable if some form of structural information is available, namely an X-ray or NMR structure, or a reliable homology model. Although labeling of multiple residues is possible and can be beneficial, the extension of labeling schemes come at a cost, both financially and spectral quality. Incorporation of additional methyl labels into the protein increases the number of probes present, which can result in an increase in the linewidth of the peaks present in the spectrum due to the availability of additional relaxation pathways. In particular when working with larger proteins or protein complexes, this can have a detrimental effect on the spectral quality (Proudfoot et al., 2016). Also, while methyl groups of the different amino acids generally occupy distinct chemical shift areas in the spectrum, there is overlap observed between signals of threonine $C\gamma^2$ methyl groups with mainly the leucine and valine $C\delta$ and $C\gamma$ geminal pairs (Fig. 3B, D, F, H), which can make interpretation of these spectra challenging. The benefits of the increased information content obtained through additional labels must therefore be carefully balanced with the drawbacks on a case-by-case basis when deciding on the labeling scheme to be used.

2.2 Scrambling of Selective Labels

When growing *E. coli* in minimal medium, the endogenous metabolic pathways used for amino acid biosynthesis utilize simple molecules such as glucose and ammonia as carbon and nitrogen sources, respectively. In addition, intermediates of these pathways can be provided and will be used to produce the respective amino acids. Where possible, published labeling schemes preferentially use a selectively labeled precursor of the amino acid biosynthetic pathway instead of the selectively labeled amino acid itself (Goto et al., 1999; Lichtenecker, Coudevylle, Konrat, & Schmid, 2013), as this significantly reduces the cost. However, due to the presence of shared biosynthetic pathways, the products and/or intermediates from one pathway are sometimes converted into intermediates for other pathways. In addition, selective labeling schemes do not result in blocking the various amino acid biosynthetic pathways but work by providing sufficient quantities of precursors or amino acids so that they are used preferentially over endogenous stocks. This can be problematic when selectively labeling an amino acid like threonine, which can be converted into both glycine and α-ketobutyrate, a precursor in the isoleucine biosynthetic pathway (Velyvis et al., 2012)

(Fig. 1). In *E. coli*, threonine is selectively labeled through the direct incorporation of the labeled amino acid into the growth medium prior to induction. Scrambling of the selectively labeled threonine into other biosynthetic pathways, e.g., scrambling into the isoleucine $C\gamma^2$ position, will reduce the amount of amino acid available for direct incorporation, and will eventually result in the expression of protein containing both selectively labeled and unlabeled threonine. This has an impact on the quality of the resulting spectrum, as there will be both a reduction in intensity of the threonine peaks, and the scrambled selective label will appear in other amino acids throughout the spectrum. To prevent this, deuterated glycine and [3,3-$^2H_2$] ^{13}C α-ketobutyrate are supplemented into the medium at the same time as the selectively labeled threonine so they can be preferentially used over the threonine metabolites (Velyvis et al., 2012) (Fig. 1).

2.3 Discrimination of Leucine and Valine Geminal Pairs

Methyl containing amino acids make up approximately 35% of all residues in proteins, with the prochiral methyl groups of leucine and valine accounting for more than 50% of the total methyl groups (Miller, Janin, Lesk, & Chothia, 1987). When working with large proteins, this can result in significant overlap in the leucine–valine region of the resulting spectrum. Leucine Cδ and valine Cγ geminal pairs are simultaneously labeled through the addition of [3-^{2}H] α-ketoisovalerate to *E. coli* growth medium prior to induction (Fig. 4A). In the case of valine, the precursor is directly converted into the amino acid, however α-ketoisovalerate is converted into α-ketoisocaproate to be utilized as a precursor for leucine biosynthesis (Kerfah, Plevin, Sounier, Gans, & Boisbouvier, 2015). Since these resonances have overlapping chemical shifts ranges, it is often desirable to differentiate between the two sets of peaks through the selective labeling of the individual residues. Lichtenecker et al. developed a methodology for the synthesis of selectively labeled α-ketoisocaproate and showed that *E. coli* can directly incorporate this into leucine residues only (Lichtenecker et al., 2013). Although this was shown to be very effective, such selectively labeled precursors are often not commercially available and require the user to synthesize them in-house. An alternative approach to differentiate between the two sets of resonance pairs is to selectively label the valine Cγ geminal pair only, which can be achieved through the addition of nutrient-rich supplements

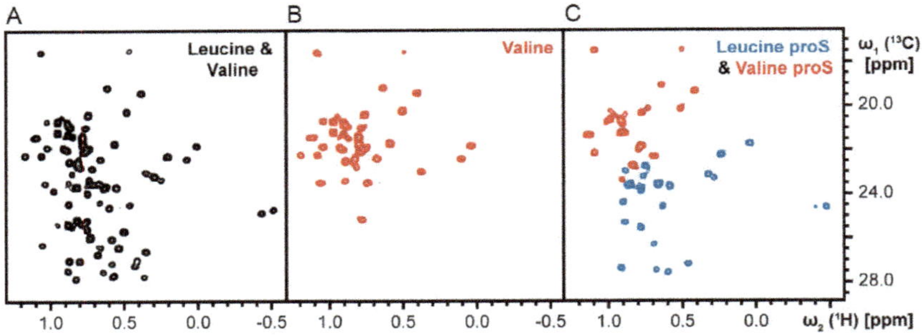

Fig. 4 2D [^{13}C,^{1}H]-SOFAST HMQC of the viral 150 kDa, truncated homopentamer BK VP1 protein. Spectra of VP1 expressed with (A) [3-^{2}H] α-ketoisovalerate to selectively label leucine Cδ and valine Cγ methyl groups (B) [3-^{2}H] α-ketoisovalerate and 20% of the nutrient-rich Bioexpress medium to suppress incorporation of label into leucine Cδ methyl groups and (C) (*S*)-2-acetolactate to selectively label *pro-S* leucine Cδ and valine Cγ methyl groups (obtained from NMR-Bio #DLAM-LVproS and used according to supplier instructions).

(e.g., Bioexpress or Isogro) into the medium. Tzeng et al. showed that incorporation of the unlabeled leucine present in such supplements was sufficient to suppress incorporation of [3-^{2}H] α-ketoisovalerate into the leucine biosynthesis pathway, resulting in a sample with only the valine Cγ geminal methyl groups labeled (Fig. 4B) (Tzeng, Pai, & Kalodimos, 2012). While the addition of 10% Bioexpress (volume percent of the medium volume) to the growth medium was reported by Tzeng et al. to be sufficient to suppress leucine Cδ labeling, our experience suggests that volumes up to 20% may sometimes be required to fully achieve this.

An alternative approach to reduce the complexity of the leucine/valine spectrum is to selectively label only one of the *pro-R* or *pro-S* methyl groups of the geminal pairs. Supplementing the medium with deuterated α-ketoisovalerate that is protonated at a single methyl group produces a racemic mixture of *pro-R*- and *pro-S*-labeled leucine and valine residues. This eliminates interactions between the prochiral methyls of the isopropyl group and thus increases the resolution and sensitivity of the methyl-TROSY spectra while maintaining all peaks in the spectrum. Use of specifically methyl-labeled acetolactate enables the stereospecific labeling of a single geminal methyl group of leucine and valine residues, which reduces complexity in the spectrum and has an increase in sensitivity of a factor of 1.6 ± 0.4 (Fig. 4C) (Gans et al., 2010).

3. EXPERIMENTAL SECTION

3.1 General and Practical Considerations for Selective Methyl Labeling of Proteins

Primary considerations for the choice of the appropriate labeling scheme are the presence and location of methyl containing amino acids in the area of interest as well as the required degree of isotope enrichment. Certain components of the medium used for selective labeling of methyl groups can be quite costly, for example the cost of 1 L medium for expression of a protein which is fully deuterated and selectively labeled on all methyl containing amino acids (MILVAT) is approximately $2600 at the time of publication (2018). Such labeling schemes are beneficial when performing very detailed studies, such as the comprehensive assignment of methyl groups and the identification of a maximum number of intermolecular NOEs between a labeled protein and its binding partner. For all other types of applications, compromises can be made on the degree of deuteration of supplemented amino acids as well as the type of carbon source used. The use of ^{12}C-deuterated glucose contributes $1000 to the total cost, and although protein production must be performed in a deuterated medium, accepting some level of proton incorporation into other parts of the protein (Fig. 5) can significantly reduce the financial burden.

Most recipes for minimal media suggest the use of glucose as the sole carbon source and we have successfully produced protein preparations with all variations (^{13}C/^{12}C, ^{1}H/^{2}H) of the sugar. As a consequence of the direct metabolism of glucose by *E. coli*, specific atoms of the molecule are incorporated into defined positions of the produced amino acids, including the α and β positions along with certain positions on the aromatic rings. Therefore, the use of protonated or deuterated glucose has a significant impact on the observed NMR spectrum. This is exemplified in Fig. 5A, which shows the 1D ^{1}H NMR spectrum of selectively MILVAT-labeled protein expressed with unlabeled glucose (black) and ^{12}C, ^{2}H glucose (red).

In addition to the peaks corresponding to the selectively labeled methyl groups, the protein prepared with nondeuterated glucose contains a considerable number of resonances corresponding to protons attached to aliphatic and aromatic carbons, which appear in the spectrum between 2.5 and 4.5, and 6 and 8 ppm, respectively. The impact of this incorporation into undesired positions is shown in Fig. 5B and C, which show the $[\omega_2(^{13}C),\omega_3(^{1}H)]$

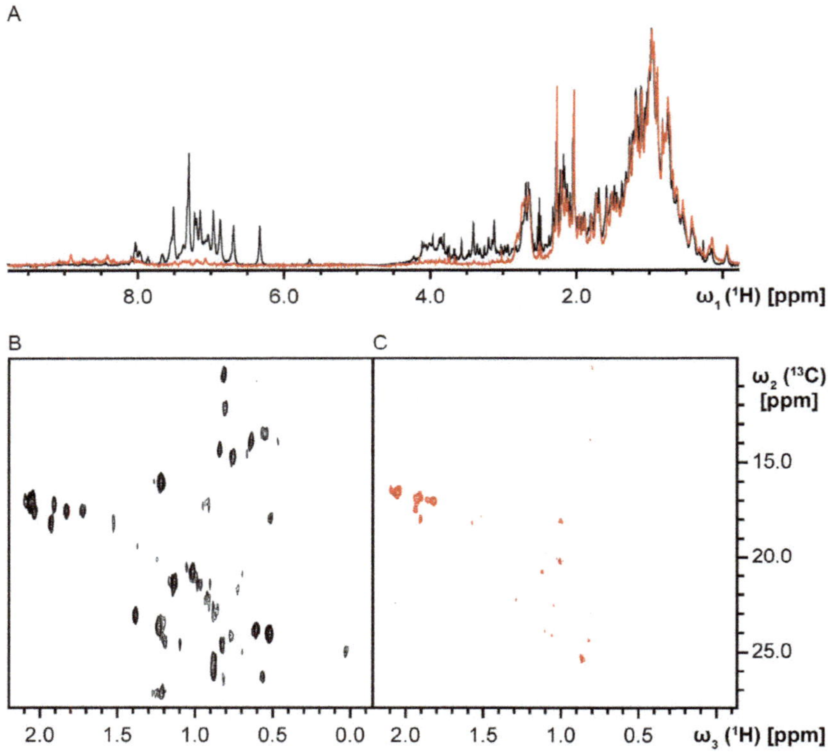

Fig. 5 NMR spectra of MILVAT-labeled *E. coli* CoaD expressed using unlabeled (*black*) and ^{12}C—^{2}H (*red*) glucose. (A) Superposition of 1D ^{1}H NMR spectra of apo CoaD. (B/C) 2D [$\omega_2(^{13}C),\omega_3(^{1}H)$] planes from the F_2-edited, F_3-filtered ^{13}C-resolved [$^{1}H,^{1}H$]-NOESY acquired in the presence of an unlabeled small molecule using CoaD expressed with (B) unlabeled (*black*) or (C) ^{12}C—^{2}H (*red*) glucose.

planes from F_2-edited, F_3-filtered ^{13}C-resolved [$^{1}H,^{1}H$]-NOESY spectra acquired with the two differently expressed MILVAT-labeled proteins in the presence of an unlabeled small molecule. In addition to observing intermolecular NOEs to the unlabeled small molecule (red), the sample produced with protonated glucose also exhibits intra- and inter-residual NOEs between the labeled methyl groups and other unlabeled positions within the protein. Since many of the chemical shifts of protons in small molecules fall within the 2.5–4.5 and 6.0–8.0 ppm ranges, this makes differentiating between the intermolecular NOEs arising from protein and small molecule resonances near impossible due to the overlap.

As mentioned earlier, the expression of selectively labeled protein is performed in a deuterated environment to maximize the incorporation of deuterium into the sample. While *E. coli* cells are able to grow in a fully deuterated medium despite the fact that water plays such an essential role in biochemical reactions, the cells have to be first acclimated to the different environment. This adaption is usually performed over several days while stepwise increasing the amount of D_2O in the growth medium. At each step of this adaption process, *E. coli* often exhibit a long lag phase prior to entering into exponential growth and it is also common for doubling times to increase significantly when compared to those observed in protonated, nutrient-rich media. Although this process reliably results in an adapted strain of *E. coli*, the final density of cells obtained following protein induction is generally lower than that observed with nutrient-rich or unlabeled minimal media, which has a negative impact on yields. As an alternative approach, *E. coli* cells can be grown up initially in a nonlabeled medium to increase the cell density. The cells are then gently pelleted and washed to remove all traces of the unlabeled medium, before being resuspended in a minimal medium which was prepared in D_2O and contains all the relevant labeling sources and precursors. If using this technique, great care should be taken to ensure that the *E. coli* cells have depleted all of their endogenous stocks of nutrient-rich amino acid precursors prior to induction, otherwise an uncontrollable combination of selectively labeled and unlabeled protein will be produced.

3.2 Equipment, Materials, and Recipes

3.2.1 Equipment

- Water bath.
- 37°C shaker and incubator.
- UV–vis spectrophotometer.
- Accurate balance.
- Lyophilizer.
- Refrigerated centrifuge.
- NMR spectrometer.

3.2.2 Materials

- Chemically competent expression strain of *E. coli*.
- SOC medium.
- LB agar plates containing appropriate antibiotics.
- LB medium containing appropriate antibiotics.
- M9 medium containing appropriate antibiotics.

- 99% (v/v) deuterium oxide.
- Selectively labeled precursors and amino acids (see Table 1).
- Isopropyl β-D-1-thiogalactopyranoside (IPTG) in D_2O.

3.2.3 Recipes for Media Components

SOC Medium

2% (w/v) tryptone, 0.5% (w/v) yeast extract, 10 m*M* NaCl, 2.5 m*M* KCl, 10 m*M* $MgCl_2$, 10 m*M* $MgSO_4$, and 20 m*M* glucose, pH 7.0.

LB Medium

1% (w/v) tryptone, 0.5% (w/v) yeast extract, 171 m*M* NaCl, pH 7.5.

Table 1 Quantity of Precursor or Amino Acid Used to Selectively Label Amino Acids

Amino Acid Labeled	Precursor or Amino Acid Used	Quantity (mg/L)	Comment and References
Methionine (Cε)	L-Methionine-(methyl-^{13}C)[a]	250	Fischer et al. (2007) and Gelis et al. (2007)
Isoleucine (Cδ^1)	α-Ketobutyric acid-4-^{13}C,3,3-D_2[b]	70	Gardner and Kay (1997)
Leucine (Cδ^1,Cδ^2)	2-Keto-3-methyl-^{13}C-butyric-4-^{13}C,3-D acid[b]	120	Labeling of leucine Cδ atoms only, can be achieved by supplementing the M9 medium with 20% Bioexpress[a] (Goto et al., 1999; Tzeng et al., 2012)
Valine (Cγ^1,Cγ^2)	2-Keto-3-methyl-^{13}C-butyric-4-^{13}C,3-D acid[b]	120	Labeling of valine and leucine occurs simultaneously using the same precursor (Goto et al., 1999)
Alanine (Cβ)	L-Alanine (3-^{13}C,2-D)[a]	100	Ayala et al. (2009)
Threonine (Cγ^2)	L-Threonine (4-^{13}C;2,3-D_2)[a]	50	Isoleucine Cδ^1 atoms need to be labeled at the same time as threonine Cγ^2 (Velyvis et al., 2012)
	Glycine (D_5)[a]	100	

[a]Selectively labeled amino acids were purchased from Cambridge Isotope Laboratory (product numbers: CLM-206, CDLM-9307, DLM-280, CGM-1000-D-S, and CDLM-8649).
[b]Selectively labeled precursors for labeling isoleucine, leucine, and valine residues were purchased from Sigma (product numbers: 589063 and 589276).
Table is adapted from Proudfoot, A., Frank, A. O., Ruggiu, F., Mamo, M., & Lingel, A. (2016). Facilitating unambiguous NMR assignments and enabling higher probe density through selective labeling of all methyl containing amino acids. *Journal of Biomolecular NMR, 65*(1), 15–27. https://doi.org/10.1007/s10858-016-0032-2.

M9 Medium

50 m*M* Na_2HPO_4, 50 m*M* KH_2PO_4, 5 m*M* Na_2SO_4, 2 m*M* $MgSO_4$, 0.3% (w/v) D-glucose,[a] 0.1% (w/v) NH_4Cl.[a] Trace elements and vitamin solution (Sigma R7256) are prepared in the correct ratio of H_2O and D_2O at a final 0.1 × concentration with respect to the stock concentrations, and subsequently sterilized by passing the solutions through a 0.2 μm filter.

1000× Trace Metals Mixture

The following is added to 36 mL water:

		MW	1× Conc.
50 mL	0.1 *M* $FeCl_3$ in 0.12 *M* HCl	270.30	50 μ*M* Fe^{3+}
2 mL	1 *M* $CaCl_2$	110.99	20 μ*M* Ca^{2+}
1 mL	1 *M* $MnCl_2$-$4H_2O$	197.91	10 μ*M* Mn^{2+}
1 mL	1 *M* $ZnSO_4$-$7H_2O$	287.56	10 μ*M* Zn^{2+}
1 mL	0.2 *M* $CoCl_2$-$6H_20$	237.95	2 μ*M* Co^{2+}
2 mL	0.1 *M* $CuCl_2$-$2H_2O$	170.486	2 μ*M* Cu^{2+}
1 mL	0.2 *M* $NiCl_2$-$6H_2O$	237.72	2 μ*M* Ni^{2+}
2 mL	0.1 *M* Na_2MoO_4-$5H_2O$	241.98	2 μ*M* Mo^{6+}
2 mL	0.1 *M* Na_2SeO_3-$5H_2O$	263.03	2 μ*M* Se^{4+}
2 mL	0.1 *M* H_3BO_3	61.83	2 μ*M* B^{3+}

All stock solutions of the individual metals should be autoclaved, except for the 0.1 *M* $FeCl_3$ in 1/100 volume concentrated HCl. Trace metal solution should be stored at room temperature.

3.3 Protocol for Selective Methyl Labeling of Proteins in *E. coli*

1. *E. coli* cells of an expression strain are transformed with 1 μL of a 50 ng/μL plasmid solution, prepared for the expression of the gene of interest, in 20 μL reaction volume. First, the reaction is incubated on ice for 2°min prior to heat shocking the cells at 42°C for 30 s. Following the heat shock, the cells are placed back on ice and subsequently incubated with 200 μL SOC medium at 37°C for 1°h. The transformed *E. coli* cells are then plated

[a] Unlabeled or labeled carbon or nitrogen sources can be used in this protocol. Unlabeled sugars and salts were obtained from Sigma, while labeled variants were purchased from Cambridge Isotope Laboratory (product numbers: NLM-467, DLM-2062, and CDLM-3813).

onto LB agar plates that contain the antibiotic which matches the selection marker of the plasmid system and incubated overnight at 37°C.

2. A single colony is picked from the agar plate and used to inoculate 5 mL of LB medium containing the appropriate antibiotic. The culture is then incubated at 37°C with agitation at 250 rpm until the cell density measured at 600 nm (OD_{600}) is 0.6. At this stage, 1 mL of the LB culture is used to inoculate 50 mL of fully protonated, unlabeled M9 minimal medium containing the appropriate antibiotic. The M9 medium starter culture is then incubated overnight at 37°C with agitation at 250 rpm.
3. 5 mL of the overnight starter culture is used to inoculate 50 mL of M9 medium containing 30% D_2O (while always keeping the appropriate antibiotic present) and incubated overnight at 37°C with agitation at 250 rpm. This process should be repeated with M9 overnight starter cultures containing 70% and 100% D_2O, with the 30% starter culture being used to inoculate the 70% growth medium, and the 70% starter culture being used to inoculate the 100% growth medium.
4. Stock solutions of IPTG and antibiotic which will be added to the 100% D_2O growth medium should be either prepared by dissolving the respective solids in D_2O, or by lyophilization of the H_2O-based stock solution and subsequent resuspension in D_2O. This can be done multiple times to remove hydration water, however only necessary if the highest levels of perdeuteration are desired.
5. Using the 100% overnight starter culture, 1 L of 100% D_2O M9 medium containing the appropriate isotopically labeled carbon and nitrogen sources (generally glucose and ammonium chloride) and relevant antibiotic is inoculated to achieve a final starting cell density at 600 nm of 0.1. Cells cultures are grown by incubation at 37°C with agitation at 250 rpm. Care should be taken to monitor the optical density of the growth overtime as *E. coli* doubling times will vary depending on carbon source and the degree of isotope enrichment.
6. The relevant quantities of precursors and amino acid precursors required for the desired labeling scheme (Table 1) are weighed out and resuspended as a mixture in D_2O.
7. When an OD_{600} of 0.6 is achieved in the large-scale growth, the temperature of the incubator is decreased to that required for induction, and the precursor solution is added to the growth medium. Afterward, the cells are incubated for an additional hour at 250 rpm, which is a crucial step to ensure that the *E. coli* cells absorb the required precursors prior to induction of the protein.

8. Protein expression is induced by adding the appropriate amount of an IPTG solution prepared in D_2O to achieve the desired final concentration (generally between 0.05 and 1 m*M*).
9. Following induction of the protein, the cells are incubated for the desired expression time (typically 4–16 h) and then harvested by centrifugation at 8000 rpm for 15 min. Long expression times can result in scrambling since the provided precursors and/or amino acids are depleted overtime. The cell pellet is then usually frozen until the protein is purified. Following protein expression, only labile protons can exchange with the solvent in the aqueous environment, so buffers used for protein purification do not need to be deuterated.

4. APPLICATIONS

4.1 Resonance Assignment

As generally true for NMR-based methods, the availability of resonance assignments is often key to establish a full understanding of the studies performed with selectively labeled proteins. When working with larger proteins, the assignment strategies pursued for small proteins are usually not applicable due to fast signal relaxation and overlap or can take a significant investment of time and resources. The exception to this is the situation when the large protein or complex can be broken down into smaller subunits and assigned using the "divide and conquer" approach (Sprangers & Kay, 2007). The obtained assignments are then transferred back to the larger complex with the assumption that chemical shifts in the core of the smaller subunits will not be affected by the disruption of the complex, and that they are structurally very similar when present in the complex or in isolation. Furthermore, in cases where backbone assignments are available or can be achieved, the methyl groups can be assigned by through-bond experiments correlating the methyl groups of amino acids to the backbone atoms via the aliphatic side chains (Kerfah, Hamelin, Boisbouvier, & Marion, 2015; Tugarinov & Kay, 2003a, 2003b). However, having partial assignments of a protein or even just knowledge of the amino acid type of an observed methyl peak is often sufficient to greatly enhance the information obtained from the experiment. Since methyl groups from different amino acids occupy distinct chemical shift ranges (Fig. 3B, D, F, H), determination of the amino acid type can be achieved through the production of individually or combinatorically labeled protein preparations and the comparison of the measured chemical shifts to available assignments in the BMRB. In addition, methods have also been developed for the partial assignment of resonances by mutagenesis or using methyl–methyl NOESY data combined with manual or automatic analysis (Monneau et al., 2017; Pritišanac et al., 2017; Proudfoot et al., 2016; Velyvis et al., 2009). Since assignments are providing such valuable information, these approaches will be covered in the following paragraphs in more detail.

4.1.1 Assignment Using Mutagenesis

One strategy that has been widely adopted for generating partial assignments of methyl groups is the individual removal of methyl groups by introducing point mutations (Velyvis et al., 2009). The premise is that mutation of a single methyl containing amino acid should remove a single peak or two peaks of a geminal methyl pair from the spectrum, which can then be assigned simply through comparison with the wild-type spectrum (Fig. 6). The decision of what the residue of interest should be mutated into can be done using a table published by Crublet et al. which was compiled based on the

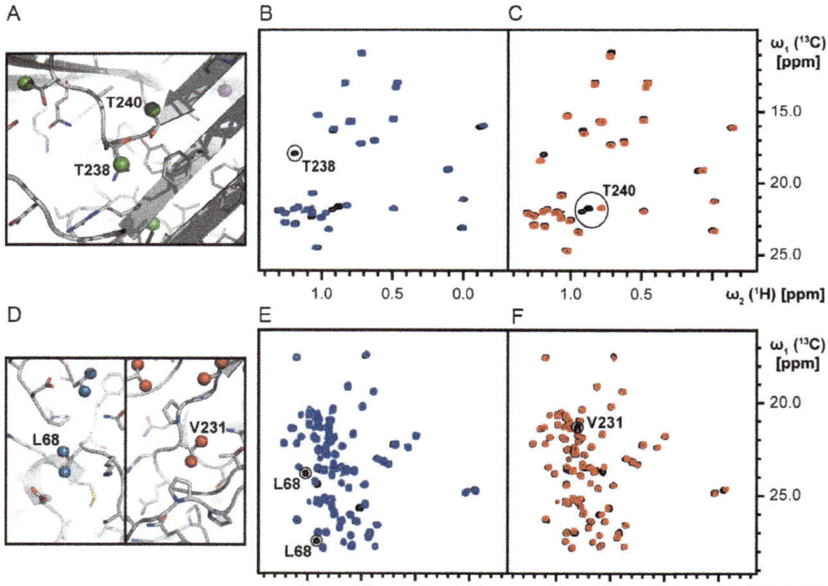

Fig. 6 Mutation studies on the viral 150 kDa, truncated homopentamer BK VP1. (A and D) 3D crystallographic structures of apo VP1 (PDB: 4MJ1) depicting the regions where mutations were made. Panels show the superposition of 2D [^{13}C,^{1}H]-SOFAST HMQC spectra of IT- or LV-labeled wild-type (*black*) VP1 with (B) T238S (*blue*), (C) T240S (*red*), (E) L68A (*blue*), and (F) V231I (*red*) mutant variants.

BLOSUM 62 substitution matrix (Crublet et al., 2014). In our experience, this table provides generally a good guideline as we have had successes mutating all six methyl containing amino acids in multiple proteins following these recommendations. Using modern day cloning techniques, the insertion and verification of point mutations into genes of interest can be achieved in only a couple of days. However, since every mutant must be individually progressed through transformation, cell growth adaption, expression, and purification, it is reasonable to expect production and purification of a set of mutant proteins to take several weeks. Even if the proteins are all produced in parallel, this process represents a significant investment of resource and time. While the approach is conceptually and experimentally straightforward, the success rates for individual mutants are variable. Methyl groups for which assignments are desired often reside in functionally important areas of the protein, and as a result the sidechain might be involved in critical interactions. Mutations of such residues might therefore not be tolerated by the protein or can negatively affect the expression level. In cases where protein expression is unaffected, other problems can result in the failure to produce unambiguous assignments. Mutation of residues within a protein can result in significant conformational changes in addition to the expected local perturbations, which will then affect the chemical shift of multiple peaks within the spectrum, making the unambiguous assignment of a single peak difficult if not impossible. Finally, if by chance the resonance of the mutated methyl group is located in a heavily overlapped region of the spectrum, the identification of the affected signal might be challenging. If structural information is available, the risk of undesired effects of the mutations can be mitigated to a certain extent by carefully selecting and prioritizing key methyl containing amino acids of interest based on their location within the protein. It is best to mutate only a single residue at a time in order to minimize the effects of potential secondary CSPs caused by conformational changes.

Fig. 6 shows examples of how this strategy has been used to generate assignments for several residues of the 150 kDa, truncated homopentameric viral protein BK VP1. Threonine 238 was mutated to a serine residue and the resulting mutant (blue) and wild-type (black) IT spectra were compared (Fig. 6B). With only a single peak missing from the spectrum and all other resonances exhibiting no significant changes in chemical shift, threonine 238 could be unambiguously assigned. A similar observation was made when mutating leucine 68 to an alanine residue (Fig. 6E). Comparison of the wild-type (black) and mutant (blue) LV spectra showed that two peaks disappeared from the spectrum. In addition, additional minor CSPs were observed throughout the spectrum of the mutant protein and most likely corresponded to methyl containing residues that are in close vicinity to the mutated leucine 68 (Fig. 6D). This in itself can also prove to be useful information as tentative assignments can be derived based on those peaks that experience CSPs due to being close by in the structure.

In contrast, when threonine 240 was mutated to a serine residue, the corresponding IT spectrum (red) shows a deficit of one threonine peak and a large CSP of another peak in close proximity of each other in the wild-type (black) spectrum (Fig. 6C, circled area). In this case, it is unclear which peak from the wild-type spectrum disappeared and which experienced the CSP. Based on analysis of the structure of VP1 (Fig. 6A), it is visible that threonine 240 is surrounded by hydrophobic residues which explains the more pronounced spectral changes observed, while threonine 238 is mainly solvent exposed. Lastly, when mutating valine 231 to an isoleucine residue, the assignment is only partial because only a single peak of the expected two disappeared from the spectrum (Fig. 6D). This assignment was later obtained through other methods and we were able to determine that the second Cγ resonance was in an overlapped region of the spectrum.

In successful cases, the process of assigning methyl groups through mutagenesis is simple to perform and easy to analyze. However, as described earlier, this process can also have variable results, with often the success of the experiment not being known until the mutant spectra are collected. In the case of BK VP1, out of the 12 single point mutations that were created, 5 provided unambiguous assignments (42%), while 3 failed to provide any information (25%), and 4 produced ambiguous assignments (33%), which were later confirmed to be correct through the use of additional assignment methods.

4.1.2 "Methyl Walk" Assignment Strategy

With the availability of highly methyl-labeled protein, it has also become feasible to directly assign methyl groups by observing and analyzing methyl–methyl NOE networks without the need for additional side chain correlations. The high probe density in such proteins reduces the average distance between individual methyl groups and thereby increases the number of observable NOEs, which enable assignment by following NOEs from one methyl to others that are in spatial proximity (Proudfoot et al., 2016). This approach is generally called "methyl walk." The requirements for this method are an X-ray structure to evaluate the expected distances, a high quality multidimensional NOESY spectrum (e.g., 3D or 4D HMQC–NOESY–HMQC), and an unique chemical shift pattern in the protein area of interest that serves as a starting point. An example of using this approach for assigning selectively labeled methyl groups in CoaD is shown in Fig. 7. By analyzing all methyl resonances observed in the various indirect HMQC planes, it is possible to generate a network of interactions observed by each

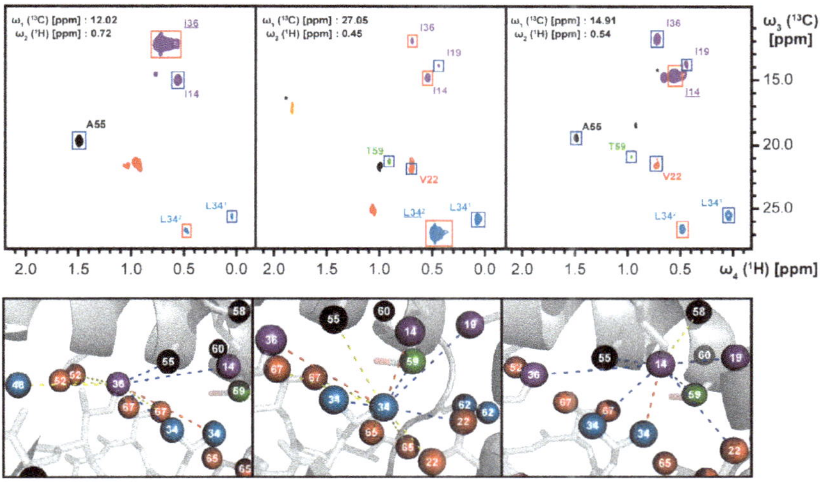

Fig. 7 *Upper panel* contains 2D [^{13}C,^{1}H] planes of the 4D methyl–methyl ^{13}C-resolved HMQC–NOESY–HMQC experiment recorded with MILVAT-labeled CoaD. Peaks are colored according to their amino acid type using the color scheme used throughout the manuscript. *Red boxes* are drawn around resonance peaks used in the methyl walk shown, whereas *blue boxes* highlight redundant resonance peaks that appear in multiple planes which could be used to generate unambiguous assignments. Assignments that are underlined correspond to peaks that have the ^{1}H, ^{13}C chemical shifts detailed at the top left of each spectrum for which NOEs are being observed to. Peaks marked with an asterisk correspond to those that have their maximum intensity in a different plane and for which residual intensity is observed in the plane shown. Peaks that are not labeled correspond to resonances that could not be assigned using the information present in the methyl walk shown. *Lower panel* contains cartoon representations of the structure of apo CoaD (PDB entry 5JBN) centered on methyl groups that have underlined assignments. Methyl groups within a 7.5 Å radius of the centered atom are connected by *colored lines*. *Red* and *blue lines* correspond to the *boxes* in the upper part of the figure, while *yellow lines* correspond to methyl groups within 7.5 Å that could not be unambiguously assigned in the methyl walk shown. *Figure adapted from Proudfoot, A., Frank, A. O., Ruggiu, F., Mamo, M., & Lingel, A. (2016). Facilitating unambiguous NMR assignments and enabling higher probe density through selective labeling of all methyl containing amino acids.* Journal of Biomolecular NMR, 65*(1), 15–27. https://doi.org/10.1007/s10858-016-0032-2.*

methyl group. Comparison of the observed and expected NOE networks is then performed to identify NOE patterns that are unique to a particular methyl group. Subsequently, such a methyl group serves as a starting point to conduct a series of methyl walks to assign the surrounding methyl groups. Alternatively, a methyl group assigned by point mutation can be used as a starting point. Although knowledge of the types of amino acids that each peak corresponds to is not a prerequisite for this process to be successful, it can greatly simplify the process and increase the number of assignments that can be made.

We have used this approach successfully in our laboratory to assign the relevant regions of several proteins of various sizes. The advantage of this simple method is that only a single experiment on a single sample is needed (provided that the amino acid types are already known), however the number and spatial range of assignments are dependent on the probe density and individual distances between methyl groups. In addition, signal dispersion has to be sufficient to unambiguously assign the signals in the indirect dimensions. NOE-based information can also be combined with paramagnetic relaxation enhancement (PREs) and pseudocontact shift information (Velyvis et al., 2009; Venditti, Fawzi, & Clore, 2011) to enhance the process and increase the number of assignments.

4.1.3 Automated Assignment Procedures

Following the previously described principle of assigning methyl groups by using existing crystal structures and identifying proton–proton distances, which can be detected in NOE experiments, multiple semi- and fully automated procedures have been developed. Automated assignment of methyl peaks in an ILV spectrum was first shown by Xu et al. and Chao et al. with the development of MAP-XS and FLAMEnGO (Chao et al., 2014; Xu et al., 2009; Xu & Matthews, 2013). Using a preexisting crystal structure of the protein, these programs facilitate the assignment of methyl peaks in the spectrum based on predicted chemical shifts, or an expected methyl–methyl NOESY cross-peak network. In recent years, new automated assignment protocols have been developed that use amended or different sampling algorithms and allow additional input data, including PRE and through-bond TOCSY data, to be used to maximize the number and accuracy of the assignments made (Chao et al., 2014; Monneau et al., 2017).

A recent publication from Pritišanac et al. introduced the MAGMA (methyl assignment by graph matching) protocol for the automated assignment of methyl groups (Pritišanac et al., 2017). Using a crystal structure of the protein being investigated, in combination with NOE peak lists derived from high dimensional solution-state or 4D DREAM solid-state NMR experiments, MAGMA applies a graph matching protocol to assign peaks and will produce a list and a confidence level of all possible assignments for each methyl resonance in the spectrum. In this study, the authors compare the functionality of MAGMA against the preexisting automated assignment protocols using proteins ranging in monomer size from 8.6 to 360 kDa which are labeled using ILV or ILVA labeling schemes and claim that their methodology produces more confident assignments than the other protocols tested. Although the performance of MAGMA was tested on ILV and ILVA labeling schemes, the authors state that MAGMA can handle and may in fact benefit from the incorporation of additional selectively labeled methyl groups, but also warn against the potential drawbacks that we discussed earlier. In circumstances where peak assignments are ambiguous, MAGMA also boasts the ability to predict which sites should be selected for site-directed mutagenesis to maximize the number of new assignments obtained (Pritišanac et al., 2017).

4.2 Applications to Drug Discovery

When used for the purposes of drug discovery, NMR is routinely applied in hit finding and validation, lead generation as well as structural characterization of target–ligand interactions. The availability of methyl-labeled protein enables numerous higher dimensional, protein-observed experiments to be acquired, which are generally preferred due to the increased extent of information that can be obtained. Multiple examples of such experiments are outlined in the following paragraphs.

4.2.1 Hit Validation

One of the most common applications of biomolecular NMR is the identification of fragment hits and the validation of ligands which were previously identified to interact with a target by other hit finding activities, including biochemical or biophysical screens. While a collection of ligand-observed experiments exist to approach this, the confidence level is significantly higher when observing changes in protein signals upon the addition of a ligand. While for small proteins perturbations in one-dimensional spectra can be sufficient to detect and quantify binding events, the use of isotope-labeled protein in combination with 2D correlation spectra allows such changes to be easily observed for larger proteins or protein complexes. In its simplest form, it can be sufficient to label a single amino acid in a nondeuterated

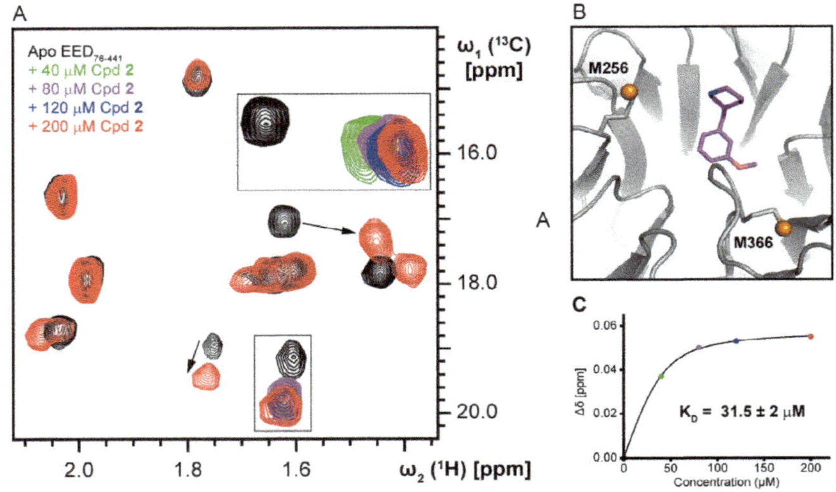

Fig. 8 Binding of compound **2** to EED. (A) Superposition of 2D [^{13}C,^{1}H]-SOFAST HMQC spectra of selectively methionine-labeled apo EED76-441 (*black*) and in the presence of 200 μ*M* compound **2** (*red*). *Insets* show the chemical shifts of two methionine peaks in the presence of increasing amounts of compound **2**. The spectra are plotted in colors corresponding to the amount of ligand added, according to the label in the *upper left corner*. (B) Costructure of EED76-441 in the presence of compound **2** showing the location of the methionine residues in the compound binding site (PDB: 5U5K). (C) K_D titration curve of the weighted chemical shift change ($\Delta\delta$) plotted vs the compound concentration calculated using the chemical shift perturbations observed in (A). *Figure adapted from Lingel, A., Sendzik, M., Huang, Y., Shultz, M. D., Cantwell, J., Dillon, M. P., … Bussiere, D. E. (2017). Structure-guided design of EED binders allosterically inhibiting the epigenetic polycomb repressive complex 2 (PRC2) methyltransferase.* Journal of Medicinal Chemistry, 60*(1), 415–427. https://doi.org/10.1021/acs.jmedchem.6b01473.*

background, as exemplified in Fig. 8. The protein EED, a subunit of the epigenetic PRC2 methyltransferase complex and evaluated as cancer drug target (Huang et al., 2017; Lingel et al., 2017; Qi et al., 2017), was produced with only the ε-methyl group of methionine sidechains ^{13}C labeled. The resulting protein yields a very simple methyl spectrum, in which CSPs induced by ligand interaction can be recognized easily (Fig. 8A) due to the presence of several methionine residues located close to the desired compound binding site (Fig. 8B). Even in the absence of assignments, the perturbations can be utilized for K_D determinations (Fig. 8C) and discrimination of different binding sites. Methionine labeling is used frequently for hit validation efforts, as the material can be produced at low cost and typically with high yields, which enable the assessment of a large number of compounds. In addition, the ε-methyl group of a methionine sidechain provides excellent signal to noise, and the peaks are found in a well-separated part of the spectrum. However, any of the aforementioned labeling schemes can be utilized in such experiments, with the exact choice often being dependent on the amino acid distribution in the region of interest.

4.2.2 Binding Site Mapping and Hit Triaging

The observation of CSPs not only confirms that the ligands tested interact with the target of interest but can also be used in the absence of peak assignments to classify the sites on the protein they interact with. Basic binning of hits can be achieved simply through comparison of the ligand-induced CSP patterns, where compounds that share a similar pattern or fingerprint (i.e., with the same peaks shifted and in the same direction) most likely bind at the same location and with a similar orientation. At this stage, knowledge of the assignments on the methyl peaks experiencing a shift can provide valuable information on the location of multiple pockets, as peaks exhibiting a CSP can be mapped onto an apo structure to provide insights into the area of compound binding.

An example of this approach is given in Fig. 9 where the superposition of 2D [^{13}C,^{1}H]-SOFAST HMQC spectra, acquired in the presence of two compounds identified by orthogonal methods to interact with the ILVT-labeled viral protein BK VP1, are shown. Compound **3** (green) had been identified by an SPR FBS campaign, and validation with 2D NMR confirmed that the compound induces CSPs of peaks corresponding to residues T118, T224, T226, V234, and T238 in the 2D [^{13}C,^{1}H]-SOFAST HMQC spectrum (Fig. 9A). A cocrystal structure of VP1 in the presence of **3** was soon obtained and showed that the ligand is bound in the pore formed by the five monomers of VP1, and in close vicinity of the assigned residues that experienced a CSP (Fig. 9C). Compound **4** (magenta) was identified through hit expansion and the selection of an analog set based on compound **3**. To our surprise, the 2D [^{13}C,^{1}H]-SOFAST HMQC spectrum of VP1 in the presence of compound **4** revealed a very different pattern of CSPs. None of the peaks previously experiencing a CSP were observed to move in the presence of compound **4**. Instead, two other peaks, identified as I121 and T192, were perturbed in the spectrum (Fig. 9B). This indicated that compound **4** was bound in a different location and was subsequently confirmed to interact with the protein outside of the main pore, in a region close to a calcium binding site (Fig. 9D). The use of this 2D NMR assay format subsequently enabled the efficient binning of follow-up compounds with respect to the identified binding sites, simply based on their CSP pattern.

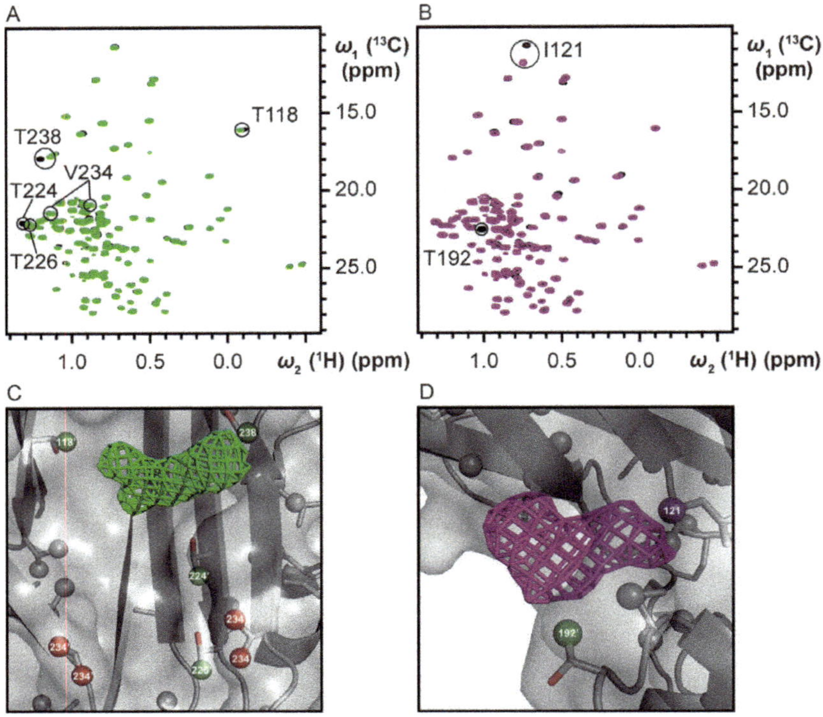

Fig. 9 Binding site mapping and binning for compounds **3** and **4** on the viral protein BK VP1. Superposition of 2D [^{13}C,^{1}H]-SOFAST HMQC spectra of the selectively ILVT-labeled viral protein VP1 acquired in the absence (*black*) and presence of (A) 1 m*M* compound **3** (*green*) and (B) 1 m*M* compound **4** (*magenta*). Assignments of peaks experiencing a chemical shift perturbations are indicated on the spectra. Cocrystal structures of VP1 in complex with (C) compound **3** and (D) compound **4**. Methyl groups experiencing chemical shift perturbations are labeled on all structures. The corresponding electron density present for the ligands in the 2FoFc map at a level of **1** sigma is displayed in the binding pockets.

In addition to the selective labeling of methyl groups, our laboratory has also selectively labeled whole amino acids. The MAP kinase ERK2 is a 41 kDa monomeric protein that is of great interest in pharmaceutical research due to its prominent role in tumorigenesis (Dhillon, Hagan, Rath, & Kolch, 2007). In this case, although the protein can be produced in *E. coli* and a high quality TROSY spectrum can be obtained when over-expressed using ^{2}H, ^{15}N isotope enrichment, the yield of purified ERK2 is usually so low (<3 mg/L) that it is prohibitively expensive to produce sufficient protein to perform a protein-observed screening or hit validation campaign. This is especially true when working with the double-phosphorylated (activated) form of the kinase as protein yields are usually further diminished by either in vivo or in vitro phosphorylation. When working with proteins that show low levels of overexpression in *E. coli*, yields can often be increased by switching to an insect or mammalian cell-based expression system (Skora et al., 2015). For ERK2, using media that was supplemented with ^{13}C-leucine, insect cells were used to produce selectively ^{13}C-leucine-labeled ERK2 with a final yield of 38 mg/L. The cost of expressing 1 L of such protein at time of press (2018) is approximately $2400, which is comparable to the cost of a fully labeled MILVAT protein preparation and was sufficient to enable protein-observed validation studies of ligands that were of high interest (Fig. 10).

As seen for selectively methyl-labeled proteins, such uniform ^{13}C-leucine-labeled protein can be assessed by 2D [^{13}C,^{1}H] HMQC experiments and yield high quality spectra (Fig. 10). In the absence of protein assignments, the natural ligand ATP was titrated into a sample of ^{13}C-leucine-labeled ERK2 to identify leucine Cδ methyl resonances that could be used as probes for compound binding to the nucleotide binding pocket. The ambiguous assignment of those resonances could then be used to differentiate between ligands that bind to the nucleotide binding site, and those that interact with the protein elsewhere.

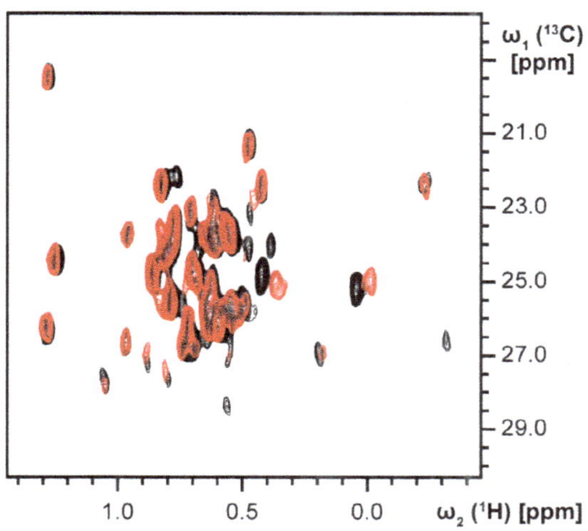

Fig. 10 Superposition of 2D [^{13}C,^{1}H]-SOFAST HMQC spectra of ^{13}C-leucine-labeled ERK2 in the absence (*black*) and presence (*red*) of saturating amounts of ATP. The easily detectable chemical shift perturbations of certain leucine methyl groups are evidence of binding of the ligand to the kinase. The nucleotide-induced CSP pattern enables the prediction of the binding location of newly discovered ligands in the absence of higher resolution structural data.

4.2.3 Docking of Small Molecules to Protein Binding Sites

In addition to validation and binning of the binding event, detailed structural information on the protein–ligand complex is often key to progress a compound into chemistry optimization. Since many ligands cannot be successfully visualized in costructures by X-ray crystallography, in particular weak binders and molecules with limited solubility in the crystallization condition, several NMR-based methods were developed to address this need (Guan et al., 2013; Hajduk et al., 2004; Mohanty et al., 2016; Orts et al., 2016; Proudfoot et al., 2017; Wälti, Riek, & Orts, 2017; Wang, Westerhoff, & Merz, 2007). Among them, the use of selectively methyl-labeled proteins combined with the determination of intermolecular NOEs applied subsequently in docking provides an efficient and robust approach to obtain high quality models of the protein–ligand complexes (Proudfoot et al., 2017; Shah et al., 2012). After assigning the observable protons of the ligand (Fig. 11A) and the protein methyl groups in the ligand binding site, unambiguous ligand proton–protein methyl distances can be obtained by acquiring an edited–filtered NOESY experiment (Fig. 11B) and subsequently employed as restraints in a protein–ligand docking approach (Fig. 11C). While the method can be applied to any ^{13}C-labeled protein, the advantage of using selectively methyl-labeled and otherwise deuterated protein is the extended observable distance as well as the better performance of intermolecular NOE experiments. In a recent study, we have shown that by increasing the coverage of methyl groups by using highly methyl-labeled protein (e.g., MILVAT-labeled), a large number and redundant set of intermolecular distances can be obtained. As a consequence, the resulting models closely recapitulate the corresponding X-ray structures and are of sufficient accuracy to be used for compound optimization in a prospective way in cases where X-ray structures are not available (Proudfoot et al., 2017).

4.3 Dynamics

Selectively methyl-labeled protein can also be used to obtain detailed information about dynamics of macromolecules (Tugarinov & Kay, 2005). In most cases described in the literature, data about intramolecular motion have been extracted from methyl groups in an indirect manner, i.e., through the use of methyl relaxation dispersion experiments (Yuwen, Vallurupalli, & Kay, 2016), by matching simulated and measured methyl residual dipolar couplings (Long, Li, Walter, Griesinger, & Brüschweiler, 2011), or through

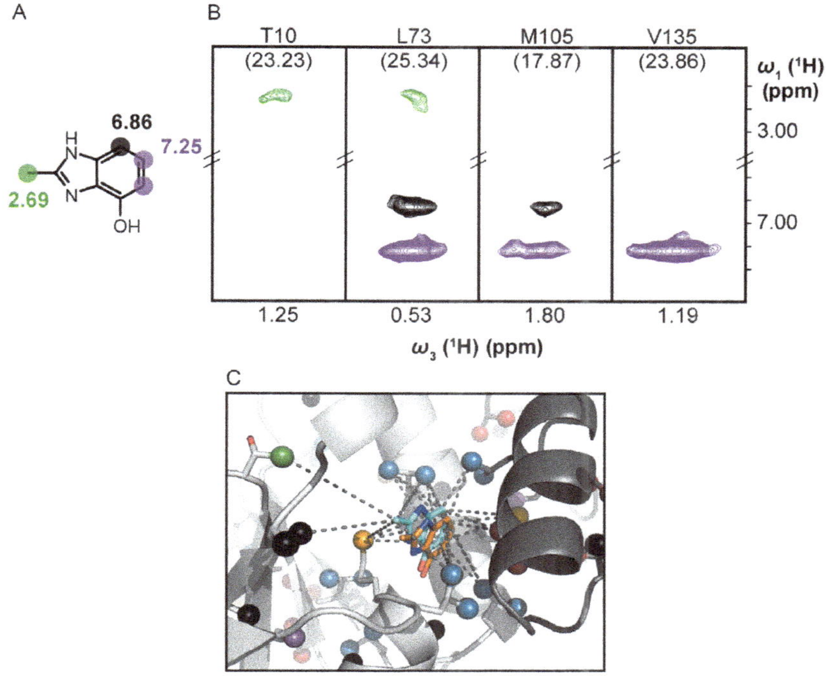

Fig. 11 NMR analysis and docking of compound **1**. (A) Chemical structure of compound **1** showing the chemical shift values of the protons present in the small molecule. (B) $[\omega_1(^1H),\omega_3(^1H)]$ strips from the F_2-edited, F_3-filtered ^{13}C-resolved $[^1H,^1H]$-NOESY showing intermolecular NOEs observed between labeled methyl groups and compound **1**. (C) Superposition of compound **1** in the docking model with the lowest docking score (*turquois*) and the crystal structure (*orange*). All observed intermolecular NOEs used to derive the model are indicated as *dotted lines. Figure adapted from Proudfoot, A., Bussiere, D. E., & Lingel, A. (2017). High-confidence protein–ligand complex modeling by NMR-guided docking enables early hit optimization.* Journal of the American Chemical Society, 139*(49), 17824–17833. https://doi.org/10.1021/jacs.7b07171.*

the use of $^{13}C_{methyl}$ chemical exchange saturation transfer or dark state exchange saturation transfer studies (Bouvignies & Kay, 2012; Fawzi, Ying, Ghirlando, Torchia, & Clore, 2011). Several reviews providing details about the use of methyl groups for determining macromolecular dynamics can be found in the literature (Rosenzweig & Kay, 2014; Ruschak & Kay, 2010; Tugarinov & Kay, 2005).

Very slow dynamics on the NMR chemical shift timescale, however, can also often be observed directly by the presence of doubled peak sets for the amino acids in motion. A nice example of a system in slow exchange was published by Xiao et al. where based on the detection of doubled signals for several isoleucine, valine, and leucine methyl groups, the authors could show that the active, double-phosphorylated form of the human kinase ERK2 samples two states (20%:80% at 298 K), the minor of which represents the inactive, nonphosphorylated form (Xiao et al., 2014). We observed similar protein dynamics for a 67 kDa complex of a different selectively (ILVT) labeled human kinase bound to an unlabeled natural partner protein (Fig. 12).

As shown in Fig. 12A (black spectrum) certain isoleucine and valine amino acids of the unphosphorylated protein complex are represented by a doubled NMR signal set (ratio of ~40%:60% at 305 K) in the [^{1}H,^{3}C]-HMQC spectrum. Changing the sample temperature changes the ratio between the two populations, indicating that the protein samples two conformational states. When the phosphorylated (activated) form of the kinase complex was produced the second signal set, representing the major form in the inactive (unphosphorylated) complex, almost completely disappeared (Fig. 12A, red spectrum, ratio ~95%:5%). It became clear that the protein complex samples two slowly exchanging conformations in the unphosphorylated state and that the minor form corresponds to the activated form of the kinase.

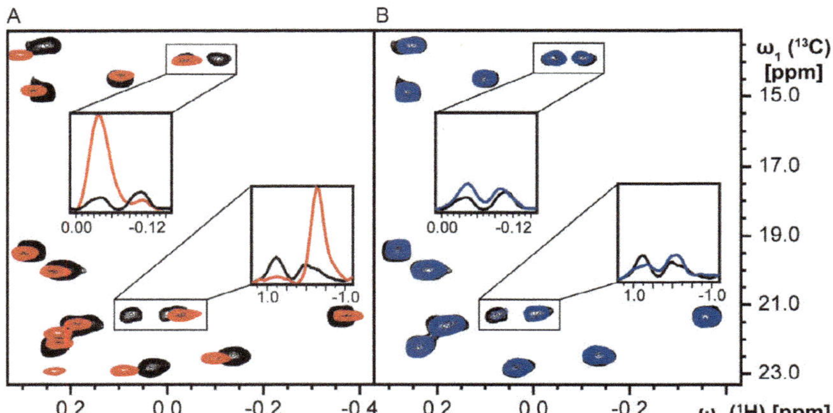

Fig. 12 Superposition of 2D [^{13}C,^{1}H]-SOFAST HMQC spectra of a 67 kDa complex of a human kinase (ILVT-labeled) bound to an unlabeled partner protein. (A) *Black spectrum*: the unphosphorylated form of the kinase complex samples two states which are represented by a doubled signal set for several methyl groups. The minor peak set has highly similar chemical shifts to the signals of the protein complex in the phosphorylated state (*red spectrum*). (B) Upon binding (*blue spectrum*) to the inactive kinase complex (*black spectrum*) a small molecule ligand shifts the dynamic equilibrium between the two conformations toward the form presumably representing the active state.

Following on from this, the selectively labeled protein complex was used for NMR-based compound binding studies. One of the small molecules which was validated to interact with the unphosphorylated protein complex appeared to reverse the conformational equilibrium upon binding and promoted formation of the active form (60:40) of the kinase domain (Fig. 12B). As no significant CSPs were observed in addition to the detected changes in signal intensities, it was speculated that the compound may interact with the unlabeled partner protein in complex with the kinase. This hypothesis was confirmed by an X-ray costructure that showed a close analog of the compound bound more than 20 Å away from the closest amino acid of the kinase. This result suggests that the small molecule has an impact on the dynamics sampled by the kinase protein and that an allosteric mechanism is potentially exploited to modulate the activity.

5. CONCLUSION

In summary, selective methyl labeling of proteins has become a highly impactful approach in the biomolecular NMR field and has enabled and greatly simplified the investigation of proteins and protein complexes beyond the molecular size limit and spectral complexity presented by uniformly labeled proteins. A range of labeling schemes is now available, ranging from simple, cost-effective methionine labeling to full MILVAT labeling for detailed determination of protein–ligand interactions. Owing to the beneficial features of methyl-labeled proteins, their use has been expanded to other challenging areas of biomolecular NMR, such as the observation of proteins by in cell NMR (Pastore & Temussi, 2017). For example, Serber et al. could detect signals of calmodulin expressed in live cells (Serber et al., 2004) and were able to monitor CSPs induced by binding of a drug molecule to the protein inside of living cells. Furthermore, selectively labeled proteins are suited for the characterization of membrane proteins (Kurauskas, Schanda, & Sounier, 2017) where observation of the corresponding methyl groups was successfully applied to study the voltage-dependent anion channel (Hiller et al., 2008) and the prokaryotic pH-dependent potassium channel KcsA (Imai, Osawa, Takeuchi, & Shimada, 2010), among others. Observing methyl groups in selectively labeled proteins are not only of great value when working with large proteins or macromolecules with heavy signal overlap but also when protein yields are low, measurement time is limited, or expression of uniform or more complex labeling patterns are not feasible or too expensive.

In our laboratory, selective methyl labeling is a cornerstone application for drug discovery, as it not only enables robust hit validation and characterization but is also instrumental to deciphering allosteric protein pockets and the impact of ligands on the equilibrium of exchanging conformational states of several disease relevant macromolecules.

ACKNOWLEDGMENTS

We thank Mark Knapp for providing VP1 X-ray costructures and Charles Wartchow for discussions on the VP1 project; Ken Crawford, Eric Fang, John Fuller, and Jacob Shaul for protein production; as well as Karin Briner, Dirksen Bussiere, Donald Ganem, Mariela Jaskelioff, Jennifer Leeds, Heinz Moser, Trixie Wagner, Kelly Wong, René Wyler, and Isabel Zaror for supporting this work.

REFERENCES

Ayala, I., Sounier, R., Usé, N., Gans, P., & Boisbouvier, J. (2009). An efficient protocol for the complete incorporation of methyl-protonated alanine in perdeuterated protein. *Journal of Biomolecular NMR*, *43*(2), 111–119. https://doi.org/10.1007/s10858-008-9294-7.

Bouvignies, G., & Kay, L. E. (2012). A 2D ^{13}C-CEST experiment for studying slowly exchanging protein systems using methyl probes: An application to protein folding. *Journal of Biomolecular NMR*, *53*(4), 303–310. https://doi.org/10.1007/s10858-012-9640-7.

Chao, F.-A., Kim, J., Xia, Y., Milligan, M., Rowe, N., & Veglia, G. (2014). FLAMEnGO 2.0: An enhanced fuzzy logic algorithm for structure-based assignment of methyl group resonances. *Journal of Magnetic Resonance*, *245*, 17–23. https://doi.org/10.1016/j.jmr.2014.04.012.

Crublet, E., Kerfah, R., Mas, G., Savoye, M. N., Lantez, V., Vernet, T., et al. (2014). A cost-effective protocol for the parallel production of libraries of $^{13}CH_3$-specifically labeled mutants for NMR studies of high molecular weight proteins. *Structural Genomics*, *1091*, 229–244. https://doi.org/10.1007/978-1-62703-691-7.

Dhillon, A. S., Hagan, S., Rath, O., & Kolch, W. (2007). MAP kinase signaling pathways in cancer. *Oncogene*, *26*(22), 3279–3290. https://doi.org/10.1038/sj.onc.1210421.

Fawzi, N. L., Ying, J., Ghirlando, R., Torchia, D. A., & Clore, G. M. (2011). Atomic-resolution dynamics on the surface of amyloid-β protofibrils probed by solution NMR. *Nature*, *480*(7376), 268–272. https://doi.org/10.1038/nature10577.

Fischer, M., Kloiber, K., Häusler, J., Ledolter, K., Konrat, R., & Schmid, W. (2007). Synthesis of a ^{13}C-methyl-group-labeled methionine precursor as a useful tool for simplifying protein structural analysis by NMR spectroscopy. *ChemBioChem*, *8*(6), 610–612. https://doi.org/10.1002/cbic.200600551.

Gans, P., Hamelin, O., Sounier, R., Ayala, I., Durá, M. A., Amero, C. D., et al. (2010). Stereospecific isotopic labeling of methyl groups for NMR spectroscopic studies of high-molecular-weight proteins. *Angewandte Chemie, International Edition*, *49*(11), 1958–1962. https://doi.org/10.1002/anie.200905660.

Gardner, K. H., & Kay, L. E. (1997). Production and incorporation of ^{15}N, ^{13}C, ^{2}H (^{1}H-δ^1 methyl) isoleucine into proteins for multidimensional NMR studies. *Journal of the American Chemical Society*, *119*(32), 7599–7600. https://doi.org/10.1021/ja9706514.

Gardner, K. H., & Kay, L. E. (1998). The use of ^{2}H, ^{13}C, ^{15}N multidimensional NMR to study the structure and dynamics of proteins. *Annual Review of Biophysics and Biomolecular Structure*, *27*(1), 357–406. https://doi.org/10.1146/annurev.biophys.27.1.357.

Garrett, D. S., Seok, Y. J., Liao, D. I., Peterkofsky, A., Gronenborn, A. M., & Clore, G. M. (1997). Solution structure of the 30 kDa N-terminal domain of enzyme I of the *Escherichia coli* phosphoenolpyruvate:sugar phosphotransferase system by multidimensional NMR. *Biochemistry*, *36*(9), 2517–2530. https://doi.org/10.1021/bi962924y.

Gelis, I., Bonvin, A. M. J. J., Keramisanou, D., Koukaki, M., Gouridis, G., Karamanou, S., et al. (2007). Structural basis for signal-sequence recognition by the translocase motor SecA as determined by NMR. *Cell*, *131*(4), 756–769. https://doi.org/10.1016/j.cell.2007.09.039.

Godoy-Ruiz, R., Guo, C., & Tugarinov, V. (2010). Alanine methyl groups as NMR probes of molecular structure and dynamics in high-molecular-weight proteins. *Journal of the American Chemical Society*, *132*(51), 18340–18350. https://doi.org/10.1021/ja1083656.

Goto, N. K., Gardner, K. H., Mueller, G. A., Willis, R. C., & Kay, L. E. (1999). A robust and cost-effective method for the production of Val, Leu, Ile (δ^1) methyl-protonated ^{15}N-, ^{13}C-, ^{2}H-labeled proteins. *Journal of Biomolecular NMR*, *13*(4), 369–374. https://doi.org/10.1023/A:1008393201236.

Gross, J. D., Gelev, V. M., & Wagner, G. (2003). A sensitive and robust method for obtaining intermolecular NOEs between side chains in large protein complexes. *Journal of Biomolecular NMR*, *25*(3), 235–242. https://doi.org/10.1023/A:1022890112109.

Guan, J. Y., Keizers, P. H. J., Liu, W. M., Löhr, F., Skinner, S. P., Heeneman, E. A., et al. (2013). Small-molecule binding sites on proteins established by paramagnetic NMR spectroscopy. *Journal of the American Chemical Society*, *135*(15), 5859–5868. https://doi.org/10.1021/ja401323m.

Hajduk, P. J., Mack, J. C., Olejniczak, E. T., Park, C., Dandliker, P. J., & Beutel, B. A. (2004). SOS-NMR: A saturation transfer NMR-based method for determining the structures of protein–ligand complexes. *Journal of the American Chemical Society*, *126*(8), 2390–2398. https://doi.org/10.1021/ja039480v.

Hiller, S., Garces, R. G., Malia, T. J., Orekhov, V. Y., Colombini, M., & Wagner, G. (2008). Solution structure of the integral human membrane protein VDAC-1 in detergent micelles. *Science*, *321*(5893), 1206–1210. https://doi.org/10.1126/science.1161302.

Huang, Y., Zhang, J., Yu, Z., Zhang, H., Wang, Y., Lingel, A., et al. (2017). Discovery of first-in-class, potent, and orally bioavailable embryonic ectoderm development (EED) inhibitor with robust anticancer efficacy. *Journal of Medicinal Chemistry*, *60*(6), 2215–2226. https://doi.org/10.1021/acs.jmedchem.6b01576.

Imai, S., Osawa, M., Takeuchi, K., & Shimada, I. (2010). Structural basis underlying the dual gate properties of KcsA. *Proceedings of the National Academy of Sciences*, *107*(14), 6216–6221. https://doi.org/10.1073/pnas.0911270107.

Isaacson, R. L., Simpson, P. J., Liu, M., Cota, E., Zhang, X., Freemont, P., et al. (2007). A new labeling method for methyl transverse relaxation-optimized spectroscopy NMR spectra of alanine residues. *Journal of the American Chemical Society*, *129*(50), 15428–15429. https://doi.org/10.1021/ja0761784.

Jiang, Y., & Kalodimos, C. G. (2017). NMR studies of large proteins. *Journal of Molecular Biology*, *429*(17), 2667–2676. https://doi.org/10.1016/j.jmb.2017.07.007.

Kerfah, R., Hamelin, O., Boisbouvier, J., & Marion, D. (2015). CH_3-specific NMR assignment of alanine, isoleucine, leucine and valine methyl groups in high molecular weight proteins using a single sample. *Journal of Biomolecular NMR*, *63*, 389–402. https://doi.org/10.1007/s10858-015-9998-4.

Kerfah, R., Plevin, M. J., Sounier, R., Gans, P., & Boisbouvier, J. (2015). Methyl-specific isotopic labeling: A molecular tool box for solution NMR studies of large proteins. *Current Opinion in Structural Biology*, *32*, 113–122. https://doi.org/10.1016/j.sbi.2015.03.009.

Kurauskas, V., Schanda, P., & Sounier, R. (2017). In J.-J. Lacapere (Ed.), *Vol. 1635. Membrane protein structure and function characterization*. New York, NY: Springer New York. https://doi.org/10.1007/978-1-4939-7151-0.

Lichtenecker, R. J., Coudevylle, N., Konrat, R., & Schmid, W. (2013). Selective isotope labelling of leucine residues by using α-ketoacid precursor compounds. *ChemBioChem*, *14*(7), 818–821. https://doi.org/10.1002/cbic.201200737.

Lichtenecker, R., Ludwiczek, M. L., Schmid, W., & Konrat, R. (2004). Simplification of protein NOESY spectra using bioorganic precursor synthesis and NMR spectral editing. *Journal of the American Chemical Society*, *126*(17), 5348–5349. https://doi.org/10.1021/ja049679n.

Lingel, A., Sendzik, M., Huang, Y., Shultz, M. D., Cantwell, J., Dillon, M. P., et al. (2017). Structure-guided design of EED binders allosterically inhibiting the epigenetic polycomb repressive complex 2 (PRC2) methyltransferase. *Journal of Medicinal Chemistry*, *60*(1), 415–427. https://doi.org/10.1021/acs.jmedchem.6b01473.

Long, D., Li, D. W., Walter, K. F. A., Griesinger, C., & Brüschweiler, R. (2011). Toward a predictive understanding of slow methyl group dynamics in proteins. *Biophysical Journal*, *101*(4), 910–915. https://doi.org/10.1016/j.bpj.2011.06.053.

Miller, S., Janin, J., Lesk, A. M., & Chothia, C. (1987). Interior and surface of monomeric proteins. *Journal of Molecular Biology*, *196*(3), 641–656. https://doi.org/10.1016/0022-2836(87)90038-6.

Mohanty, B., Williams, M., Doak, B., Vazirani, M., Ilyichova, O., Wang, G., et al. (2016). Determination of ligand binding modes in weak protein–ligand complexes using sparse NMR data. *Journal of Biomolecular NMR*, *66*(3), 195–208. https://doi.org/10.1007/s10858-016-0067-4.

Monneau, Y. R., Rossi, P., Bhaumik, A., Huang, C., Jiang, Y., Saleh, T., et al. (2017). Automatic methyl assignment in large proteins by the MAGIC algorithm. *Journal of Biomolecular NMR*, *69*(4), 215–227. https://doi.org/10.1007/s10858-017-0149-y.

Orts, J., Walti, M. A., Marsh, M., Vera, L., Gossert, A. D., Guntert, P., et al. (2016). NMR-based determination of the 3D structure of the ligand–protein interaction site without protein resonance assignment. *Journal of the American Chemical Society*, *138*(13), 4393–4400. https://doi.org/10.1021/jacs.5b12391.

Pastore, A., & Temussi, P. A. (2017). The Emperor's new clothes: Myths and truths of in-cell NMR. *Archives of Biochemistry and Biophysics*, *628*, 114–122. https://doi.org/10.1016/j.abb.2017.02.008.

Pritišanac, I., Degiacomi, M. T., Alderson, T. R., Carneiro, M. G., Ab, E., Siegal, G., et al. (2017). Automatic assignment of methyl-NMR spectra of supramolecular machines using graph theory. *Journal of the American Chemical Society*, *139*(28), 9523–9533. https://doi.org/10.1021/jacs.6b11358.

Proudfoot, A., Bussiere, D. E., & Lingel, A. (2017). High-confidence protein–ligand complex modeling by NMR-guided docking enables early hit optimization. *Journal of the American Chemical Society*, *139*(49), 17824–17833. https://doi.org/10.1021/jacs.7b07171.

Proudfoot, A., Frank, A. O., Ruggiu, F., Mamo, M., & Lingel, A. (2016). Facilitating unambiguous NMR assignments and enabling higher probe density through selective labeling of all methyl containing amino acids. *Journal of Biomolecular NMR*, *65*(1), 15–27. https://doi.org/10.1007/s10858-016-0032-2.

Qi, W., Zhao, K., Gu, J., Huang, Y., Wang, Y., Zhang, H., et al. (2017). An allosteric PRC2 inhibitor targeting the H3K27me3 binding pocket of EED. *Nature Chemical Biology*, *13*(4), 381–388. https://doi.org/10.1038/nchembio.2304.

Rosenzweig, R., & Kay, L. E. (2014). Bringing dynamic molecular machines into focus by methyl-TROSY NMR. *Annual Review of Biochemistry*, *83*(1), 291–315. https://doi.org/10.1146/annurev-biochem-060713-035829.

Ruschak, A. M., & Kay, L. E. (2010). Methyl groups as probes of supra-molecular structure, dynamics and function. *Journal of Biomolecular NMR*, *46*(1), 75–87. https://doi.org/10.1007/s10858-009-9376-1.

Saio, T., Guan, X., Rossi, P., Economou, A., & Kalodimos, C. G. (2014). Structural basis for protein antiaggregation activity of the trigger factor chaperone. *Science*, *344*(6184), 1250494. https://doi.org/10.1126/science.1250494.

Serber, Z., Straub, W., Corsini, L., Nomura, A. M., Shimba, N., Craik, C. S., et al. (2004). Methyl groups as probes for proteins and complexes in in-cell NMR experiments. *Journal of the American Chemical Society*, *126*(22), 7119–7125. https://doi.org/10.1021/ja049977k.

Shah, D., AB, E., Diercks, T., Hass, M., van Nuland, N., & Siegal, G. (2012). Rapid protein–ligand costructures from sparse NOE data. *Journal of Medicinal Chemistry*, *55*(23), 10786–10790. https://doi.org/10.1021/jm301396d.

Skora, L., Shrestha, B., & Gossert, A. D. (2015). Isotope labeling of proteins in insect cells. In *Vol. 565. Methods in enzymology.* (1st ed.): Elsevier. https://doi.org/10.1016/bs.mie.2015.05.013.

Sprangers, R., & Kay, L. E. (2007). Quantitative dynamics and binding studies of the 20S proteasome by NMR. *Nature*, *445*(7128), 618–622. https://doi.org/10.1038/nature05512.

Stockman, B. J., & Dalvit, C. (2002). NMR screening techniques in drug discovery and drug design. *Progress in Nuclear Magnetic Resonance Spectroscopy*, *41*(3–4), 187–231. https://doi.org/10.1016/S0079-6565(02)00049-3.

Stoffregen, M. C., Schwer, M. M., Renschler, F. A., & Wiesner, S. (2012). Methionine scanning as an NMR tool for detecting and analyzing biomolecular interaction surfaces. *Structure*, *20*(4), 573–581. https://doi.org/10.1016/j.str.2012.02.012.

Tugarinov, V., Hwang, P. M., Ollerenshaw, J. E., & Kay, L. E. (2003). Cross-correlated relaxation enhanced ^{1}H–^{13}C NMR spectroscopy of methyl groups in very high molecular weight proteins and protein complexes. *Journal of the American Chemical Society*, *125*(34), 10420–10428. https://doi.org/10.1021/ja030153x.

Tugarinov, V., & Kay, L. E. (2003a). Ile, Leu, and Val methyl assignments of the 723-residue malate synthase G using a new labeling strategy and novel NMR methods. *Journal of the American Chemical Society*, *125*(45), 13868–13878. https://doi.org/10.1021/ja030345s.

Tugarinov, V., & Kay, L. E. (2003b). Side chain assignments of Ile δ^1 methyl groups in high molecular weight proteins: An application to a 46 ns tumbling molecule. *Journal of the American Chemical Society*, *125*(19), 5701–5706. https://doi.org/10.1021/ja021452+.

Tugarinov, V., & Kay, L. E. (2005). Methyl groups as probes of structure and dynamics in NMR studies of high-molecular-weight proteins. *ChemBioChem*, *6*(9), 1567–1577. https://doi.org/10.1002/cbic.200500110.

Tzeng, S.-R., Pai, M.-T., & Kalodimos, C. G. (2012). NMR studies of large protein systems. In A. Shekhtman & D. S. Burz (Eds.), *Methods in molecular biology: Vol. 831. Protein NMR techniques* (pp. 133–140). Totowa, NJ: Humana Press. https://doi.org/10.1007/978-1-61779-480-3_8.

Velyvis, A., Ruschak, A. M., & Kay, L. E. (2012). An economical method for production of ^{2}H, $^{13}CH_3$-threonine for solution NMR studies of large protein complexes: Application to the 670 kDa proteasome. *PLoS One*, 7(9), e43725. https://doi.org/10.1371/journal.pone.0043725.

Velyvis, A., Schachman, H. K., & Kay, L. E. (2009). Assignment of Ile, Leu, and Val methyl correlations in supra-molecular systems: An application to aspartate transcarbamoylase. *Journal of the American Chemical Society*, *131*(45), 16534–16543. https://doi.org/10.1021/ja906978r.

Venditti, V., Fawzi, N. L., & Clore, G. M. (2011). Automated sequence- and stereo-specific assignment of methyl-labeled proteins by paramagnetic relaxation and methyl-methyl nuclear Overhauser enhancement spectroscopy. *Journal of Biomolecular NMR*, *51*(3), 319–328. https://doi.org/10.1007/s10858-011-9559-4.

Venters, R. A., Huang, C. C., Farmer, B. T., Trolard, R., Spicer, L. D., & Fierke, C. A. (1995). High-level $^{2}H/^{13}C/^{15}N$ labeling of proteins for NMR studies. *Journal of Biomolecular NMR*, *5*(4), 339–344. https://doi.org/10.1007/BF00182275.

Wälti, M. A., Riek, R., & Orts, J. (2017). Fast NMR-based determination of the 3D structure of the binding site of protein–ligand complexes with weak affinity binders. *Angewandte Chemie, International Edition*, *56*(19), 5208–5211. https://doi.org/10.1002/anie.201612304.

Wang, B., Westerhoff, L. M., & Merz, K. M. (2007). A critical assessment of the performance of protein–ligand scoring functions based on NMR chemical shift perturbations. *Journal of Medicinal Chemistry*, *50*(21), 5128–5134. https://doi.org/10.1021/jm070484a.

Xiao, Y., Lee, T., Latham, M. P., Warner, L. R., Tanimoto, A., Pardi, A., et al. (2014). Phosphorylation releases constraints to domain motion in ERK2. *Proceedings of the National Academy of Sciences*, *111*(7), 2506–2511. https://doi.org/10.1073/pnas.1318899111.

Xu, Y., Liu, M., Simpson, P. J., Isaacson, R., Cota, E., Marchant, J., et al. (2009). Automated assignment in selectively methyl-labeled proteins. *Journal of the American Chemical Society*, *131*(27), 9480–9481. https://doi.org/10.1021/ja9020233.

Xu, Y., & Matthews, S. (2013). TROSY NMR spectroscopy of large soluble proteins. *Topics in Current Chemistry*, *335*(1), 97–120. https://doi.org/10.1007/128_2011_228.

Yamazaki, T., Lee, W., Arrowsmith, C. H., Muhandiram, D. R., & Kay, L. E. (1994). A suite of triple resonance NMR experiments for the backbone assignment of ^{15}N, ^{13}C, ^{2}H labeled proteins with high sensitivity. *Journal of the American Chemical Society*, *116*, 11655–11666. https://doi.org/10.1021/ja00105a005.

Yuwen, T., Vallurupalli, P., & Kay, L. E. (2016). Enhancing the sensitivity of CPMG relaxation dispersion to conformational exchange processes by multiple-quantum spectroscopy. *Angewandte Chemie, International Edition*, *55*(38), 11490–11494. https://doi.org/10.1002/anie.201605843.

CHAPTER TWO

Isotopic Labeling of Eukaryotic Membrane Proteins for NMR Studies of Interactions and Dynamics

Igor Dikiy*[,2], Lindsay D. Clark[†,‡,2], Kevin H. Gardner*[,§,¶,1], Daniel M. Rosenbaum[†,‡,1]

*Structural Biology Initiative, CUNY Advanced Science Research Center, New York, NY, United States
[†]Department of Biophysics, The University of Texas Southwestern Medical Center, Dallas, TX, United States
[‡]Molecular Biophysics Graduate Program, The University of Texas Southwestern Medical Center, Dallas, TX, United States
[§]Department of Chemistry and Biochemistry, City College of New York, New York, NY, United States
[¶]Biochemistry, Chemistry and Biology Ph.D. Programs, Graduate Center, City University of New York, New York, NY, United States
[1]Corresponding authors: e-mail address: kevin.gardner@asrc.cuny.edu; dan.rosenbaum@utsouthwestern.edu

Contents

[2] These authors contributed equally.

Methods in Enzymology, Volume 614
ISSN 0076-6879
https://doi.org/10.1016/bs.mie.2018.08.030

Abstract

Membrane proteins, and especially G-protein coupled receptors (GPCRs), are increasingly important targets of structural biology studies due to their involvement in many biomedically critical pathways in humans. These proteins are often highly dynamic and thus benefit from studies by NMR spectroscopy in parallel with complementary crystallographic and cryo-EM analyses. However, such studies are often complicated by a range of practical concerns, including challenges in preparing suitably isotopically labeled membrane protein samples, large sizes of protein/detergent or protein/lipid complexes, and limitations on sample concentrations and stabilities. Here we describe our approach to addressing these challenges via the use of simple eukaryotic expression systems and modified NMR experiments, using the human adenosine A_{2A} receptor as an example. Protocols are provided for the preparation of U-^{2}H (^{13}C,^{1}H-Ile δ1)-labeled membrane proteins from overexpression in the methylotrophic yeast *Pichia pastoris*, as well as techniques for studying the fast ns–ps sidechain dynamics of the methyl groups of such samples. We believe that, with the proper optimization, these protocols should be generalizable to other GPCRs and human membrane proteins.

1. INTRODUCTION

Integral membrane proteins make up a large proportion of the genomes of many organisms—approximately 25% of the human genome—and perform a diverse range of functions, including key steps in the communication of a cell with its environment. Because of their biological and therapeutic importance (Almén, Nordström, Fredriksson, & Schiöth, 2009), membrane proteins are the focus of fundamental and applied biophysical research to characterize three-dimensional structures, dynamics, and interactions in native-like environments.

Solution-state NMR spectroscopy has played a critical role in membrane protein biophysical studies, as the site-specific dynamic and interaction information provided by such approaches nicely complements structural data obtained from X-ray diffraction, cryo-EM, and computational analyses (Cuniasse, Tavares, Orlova, & Zinn-Justin, 2017; Opella & Marassi, 2017). Fundamental to such studies are several 2D "fingerprint spectra," most often ^{15}N/^{1}H HSQC (heteronuclear single-quantum coherence) spectra (for backbone amide plus Trp, Asn, and Gln sidechains) or methyl ^{13}C/^{1}H HMQC (heteronuclear multiple-quantum coherence) spectra for sidechain methyl groups (Pellecchia et al., 2008). These methyl-directed experiments are especially advantageous for large, slow-tumbling membrane protein/lipid complexes; experiments directed to other sidechain and mainchain sites have been successfully applied as well. NMR experiments can provide information about protein dynamics over many timescales, from fast (ps–ns) sidechain motions to slow conformational changes (μs–ms) (Kasinath, Sharp, & Wand, 2013; Liang & Tamm, 2016; Palmer, 2012; Wand, Moorman, & Harpole, 2013). Many of these dynamics experiments, often using sidechain methyl groups as probes, have been adapted and developed for large biomolecular systems and can be used for membrane proteins (Rosenzweig & Kay, 2014; Sun, Kay, & Tugarinov, 2011; Tugarinov, Hwang, Ollerenshaw, & Kay, 2003).

A particular advantage of solution-state NMR is that proteins are studied in a native-like solution state where they can interconvert among multiple conformations. However, membrane proteins must be solubilized in a suitable membrane mimetic that maintains native structure and dynamics. Different options include detergent micelles, amphipols, bicelles, nanodiscs, SMALPs, and lipid vesicles, each having their own benefits and drawbacks (Liang & Tamm, 2016, 2018; Zhou & Cross, 2013). It is often necessary to test different solubilization strategies for a given protein sample for stability, signal intensity and resolution, and native structure/activity. Two important considerations for all membrane mimetics are (1) a uniform and small particle size and (2) a high extent of deuteration.

High-level deuteration, both within the membrane mimetic and protein itself, is critical to reduce the number of ^{1}H signals present in spectra (including those from lipids, which can be intense) and to improve the relaxation characteristics of the remaining NMR-active spins in the sample. While deuteration is possible for the membrane mimetic through the purchase/synthesis of deuterated compounds, replacing ^{1}H with ^{2}H in proteins requires biosynthetic incorporation. For backbone experiments in eukaryotic expression systems, one can label uniformly with ^{15}N to observe all amides (Eddy et al., 2018; Opitz, Isogai, & Grzesiek, 2015) or through addition of specifically labeled amino acids (Isogai et al., 2016). For methyl groups, one can provide either appropriately labeled amino acids or amino acid precursors (particularly alpha-keto acids) to growth media to access various labeling patterns in the sidechains of several amino acids (Kofuku et al., 2014, 2018). We find isoleucine δ1 methyl groups particularly useful given (1) the abundance of Ile residues in integral membrane proteins including GPCRs (Ulmschneider & Sansom, 2001), (2) the far upfield ^{13}C shift of isoleucine δ1 methyl groups [average 13.5 ± 3.6 ppm ^{13}C according to BioMagResBank (Ulrich et al., 2008)], putting them in a particularly uncrowded region of 2D ^{13}C/^{1}H spectra, (3) the lack of need to stereospecifically assign these methyl groups, unlike Val and Leu, and (4) the presence of multiple, freely rotatable bonds between the methyl group and protein backbone, providing substantial independence of dynamics at these sites (Kasinath et al., 2013).

While many of the aforementioned labeling strategies have been well developed for *E. coli*, many integral membrane proteins can only be expressed at high levels in eukaryotic hosts. Among these, the methylotrophic yeast *Pichia pastoris* is a convenient host for heterologous expression and isotopic labeling of eukaryotic membrane proteins (Clark, Dikiy, Rosenbaum, & Gardner, 2018). Advantages of *Pichia* include rapidity of genetic manipulation, high yields of recombinant protein, existence of post-translational modification (PTM) and chaperone machinery necessary for eukaryotic membrane proteins, and ability to grow on defined minimal media allowing for perdeuteration (Cereghino & Cregg, 2000; Morgan, Kragt, & Feeney, 2000). Uniform isotopic labeling in *Pichia* has been well established (Morgan et al., 2000; Pickford & O'Leary, 2004). We have extended this work by demonstrating the ^{13}C, ^{1}H labeling of isoleucine δ1-methyl groups in a perdeuterated background by adding labeled α-ketobutyrate (~50% labeling, ~90% deuteration) to highly deuterated growth media (Clark et al., 2017, 2015). In contrast, simultaneous labeling of leucine δ- and valine γ-methyl groups with α-ketoisovalerate is inefficient but can be achieved by adding labeled valine directly to the growth media or modifying culture conditions (Clark et al., 2015; Suzuki et al., 2018; Zhang et al., 2017). *Pichia* can readily take up additional amino acids from media, with a general correlation between uptake efficiency and the energetic cost to synthesize that amino acid type de novo (Heyland, Fu, Blank, & Schmid, 2011). However, after uptake into cells, labeled amino acids can be fed into metabolic pathways (Solà, Maaheimo, Ylonen, Ferrer, & Szyperski, 2004), diluting signal of desired amino acids and complicating data analysis by isotopic scrambling. Alternatively, auxotrophic strains can be developed for labeling a specific amino acid; however, care must be taken to confirm that off-target effects in other metabolic pathways do not arise (Whittaker, 2007).

Here we provide detailed protocols needed to generate such U-^{2}H (^{13}C, ^{1}H-Ile δ1 methyl)-labeled integral membrane proteins by overexpression in *Pichia*, using the human adenosine A_{2A} receptor [A_{2A}R] as a model system. We further detail how such samples can be used in solution NMR studies, from acquiring simple ^{13}C/^{1}H HMQC spectra, through chemical shift assignments by site-directed mutagenesis, to analyses of ^{1}H–^{1}H cross-relaxation measurements of fast sidechain dynamics.

2. EXPRESSION AND ISOTOPIC LABELING IN *PICHIA*

The overall strategy for expression of membrane proteins in *Pichia* is outlined below. After cloning the gene of interest into a *Pichia* expression vector [see *Note 1* below and several literature reviews where such reagents are comprehensively discussed (Byrne, 2015; Cos, Ramon, Montesinos, & Valero, 2006; Fan, Emami, Munro, Ladizhansky, & Brown, 2015)], the total process to obtaining initial expression cultures takes about a week. While *Pichia* can be grown in large-scale fermenters (Cereghino, Cereghino, Ilgen, & Cregg, 2002), this protocol describes growth in shake flasks, as this is the most straightforward and cost-efficient way to initiate *Pichia* expression at levels sufficient for structural biology work. Although the process outlined here is a good general starting point for membrane protein expression, it is likely that each sample will require optimization to achieve high levels of expression, and pertinent notes are indicated after each step.

2.1 Equipment

1. Plate incubator (28°C)
2. DNA agarose gel apparatus
3. Electroporator (e.g., Bio-Rad Micropulser) and cuvettes
4. Temperature-controlled shake flask incubator, ideally dedicated for *Pichia* expression

2.2 Required Buffers and Media (All Solutions 0.2 μm Sterile-Filtered, Store All at 4°C)

1. 1 *M* sorbitol
2. 20% w/v dextrose (use as 10 × stock)
3. 10 × YNB (per 1 L: 34 g yeast nitrogen base without amino acids or nitrogen, 100 g ammonium sulfate)
4. 1 *M* Potassium phosphate pH 6.0 (use as 10 × stock)
5. 2000 × biotin (0.08% in 0.05 *M* NaOH)
6. 200 × histidine (0.8% in water)
7. MD agar (1.5% agar, 1 × YNB, 1 × dextrose, 1 × biotin, 1 × histidine if required ("MDH" plates); add all stock solutions after autoclaving agar and water)
8. 10% w/v glycerol (use as 10 × stock)

9. YPD (1% yeast extract, 2% peptone, 1 × dextrose; dextrose solution added after autoclaving other components)
10. YPDS (YPD + 2% agar + 1 *M* sorbitol; sorbitol added after autoclaving)
11. BMG (1 × glycerol, 1 × potassium phosphate pH 6.0, 1 × YNB, 1 × biotin, and 1 × histidine if required)
12. BMM without methanol (1 × potassium phosphate pH 6.0, 1 × YNB, 1 × biotin, 1 × histidine if required)

2.3 Procedure

1. Clone the gene of interest into the desired vector for expression in *Pichia*. The cDNA for wild-type human ADORA2A adenosine A_{2A} receptor with C-terminal His_6 and Protein C affinity tags was cloned into the pPICZ vector containing a modified MFα secretion signal (see *Note 1* below). Molecular biology steps were carried out using standard techniques and utilizing Zeocin selection.
   a. *Note 1*: Although some membrane proteins have been successfully purified when expressing intracellularly (Byrne, 2015; Lee et al., 2016), we have observed that the best expression of $A_{2A}R$ occurred when the gene was preceded by the MFα signal sequence from *Saccharomyces cerevisiae*. Furthermore, the highest amount of fully processed (e.g., plasma membrane localized) receptor occurred when the signal sequence was modified for more efficient processing. These modifications were based on previous literature precedents (Lin-Cereghino et al., 2013; Rakestraw, Sazinsky, Piatesi, Antipov, & Wittrup, 2009), comprising the V22A, G40D, L42S, V50A, V52A, and F55L mutations plus the deletion of residues 57–70 (Fig. 1).
   b. *Note 2*: GPCRs and other membrane proteins commonly contain PTMs and/or large flexible regions that are problematic for NMR or other structural analyses (Yurugi-Kobayashi et al., 2009). PTM target residues can be removed though point mutations and flexible regions can be truncated during this stage. While *Pichia* does not hyperglycosylate overexpressed proteins to the extent of other yeast such as *S. cerevisiae* (Pickford & O'Leary, 2004), a simple way to minimize this possibility is to remove targeted asparagine residues. To avoid such issues with $A_{2A}R$, we mutated an N-glycosylation site at Asn154 to glutamine and truncated the construct at residue 316 as previously described (Jaakola et al., 2008).

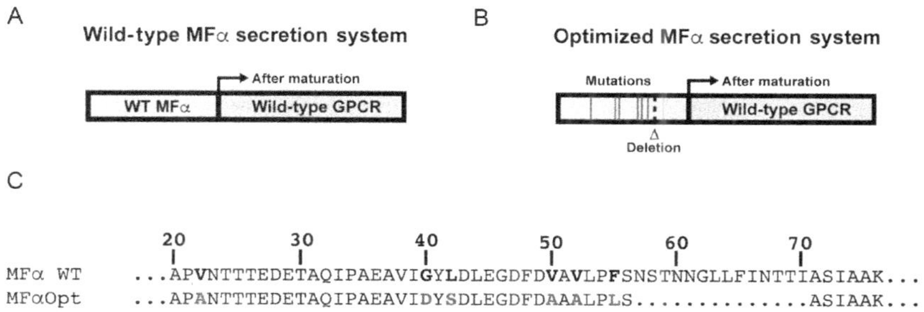

Fig. 1 Optimization of MFα signal sequence. (A, B) General schematics of WT and optimized MFα signal sequences prior to start of WT GPCR gene of interest. As shown in alignments in (C), six mutations (V22A, G40D, L42S, V50A, V52A, and F55L) plus a deletion of residues 57–70 generate an optimized version of the MFα signal sequence ("MFαOpt") that was used for efficient $A_{2A}R$ expression at the plasma membrane.

2. Prepare a fresh batch of competent KM71H cells. Briefly, cells should be grown from a freshly streaked MD plate in YPD to an OD_{600} of 1, then spun down under sterile conditions at 4000 rcf for 10 min. Cells were washed and repelleted in ice-cold, sterile Milli-Q water twice, followed by a wash in ice-cold 1 *M* sorbitol and resuspension in a minimal volume of 1 *M* sorbitol to create a thick, but pipettable, cell paste (e.g., ~500 μL for 500 mL pellet). Competent cells may be kept on ice at this point until electroporation.
   a. *Note 1*: It is important that the competent cells are grown and prepared fresh for efficient transformation, preferably no more than 24 h prior to electroporation. Sterility must be maintained as much as possible.
   b. *Note 2*: The popularity of *Pichia* as an expression host has expanded the options for expression and culturing conditions. *Pichia* is commercially available in various cell strains from several vendors including Invitrogen, including the most commonly used strains for MP expression (KM71/H, GS115, SMD1163/8, and wild-type X-33). KM71H is the strain that gave us the highest expression levels of $A_{2A}R$, although other membrane proteins have been successfully overexpressed and purified from alternative strains for structural studies (Byrne, 2015).
3. Purify 10–20 μg of plasmid prior to linearization. Linearize with *PmeI* or appropriate restriction enzyme and use a small amount of plasmid to verify efficient digestion with a DNA agarose gel. Repurify linearized plasmid with PCR Clean-up Kit (Qiagen), with the final elution step carried out in a minimal volume of sterile Milli-Q water.

a. *Note*: It is important to use a large amount of linearized plasmid (5 μg is ideal) for each electroporation to maximize efficiency of incorporation into the *Pichia* genome. If screening multiple *Pichia* strains, it is helpful to increase the scale of plasmid linearization and divide the resultant DNA between strains.

4. Add linearized plasmid to ~100 μL of competent *Pichia*, and transfer to an electroporation cuvette with a 0.2 cm gap. Include an extra cuvette containing cells and an equal volume of sterile water instead of plasmid as a negative control. Process each cuvette with 1 pulse of 2.00 kV ("Pic" program on Bio-Rad MicroPulser system), and immediately transfer cuvette back to ice. Add 1 mL of ice-cold 1 *M* sorbitol and transfer the contents of the cuvette to a 14 mL round-bottom culture tube.
   a. *Note*: Arcing during electroporation is rare but can happen if there is salt in the sample. This risk can be minimized by ensuring that the paste of competent cells has been thoroughly rinsed of media and eluting linearized plasmid with water instead of elution buffer.
5. Incubate cells at 28°C for 30–60 min without shaking, and then add 1 mL YPD to each tube and shake at 200 rpm for 4 h at 28°C.
6. Plate cells on YPDS plates containing increasing amounts of Zeocin (100–1000+ μg/mL) to check for efficiency of gene integration. Incubate plates at 28°C for 3–4 days. "Pinprick" colonies should appear after 48 h and individual colonies should be obvious (~1 mm diameter) after 72 h.
7. Individual colonies should be expanded and screened for expression levels, generally through western blots, as simply using the copy number as a proxy for expression levels is not advisable (Aw & Polizzi, 2013). Typically, 10–12 clones from each sample are selected mostly (but not solely) from the plates containing the higher levels of Zeocin and easily expanded on a single MD (or MDH) plate. Individually pick colonies with a sterile pipet tip and gently spread each colony on a labeled area of the plate. After 24 h of incubation at 28°C, each colony will grow into a small "patch" that can be easily looped for screening cultures and subsequent preservation (see *Step 10* below).
8. Grow a small culture (~10 mL) of each clone by inoculating a loopful of yeast into sterile BMG media and incubating at 200 rpm/28°C. After growing to saturation (OD_{600} ~20–30), spin down cells at 4000 rcf for 5 min. While maintaining sterility, remove media and replace with an equal volume of BMM media without methanol. After continued

shaking at 28°C for 4–6 h to allow for metabolism of residual glycerol, induce expression with the addition of 0.5% methanol. Additional aliquots of 0.5% methanol should be added every 12 h to maintain expression, and cells can be harvested and subjected to western blots after 48 h.

a. *Note 1*: Although each sample will likely require individual optimization to achieve the best expression conditions, the strategy outlined above can be viewed as a general starting point. Common additives to increase expression levels are DMSO (André et al., 2006), histidine (even if not required by cell strain as it is an energetically "expensive" amino acid to synthesize; Heyland et al., 2011), and individual ligands, especially in the case of GPCRs (André et al., 2006; Clark et al., 2017; Hino et al., 2012; Shimamura et al., 2011). It is also common to use unbuffered media (Lee et al., 2016), BMG/BMM at different pH levels (known to influence expression and activity levels of various GPCRs (André et al., 2006; Yurugi-Kobayashi et al., 2009)), or alter the temperature and length of expression (André et al., 2006; Yurugi-Kobayashi et al., 2009). In all cases, effects on protein expression can be assayed for with western blots.

b. *Note 2*: If methanol is not suitable for expression induction due to toxicity or other sample constraints, proteins can be expressed constitutively through use of the commercially available pGAPZ vectors (Invitrogen).

9. Once the best expression clone is identified, it can be stored as a glycerol stock at −80°C and streaked onto fresh MD (or MDH) plates as needed. To prepare a culture for a glycerol stock, inoculate a loopful of the colony "patch" in YPD and shake for 24 h at 28°C. Store culture in cryovials with a final glycerol concentration of 25%.

10. To grow deuterated and isotopically labeled cultures for NMR experiments, use a loopful of freshly streaked cells from a MD (or MDH) plate and inoculate 50 mL of BMG media made with 90% D_2O/10% H_2O. Shake the culture at 28°C until it reaches an OD_{600} of ~10 (approximately 36–48 h). Use 100 μL of this culture to inoculate 50 mL of BMG media made with 100% D_2O. Shake the culture at 28°C until it reaches an OD_{600} of ~10 (approximately 36–48 h). Use 100 μL of this culture to inoculate 50 mL of BMG media made with 100% D_2O, but replacing the 1% glycerol with 1% d_8-glycerol. Shake the culture until reaching an OD_{600} of ~10, and then use the culture in its entirety to inoculate large-scale (1 − 2 L) cultures of identical media.

Shake the large-scale cultures at 28°C until reaching saturation (OD_{600} of ~20–30), and then pellet cells in sterile 1 L bottles at 4000 rcf for 30 min. Remove the media and gently resuspend cell pellets in BMM media without methanol. Continue to shake flasks at 28°C for ~12 more hours to metabolize any remaining d_8-glycerol. One hour prior to induction, add 200 mg/L of labeled α-ketobutyric acid (methyl-^{13}C, 99%; 3,3-D2, 98%; Cambridge Isotope Laboratories). For $A_{2A}R$ cultures, add 4 m*M* dry theophylline (low-affinity antagonist) 10 min prior to induction, and induce with 0.5% d_4-methanol. Reduce the temperature to 20°C after induction. As with the small test cultures, add additional aliquots of 0.5% d_4-methanol roughly every 12 h to maintain expression levels. Harvest cells after 36–48 h and store at −80°C. See Fig. 2 for a schematic of the deuterated media adaptation procedure.

a. *Note 1*: Sterility should be maintained as much as possible during adaptation due to the lack of antibiotics in the media. Typically, this requires a dedicated shaker for *Pichia* cultures and a clean flame area for inoculations and media transfers.

b. *Note 2*: Instead of d_8-glycerol, deuterated glucose can be used as an alternative (and somewhat less expensive) carbon source in *Pichia* cultures (Emami, Fan, Munro, Ladizhansky, & Brown, 2013; Fan et al., 2015; Fan, Shi, Ladizhansky, & Brown, 2011). However, glucose is a very strong repressor of the AOX promoter; thus, any residual glucose in the cells or media can result in toxic conditions during induction (Wood & Komives, 1999).

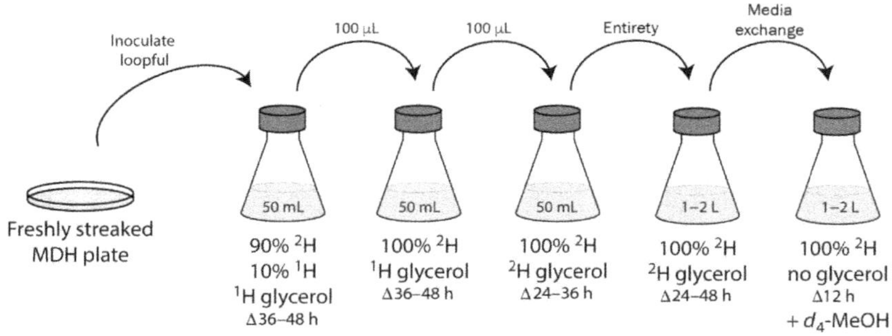

Fig. 2 Adaptation of *Pichia pastoris* to deuterated media. Summary of schemes used to adapt *Pichia* cultures to increasingly deuterated media to ensure efficient deuteration levels and protein expression. See Section 2 for more details.

c. *Note 3*: Deuterated media can be saved and recycled to offset some of the costs of deuteration. Recycling can be done in-house by flash distillation (Moore, 1979) or shipped to Cambridge Isotope Laboratories for their D_2O recovery program (https://www.isotope.com/sales-technical-support/D2O.cfm).

3. PURIFICATION/NMR SAMPLE PREPARATION

During optimization of culture and expression conditions (prior to purification), it is important to validate expression of $A_{2A}R$ or other target proteins via western blots using specific antibodies to the protein or affinity tag(s). In our experience, a 10–20 μL volume of cells from the final harvested culture is typically sufficient to obtain a robustly visible signal with an anti-His_6 antibody (Pierce) for $A_{2A}R$ or other His-tagged membrane proteins. Once an optimized level of expression is demonstrated, the receptor can be purified from *Pichia* pellets following the procedure described below. Ideally, the purification should be "straight-through," without any substantial delays from cell disruption to final sample; if required, the affinity-purified receptor can be flash-frozen in liquid N_2 prior to SEC with minimal (10%–15%) losses of final yield.

3.1 Equipment

Stir plate
Glass Dounce homogenizer with loose and tight plungers
Refrigerated ultracentrifuge (100,000 + rcf)
Orbital shaker
FPLC with Superdex 200 or comparable size exclusion column

3.2 Buffers

1. Lysis buffer: PBS containing 10% glycerol, 4 m*M* theophylline (low-affinity antagonist for $A_{2A}R$), 2 m*M* EDTA, and protease inhibitors (160 μg/mL benzamidine, 2.5 μg/mL leupeptin, 1 m*M* PMSF, 1 μ*M* E-64).
2. Hypotonic buffer: 10 m*M* HEPES pH 7.5, 2 m*M* EDTA, 4 m*M* theophylline, protease inhibitors as above.
3. Solubilization buffer: 500 m*M* NaCl, 50 m*M* HEPES pH 7.5, 20% glycerol, 1% w/v DDM (Anatrace), 4 m*M* theophylline, and protease inhibitors as above.

4. TALON equilibration buffer: 250 m*M* NaCl, 50 m*M* HEPES pH 7.5, 0.05% w/v protonated DDM, 5% glycerol, 4 m*M* theophylline, and 30 m*M* imidazole.
5. Buffers A–E: 250 m*M* NaCl, 50 m*M* HEPES pH 7.5, 5% glycerol, 20 m*M* imidazole, ligand of interest (e.g., 10 μ*M* of the $A_{2A}R$ antagonist ZM241385 or 20 μ*M* of the agonist NECA), made in D_2O, and containing 0.05% w/v DDM in the following protonated:deuterated ratios: (A) 4:0; (B) 3:1; (C) 2:2; (D) 1:3; (E) 0:4.
6. SEC buffer: 150 m*M* NaCl, 20 m*M* HEPES pH 7.5, 0.05% w/v deuterated DDM, ligand of interest (e.g., 10 μ*M* ZM241385 or 20 μ*M* NECA), 99% D_2O. Store at 4°C after 0.2 μm sterile filtering.

3.3 Procedure

Day 1

1. Thaw cell pellets and resuspend in 200–300 mL cold lysis buffer. All volumes in this protocol are given for a culture of 1 L, and should be scaled appropriately for different volumes.
2. Lyse cells with a high-pressure microfluidizer (Microfluidics M-110P or similar), using three passes at 24,000 psi, allowing the lysate to cool between passes. Pack ice or ice packs around microfluidizer coils to minimize sample heating.
3. To digest yeast cell wall glycans, add zymolyase (LongLife Zymolyase from G-Bio Sciences or similar) to 15 U/mL and incubate at 37°C for 1 h with stirring.
4. Separate out total membranes by centrifugation at 140,000 rcf for 30 min. Carefully decant the supernatant and keep membrane pellets on ice.
5. To wash membranes, resuspend the pellets in 200–300 mL of cold lysis buffer using a Dounce homogenizer (loose, 10 passes), then centrifuge at 140,000 rcf for 30 min again. Carefully decant the supernatant and keep membrane pellets on ice.
6. To lyse any intact spheroplasts and further wash the membranes, resuspend the pellets in 200–300 mL of cold hypotonic buffer using a Dounce homogenizer (loose, 10 passes). Incubate at 4°C for 30 min with stirring, then centrifuge at 140,000 rcf for 30 min. Carefully decant the supernatant and keep membrane pellets on ice.
7. To solubilize membranes, resuspend the pellets in cold solubilization buffer using a Dounce homogenizer (loose, 10 passes; tight, 30 passes)

and incubate at 4°C for 2h with stirring. Centrifuge at 140,000 rcf for 30 min to remove insoluble material.

a. *Note*: The addition of cholesteryl hemisuccinate (CHS) to the micelles as a cholesterol analog has been demonstrated to increase the stability and activity of $A_{2A}R$ and other GPCRs (Liu et al., 2012; Weiss & Grisshammer, 2002). We initially solubilized and purified $A_{2A}R$ in DDM buffer with CHS included at a concentration of 0.2% and 0.01% w/v, respectively (Fig. 3A). While the inclusion of CHS improved NMR sample stability (Fig. 3B), it gave rise to an intense methyl peak at around 25 ppm ^{13}C/0.85 ppm ^{1}H that interfered with analyses of NMR spectra (particularly relaxation measurements detailed below, which were extremely difficult to analyze until CHS was removed from the preparation). We note that a small peak at the same chemical shift remained in our CHS-free samples (Fig. 3A), suggesting that the receptor copurified with a cholesterol-like molecule.

8. During the final centrifugation step, preequilibrate 10–15 mL of fresh immobilized cobalt resin (such as TALON resin from Clontech) in TALON equilibration buffer. After centrifugation, decant supernatant into 250 mL tubes and add 30 m*M* imidazole to reduce background nonspecific binding. Incubate the supernatant with the equilibrated resin and bind overnight at 4°C with gentle agitation.

Day 2

1. Spin down loose TALON resin at 2000 rcf for 10 min, remove the majority of the unbound material, and transfer to a gravity flow column. Through a series of wash steps, exchange the protonated DDM for deuterated DDM and simultaneously exchange theophylline (low-affinity antagonist) for a high-affinity ligand of interest (or no ligand). This is achieved by washing resin sequentially with 4 column volumes (CV) each of buffers A–E. Wash steps should be carried out at 4°C and at a flow rate of 1–1.5 mL/min to ensure efficient exchange.
2. Elute tagged protein with Buffer E + 250 m*M* imidazole. $A_{2A}R$ should elute within 2 CV. If necessary, purified $A_{2A}R$ can be snap frozen in liquid N_2 in a buffer containing glycerol at a final concentration of 15% (e.g., addition of 10% glycerol to eluate).

   a. *Note*: Another option for solution-state NMR is reconstituting the proteins into nanodiscs, in which the protein is surrounded by lipids held together by an apolipoprotein band (Bayburt & Sligar, 2002, 2010; Denisov, Grinkova, Lazarides, & Sligar, 2004; Puthenveetil,

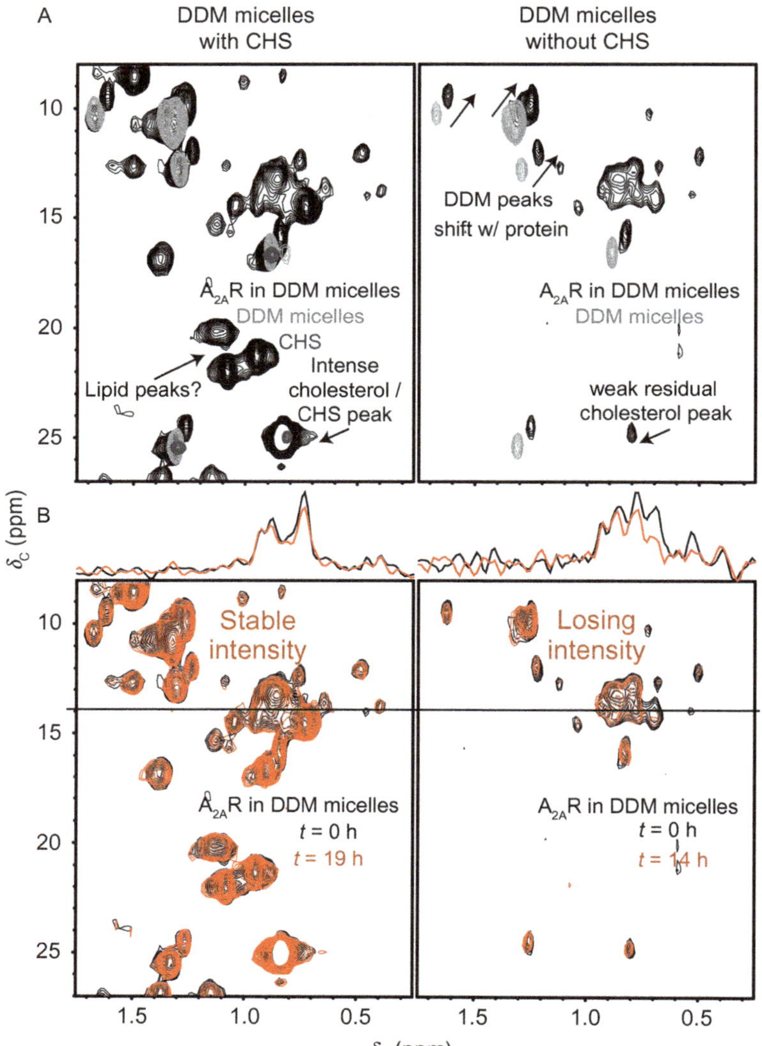

Fig. 3 Receptors reconstituted with CHS are more stable but have crowded spectra. (A) $^{13}C/^{1}H$ HMQC spectra of $A_{2A}R$ in DDM micelles with CHS (*left*) and without added CHS (*right*). Reconstitution with CHS resulted in spectra with intense extraneous signals, likely arising from lipids, as well as a very intense cholesteryl methyl peak (*dark gray*). In DDM micelles without added CHS, the dispersion of protein-derived peaks is comparable, but there are fewer extraneous peaks in the region. A weak signal at the CHS/cholesterol methyl chemical shift suggests that some sterol was copurified with the protein. (B) $^{13}C/^{1}H$ HMQC spectra of $A_{2A}R$ in DDM micelles with CHS (*left*) and without added CHS (*right*) collected after different lengths of time at 30°C. Several protein signals in the DDM micelles without CHS drop significantly after 14 h (*dark orange*).

Nguyen, & Vinogradova, 2017). We carried out experiments in which $A_{2A}R$ was reconstituted into deuterated DMPC nanodiscs containing 10% cholesterol. While these samples exhibited a substantially improved stability under NMR conditions, this advantage was negated by the presence of many intense lipid-derived peaks in the methyl region which significantly complicated spectral analysis. As a result, we decided to use deuterated DDM micelles as our membrane mimetic. For further discussion of nanodisc use in NMR, see also chapter "Synthetic biology-based solution NMR studies on membrane proteins in lipid environments" by Henrich et al.

3. Concentrate eluted protein using 100 kDa MWCO centrifugal concentrators to 1 mL for injection on a Superdex 200 or comparable column (e.g., S200 GL Increase 10/300). As with many other membrane proteins, it is important to frequently resuspend the sample during concentration and spin at a slow speed (<3500 rcf) to reduce aggregation.
4. Preequilibrate Superdex 200 column with SEC buffer, inject concentrated protein, and collect elution peak. The peak should elute at a volume commensurate with monomeric receptor with bound micelle.
   a. *Note*: Sodium has been shown to be an allosteric modulator of $A_{2A}R$ and other Class A GPCRs (Gao & Ijzerman, 2000; Liu et al., 2012). We obtained high-quality spectra of $A_{2A}R$ in the presence of 150 m*M* NaCl, although the minimal chemical shift differences observed between agonist- and antagonist-loaded samples suggest that cation choice has a substantial effect in this system. In contrast, purification in 150 m*M* KCl led to more heterogeneous spectra and less stable samples that were resistant to relaxation experiments and analysis, although there was a clear difference between agonist- and antagonist-bound spectra under these conditions (Clark et al., 2017).
5. Concentrate eluted protein using 100 kDa MWCO centrifugal concentrators to ~100 μL for NMR samples. Frequently resuspend sample and spin at a slow speed (<3500 rcf) to reduce aggregation. Using 3 mm Shigemi tubes (100 μL volume) allows us to concentrate samples to a concentration of ~100 μ*M*, leading to adequate signal-to-noise ratios for relaxation data.

Final Note: After establishing a purification protocol, it is important to verify that the protein sample is functional. In the case of GPCRs, the standard readout is coupling to G-protein. This is achieved by liposome reconstitution and G-protein activation assays (Cerione & Ross, 1991) and/or G-protein complex formation in detergent followed by subsequent gel filtration chromatography steps (Rasmussen et al., 2011). For $A_{2A}R$, we verified functionality with both methods (Clark et al., 2017) and found that the purification yielded a high level of functional receptor that can efficiently couple to the G_s heterotrimer.

4. RUNNING $^{13}C/^{1}H$ HMQC AND METHYL RELAXATION EXPERIMENTS

Once the protein sample is generated, the protein can be analyzed by NMR. A $^{13}C/^{1}H$ HMQC experiment reports on overall folding and stability of the protein sample, as well as the effects of any perturbations, such as addition of binding partners. To be able to interpret the NMR spectra, some or all of the methyl peaks must be assigned to the residues that they arise from. Finally, residue-specific dynamics at various time scales can be measured by specialized NMR experiments. Here, we focus on our application of a methyl relaxation experiment measuring the fast (ps–ns) sidechain motions (Sun et al., 2011; Tugarinov, Sprangers, & Kay, 2007) and adaption to short-lived membrane protein samples.

4.1 Equipment and Software

High-field NMR spectrometer with cryoprobe (e.g., Bruker AVANCE III HD 800 MHz with TCI probe)
3 mm solvent-matched Shigemi NMR tube
NMRpipe software (Delaglio et al., 1995)
Python installation with NumPy and SciPy modules

4.2 Procedure

1. Load the sample into an NMR tube (e.g., 3 mm solvent-matched Shigemi tube). Insert tube into spinner and spinner into high-field NMR spectrometer with cryogenically cooled probe.
2. Set temperature to 30°C and allow to equilibrate for ~5 min.
3. Lock and shim the sample, tune the probe head for ^{1}H and ^{13}C, and determine the ^{1}H 90° pulse width as per usual NMR protocols.
4. Acquire a 1D ^{1}H spectrum to look for detectable isoleucine methyl signals (approx. −0.5 to 1.2 ppm) that are not swamped by other signals (buffer, detergent, lipids).
5. Acquire a $^{13}C/^{1}H$ HMQC (Ollerenshaw, Tugarinov, & Kay, 2003) spectrum centered on the isoleucine methyl region to assess sample quality, stability, and degree of secondary and tertiary structure.

Table 1 Typical Spectral Parameters for $^{13}C/^{1}H$ HMQC Spectra of U-^{2}H ($^{13}C,^{1}H$ Ile δ1 CH_3)-Labeled GPCR Samples

Nucleus	Center Frequency (ppm)	Spectral Width (ppm)	Number of Points	Acquisition Time (ms)
^{1}H	4.7	13.35	1024	7
^{13}C	14	22	64	48

a. *Note 1*: The optimal spectral parameters may vary a bit from sample to sample. The ^{13}C and ^{1}H carrier frequencies, spectral widths, and number of points collected for our $A_{2A}R$ samples are shown in Table 1. For a new sample, it is recommended to collect a spectrum with increased spectral width in the ^{13}C dimension (e.g., 40 ppm centered at 20 ppm) to determine where all isoleucine δ1 methyl peaks appear and where other signals arising from the sample (e.g., lipid, detergent, buffer, ligand) may fall. This initial spectrum can then be used to design a combination of spectral width and center frequency that avoids aliasing in contaminating peaks over the peaks of interest.

b. *Note 2*: Wild-type eukaryotic membrane proteins, and especially GPCRs, have limited stability in the micelle-solubilized form required for obtaining well-resolved spectra in solution-state NMR studies. As shown in Fig. 3B, the intensities of several $A_{2A}R$ signals dropped significantly at ~12 h. With this in mind, it is highly recommended to collect consecutive somewhat short (1 − 2 h) HMQC experiments, process them independently, and look whether peaks are decreasing or changing radically with time. Those experiments that result in similar spectra can then be summed and processed together to increase the signal-to-noise ratio. NMRpipe (Delaglio et al., 1995) offers simple commands ("addNMR") for summing NMR data.

6. To assign isoleucine δ1 methyl peaks, generate mutants in which a single isoleucine has been mutated to a different residue type, then express, purify, and collect a $^{13}C/^{1}H$ HMQC spectrum on each as described above. Ideally, one peak will disappear and that peak can be assigned to the mutated residue.

a. *Note*: This method relies on the mutation perturbing as little of the overall protein structure as possible. Thus, it is important to choose residues to mutate (and thus assign) that are biologically interesting but not indispensable for the proper folding of the protein. The

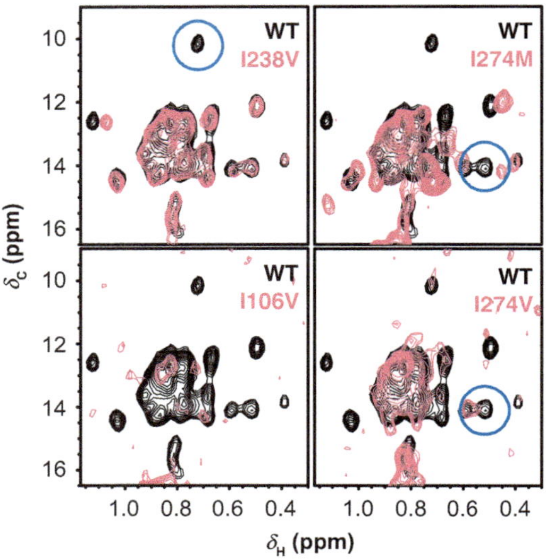

Fig. 4 Assignment of selected $A_{2A}R$ Ile-δ1 methyl peaks by site-directed mutagenesis. $^{13}C/^{1}H$ HMQC spectra of wild-type $A_{2A}R$ (*black*) are shown overlaid with four mutants (*purple*). In some cases, such as I238, only one peak disappeared and was assigned (*blue circle*). In others, such as I274, manifold spectral changes were present, and multiple mutants were needed, along with solvent PRE data (Clark et al., 2017) to make the assignment (*blue circle*). In other cases, such as I106, the mutation completely abrogated protein folding or stability.

identity of the mutant also plays a large role; we have found that in general isoleucine to valine mutations work well; however, in certain cases it may be necessary to test multiple mutants, such as leucine or methionine (Fig. 4). This is often the most time- and cost-limiting step of a study. It is recommended to limit assignments to the most interesting or important residues.

7. On the same, or ideally on another freshly prepared sample, acquire relaxation measurements. Here we focus on the relaxation-violated coherence transfer experiment and its measurements of methyl order parameters (S^2_{axis}) which report on fast (ps–ns) motions of the sidechains (Sun et al., 2011; Tugarinov et al., 2007). The experiment and analysis are outlined in Fig. 5A. Additional discussion of the practical aspects of collection and analysis of methyl relaxation data is presented in Chapter "Characterization of internal protein dynamics and conformational entropy by NMR relaxation" by Stetz et al. Other experiments for measuring slower (μs–ms) dynamics exist, but will not be detailed here (Korzhnev, Kloiber, Kanelis, Tugarinov, & Kay, 2004).

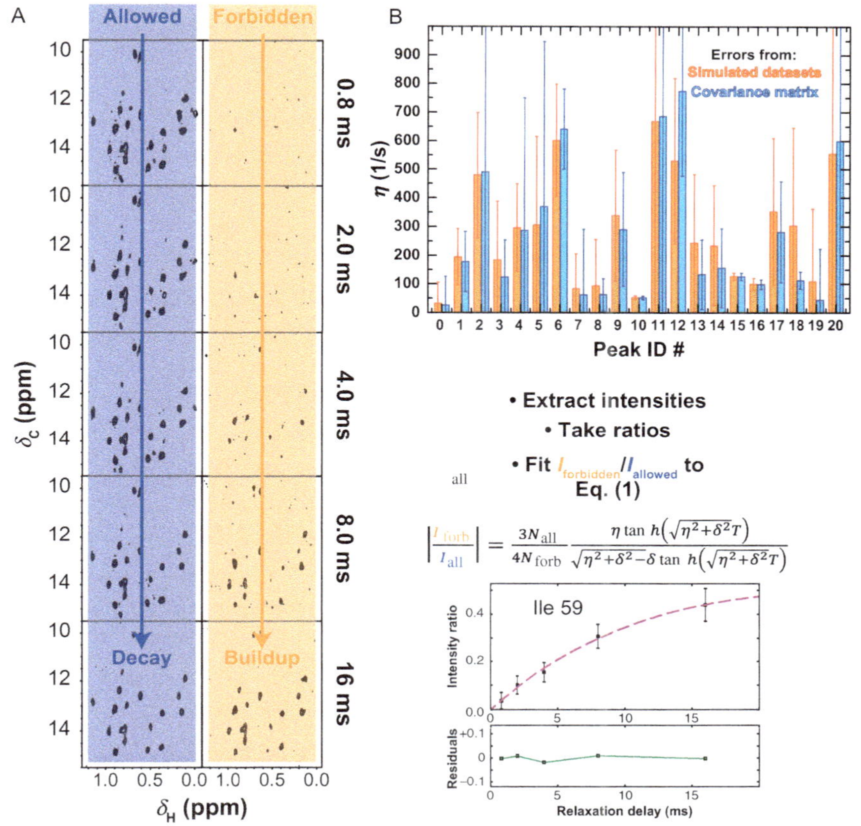

Fig. 5 Schematic of fast motions relaxation experiment. (A) Sample $^{13}C/^{1}H$ planes at the indicated relaxation delays for *E. coli* MBP are shown to illustrate the decay of allowed signal intensity and buildup of forbidden signal intensity. To derive relaxation rate η, peak intensities are extracted and the forbidden:allowed ratio is fit to Eq. (1) as a function of relaxation delay. (B) Plot of extracted η relaxation rates and their errors using either a Monte Carlo bootstrapping approach with 1000 selected datasets (*red/yellow*) or covariance matrix of the fit (*blue/cyan*) for $A_{2A}R$. Ile-δ1 methyl peaks are listed by an arbitrary peak ID number.

a. *Note 1*: Relaxation experiments rely on collecting multiple spectra from experiments which differ solely by the duration of a relaxation delay and fitting spectral peak intensities as a function of this delay to the appropriate equation to extract relaxation parameters which in turn provide insights into dynamics. It is important to collect sufficient data points to generate a good fit (number of delays) and measure a long enough delay for the (usually exponential) function to decay (longest delay). These values vary based on the size of the particle and rotational tumbling time. For the class of experiments described here, a recommended set of delays for ~100 kDa particle is 0.8, 2, 3, 4, 6, 8, 10, 14 ms (Sun et al., 2011). However, since a pair of forbidden and allowed experiments needs to be collected for each delay in this class of relaxation experiment, this large number of time delays can lead to long experimental times, which may be problematic for samples with relatively short lifetimes (see below).

b. *Note 2*: In our experience, the combination of sample concentration and stability of the membrane protein sample significantly limited our ability to collect a full set of 16 spectra as described in *Note 1*. As detailed above (*Note 2* on *Step 5* and Fig. 3B), our samples had a useful lifetime of 12–14 h in the spectrometer at 30°C, limiting the number of relaxation delays that could be tested. To overcome this, we modified the experimental setup to collect allowed experiments (the higher signal-to-noise experiment) at five relaxation delays and a forbidden experiment with five times the number of scans at one relaxation delay. The choice of forbidden relaxation delay is important; an intermediate value (for us, 8 ms) that has good peak intensity for both forbidden and allowed spectra is highly preferable. The pulse sequence was modified accordingly and the experiment was processed using a custom-written NMRpipe (Delaglio et al., 1995) macro to appropriately split the experiments. Both pulse sequence and macro are available upon request. This change requires modifying the analysis and fitting protocol (see *Note 2* on *Step 10* below).

8. This relaxation experiment is implemented as a pseudo-4D experiment with alternating forbidden/allowed experiments and relaxation delays as the third and fourth dimensions, respectively. To process relaxation experiments, either extract the 2D spectrum at each relaxation delay (allowed and forbidden) using the Bruker TopSpin *rser2d* command and process each dataset as an individual 2D spectrum, or use a NMRpipe script that iterates over the relaxation delays to simultaneously split and process (a sample can be provided on request). NMRpipe (Delaglio et al., 1995) and NMRFx Processor (Norris, Fetler, Marchant, & Johnson, 2016) are both good options to process extracted 2D datasets.

9. To analyze the relaxation data, first extract peak intensities for each peak to be analyzed for each relaxation delay (allowed and forbidden) using an analysis program such as NMRViewJ (Johnson, 2004). To estimate peak intensity errors in NMRViewJ, either select an empty region of the spectrum and measure the standard deviation of signal across it (in the *Analysis* tab in NMRViewJ), or use the command *nv_dataset noise data_name.nv* in the NMRViewJ console. This noise estimate may be different for each plane or spectrum and will be used for error analysis of the fit.
10. To extract relaxation parameters, fit the ratio of forbidden to allowed peak intensities to Eq. (1) (Sun et al., 2011), where N is the number of scans for each experiment, T is the relaxation delay, δ (<0) is a term for the coupling between rapidly and slowly decaying single-quantum coherences, and η is a relaxation rate:

$$\left|\frac{I_{\text{forb}}}{I_{\text{all}}}\right| = \frac{3N_{\text{all}}}{4N_{\text{forb}}}\frac{\eta \tan h\left(\sqrt{\eta^2+\delta^2}T\right)}{\sqrt{\eta^2+\delta^2}-\delta \tan h\left(\sqrt{\eta^2+\delta^2}T\right)} \tag{1}$$

The two fit parameters are η and δ, which are constrained to be >0 and <0, respectively, within the fit.

a. *Note 1*: There are multiple approaches to estimate errors in the fit parameters. One is to propagate noise estimates into the intensity ratios, then use those errors to weight the data points during the least-squares minimization and determine the errors of the fit parameters from the covariance matrix. The *optimize.curve_fit* function in the SciPy Python module provides reasonably simple functionality for this process (Oliphant, 2007). Another approach is a *nonparametric Monte Carlo bootstrap* (Efron, 1981), in which for each set of datapoints to fit, many (100 − 1000) artificial datasets are simulated centered on the measured peak intensities (and within error estimates from the noise) and then fit individually. The mean and standard deviation of the fit parameters provide the η and δ values and errors. A comparison of these two methods applied to data from $A_{2A}R$ is shown in Fig. 5B; the error estimates in the bootstrap approach are sometimes larger and sometimes smaller than those from the covariance matrix.

b. *Note 2*: For the modified relaxation experiment with only one forbidden relaxation delay (together with a complement of several allowed relaxation spectra) described above in *Note 2* on *Step 7*, the processing step (*Step 8*) should be modified to correctly split allowed experiments from the one forbidden experiment. A sample NMRpipe script can be provided upon request. Peak intensity quantification and error estimates (*Step 9*) are unaffected. The analysis (*Step 10*) involves simultaneous fitting of the single forbidden to allowed intensity ratio to Eq. (1) and all allowed peak intensities to Eq. (2) (Sun et al., 2011), where A is a constant, T is the relaxation delay, and $R_{2,\mathrm{H}}^{\mathrm{S}}$ and $R_{2,\mathrm{H}}^{\mathrm{F}}$ are the slow and fast relaxation rates of methyl proton single-quantum transitions:

$$I_{\mathrm{all}} = A\frac{3}{2}\left[\exp\left(-R_{2,\mathrm{H}}^{S}T\right) + \exp\left(-R_{2,\mathrm{H}}^{F}T\right)\right] \tag{2}$$

Unfortunately, this approach introduces more fit parameters: A, $R_{2,\mathrm{H}}^{\mathrm{S}}$, and $R_{2,\mathrm{H}}^{\mathrm{F}}$ (all of which were constrained to be >0). Using data collected on *E. coli* maltose binding protein as a test case, we achieved good correspondence between η values calculated from five pairs of allowed and forbidden experiments and from five allowed and one forbidden experiment (Clark et al., 2017). In the case of $A_{2A}R$, the isoleucine residue in the flexible C-terminal Protein C tag served as a serendipitous control with markedly reduced relaxation rate than the other residues.

11. The value of η can be used to extract the S_{axis}^2 order parameter of the methyl group using Eq. (3) (Sun et al., 2011), where τ_c is the molecular rotational correlation time, μ_0 is the vacuum permittivity constant, γ_{H} is the proton gyromagnetic ratio, r_{HH} is the distance between pairs of methyl protons (estimated as 1.813 Å), the function $P_2(x) = 1/2\,(3x^2 - 1)$, and $\theta_{\mathrm{axis,HH}}$ is the angle between the methyl axis and a line between two methyl ^{1}H nuclei (90°) (Tugarinov & Kay, 2006):

$$\eta = \frac{9}{10}\left(\frac{\mu_0}{4\pi}\right)^2 \left[P_2\left(\cos\theta_{\mathrm{axis,HH}}\right)\right]^2 \frac{S_{\mathrm{axis}}^2 \gamma_{\mathrm{H}}^4 \hbar^2 \tau_c}{r_{\mathrm{HH}}^6} \tag{3}$$

The rotational correlation time τ_c can be measured using a ^{15}N/^{1}H-TRACT experiment (Lee, Hilty, Wider, & Wuthrich, 2006) or ^{15}N T_1/T_2 relaxation analysis (Farrow et al., 1994). However, like all global estimates of tumbling time, these methods may underestimate the rotational correlation time due to contributions from fast internal dynamics.

Final Note: Pulse sequence, macros, and scripts all available on request.

5. CONCLUDING REMARKS

As detailed here, we have developed a protocol using α-ketobutyrate (the keto acid precursor for isoleucine) supplementation of the growth media for *P. pastoris,* to prepare Ile δ1 methyl-labeled GPCRs (Clark et al., 2017, 2018, 2015). This protocol is analogous to methods previously developed for expression and labeling in *E. coli* (Gardner & Kay, 1997); however, it allows access to integral membrane proteins such as GPCRs that cannot be efficiently expressed in a prokaryotic host. The samples we have generated are suitable for a range of solution NMR measurements, including relaxation-based determination of ligand-dependent changes in receptor dynamics. These data complement other NMR-based measurements of receptor structure, dynamics, and interactions using probes at other ^{13}C methyl groups (Kofuku et al., 2014, 2018; Solt et al., 2017), exogenously added ^{19}F probes (Ye et al., 2018; Ye, Van Eps, Zimmer, Ernst, & Prosser, 2016), or $^{15}N/^{1}H$ pairs in receptors stabilized by point mutations or insertion into covalently circularized nanodiscs (Isogai et al., 2016; Nasr et al., 2017).

Taken together, these studies show the rich diversity of GPCR conformations and dynamics, taking advantage of the power of solution NMR to get site-specific information from receptors in equilibrium and complementing structural studies from X-ray diffraction and cryo-EM. In addition, we believe that, properly modified, these approaches can be applied to study the dynamics and interactions of other eukaryotic membrane proteins as well. We anticipate that further technological developments, including those in sample preparation [e.g., additional labeling strategies (Robson et al., 2018), ways to improve sample stability or concentration via improved membrane mimetics such as smaller, more "NMR-friendly" nanodiscs (Chien et al., 2017; Hagn, Etzkorn, Raschle, & Wagner, 2013)], NMR data acquisition [e.g., ultrahigh-field NMR systems (Quinn, Wang, & Polenova, 2018), improved $^{13}C/^{15}N$ direct detection methods (Takeuchi, Heffron, Sun, Frueh, & Wagner, 2010)], and analysis [e.g., improved methods for chemical shift assignment through PRE or NOESY experiments (Lescanne et al., 2017)], will only further improve such prospects in the future.

ACKNOWLEDGMENTS

Research detailed herein was supported by the National Science Foundation (Grant 1000136529 to L.D.C.), the American Heart Association (Grant 16PRE27200004 to L.D.C., the Welch Foundation (I-1770 to D.M.R), a Packard Foundation Fellowship (to D.M.R.), the Mallinckrodt Foundation (to D.M.R), and the National Institutes of Health (F32 GM119311 to I.D., R01 GM113050 to D.M.R., and R01 GM106239 to K.H.G.). Reagents for NMR spectroscopy were generously provided by a research award from Cambridge Isotope Laboratories, Inc.

REFERENCES

Almén, M. S., Nordström, K. J., Fredriksson, R., & Schiöth, H. B. (2009). Mapping the human membrane proteome: A majority of the human membrane proteins can be classified according to function and evolutionary origin. *BMC Biology*, 7, 50. https://doi.org/10.1186/1741-7007-7-50.

André, N., Cherouati, N., Prual, C., Steffan, T., Zeder-Lutz, G., Magnin, T., et al. (2006). Enhancing functional production of G protein-coupled receptors in Pichia pastoris to levels required for structural studies via a single expression screen. *Protein Science*, *15*(5), 1115–1126. https://doi.org/10.1110/ps.062098206.

Aw, R., & Polizzi, K. M. (2013). Can too many copies spoil the broth? *Microbial Cell Factories*, *12*, 128. https://doi.org/10.1186/1475-2859-12-128.

Bayburt, T. H., & Sligar, S. G. (2002). Single-molecule height measurements on microsomal cytochrome P450 in nanometer-scale phospholipid bilayer disks. *Proceedings of the National Academy of Sciences of the United States of America*, *99*(10), 6725–6730. https://doi.org/10.1073/pnas.062565599.

Bayburt, T. H., & Sligar, S. G. (2010). Membrane protein assembly into nanodiscs. *FEBS Letters*, *584*(9), 1721–1727. https://doi.org/10.1016/j.febslet.2009.10.024.

Byrne, B. (2015). Pichia pastoris as an expression host for membrane protein structural biology. *Current Opinion in Structural Biology*, *32*, 9–17. https://doi.org/10.1016/j.sbi.2015.01.005.

Cereghino, G. P., Cereghino, J. L., Ilgen, C., & Cregg, J. M. (2002). Production of recombinant proteins in fermenter cultures of the yeast Pichia pastoris. *Current Opinion in Biotechnology*, *13*(4), 329–332.

Cereghino, J. L., & Cregg, J. M. (2000). Heterologous protein expression in the methylotrophic yeast Pichia pastoris. *FEMS Microbiology Reviews*, *24*(1), 45–66.

Cerione, R. A., & Ross, E. M. (1991). Reconstitution of receptors and G proteins in phospholipid vesicles. *Methods in Enzymology*, *195*, 329–342.

Chien, C. H., Helfinger, L. R., Bostock, M. J., Solt, A., Tan, Y. L., & Nietlispach, D. (2017). An adaptable phospholipid membrane mimetic system for solution NMR studies of membrane proteins. *Journal of the American Chemical Society*, *139*(42), 14829–14832. https://doi.org/10.1021/jacs.7b06730.

Clark, L., Dikiy, I., Chapman, K., Rodstrom, K. E., Aramini, J., LeVine, M. V., et al. (2017). Ligand modulation of sidechain dynamics in a wild-type human GPCR. *eLife*, *6*, e28505. https://doi.org/10.7554/eLife.28505.

Clark, L., Dikiy, I., Rosenbaum, D. M., & Gardner, K. H. (2018). On the use of Pichia pastoris for isotopic labeling of human GPCRs for NMR studies. *Journal of Biomolecular NMR*, *71*(4), 203–211.

Clark, L., Zahm, J. A., Ali, R., Kukula, M., Bian, L., Patrie, S. M., et al. (2015). Methyl labeling and TROSY NMR spectroscopy of proteins expressed in the eukaryote Pichia pastoris. *Journal of Biomolecular NMR*, *62*(3), 239–245. https://doi.org/10.1007/s10858-015-9939-2.

Cos, O., Ramon, R., Montesinos, J. L., & Valero, F. (2006). Operational strategies, monitoring and control of heterologous protein production in the methylotrophic yeast Pichia pastoris under different promoters: A review. *Microbial Cell Factories*, *5*, 17. https://doi.org/10.1186/1475-2859-5-17.

Cuniasse, P., Tavares, P., Orlova, E. V., & Zinn-Justin, S. (2017). Structures of biomolecular complexes by combination of NMR and cryoEM methods. *Current Opinion in Structural Biology*, *43*, 104–113. https://doi.org/10.1016/j.sbi.2016.12.008.

Delaglio, F., Grzesiek, S., Vuister, G. W., Zhu, G., Pfeifer, J., & Bax, A. (1995). NMRPipe: A multidimensional spectral processing system based on UNIX pipes. *Journal of Biomolecular NMR*, *6*(3), 277–293. https://doi.org/10.1007/BF00197809.

Denisov, I. G., Grinkova, Y. V., Lazarides, A. A., & Sligar, S. G. (2004). Directed self-assembly of monodisperse phospholipid bilayer nanodiscs with controlled size. *Journal of the American Chemical Society*, *126*(11), 3477–3487. https://doi.org/10.1021/ja0393574.

Eddy, M. T., Lee, M. Y., Gao, Z. G., White, K. L., Didenko, T., Horst, R., et al. (2018). Allosteric coupling of drug binding and intracellular signaling in the A2A adenosine receptor. *Cell*, *172*(1–2), 68–80. e12. https://doi.org/10.1016/j.cell.2017.12.004.

Efron, B. (1981). Nonparametric estimates of standard error: The jackknife, the bootstrap and other methods. *Biometrika*, *68*(3), 589–599. https://doi.org/10.1093/biomet/68.3.589.

Emami, S., Fan, Y., Munro, R., Ladizhansky, V., & Brown, L. S. (2013). Yeast-expressed human membrane protein aquaporin-1 yields excellent resolution of solid-state MAS NMR spectra. *Journal of Biomolecular NMR*, *55*(2), 147–155. https://doi.org/10.1007/s10858-013-9710-5.

Fan, Y., Emami, S., Munro, R., Ladizhansky, V., & Brown, L. S. (2015). Isotope labeling of eukaryotic membrane proteins in yeast for solid-state NMR. *Methods in Enzymology*, *565*, 193–212. https://doi.org/10.1016/bs.mie.2015.05.010.

Fan, Y., Shi, L., Ladizhansky, V., & Brown, L. S. (2011). Uniform isotope labeling of a eukaryotic seven-transmembrane helical protein in yeast enables high-resolution solid-state NMR studies in the lipid environment. *Journal of Biomolecular NMR*, *49*(2), 151–161. https://doi.org/10.1007/s10858-011-9473-9.

Farrow, N. A., Muhandiram, R., Singer, A. U., Pascal, S. M., Kay, C. M., Gish, G., et al. (1994). Backbone dynamics of a free and a phosphopeptide-complexed Src homology 2 domain studied by 15N NMR relaxation. *Biochemistry*, *33*(19), 5984–6003. https://doi.org/10.1021/bi00185a040.

Gao, Z. G., & Ijzerman, A. P. (2000). Allosteric modulation of A(2A) adenosine receptors by amiloride analogues and sodium ions. *Biochemical Pharmacology*, *60*(5), 669–676.

Gardner, K. H., & Kay, L. E. (1997). Production and incorporation of 15N, 13C, 2H (1H-δ1 methyl) isoleucine into proteins for multidimensional NMR studies. *Journal of the American Chemical Society*, *119*(32), 7599–7600. https://doi.org/10.1021/ja9706514.

Hagn, F., Etzkorn, M., Raschle, T., & Wagner, G. (2013). Optimized phospholipid bilayer nanodiscs facilitate high-resolution structure determination of membrane proteins. *Journal of the American Chemical Society*, *135*(5), 1919–1925. https://doi.org/10.1021/ja310901f.

Heyland, J., Fu, J., Blank, L. M., & Schmid, A. (2011). Carbon metabolism limits recombinant protein production in Pichia pastoris. *Biotechnology and Bioengineering*, *108*(8), 1942–1953. https://doi.org/10.1002/bit.23114.

Hino, T., Arakawa, T., Iwanari, H., Yurugi-Kobayashi, T., Ikeda-Suno, C., Nakada-Nakura, Y., et al. (2012). G-protein-coupled receptor inactivation by an allosteric inverse-agonist antibody. *Nature*, *482*(7384), 237–240. https://doi.org/10.1038/nature10750.

Isogai, S., Deupi, X., Opitz, C., Heydenreich, F. M., Tsai, C. J., Brueckner, F., et al. (2016). Backbone NMR reveals allosteric signal transduction networks in the beta1-adrenergic receptor. *Nature*, *530*(7589), 237–241. https://doi.org/10.1038/nature16577.

Jaakola, V. P., Griffith, M. T., Hanson, M. A., Cherezov, V., Chien, E. Y., Lane, J. R., et al. (2008). The 2.6 angstrom crystal structure of a human A2A adenosine receptor bound to an antagonist. *Science*, *322*(5905), 1211–1217. https://doi.org/10.1126/science.1164772.

Johnson, B. A. (2004). Using NMRView to visualize and analyze the NMR spectra of macromolecules. *Methods in Molecular Biology*, *278*, 313–352. https://doi.org/10.1385/1-59259-809-9:313.

Kasinath, V., Sharp, K. A., & Wand, A. J. (2013). Microscopic insights into the NMR relaxation-based protein conformational entropy meter. *Journal of the American Chemical Society*, *135*(40), 15092–15100. https://doi.org/10.1021/ja405200u.

Kofuku, Y., Ueda, T., Okude, J., Shiraishi, Y., Kondo, K., Mizumura, T., et al. (2014). Functional dynamics of deuterated beta2-adrenergic receptor in lipid bilayers revealed by NMR spectroscopy. *Angewandte Chemie (International ed. in English)*, *53*(49), 13376–13379. https://doi.org/10.1002/anie.201406603.

Kofuku, Y., Yokomizo, T., Imai, S., Shiraishi, Y., Natsume, M., Itoh, H., et al. (2018). Deuteration and selective labeling of alanine methyl groups of beta2-adrenergic receptor expressed in a baculovirus-insect cell expression system. *Journal of Biomolecular NMR*, *71*(3), 185–192. https://doi.org/10.1007/s10858-018-0174-5.

Korzhnev, D. M., Kloiber, K., Kanelis, V., Tugarinov, V., & Kay, L. E. (2004). Probing slow dynamics in high molecular weight proteins by methyl-TROSY NMR spectroscopy: Application to a 723-residue enzyme. *Journal of the American Chemical Society*, *126*(12), 3964–3973. https://doi.org/10.1021/ja039587i.

Lee, D., Hilty, C., Wider, G., & Wuthrich, K. (2006). Effective rotational correlation times of proteins from NMR relaxation interference. *Journal of Magnetic Resonance*, *178*(1), 72–76. https://doi.org/10.1016/j.jmr.2005.08.014.

Lee, J. Y., Kinch, L. N., Borek, D. M., Wang, J., Wang, J., Urbatsch, I. L., et al. (2016). Crystal structure of the human sterol transporter ABCG5/ABCG8. *Nature*, *533*(7604), 561–564. https://doi.org/10.1038/nature17666.

Lescanne, M., Skinner, S. P., Blok, A., Timmer, M., Cerofolini, L., Fragai, M., et al. (2017). Methyl group assignment using pseudocontact shifts with PARAssign. *Journal of Biomolecular NMR*, *69*(4), 183–195. https://doi.org/10.1007/s10858-017-0136-3.

Liang, B., & Tamm, L. K. (2016). NMR as a tool to investigate the structure, dynamics and function of membrane proteins. *Nature Structural & Molecular Biology*, *23*(6), 468–474. https://doi.org/10.1038/nsmb.3226.

Liang, B., & Tamm, L. K. (2018). Solution NMR of SNAREs, complexin and alpha-synuclein in association with membrane-mimetics. *Progress in Nuclear Magnetic Resonance Spectroscopy*, *105*, 41–53. https://doi.org/10.1016/j.pnmrs.2018.02.001.

Lin-Cereghino, G. P., Stark, C. M., Kim, D., Chang, J., Shaheen, N., Poerwanto, H., et al. (2013). The effect of alpha-mating factor secretion signal mutations on recombinant protein expression in Pichia pastoris. *Gene*, *519*(2), 311–317. https://doi.org/10.1016/j.gene.2013.01.062.

Liu, W., Chun, E., Thompson, A. A., Chubukov, P., Xu, F., Katritch, V., et al. (2012). Structural basis for allosteric regulation of GPCRs by sodium ions. *Science*, *337*(6091), 232–236. https://doi.org/10.1126/science.1219218.

Moore, P. B. (1979). The preparation of deuterated ribosomal materials for neutron scattering. *Methods in Enzymology*, *59*, 639–655.

Morgan, W. D., Kragt, A., & Feeney, J. (2000). Expression of deuterium-isotope-labelled protein in the yeast Pichia pastoris for NMR studies. *Journal of Biomolecular NMR*, *17*(4), 337–347. https://doi.org/10.1023/A:1008313530207.

Nasr, M. L., Baptista, D., Strauss, M., Sun, Z. J., Grigoriu, S., Huser, S., et al. (2017). Covalently circularized nanodiscs for studying membrane proteins and viral entry. *Nature Methods*, *14*(1), 49–52. https://doi.org/10.1038/nmeth.4079.

Norris, M., Fetler, B., Marchant, J., & Johnson, B. A. (2016). NMRFx processor: A cross-platform NMR data processing program. *Journal of Biomolecular NMR*, *65*(3–4), 205–216. https://doi.org/10.1007/s10858-016-0049-6.

Oliphant, T. E. (2007). Python for scientific computing. *Computing in Science & Engineering*, *9*(3), 10–20. https://doi.org/10.1109/mcse.2007.58.

Ollerenshaw, J. E., Tugarinov, V., & Kay, L. E. (2003). Methyl TROSY: Explanation and experimental verification. *Magnetic Resonance in Chemistry*, *41*(10), 843–852. https://doi.org/10.1002/mrc.1256.

Opella, S. J., & Marassi, F. M. (2017). Applications of NMR to membrane proteins. *Archives of Biochemistry and Biophysics*, *628*, 92–101. https://doi.org/10.1016/j.abb.2017.05.011.

Opitz, C., Isogai, S., & Grzesiek, S. (2015). An economic approach to efficient isotope labeling in insect cells using homemade 15N-, 13C- and 2H-labeled yeast extracts. *Journal of Biomolecular NMR*, *62*(3), 373–385. https://doi.org/10.1007/s10858-015-9954-3.

Palmer, A. G. (2012). NMR spectroscopy: NMR relaxation methods. In H. J. Dyson (Ed.), *Biophysical techniques for structural characterization of macromolecules* (pp. 216–244). Oxford: Academic Press.

Pellecchia, M., Bertini, I., Cowburn, D., Dalvit, C., Giralt, E., Jahnke, W., et al. (2008). Perspectives on NMR in drug discovery: A technique comes of age. *Nature Reviews. Drug Discovery*, 7(9), 738–745. https://doi.org/10.1038/nrd2606.

Pickford, A. R., & O'Leary, J. M. (2004). Isotopic labeling of recombinant proteins from the methylotrophic yeast Pichia pastoris. *Methods in Molecular Biology*, *278*, 17–33. https://doi.org/10.1385/1-59259-809-9:017.

Puthenveetil, R., Nguyen, K., & Vinogradova, O. (2017). Nanodiscs and solution NMR: Preparation, application and challenges. *Nanotechnology Reviews*, *6*(1), 111–126. https://doi.org/10.1515/ntrev-2016-0076.

Quinn, C. M., Wang, M., & Polenova, T. (2018). NMR of macromolecular assemblies and machines at 1 GHz and beyond: New transformative opportunities for molecular structural biology. *Methods in Molecular Biology*, *1688*, 1–35. https://doi.org/10.1007/978-1-4939-7386-6_1.

Rakestraw, J. A., Sazinsky, S. L., Piatesi, A., Antipov, E., & Wittrup, K. D. (2009). Directed evolution of a secretory leader for the improved expression of heterologous proteins and full-length antibodies in Saccharomyces cerevisiae. *Biotechnology and Bioengineering*, *103*(6), 1192–1201. https://doi.org/10.1002/bit.22338.

Rasmussen, S. G., DeVree, B. T., Zou, Y., Kruse, A. C., Chung, K. Y., Kobilka, T. S., et al. (2011). Crystal structure of the beta2 adrenergic receptor-Gs protein complex. *Nature*, *477*(7366), 549–555. https://doi.org/10.1038/nature10361.

Robson, S. A., Takeuchi, K., Boeszoermenyi, A., Coote, P. W., Dubey, A., Hyberts, S., et al. (2018). Mixed pyruvate labeling enables backbone resonance assignment of large proteins using a single experiment. *Nature Communications*, *9*(1), 356. https://doi.org/10.1038/s41467-017-02767-8.

Rosenzweig, R., & Kay, L. E. (2014). Bringing dynamic molecular machines into focus by methyl-TROSY NMR. *Annual Review of Biochemistry*, *83*, 291–315. https://doi.org/10.1146/annurev-biochem-060713-035829.

Shimamura, T., Shiroishi, M., Weyand, S., Tsujimoto, H., Winter, G., Katritch, V., et al. (2011). Structure of the human histamine H1 receptor complex with doxepin. *Nature*, *475*(7354), 65–70. https://doi.org/10.1038/nature10236.

Solà, A., Maaheimo, H., Ylonen, K., Ferrer, P., & Szyperski, T. (2004). Amino acid biosynthesis and metabolic flux profiling of Pichia pastoris. *European Journal of Biochemistry*, *271*(12), 2462–2470. https://doi.org/10.1111/j.1432-1033.2004.04176.x.

Solt, A. S., Bostock, M. J., Shrestha, B., Kumar, P., Warne, T., Tate, C. G., et al. (2017). Insight into partial agonism by observing multiple equilibria for ligand-bound and Gs-mimetic nanobody-bound beta1-adrenergic receptor. *Nature Communications*, *8*(1), 1795. https://doi.org/10.1038/s41467-017-02008-y.

Sun, H., Kay, L. E., & Tugarinov, V. (2011). An optimized relaxation-based coherence transfer NMR experiment for the measurement of side-chain order in methyl-protonated, highly deuterated proteins. *The Journal of Physical Chemistry B*, *115*(49), 14878–14884. https://doi.org/10.1021/jp209049k.

Suzuki, R., Sakakura, M., Mori, M., Fujii, M., Akashi, S., & Takahashi, H. (2018). Methyl-selective isotope labeling using alpha-ketoisovalerate for the yeast Pichia pastoris recombinant protein expression system. *Journal of Biomolecular NMR*, *71*(4), 213–223. https://doi.org/10.1007/s10858-018-0192-3.

Takeuchi, K., Heffron, G., Sun, Z. Y., Frueh, D. P., & Wagner, G. (2010). Nitrogen-detected CAN and CON experiments as alternative experiments for main chain NMR resonance assignments. *Journal of Biomolecular NMR*, *47*(4), 271–282. https://doi.org/10.1007/s10858-010-9430-z.

Tugarinov, V., Hwang, P. M., Ollerenshaw, J. E., & Kay, L. E. (2003). Cross-correlated relaxation enhanced 1H-13C NMR spectroscopy of methyl groups in very high molecular weight proteins and protein complexes. *Journal of the American Chemical Society*, *125*(34), 10420–10428. https://doi.org/10.1021/ja030153x.

Tugarinov, V., & Kay, L. E. (2006). A 2H NMR relaxation experiment for the measurement of the time scale of methyl side-chain dynamics in large proteins. *Journal of the American Chemical Society*, *128*(38), 12484–12489. https://doi.org/10.1021/ja063071s.

Tugarinov, V., Sprangers, R., & Kay, L. E. (2007). Probing side-chain dynamics in the proteasome by relaxation violated coherence transfer NMR spectroscopy. *Journal of the American Chemical Society*, *129*(6), 1743–1750. https://doi.org/10.1021/ja067827z.

Ulmschneider, M. B., & Sansom, M. S. (2001). Amino acid distributions in integral membrane protein structures. *Biochimica et Biophysica Acta*, *1512*(1), 1–14. https://doi.org/10.1016/S0005-2736(01)00299-1.

Ulrich, E. L., Akutsu, H., Doreleijers, J. F., Harano, Y., Ioannidis, Y. E., Lin, J., et al. (2008). BioMagResBank. *Nucleic Acids Research*, *36*(Database issue), D402–D408. https://doi.org/10.1093/nar/gkm957.

Wand, A. J., Moorman, V. R., & Harpole, K. W. (2013). A surprising role for conformational entropy in protein function. *Topics in Current Chemistry*, *337*, 69–94. https://doi.org/10.1007/128_2012_418.

Weiss, H. M., & Grisshammer, R. (2002). Purification and characterization of the human adenosine A(2a) receptor functionally expressed in Escherichia coli. *European Journal of Biochemistry*, *269*(1), 82–92.

Whittaker, J. W. (2007). Selective isotopic labeling of recombinant proteins using amino acid auxotroph strains. *Methods in Molecular Biology*, *389*, 175–188. https://doi.org/10.1007/978-1-59745-456-8_13.

Wood, M. J., & Komives, E. A. (1999). Production of large quantities of isotopically labeled protein in Pichia pastoris by fermentation. *Journal of Biomolecular NMR*, *13*(2), 149–159.

Ye, L., Neale, C., Sljoka, A., Lyda, B., Pichugin, D., Tsuchimura, N., et al. (2018). Mechanistic insights into allosteric regulation of the A2A adenosine G protein-coupled receptor by physiological cations. *Nature Communications*, *9*(1), 1372. https://doi.org/10.1038/s41467-018-03314-9.

Ye, L., Van Eps, N., Zimmer, M., Ernst, O. P., & Prosser, R. S. (2016). Activation of the A2A adenosine G-protein-coupled receptor by conformational selection. *Nature*, *533*(7602), 265–268. https://doi.org/10.1038/nature17668.

Yurugi-Kobayashi, T., Asada, H., Shiroishi, M., Shimamura, T., Funamoto, S., Katsuta, N., et al. (2009). Comparison of functional non-glycosylated GPCRs expression in Pichia pastoris. *Biochemical and Biophysical Research Communications*, *380*(2), 271–276. https://doi.org/10.1016/j.bbrc.2009.01.053.

Zhang, M., Yu, X. W., Xu, Y., Jouhten, P., Swapna, G. V. T., Glaser, R. W., et al. (2017). (13) C metabolic flux profiling of Pichia pastoris grown in aerobic batch cultures on glucose revealed high relative anabolic use of TCA cycle and limited incorporation of provided precursors of branched-chain amino acids. *The FEBS Journal*, *284*(18), 3100–3113. https://doi.org/10.1111/febs.14180.

Zhou, H. X., & Cross, T. A. (2013). Influences of membrane mimetic environments on membrane protein structures. *Annual Review of Biophysics*, *42*, 361–392. https://doi.org/10.1146/annurev-biophys-083012-130326.

CHAPTER THREE

Optimal Isotope Labeling of Aromatic Amino Acid Side Chains for NMR Studies of Protein Dynamics

Ulrich Weininger[1]
Institute of Physics, Biophysics, Martin-Luther-University Halle-Wittenberg, Halle, Germany
[1]Corresponding author: e-mail address: ulrich.weininger@physik.uni-halle.de

Contents

Abstract

Aromatic side chains in proteins are often directly evolved in stabilizing the hydrophobic core, protein binding, or enzymatic activity. They are also responsible for specific local dynamic processes, such as histidine tautomerization or ring flips. Despite their importance, they are often not targeted directly by NMR spectroscopy, because of

Methods in Enzymology, Volume 614
ISSN 0076-6879
https://doi.org/10.1016/bs.mie.2018.08.028

spectroscopic complications and challenges. This chapter addresses state-of-the-art site-selective ^{13}C-labeling methods for aromatic side chains, and describes how they solve several of the spectroscopic issues. A special emphasis is put on thereby enabled protein dynamics experiments of aromatic side chains.

ABBREVIATIONS

COSY correlation spectroscopy
CSA chemical shielding anisotropy
HSQC hetero single quantum coherence
NMR nuclear magnetic resonance
TOCSY total correlated spectroscopy
TROSY transverse relaxation optimized spectroscopy

1. INTRODUCTION

Aromatic side chains are a prominent and versatile subgroup within the 20 amino acids and contain a lot of interesting features. They are bulky and contribute a significant part (roughly 25% of the volume in average) of the hydrophobic core, where they often pair in clusters. Further, they are overrepresented in protein binding interfaces (Lo Conte, Chothia, & Janin, 1999), where they contribute a significant part of the binding free energy (Bogan & Thorn, 1998). Because of their quadrupolar electrostatic character, they can be engaged in specific interactions with other aromatic groups (Burley & Petsko, 1985, 1989), cations (Berry, Elvekrog, & Tommos, 2007; Mahadevi & Sastry, 2013), or sulfur (Valley et al., 2012). Tyr and especially His play an important role in enzyme activity (Bartlett, Porter, Borkakoti, & Thornton, 2002). The pK_a value of His is close to physiological pH, and it can exist in three different tautomeric forms (Reynolds, Peat, Freedman, & Lyerla, 1973). Thus it can act as a proton acceptor/donor (Fersht, 1977), or shuttle (Lindskog, 1997), as well as a hydrogen bond donor, acceptor, or switch (Krishna Deepak & Sankararamakrishnan, 2016; Preimesberger, Majumdar, Rice, Que, & Lecomte, 2015). Additionally many Phe and Tyr residues undergo frequent 180 degree rotations (ring flips) of the χ_2 dihedral angle (around the C_β–C_γ–C_ζ axis). Thereby they can provide insights into aromatic side-chain packing and local transient protein breathing motions. Taken together, they are extremely interesting probes that allow to gain specific insights into structure, thermodynamics, and conformational dynamics of proteins.

Despite their importance and early groundbreaking work on ring flips (Wagner, Demarco, & Wuthrich, 1976), aromatic side chains have not been in the main focus of nuclear magnetic resonance (NMR) spectroscopy for a long time. There are a number of reasons for this. First, they are difficult to assign since they are far (in terms of J-coupling steps) from the usually studied backbone, and are disconnected in $^{1}H^{1}H$-COSY/TOCSY experiments by the absence of a proton in Cγ and in $^{13}C^{13}C$-COSY/TOCSY experiments by different ^{13}C chemical shifts. Second, their spectra are affected by the fast relaxation properties of aromatic ^{1}H and ^{13}C. Further the spectra can suffer greatly from strong ^{13}C–^{13}C J-couplings. This has been changed by site-selective ^{13}C labeling of aromatic side chains (Teilum, Brath, Lundstrom, & Akke, 2006) that overcomes the strong coupling problem, enables the acquirement of well-resolved spectra, and further makes advanced experiments for studying protein dynamics by NMR (Weininger, Diehl, & Akke, 2012; Weininger, Respondek, & Akke, 2012) possible.

2. NMR PROPERTIES OF AROMATIC SIDE CHAINS

There are four different aromatic amino acids in proteins: tryptophan, phenylalanine, tyrosine, and histidine (Fig. 1). The NMR properties of the aromatic side chains differ significantly from aliphatic side chains. The J-couplings of interest (^{1}J ^{1}H–^{13}C, ^{1}J ^{13}C–^{13}C, and ^{3}J ^{1}H–^{1}H) are summarized in Fig. 2. First, J-couplings are larger. All other higher order proton and carbon J-couplings are too small to matter in experiments involving proteins. The ^{1}J ^{1}H–^{13}C couplings are heterogeneous between the different amino acids, in contrast to the more uniform amide ^{1}J ^{1}H–^{15}N couplings. Second, ^{1}H and ^{13}C chemical shifts within an aromatic side chain can be very similar. This is usually the case in Phe, but can also occur in the six-ring moiety of Trp. This leads to reduced spectral resolution that can, in combination with high homonuclear J-couplings, result in strong couplings.

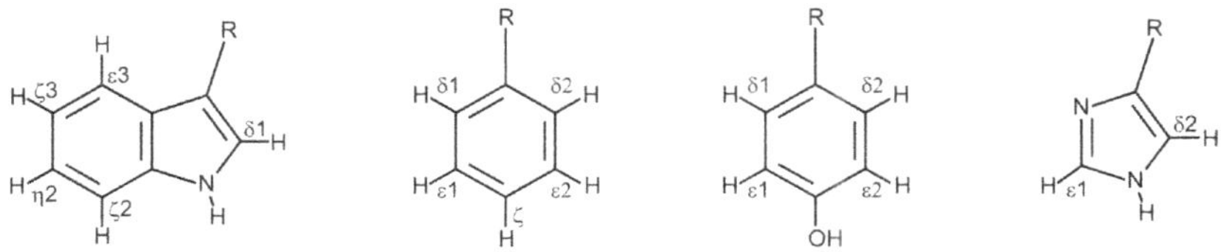

Fig. 1 Nomenclature of ^{1}H–^{13}C positions in aromatic side chains. From *left* to *right*: Trp, Phe, Tyr, His. δ1, δ2 and ε1, ε2 in Phe and Tyr are usually averaged to δ* and ε* because of fast ring flips. His is shown in its most stable neutral tautomeric form.

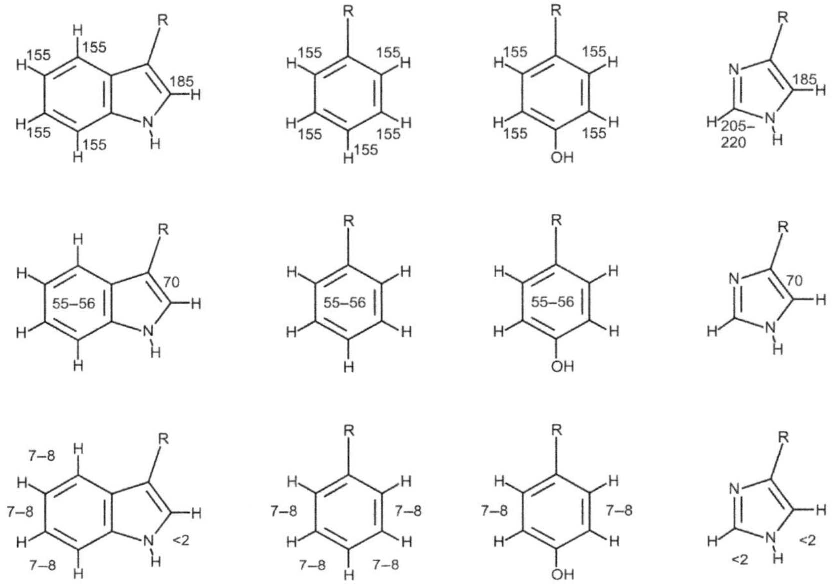

Fig. 2 Relevant J-couplings (in Hz) in aromatic side chains. *First row*: ^1J ^1H–^{13}C, *second row*: ^1J ^{13}C–^{13}C, *third row*: ^3J ^1H–^1H. From *left* to *right*: Trp, Phe, Tyr, His. His is shown in its HNε2 tautomeric form. The ^1J ^1H–^{13}C coupling in His ε1 varies between the different tautomeric forms.

Strong J-couplings occur, if the difference in frequency between two coupling atoms is not sufficiently larger than the J-coupling. Strong couplings are rather common in Phe and cause (in case of ^1J ^{13}C–^{13}C) heavily disturbed and weakened signals. Additionally, the similarity of chemical shifts within an aromatic side chain makes selective pulsing, and thereby refocusing of transverse magnetization, extremely hard and often practically impossible. Third, transverse relaxation in aromatic side chains is fast, especially for ^{13}C, due to large contributions from shift anisotropy (CSA). ^1H–^{13}C spectra of uniformly ^{13}C-labeled aromatic side chains are best run in a constant time approach in the indirect ^{13}C dimension where ^1J ^{13}C–^{13}C couplings are self-refocused. However large J-couplings in combination with rapid ^{13}C transverse relaxation only allow for a relatively short constant time delay. As a result the accessible resolution is limited and intensity losses during the constant time delay are significant nevertheless. Fast transverse relaxation can mitigated by the use of the TROSY approach and site-selective ^{13}C labeling avoids the need for the constant time approach, eliminates strong ^1J ^{13}C–^{13}C couplings, improves the spectral resolution, and makes refocusing of transverse ^{13}C magnetization possible.

3. SITE-SELECTIVE ^{13}C LABELING OF AROMATIC SIDE CHAINS

3.1 Principles

The key reason for site-selective ^{13}C labeling of aromatic side chains is to get rid of the ^{1}J ^{13}C–^{13}C couplings. This brings the following advantages: constant time detection of ^{13}C is no longer needed, spectra are not suffering from strong ^{13}C–^{13}C couplings, and relaxation experiments become possible, by being able to refocus all remaining J-couplings (and thus acquiring artifact-free R_2). To achieve this, it is not necessary (at least at the current experimental performance in terms of S/N) to have 100% ^{12}C incorporated in the neighboring carbon positions (^{13}C depleted), but to have it at a low rate. 1% ^{13}C incorporation (natural abundance) is working fine. Any labeling approach achieving this is sufficient for relaxation studies. So is, in principle, using regular glucose with a natural abundance isotope pattern. Because of low S/N in NMR this is, however, absolutely not practical in the vast majority of cases. For sensitivity reasons the optimal labeling would result in a high ^{13}C incorporation (ideally 99%) in the positions of interest, while maintaining low ^{13}C incorporation (1%) in the neighboring positions in order to avoid ^{13}C–^{13}C couplings. Other requirements on the labeling are more customized. Does one want to study as many positions as possible with the same sample and thereby risking potential overlap and more complicated spectra. Or does one want to study just one specific type of position, resulting in few signals in the spectra and minimizing the chance of overlap. In the following sections currently available practical labeling approaches are discussed. Resulting ^{13}C incorporation in the aromatic side chains is summarized in Table 1.

3.2 Site-Selective Labeling Using Glucose

Site-selective ^{13}C incorporation, in its most simple and robust form, uses site-selectively ^{13}C-enriched glucose as the carbon source in the expression medium, instead of regular glucose or uniformly ^{13}C-enriched ($^{13}C_6$) glucose. The most prominent examples are 1-^{13}C (Teilum et al., 2006) and 2-^{13}C (Lundström et al., 2007) glucose. By diluting ^{13}C, a ^{13}C incorporation in neighboring positions, which would lead to a ^{13}C–^{13}C pattern, is completely avoided in aromatic side chains. 1-^{13}C glucose is labeling the following positions, which contain a directly attached proton: Phe/Tyr δ, Trp δ1 and ε3, His δ2 and ε1. 2-^{13}C glucose is labeling Phe/Tyr ε, Trp δ1, ζ3

Table 1 Summary of ^{13}C-Labeled Positions With Attached Proton in Aromatic Side Chains Available by Current In Vivo Labeling Approaches

	glc	glc	gly/pyr	pyr	ery	ery	ery	ery	rib	phenylpyr			indolepyr		anthr		hisp
	1-^{13}C	2-^{13}C	2-^{13}C	3-^{13}C	1-^{13}C	2-^{13}C	3-^{13}C	4-^{13}C	1-^{13}C	(1)	(2)	(3)	(1)	(2)	(1)	(2)	(1)
Fδ	^{13}C			^{13}C				^{13}C									
Fε		^{13}C	^{13}C		^{13}C		^{13}C			^{13}C							
Fζ						^{13}C					^{13}C						
Yδ	^{13}C			^{13}C				^{13}C									
Yε		^{13}C	^{13}C		^{13}C		^{13}C					^{13}C					
Wδ1	^{13}C	^{13}C		^{13}C					^{13}C								
Wε3	^{13}C		^{13}C	^{13}C				^{13}C							^{13}C		
Wζ3		^{13}C					^{13}C						^{13}C			^{13}C	
Wη2						^{13}C								^{13}C	^{13}C		
Wζ2		^{13}C			^{13}C								^{13}C				
Hδ2	^{13}C	^{13}C		^{13}C					^{13}C								
Hε1	^{13}C			^{13}C													^{13}C

anthr, anthranilate (Schörghuber et al., 2017); *ery*, erythrose (Weininger, 2017a); *indolepyr*, indolepyruvate (Schörghuber et al., 2015); *glc*, glucose (Lundström et al., 2007); *gly*, glycerol; *pyr*, pyruvate (Milbradt et al., 2015); *hisp*, histidine precursor (Schörghuber, Geist, Platzer, Konrat, & Lichtenecker, 2017); *phenylpyr*, phenylpyruvate (Lichtenecker, 2014; Lichtenecker, Weinhaupl, Schmid, & Konrat, 2013); *rib*, ribose (Weininger, 2017b); x-^{13}C, site-selectively ^{13}C-enriched position; (x) arbitrary number of precursor.

and ζ2, His δ2. Phe ζ and Trp η2 are not labeled by 1-^{13}C or 2-^{13}C glucose. ^{13}C incorporation levels are 20%–50%. 2-^{13}C glucose yields a higher incorporation in Trp δ1 and His δ2. The carbon of His ε1 is isolated per default, since it is located between two nitrogens. The highest ^{13}C incorporation is obviously achieved by uniformly ^{13}C-enriched glucose, which therefore is a perfect way of labeling, if one is solely interested in this position (no overlap with signals from other positions). In order to label the above listed positions most quantitatively and exclusively, it is best to avoid long expression times. For 5 h expression after induction hardly any scrambling is seen in the spectra, while in much longer expressions other positions than mentioned before can become labeled to some amount. While this leads to less clean spectra, it does not cause any detectable ^{13}C–^{13}C bonds.

3.3 Nonglucose Custom Precursors

Instead of glucose as the solely carbon source, glycerol or pyruvate could be used. With one out of three carbons ^{13}C enriched compared to one out of six for glucose, they are expected to double the incorporation level, but also increase the risk of creating ^{13}C–^{13}C bonds as demonstrated for Cα (Ahlner, Andresen, Khan, Kay, & Lundström, 2015). 2-^{13}C glycerol and 2-^{13}C pyruvate are labeling Phe/Tyr ε and Trp ε3. 3-^{13}C pyruvate is labeling Phe/Tyr δ, Trp δ1 and ε3, His δ2 and ε1, the same pattern as 1-^{13}C glucose (Milbradt et al., 2015).

3.4 Adding Custom Precursors to Unlabeled Glucose

Instead of using site-selectively ^{13}C-enriched glucose as the energy source for bacteria, and the carbon source for aliphatic and aromatic side chains, one can use unlabeled glucose for the first two objectives and add common site-selectively ^{13}C-enriched precursors closer to the product as the carbon source for aromatic side chains. These precursors can be added directly in the expression medium at the start. This has been established for commercially available site-selectively ^{13}C-enriched erythrose (Weininger, 2017a) and ribose (Weininger, 2017b). In general these approaches are more selective than the glucose approach. 1-^{13}C erythrose is labeling Phe/Tyr ε and Trp ζ2, 2-^{13}C erythrose is labeling Phe/Tyr ζ and Trp η2, 3-^{13}C erythrose is labeling Phe/Tyr ε and Trp ζ3, and 4-^{13}C erythrose is labeling Phe/Tyr δ and Trp ε3. ^{13}C incorporation levels are 20%–50% and comparable to glucose for Phe/Tyr δ and ε. However ^{13}C incorporation in Trp is doubled. Phe ζ and Trp η2 become effectively labeled and available for relaxation studies. 1-^{13}C ribose is labeling Trp δ1 and His δ2 exclusively. The ^{13}C incorporation level is in between the levels achieved by 1-^{13}C and 2-^{13}C glucose. A combination of these approaches with site-selectively ^{13}C-enriched glucose is possible and results in higher ^{13}C incorporation and less selectivity of the labeling. 4-^{13}C erythrose has also been used in combination with unlabeled pyruvate instead of glucose, labeling Phe/Tyr δ and Trp ε3 (Kasinath, Valentine, & Wand, 2013).

Alternatively chemically synthesized site-selectively ^{13}C-enriched precursors can be added to unlabeled glucose. This approach is very similar to the one described above, but potentially of advantage, because the precursors are even closer to the product. Right now, these precursors are not commercially available, but chemical synthesis is required. Precursors have been made for labeling Phe ε (Lichtenecker et al., 2013), Tyr ε, Phe ζ (Lichtenecker, 2014), Trp ζ3 and ζ2, η2 (Schörghuber et al., 2015), ζ3, ε3 and η2 (Schörghuber, Geist, Bisaccia, et al., 2017), as well as for His ε1 (Schörghuber, Geist, Platzer, et al., 2017). The labeling in Phe/Tyr and Trp allows for additional deuteration in sites neighbored to the ^{1}H–^{13}C sites of interest. His ε1 is labeled selectively and quantitatively by uniformly ^{13}C glucose. The synthetic precursor, however, allows a combination with other site-selectively labeled sites in one sample.

3.5 Cell-Free Labeling

Chemically synthesized site-selectively ^{13}C-enriched precursors are also available for in vitro expression systems that allow for the highest control of the product. They have been developed for Phe/Tyr (Takeda, Ono, Terauchi, & Kainosho, 2010) and Trp (Miyanoiri et al., 2011). By this approach all sites neighbored to the ^{13}C-labeled positions are deuterated.

3.6 Deuteration

Some of the approaches mentioned above enable additional deuteration in the aromatic side chains, some do not. Here the effect of deuteration is discussed. First, it should be noted that it is not 100% possible to maintain a ^{1}H/^{2}H pattern in an in vivo expression. Some percentage will exchange with the solvent (H_2O/D_2O). By expressing in D_2O some percentage of the wanted protons will be replaced with deuterium, thus reducing the signal. By expressing in H_2O some percentage of protons will be reintroduced in deuterated neighboring sites. In Trp δ1, and His δ2 and ε1, the neighboring protons are attached to nitrogens and exchange with the solvent. The level of deuteration on these sites is equal to the level of deuteration of the solvent. However the coupling to this protons is too weak to matter anyway.

Since in all approaches neighbored sites are ^{13}C at natural abundance, they will give rise to weak signals in the spectra. Having this sites additionally deuterated is reducing these signals. By deuteration one therefore is getting cleaner spectra, but usually the weak signals do not interfere. In CPMG (Weininger, Respondek, et al., 2012) and $R_{1\rho}$ (Weininger, Brath, Modig, Teilum, & Akke, 2014) relaxation dispersion experiments deuteration is not needed in order to extract the correct parameters, if strong ^{1}H–^{1}H couplings do not occur. In relaxation experiments for determining order parameters, things are more complex, and require more research to sort out all the details. Strictly spoken eliminating protons, and their impact on dipolar relaxation is better. However the impact is strongly distance dependent ($\sim r^{-6}$) and thereby reduced to 1.7% for vicinal aromatic protons (Weininger, Diehl, et al., 2012) or negligible for more distanced protons ($\geq$3 Å) (Kasinath et al., 2013). Deuteration does not seem to affect R_1 (Kasinath et al., 2013; Weininger, Diehl, et al., 2012) but impact $R_2/R_{1\rho}$ (Kasinath et al., 2013). Things might be more severe for proteins >20 kDa. As a result on this one can still get reasonable order parameters with a higher error from protonated samples. They can be used for more qualitative interpretations. For most accurate order parameters it is advisable to use deuteration (Kasinath et al., 2013).

4. NMR SPECTRA OF SITE-SELECTIVELY LABELED AROMATIC SIDE CHAINS

Using site-selective ^{13}C labeling one is able to acquire high-resolution ^{1}H–^{13}C spectra on aromatic side chains (Fig. 3). The main reasons for improvement in the spectra are: avoiding constant time detection of the ^{13}C dimension and getting rid of ^{13}C–^{13}C strong couplings. Additionally this labeling allows for the further improvement of TROSY detection (Eletsky, Atreya, Liu, & Szyperski, 2005; Weininger, Diehl, et al., 2012). Another advantage is that crowded regions with potential overlap (Fδ, Fε, Fζ, Yδ and Hδ2, Yε) are simplified. These improvements that overcompensate intensity losses from incomplete ^{13}C incorporation at the positions of interest allow to study large systems (Milbradt et al., 2015) and protein dynamics in aromatic side chains (Weininger, Brath, et al., 2014; Weininger, Diehl, et al., 2012; Weininger, Respondek, et al., 2012).

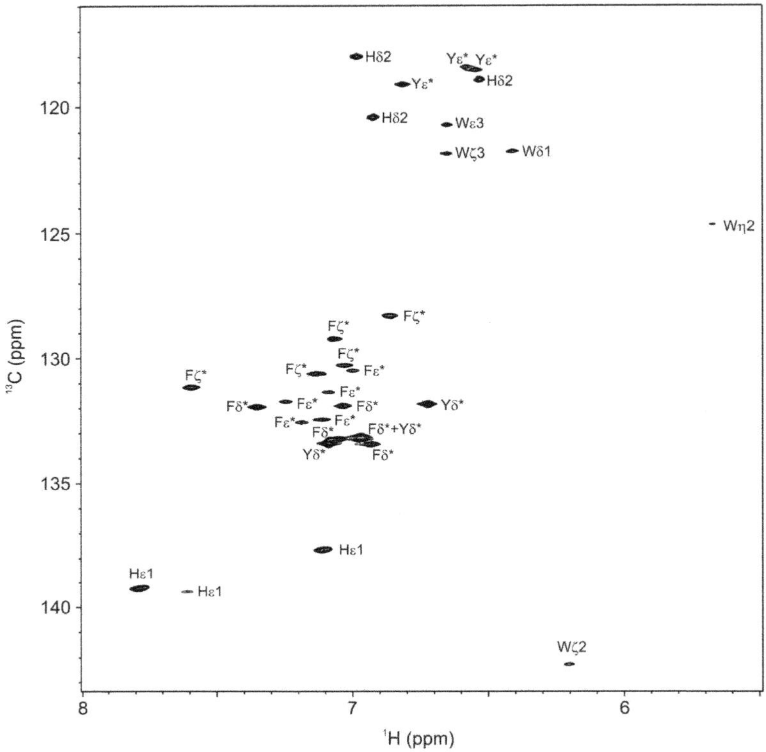

Fig. 3 Aromatic ^{1}H–^{13}C HSQC spectrum of FKBP12. The spectrum is an overlay of several experiments performed on differently site-selective ^{13}C-labeled samples. FKBP12 contains 1 Trp, 3 His, 5 Phe, and 3 Tyr. All expected signals are seen and labeled. Intensities are not fully comparable. Signals from Hε1 are not ^{15}N decoupled, Wζ2 is folded in the ^{13}C dimension.

5. RELAXATION DISPERSION EXPERIMENTS

5.1 ^{13}C Relaxation Dispersion Experiments

Site-selective ^{13}C labeling enables the use of μs–ms protein dynamic experiments in aromatic side chains, namely, ^{13}C CPMG relaxation dispersion (Weininger, Respondek, et al., 2012) and ^{13}C $R_{1\rho}$ relaxation dispersion (Weininger, Brath, et al., 2014) experiments. It eliminates large ^{1}J ^{13}C–^{13}C couplings that are impossible to refocus and thus would lead to severe artifacts. Without the ^{1}J ^{13}C–^{13}C couplings one can extract the correct parameters of dynamic processes between state A and B: the exchange rate (k_{ex}), the populations (p_A and p_B), and the chemical shift difference ($\Delta\delta$).

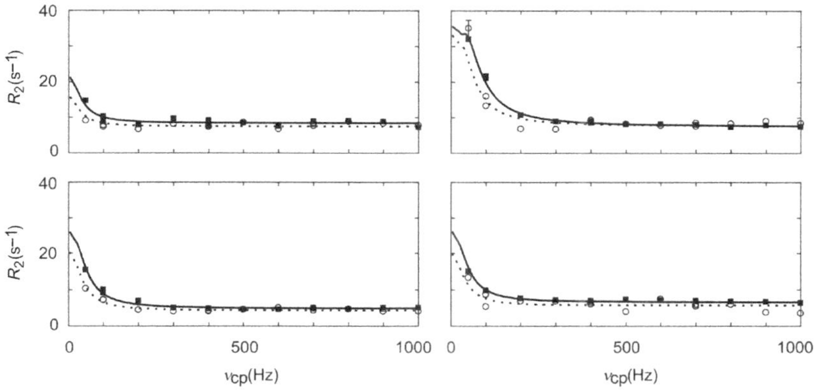

Fig. 4 Aromatic ^{13}C CPMG relaxation dispersion profiles of folding/unfolding of 2-^{13}C glucose-labeled GB1 at 40°C and static magnetic field strengths of 14.1 T (*open circles, dashed line*) and 18.8 T (*closed squares, solid line*). *First row left*: Y3ε, *right*: F30ε. *Second row left*: Y33ε, *right*: W43δ1. *Lines* represent the global fit to the experimental data with $k_{ex} = 223 \pm 8\,s^{-1}$, $p_u = 14.2 \pm 0.3\%$. Extracted chemical shift differences are 0.15 ± 0.01 ppm (Y3ε), 0.42 ± 0.02 ppm (F30ε), 0.23 ± 0.01 ppm (Y33ε), 0.21 ± 0.01 ppm (W43δ1). Chemical shift differences taken from the spectrum are 0.12 ppm (Y3ε), 0.65 ppm (F30ε), 0.27 ppm (Y33ε), 0.23 ppm (W43δ1).

This is shown in Fig. 4 on the example of folding/unfolding of GB1. Additional deuteration in the aromatic side chain is not required, if strong ^{1}H–^{1}H couplings are absent (see below). However it is strongly beneficial to run the experiments as TROSY versions (Igumenova & Palmer, 2006; Loria, Rance, & Palmer, 1999; Weininger, Brath, et al., 2014; Weininger, Respondek, et al., 2012), in order to deal with high ^{13}C R_2 rates in the aromatic side chain caused by the sizeable ^{13}C CSA.

5.2 Special Dynamic Processes in Aromatic Side Chains

Using ^{13}C relaxation dispersion experiments, one can study general dynamic processes on the μs–ms timescale, like folding/unfolding, binding, or conformational exchange between two or more states. In addition, one can study two dynamic processes special for aromatic side chains. These include tautomerization in His and ring flips or rotations in Phe and Tyr (Fig. 5). These processes can only be rationalized (in case of tautomerization), or even seen at all (in case of ring flips), using relaxation dispersion experiments that directly target the involved side chain. Both processes usually cause large effects in the dispersion profiles recorded on the side chain where they occur. In order to decipher a possible additional general dynamic process

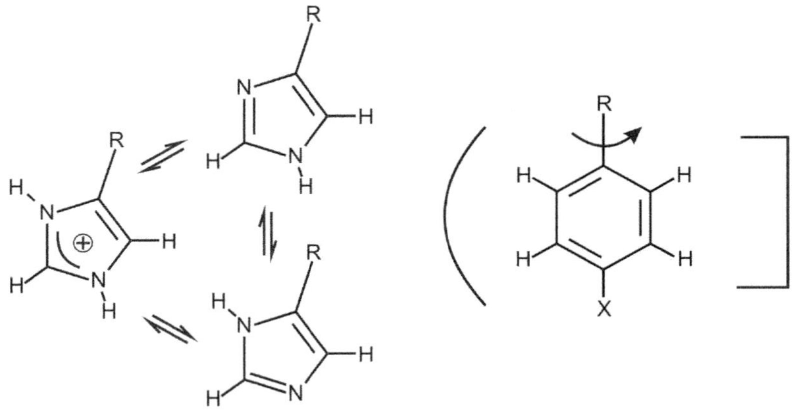

Fig. 5 Intrinsic exchange processes in aromatic side chains. *Left*: Different tautomeric forms of His. *Right*: Ring flips in Phe/Tyr. Different chemical environments on both sides are sketched.

from these one has to study multiple sites, including some, that are not affected by tautomerization or ring flips but only by the general process. His can exist in three different tautomeric forms, one protonated and two neutral (Fig. 5). Changes in the neutral tautomeric forms alter the hydrogen bonding pattern. $^{13}C\delta 2$ is very sensitive to the neutral tautomeric form, changing >8 ppm, significantly more than in different chemical environments. By identifying such a large shift change between two states in FKBP12, a change of the tautomeric states of H87 coupled to the general process could be monitored (Weininger et al., 2017). The study of dynamic processes directly in the His side chain thus is providing specific insights into chemical properties of the His. Phe and Tyr undergo frequent 180 degree rotations of the χ_2 dihedral angle. If ^{13}C shifts are different between the two structural sides the ring is placed in, a specific local two-state exchange process with fixed population $p_A = p_B = 0.5$ can be investigated (Fig. 5). Studying slow (μs–ms timescale) ring flips in 2D with advanced modern methods has led to the identification of two additional slow ring flips in BPTI (Weininger, Modig, & Akke, 2014), solving inconsistencies between NMR (Wagner, Bruhwiler, & Wuthrich, 1987; Wagner et al., 1976) and MD (Shaw et al., 2010). This technological advancement might lead to a renaissance of this special field of protein dynamics.

5.3 Strong Couplings

Things can be complicated by strong ^3J ^1H–^1H couplings of ^{13}C (position of interest) and ^{12}C (neighboring position) bound protons. They effectively increase $R_2/R_{1\rho}$ rate constants either on the low- or high-frequency side of CPMG/R_{1r} relaxation dispersion profiles. These profiles are thereby spoiled and cannot be interpreted correctly (Weininger, Respondek, Low, & Akke, 2013). The requirement for strong J-couplings is that the difference in frequency between two coupling atoms is not sufficiently larger than the J-coupling, which can occur in aromatic side chains. In order to cause an appreciable effect the J-couplings themselves have to be of a certain size. Coupling constants of 2 Hz do not cause any problems, even if in the strong coupling regime. Coupling constants of 7 Hz do (Weininger et al., 2013).

In combination with possible chemical shifts in the different positions of the aromatic side chains, we have the following situation. Hδ2, Hε1, and Wδ1 are never affected by strong couplings. Wε3, Wζ3, Wη2, and Wζ2 can potentially be. However they are not very likely to be strongly coupled and the possibility of strong coupling can directly be evaluated by the spectra. Fδ*, Fε*, Fζ, Yδ*, and Yε* can also be strongly coupled. In Phe it is in fact somewhat likely. Further, since both sides (δ and ε) in Phe and Tyr are usually in fast exchange caused by fast ring flips, they display an averaged signal. A potential strong coupling on one side thus cannot be evaluated from the spectra, unless the ring flip is very slow, which is unlikely.

However, one can still tell from the spectra, if occurring strong coupling artifacts are caused by strong couplings on both sides of the ring (similar averaged ^{1}H shift) or just on one side (different averaged ^{1}H shift). If happening just on one side, the effect is efficiently quenched by ring flips that alter the atoms between the strong coupling and the weak coupling side. The artifacts in the relaxation dispersion profiles remain only for ring flips $<100\,s^{-1}$. Since this is uncommon, ^{1}H–^{1}H strong coupling artifacts are relatively rare. On the other side one can identify slow ring flips by this phenomenon in rings with completely degenerate chemical shifts (Weininger et al., 2013). If one needs to get rid of ^{1}H–^{1}H strong coupling artifacts in Trp, slow flipping Phe or Tyr, or if they happen on both sides of Phe or Tyr, additional selective deuteration of the neighboring ^{12}C positions is needed.

6. PICOSECOND TO NANOSECOND DYNAMICS

Site-selectively ^{13}C-labeled protein samples provide isolated ^{1}H–^{13}C pairs in the aromatic side chains. The relaxation of aromatic ^{13}C is dominated by its CSA and the dipolar relaxation of the directly attached proton, a situation that is in close resemblance to backbone amide ^{15}N. Therefore it should be straightforward to derive order parameters, which would complement methyl order parameters in monitoring the protein hydrophobic core, and give unique insights into binding and active sites, where aromatic side chains often play critical roles. However there are some differences of yet unknown impact, one should keep in mind.

6.1 Remote Atoms

First, aromatic ^{1}H–^{13}C pairs are less isolated from other ^{1}H and ^{13}C than it is the case for backbone amide ^{1}H–^{15}N. Remote ^{13}C might lead to uncontrolled relaxation losses via J-couplings, while remote ^{1}H contributes to the dipolar relaxation. Higher order ^{13}C–^{13}C J-couplings, however, are small (<2 Hz) and therefore are not expected to matter. This has been experimentally confirmed by retrieving identical R_2 rate constants for differently ^{13}C-diluted samples (Weininger, Diehl, et al., 2012). The situation for a potential impact of remote ^{1}H is entirely clear yet. They matter, but their impact is strongly distance dependent ($\sim r^{-6}$). Vicinal aromatic protons are expected to contribute 1.7% to dipolar relaxation (Weininger, Diehl, et al., 2012); more distanced protons ($\geq$3 Å) (Kasinath et al., 2013) are negligible. So far there are two studies that have investigated the impact of remote protons on the actual relaxation experiments (for proteins <20 kDa). Both (Kasinath, Fu, Sharp, & Wand, 2015; Weininger, Diehl, et al., 2012) observe identical R_1 rate constants in samples with different ^{1}H dilution. The second (Kasinath et al., 2013) study observed somewhat higher $R_{1\rho}$ relaxation rate constants in protonated samples compared to samples where remote protons have been deuterated. The impact on the $\{^{1}H\}$–^{13}C NOE is not known. Virtually unaffected R_1 rate constants, but affected $R_{1\rho}$ relaxation rate constants still allow to determine similar order parameters for protonated samples, but with higher errors (Kasinath et al., 2013). In summary, if a more qualitative interpretation of order parameters is fine, one can use protonated samples; if more accurate order parameters are needed, one probably should use deuterated samples.

6.2 ^{13}C CSA in Aromatic Side Chains

While it is well known that ^{13}C CSA values in aromatic side chains are large, their exact values and differences for different sites are not well known. Till now, most interpretations are based on CSA values of benzene ($\sigma_{11}=-57$ ppm, $\sigma_{22}=39$ ppm, $\sigma_{33}=174$ ppm) (Strub et al., 1983). While benzene might be a useful model for Phe, it surely is insufficient for Tyr, His, and Trp. For instance R_1 values of Fε in GB1 are higher than R_1 values of Yε, while $\{^{1}H\text{–}\}^{13}C$ NOE values are comparable (Fig. 6), except for

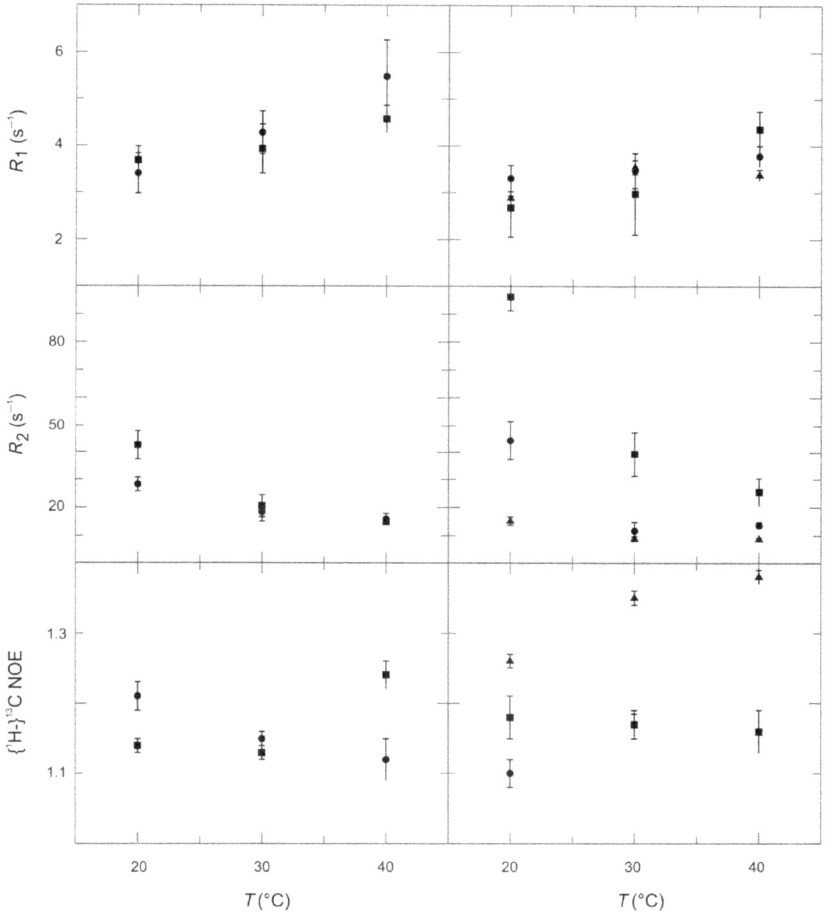

Fig. 6 Aromatic ^{13}C relaxation experiments on 2-^{13}C glucose-labeled GB1 at different temperatures. *Left column*: F30 (*circles*), F52 (*squares*). *Right column*: Y3 (*circles*), Y45 (*squares*), Y33 (*triangles*). *First row*: R_1, *second row*: R_2, *third row*: heteronuclear NOE.

the more flexible surface-exposed Y33. CSA values derived from solid-state NMR of the free amino acids (Separovic, Ashida, Woolf, Smith, & Terao, 1999; Separovic, Hayamizu, Smith, & Cornell, 1991; Ye, Fu, Hu, Hou, & Ding, 1993) allow for a more appropriate use of CSA values of individual positions model-free analysis and for somewhat realistic propagation of uncertainties to the order parameter. However, what we know of CSA values of aromatic side chain ^{13}C is far less established than for amide ^{15}N, and therefore having an impact on the precise determination of order parameters at the moment.

6.3 Ring Flips

Symmetric Phe and Tyr side chains undergo ring flips (Wagner et al., 1976). They are often very fast (at least on the NMR timescale) causing collapsed averaged signals on both sides of the ring. If sufficiently fast, their R_{ex} contribution to R_2 is zero. If slowed down in the appropriate time regime, they can cause extremely high large exchange contributions in R_2 (Fig. 6 Y3, F30 and F52 20°C, Y45 all temperatures). These contributions have to be fitted correctly to R_{ex} terms in the model-free analysis. If not, they can cause significantly wrong order parameters. Another potential complication on the other side of timescale was picked up in a temperature-dependent study of aromatic order parameters in ubiquitin (Kasinath et al., 2015). Reaching certain temperatures the derived order parameters decrease. The explanation is that the motion of the ring around χ_2 becomes faster than the correlation time of the protein and therefore is contributing to dipolar relaxation. If ring flips can become so fast (very short residence time), they might be better described as a more continuous rotation (very small to no energy barrier) than an actual flip process (high energy barrier). As a consequence the continuous rotation of the ring (or super fast flipping) becomes rather independent of pressure (no activation volume needed). In summary order parameters of Phe and Tyr residues can be affected indirectly by significant R_{ex} terms (if ring flips are slow enough), or directly if the flip (or a continuous rotation) is faster than the correlation time of the protein.

6.4 Models

Fast motions in aromatic side chains have been addressed by simple model-free analysis and more complex models (Bruschweiler, Blackledge, & Ernst, 1991). So far model-free-determined order parameters look reasonable (Boyer & Lee, 2008; Kasinath et al., 2013; Sapienza, Mauldin, & Lee, 2011), despite any complications mentioned above. However, there is the potential of a more physical interpretation of motion by analyzing multiple sites in a rigid ring. This is discussed on the example of Phe and Tyr. Assuming a rigid backbone, aromatic side chains can fluctuate around χ_1, around χ_2, or a combination of both (Levy & Sheridan, 1983). If the fluctuation around χ_2 is dominant, one can expect identical order parameters for the positions δ and ε. If the fluctuation around χ_1 is dominant, one can expect lower order parameters for ε than for δ, since the amplitude of motion is larger. Therefore studying the ps–ns dynamic of aromatic side chains can provide insights not only into unique sites of the protein but also in unique physical motions.

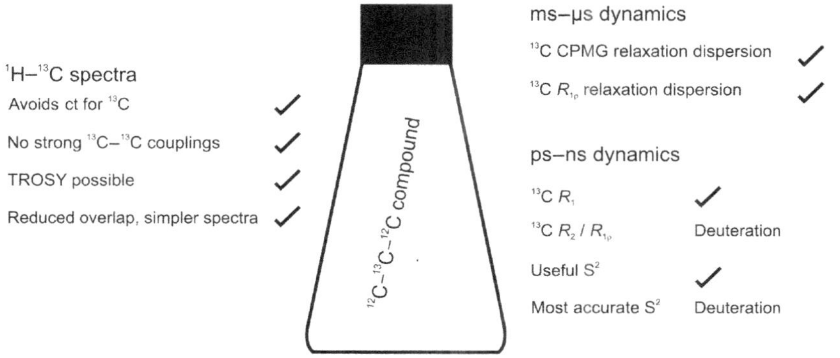

Fig. 7 Improvements and enablings by site-selective ^{13}C labeling of aromatic side chains.

7. CONCLUSIONS

Site-selective ^{13}C labeling of aromatic side chains can be achieved in multiple ways, by site-selectively ^{13}C-enriched precursors in the expression medium. It solves multiple problems and complication in the NMR spectroscopy of aromatic side chains (Fig. 7): ^{13}C–^{13}C strong couplings, high relaxation losses and limited resolution by constant time detection of the indirect ^{13}C dimension, overlap in the spectra, and the inability to refocus transverse aromatic ^{13}C magnetization when ^{1}J ^{13}C–^{13}C couplings are present. By this, one can obtain high-resolution ^{1}H–^{13}C HSQC spectra of aromatic side chains that can be further optimized by the TROSY approach. Protein dynamic on the μs–ms timescale can be studied by ^{13}C relaxation dispersion methods in a straightforward and robust way. NMR order parameters of aromatic side-chain positions are possible, but not all possible issues and complications are settled at the moment.

REFERENCES

Ahlner, A., Andresen, C., Khan, S. N., Kay, L. E., & Lundström, P. (2015). Fractional enrichment of proteins using [2-C-13]-glycerol as the carbon source facilitates measurement of excited state C-13 alpha chemical shifts with improved sensitivity. *Journal of Biomolecular NMR*, *62*(3), 341–351.

Bartlett, G. J., Porter, C. T., Borkakoti, N., & Thornton, J. M. (2002). Analysis of catalytic residues in enzyme active sites. *Journal of Molecular Biology*, *324*(1), 105–121.

Berry, B. W., Elvekrog, M. M., & Tommos, C. (2007). Environmental modulation of protein cation-pi interactions. *Journal of the American Chemical Society*, *129*(17), 5308–5309.

Bogan, A. A., & Thorn, K. S. (1998). Anatomy of hot spots in protein interfaces. *Journal of Molecular Biology*, *280*(1), 1–9.

Boyer, J. A., & Lee, A. L. (2008). Monitoring aromatic picosecond to nanosecond dynamics in proteins via C-13 relaxation: Expanding perturbation mapping of the rigidifying core mutation, V54A, in eglin c. *Biochemistry*, *47*(17), 4876–4886.

Bruschweiler, R., Blackledge, M., & Ernst, R. R. (1991). Multi-conformational peptide dynamics derived from NMR data: A new search algorithm and its application to antamanide. *Journal of Biomolecular NMR*, *1*(1), 3–11.

Burley, S. K., & Petsko, G. A. (1985). Aromatic-aromatic interaction: A mechanism of protein structure stabilization. *Science*, *229*(4708), 23–28.

Burley, S. K., & Petsko, G. A. (1989). Electrostatic interactions in aromatic oligopeptides contribute to protein stability. *Trends in Biotechnology*, 7(12), 354–359.

Eletsky, A., Atreya, H. S., Liu, G. H., & Szyperski, T. (2005). Probing structure and functional dynamics of (large) proteins with aromatic rings: L-GFT-TROSY (4,3)D HCCH NMR spectroscopy. *Journal of the American Chemical Society*, *127*(42), 14578–14579.

Fersht, A. (1977). *Enzyme structure and mechanism*. W. H. Freeman.

Igumenova, T. I., & Palmer, A. G. (2006). Off-resonance TROSY-selected R1ρ experiment with improved sensitivity for medium- and high-molecular-weight proteins. *Journal of the American Chemical Society*, *128*(25), 8110–8111.

Kasinath, V., Fu, Y. N., Sharp, K. A., & Wand, A. J. (2015). A sharp thermal transition of fast aromatic-ring dynamics in ubiquitin. *Angewandte Chemie (International ed. in English)*, *54*(1), 102–107.

Kasinath, V., Valentine, K. G., & Wand, A. J. (2013). A ^{13}C labeling strategy reveals a range of aromatic side chain motion in calmodulin. *Journal of the American Chemical Society*, *135*(26), 9560–9563.

Krishna Deepak, R. N., & Sankararamakrishnan, R. (2016). N-H...N hydrogen bonds involving histidine imidazole nitrogen atoms: A new structural role for histidine residues in proteins. *Biochemistry*, *55*(27), 3774–3783.

Levy, R. M., & Sheridan, R. P. (1983). Combined effect of restricted rotational diffusion plus jumps on nuclear magnetic-resonance and fluorescence probes of aromatic ring motions in proteins. *Biophysical Journal*, *41*(2), 217–221.

Lichtenecker, R. J. (2014). Synthesis of aromatic ^{13}C/^{2}H-alpha-ketoacid precursors to be used in selective phenylalanine and tyrosine protein labelling. *Organic & Biomolecular Chemistry*, *12*(38), 7551–7560.

Lichtenecker, R. J., Weinhaupl, K., Schmid, W., & Konrat, R. (2013). α-Ketoacids as precursors for phenylalanine and tyrosine labelling in cell-based protein overexpression. *Journal of Biomolecular NMR*, *57*(4), 327–331.

Lindskog, S. (1997). Structure and mechanism of carbonic anhydrase. *Pharmacology & Therapeutics*, *74*(1), 1–20.

Lo Conte, L., Chothia, C., & Janin, J. (1999). The atomic structure of protein-protein recognition sites. *Journal of Molecular Biology*, *285*(5), 2177–2198.

Loria, J. P., Rance, M., & Palmer, A. G. (1999). A TROSY CPMG sequence for characterizing chemical exchange in large proteins. *Journal of Biomolecular NMR*, *15*(2), 151–155.

Lundström, P., Teilum, K., Carstensen, T., Bezsonova, I., Wiesner, S., Hansen, D. F., et al. (2007). Fractional ^{13}C enrichment of isolated carbons using [1-C-13]- or [2-C-13]-glucose facilitates the accurate measurement of dynamics at backbone C-alpha and side-chain methyl positions in proteins. *Journal of Biomolecular NMR*, *38*(3), 199–212.

Mahadevi, A. S., & Sastry, G. N. (2013). Cation–π interaction: Its role and relevance in chemistry, biology, and material science. *Chemical Reviews*, *113*(3), 2100–2138.

Milbradt, A. G., Arthanari, H., Takeuchi, K., Boeszoermenyi, A., Hagn, F., & Wagner, G. (2015). Increased resolution of aromatic cross peaks using alternate C-13 labeling and TROSY. *Journal of Biomolecular NMR*, *62*(3), 291–301.

Miyanoiri, Y., Takeda, M., Jee, J., Ono, A. M., Okuma, K., Terauchi, T., et al. (2011). Alternative SAIL-Trp for robust aromatic signal assignment and determination of the chi(2) conformation by intra-residue NOEs. *Journal of Biomolecular NMR*, *51*(4), 425–435.

Preimesberger, M. R., Majumdar, A., Rice, S. L., Que, L., & Lecomte, J. T. (2015). Helix-capping histidines: Diversity of N-H...N hydrogen bond strength revealed by (2h)JNN scalar couplings. *Biochemistry*, *54*(46), 6896–6908.

Reynolds, W. F., Peat, I. R., Freedman, M. H., & Lyerla, J. R., Jr. (1973). Determination of the tautomeric form of the imidazole ring of L-histidine in basic solution by carbon-13 magnetic resonance spectroscopy. *Journal of the American Chemical Society*, *95*(2), 328–331.

Sapienza, P. J., Mauldin, R. V., & Lee, A. L. (2011). Multi-timescale dynamics study of FKBP12 along the rapamycin-mTOR binding coordinate. *Journal of Molecular Biology*, *405*(2), 378–394.

Schörghuber, J., Geist, L., Bisaccia, M., Weber, F., Konrat, R., & Lichtenecker, R. (2017). Anthranilic acid, the new player in the ensemble of aromatic residue labeling precursor compounds. *Journal of Biomolecular NMR*, *69*(1), 13–22.

Schörghuber, J., Geist, L., Platzer, G., Konrat, R., & Lichtenecker, R. J. (2017). Highly selective stable isotope labeling of histidine residues by using a novel precursor in *E. coli*-based overexpression systems. *Chembiochem*, *18*(15), 1487–1491.

Schörghuber, J., Sara, T., Bisaccia, M., Schmid, W., Konrat, R., & Lichtenecker, R. J. (2015). Novel approaches in selective tryptophan isotope labeling by using *Escherichia coli* overexpression media. *Chembiochem*, *16*(5), 746–751.

Separovic, F., Ashida, J., Woolf, T., Smith, R., & Terao, T. (1999). Determination of chemical shielding tensor of an indole carbon and application to tryptophan orientation of a membrane peptide. *Chemical Physics Letters*, *303*(5–6), 493–498.

Separovic, F., Hayamizu, K., Smith, R., & Cornell, B. A. (1991). C-13 chemical-shift tensor of L-tryptophan and its application to polypeptide structure determination. *Chemical Physics Letters*, *181*(2–3), 157–162.

Shaw, D. E., Maragakis, P., Lindorff-Larsen, K., Piana, S., Dror, R. O., Eastwood, M. P., et al. (2010). Atomic-level characterization of the structural dynamics of proteins. *Science*, *330*(6002), 341–346.

Strub, H., Beeler, A. J., Grant, D. M., Michl, J., Cutts, P. W., & Zilm, K. W. (1983). Low-temperature C-13 magnetic-resonance in solids. 2. 6 Pi-electron ring-systems—C7H7+, C6H6, and C5H5−. *Journal of the American Chemical Society*, *105*(10), 3333–3334.

Takeda, M., Ono, A. M., Terauchi, T., & Kainosho, M. (2010). Application of SAIL phenylalanine and tyrosine with alternative isotope-labeling patterns for protein structure determination. *Journal of Biomolecular NMR*, *46*(1), 45–49.

Teilum, K., Brath, U., Lundstrom, P., & Akke, M. (2006). Biosynthetic C-13 labeling of aromatic side chains in proteins for NMR relaxation measurements. *Journal of the American Chemical Society*, *128*(8), 2506–2507.

Valley, C. C., Cembran, A., Perlmutter, J. D., Lewis, A. K., Labello, N. P., Gao, J., et al. (2012). The methionine-aromatic motif plays a unique role in stabilizing protein structure. *Journal of Biological Chemistry*, *287*(42), 34979–34991.

Wagner, G., Bruhwiler, D., & Wuthrich, K. (1987). Reinvestigation of the aromatic side-chains in the basic pancreatic trypsin-inhibitor by heteronuclear two-dimensional nuclear-magnetic-resonance. *Journal of Molecular Biology*, *196*(1), 227–231.

Wagner, G., Demarco, A., & Wuthrich, K. (1976). Dynamics of aromatic amino-acid residues in globular conformation of basic pancreatic trypsin-inhibitor (Bpti). 1. H-1 NMR-studies. *Biophysics of Structure and Mechanism*, *2*(2), 139–158.

Weininger, U. (2017a). Site-selective ^{13}C labeling of proteins using erythrose. *Journal of Biomolecular NMR*, *67*(3), 191–200.
Weininger, U. (2017b). Site-selective C-13 labeling of histidine and tryptophan using ribose. *Journal of Biomolecular NMR*, *69*(1), 23–30.
Weininger, U., Brath, U., Modig, K., Teilum, K., & Akke, M. (2014). Off-resonance rotating-frame relaxation dispersion experiment for C-13 in aromatic side chains using L-optimized TROSY-selection. *Journal of Biomolecular NMR*, *59*(1), 23–29.
Weininger, U., Diehl, C., & Akke, M. (2012). C-13 relaxation experiments for aromatic side chains employing longitudinal- and transverse-relaxation optimized NMR spectroscopy. *Journal of Biomolecular NMR*, *53*(3), 181–190.
Weininger, U., Modig, K., & Akke, M. (2014). Ring flips revisited: C-13 relaxation dispersion measurements of aromatic side chain dynamics and activation barriers in basic pancreatic trypsin inhibitor. *Biochemistry*, *53*(28), 4519–4525.
Weininger, U., Modig, K., Geitner, A. J., Schmidpeter, P. A. M., Koch, J. R., & Akke, M. (2017). Dynamics of aromatic side chains in the active site of FKBP12. *Biochemistry*, *56*(1), 334–343.
Weininger, U., Respondek, M., & Akke, M. (2012). Conformational exchange of aromatic side chains characterized by L-optimized TROSY-selected C-13 CPMG relaxation dispersion. *Journal of Biomolecular NMR*, *54*(1), 9–14.
Weininger, U., Respondek, M., Low, C., & Akke, M. (2013). Slow aromatic ring flips detected despite near-degenerate NMR frequencies of the exchanging nuclei. *Journal of Physical Chemistry B*, *117*(31), 9241–9247.
Ye, C. H., Fu, R. Q., Hu, J. Z., Hou, L., & Ding, S. W. (1993). C-13 chemical-shift anisotropies of solid amino-acids. *Magnetic Resonance in Chemistry*, *31*(8), 699–704.

CHAPTER FOUR

A Quick and Colorful Method to Measure Low-Level Contaminations of Paramagnetic Ni^{2+} in Protein Samples Purified by Immobilized Metal Ion Affinity Chromatography

Starla D. Glover[2], **Cecilia Tommos**[1]
Department of Biochemistry & Biophysics, University of Pennsylvania Perelman School of Medicine, Philadelphia, PA, United States
[1]Corresponding author: e-mail address: tommos@pennmedicine.upenn.edu

Contents

[2] Present address: Department of Chemistry—Ångström Laboratories, Uppsala University, Box 523, 751 20 Uppsala, Sweden.

Methods in Enzymology, Volume 614
ISSN 0076-6879
https://doi.org/10.1016/bs.mie.2018.08.037

Abstract

Isotopic labeling of recombinantly expressed proteins is generally required for investigation by modern nuclear magnetic resonance (NMR) methods. Purification strategies of the labeled proteins often include the use of a polyhistidine affinity tag (His-tag) and immobilized metal ion affinity chromatography (IMAC). Described herein are rapid and inexpensive qualitative and quantitative assays to determine the concentration of paramagnetic Ni^{2+} in protein samples purified by IMAC. Both qualitative and quantitative colorimetric methods detect the amount of Ni^{2+} via the color change produced when a $[Ni(PAR)_n]^{2+}$ (PAR = 4-(2-pyridylazo)resorcinol, $n = 1, 2$) complex is formed. The qualitative assay provides a rapid visual test for the presence of Ni^{2+} in the low micromolar range in a sample of interest. The usefulness of the spectroscopic quantitative assay is illustrated by: (i) detecting a 12 μ*M* Ni^{2+} contamination in an NMR sample containing 950 μ*M* of the 7.5 kDa α_3W protein purified by a standard His-tag Ni^{2+}/IMAC approach and (ii) showing that the ^{15}N-HSQC spectrum of the α_3W NMR sample, containing 1 paramagnetic Ni^{2+} ion per 80 protein molecules, displays clear line broadening of both water and protein spectral lines. We also (iii) measured Ni^{2+} release during the equilibration, wash, and elution steps of three commonly used Ni^{2+}/IMAC resins when following manufacturer's protocols. The concentration of Ni^{2+} detected in elutes of the three resins ranged from 2 μ*M* to nearly 1 m*M*.

1. INTRODUCTION

Nuclear magnetic resonance (NMR) spectroscopy has an inherently low sensitivity and NMR-based structural biology studies typically require tens of milligrams of purified material. Protein production generally involves recombinantly expressed proteins and purification strategies based on affinity chromatography. Proteins engineered to contain a polyhistidine affinity tag (His-tag) and purified using immobilized metal ion affinity chromatography (IMAC) represent a major protein production strategy in protein NMR and generally across the biophysical and biochemical research fields. IMAC resins are often charged with paramagnetic metal ions. For example, paramagnetic Co^{2+}, Ni^{2+}, or Cu^{2+} are the most commonly used metal ions for purification of His-tagged proteins. All charged IMAC resins have a propensity for metal ion leakage, although it varies depending on the specific chemical properties of the IMAC system and the solution conditions used. Thus, copurification of substoichiometric amount of a metal ion nonspecifically bound to the target protein presents a significant possibility. This situation is of particular concern for the protein NMR field since even a very minor contamination of a paramagnetic species may give rise to undesirable spectral perturbations such as line broadening of water and protein resonances. Here we present quickly and easily conducted qualitative and quantitative colorimetric assays to check Ni^{2+}/IMAC-purified protein material for micromolar contaminations of Ni^{2+}.

IMAC separates proteins present in the solvent phase according to their affinity for metal ions such as Fe^{3+}, Co^{2+}, Ni^{2+}, Cu^{2+}, and Zn^{2+} that have been immobilized by a metal-chelating agent covalently bound to the stationary matrix. The IMAC approach was based on the known affinities of aqueous histidine and cysteine to coordinate transition metal ions (Hearon, 1948) and the novel idea of using immobilized metal ions to selectively bind peptides and proteins with exposed histidine and cysteine residues. In the proof-of-concept study, Porath et al. used iminodiacetate (IDA) as the chelating agent to fix Ni^{2+} and Cu^{2+} to agarose and successfully demonstrated fractionation of human serum proteins using the developed M^{2+}-IDA/agarose resins (Porath, Carlsson, Olsson, & Belfrage, 1975). The IMAC protein purification approach was validated and further developed over the following years and gained general recognition (Porath, 1992; Sulkowski, 1985). Hochuli et al. introduced nitrilotriacetate (NTA) as a metal-chelating agent and showed that the new Ni^{2+}-NTA/sepharose resin had a high specificity for peptides and proteins containing adjacent histidines (Hochuli, Döbeli, & Schacher, 1987). In a subsequent publication, mouse dihydrofolate reductase (DHFR) was engineered to contain polyhistidine tags and efficient isolation of the expressed DHFR/His-tag fusions on the Ni^{2+}-NTA/sepharose resin was shown (Hochuli, Bannwarth, Döbeli, Gentz, & Stüber, 1988). Thus, the application of recombinant techniques and the concept of engineered IMAC affinity tags (Hochuli et al., 1988; Smith, Furman, Ingolia, & Pidgeon, 1988; Smith, Furman, & Pidgeon, 1987) refined the IMAC methodology toward the well-established and widely used protein purification tool it is today.

The metal-chelating agent binds the metal ion to the supporting matrix and modulates the metal ion/protein interactions (Porath, 1992). The original IDA and NTA chelators are used in many commercial IMAC resins and protein purification kits including IDA-containing HiTrap™ products developed by Amersham Pharmacia Biotech and NTA resins from Qiagen, Thermo Fisher Scientific, Bio-Rad Laboratories, etc. Commonly used chelators are tridentate (e.g., IDA), tetradentate (e.g., NTA), or pentadentate (e.g., carboxymethylated aspartate used in Co^{2+}-TALON resins, Clontech

Laboratories). Upon charging the resin, the chelator binds the added metal ion while leaving some coordination sites available for protein ligation. In general, the tetradentate and pentadentate chelators have higher affinity for the metal ion (i.e., less metal ion leakage), but they exhibit lower protein binding capacity due to having fewer protein coordination sites. When choosing an IMAC system for a specific protein, there are multiple factors to consider including the protein binding capacity and selectivity of the IMAC resin, suitable protein elution conditions, the potential need for downstream purification steps, and cost. The degree of metal ion leakage is thus only one of several parameters to consider for optimized protein production.

An assay to screen for contaminating Ni^{2+} ions that may copurify with a target protein should ideally be rapid and easy to perform and require little protein material. Here we used commercially available 4-(2-pyridylazo)resorcinol (PAR) as the active reagent in qualitative and quantitative colorimetric assays to screen for possible Ni^{2+} contaminations in protein samples whose purification protocols include Ni^{2+}/IMAC. The sensitivity of the developed method is illustrated by checking protein NMR samples for low-level Ni^{2+} contaminations. We show that Ni^{2+} released during a typical IMAC purification step copurifies with the target protein through a subsequent reversed-phase HPLC step and affects NMR spectral properties at concentrations as low as one Ni^{2+} ion per 80 protein molecules. The usefulness of the developed method is further illustrated by measuring Ni^{2+} leakage from three commercially available and commonly used IMAC resins. Either protocol described below provides a straightforward method to check a protein purification midstream or an NMR sample and to exclude the possibility of spectral degradation due to a minor contamination of paramagnetic Ni^{2+}.

2. MATERIALS AND METHODS

CELLSTAR 50 mL light-impermeable centrifuges tubes were purchased from Greiner Bio-One via VWR (catalog number 82051-630). Sodium acetate-d_3 (99 atom % D) was purchased from Cambridge Isotope Laboratories (DLM-3126). His•Bind resin (Novagen) was purchased from EMD Millipore Corporation (69670). Ethylenediaminetetraacetate dipotassium salt extra pure (EDTA, AC209745000), His60 Ni Superflow resin (Clontech Laboratories) (NC0319531), imidazole analytical grade

(AC396745000), NaCl extra pure (AC194090010), and Tris–HCl extra pure (AC228030010) were purchased from ACROS Organics via Fisher Scientific. $NiSO_4{\bullet}6H_2O$ (N73500) and NaH_2PO_4 (S369), both ACS Reagent grade, and GE Healthcare Sephadex G-25 PD10 columns (45-000-148) were from Fisher Scientific. Acetonitrile HPLC plus (34998), cOmplete His-Tag Purification resin (Roche; 5893682001), D_2O (99.9 atom % D; 613444), 2,2-dimethyl-2-silapentane-5-sulfonate (DSS; 178837), NaCl BioXtra (S7653), 4-(2-pyridylazo)resorcinol monosodium salt monohydrate (PAR; 1075330005), and trifluoroacetic acid for HPLC (302031) were purchased from Sigma-Aldrich. 5 mm Precision NMR tubes were purchased from Wilmad via Sigma-Aldrich (Z412015) and Spectrosil quartz cuvettes with 1 (21-Q-1), 2 (21-Q-2), 5 (9-Q-5), and 10 (3-Q-10) mm pathlengths were purchased from Starna Cells Inc. α_3X proteins are expressed in fusion with thioredoxin (Dai et al., 2002) in *E. coli* BL21-CodonBlue (DE3)-RIL grown on LB or minimal media and purified as described previously (Martínez-Rivera et al., 2011).

3. METHOD TO DETERMINE MICROMOLAR NI^{2+} CONTAMINATIONS IN PROTEIN SAMPLES

3.1 Description of the Qualitative Visually Based Colorimetric Assay to Screen for Ni^{2+}

The qualitative assay described in this section utilizes the observed differences in color between two solutions: a control containing 50 μM of free PAR and a sample containing 50 μM PAR dissolved in the protein solution of interest. The colorimetric scale for this assay is shown in Fig. 1 where the Ni^{2+} concentration spans a range of 0–37.5 μM in a solution containing 50 μM PAR. As the Ni^{2+} concentration and the nickel to PAR (Ni:PAR) ratio increase, the color of the solution shifts from bright yellow to yellow–orange to coral. Each of the 12 samples shown in Fig. 1 was made by titrating together 1 mL of a 100 μM PAR stock solution and 1 mL of a buffered solution containing 0–75 μM Ni^{2+}. These 1:1 volumetric titrations result in final concentrations of 50 μM PAR and 0–37.5 μM Ni^{2+}. By following Sections 3.2.1–3.2.5 described below, protein samples can be screened for micromolar Ni^{2+} contaminations by a simple visual comparison to the colorimetric scale shown in Fig. 1.

The upper detection limit for Ni^{2+} is set by the final PAR concentration in the sample. With a Ni^{2+} to PAR binding stoichiometry of 1:2, the upper

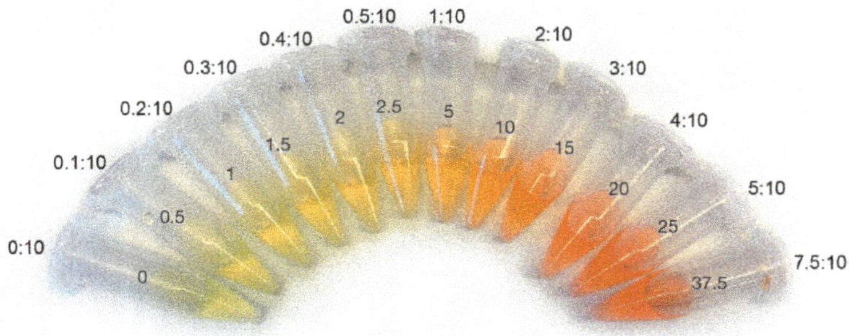

Fig. 1 Colorimetric series made from samples containing 25 m*M* NaH_2PO_4, 150 m*M* NaCl, pH 8.0, 50 μ*M* PAR, and 0–37.5 μ*M* Ni^{2+}. The Ni^{2+}/PAR titration series displays colors ranging from *bright yellow* (0 μ*M* Ni^{2+}, *far left*) to *coral* (37.5 μ*M* Ni^{2+}, *far right*). The series was made by mixing equal volumes of a 100 μ*M* PAR (in milliQ water) stock and 0–75 μ*M* Ni^{2+} (in Buffer A) solutions. The Ni^{2+} concentration in μ*M* and the Ni:PAR ratio are printed at the center and the top of each tube, respectively.

limit for Ni^{2+} detection is predicted to be around 25 μ*M* for a solution containing 50 μ*M* PAR. When the final concentration is above 25 μ*M* Ni^{2+}, essentially all of the PAR molecules are coordinated to Ni^{2+} and the samples will be coral in color. Section 3.3.1 provides a more detailed discussion of the binding behavior of PAR to Ni^{2+}.

Increasing or decreasing the concentration of the PAR stock solution will raise or lower the Ni^{2+} detection limit, respectively. There are two advantages to carry out the qualitative assay with a final concentration of 50 μ*M* PAR: (a) the intensity of the colors allows them to be readily distinguished from one another and (b) the samples from the qualitative assay are at ideal concentrations for the spectroscopic quantitative assay described in Section 3.5, should such analysis be needed.

3.1.1 Preparation of a 100 μM PAR Stock

25.55 mg of 4-(2-pyridylazo)resorcinol monosodium salt monohydrate (molecular weight 255.21 g mol^{-1}) was weighed using a micrometer balance and dissolved in 100 mL of milliQ water using a volumetric flask to give a 1.001 m*M* PAR solution. 5 mL of 1.001 m*M* PAR was diluted to 50 mL using milliQ water in a volumetric flask to give 100 μ*M* PAR. PAR stock solutions were stored in 50 mL black and opaque polypropylene tubes designed for light exclusion. No spectral changes were observed for 100 μ*M* PAR/milliQ stock solutions recorded on the day of preparation and 1 month later.

3.1.2 Tip for Measuring Small Masses

It is good practice to weigh masses that are well within the bounds of the balance you are using. If possible, use a calibrated micrometer balance for measuring out mg quantities of PAR.

3.1.3 Optional Preparations of Ni^{2+} and Ni^{2+}/PAR Solutions

1.319 g of $NiSO_4 \cdot 6H_2O$ (molecular weight 262.84 g mol^{-1}) was dissolved into 100 mL of Buffer A (50 mM NaH_2PO_4, 300 mM NaCl, pH 8.0) using a volumetric flask to give a 50.2 mM Ni^{2+} stock solution. Serial dilutions using volumetric glassware and calibrated automatic pipettors were performed to obtain a concentration series ranging from 1 to 75 μM Ni^{2+} in Buffer A.

To generate a colorimetric series like the one shown in Fig. 1, titrate 1 mL of the 100 μM PAR stock and 1 mL of a Ni^{2+} standard into a 5 mL microcentrifuge tube or glass vial. Solutions prepared in our laboratory maintain their color for several months. Importantly, $[Ni(PAR)_n]^{2+}$ ($n=1$, 2) solutions should be buffered and have a pH in the 7–10 range to prevent precipitation.

3.2 Protocol for the Qualitative Visually Based Colorimetric Assay to Screen for Ni^{2+}

3.2.1 Prepare the PAR Stock Solution

Prepare 100 μM PAR in milliQ water or in a pH 7–10 buffer (see Section 3.1.1).

3.2.2 Prepare the Free PAR Control

Titrate 1:1 volumetric ratios of the 100 μM PAR stock and the buffer used for the protein sample into a translucent vial. The final pH should be >7.0. The control should be a vibrant yellow in color.

3.2.3 Prepare the Protein-PAR Sample

Titrate 1:1 volumetric ratios of the 100 μM PAR stock and the protein sample into a translucent vial. Gently mix and let the protein-PAR sample sit for 5–10 min or until its color does not change. The final pH should be >7.0.

3.2.4 Observe Differences in Color

Visually compare the PAR control and the protein-PAR sample against a white background. The Ni^{2+} concentration in the protein sample can be estimated from Fig. 1 or using a colorimetric series prepared using a buffer system of choice (see Section 3.1.3).

3.2.5 Tips for Low and High Ni:PAR Ratios

It can be challenging to visually detect color changes associated with small Ni:PAR ratios (0.1:10 to 0.3:10, see Fig. 1). In this situation, it is better to use the quantitative assay described below for a more accurate determination of the Ni^{2+} concentration. When the Ni:PAR ratio is 4:10 or greater, the samples will be coral in color as shown in Fig. 1. This indicates that all, or nearly all, PAR molecules are coordinated to Ni^{2+}. If the instructions described in Section 3.2.4 result in coral-colored samples, dilute the sample at least 4× and repeat Sections 3.2.3 and 3.2.4 until the protein-PAR sample is yellow to yellow–orange.

3.3 Description of the Spectroscopic Assay to Quantitate Possible Ni^{2+} Contaminations

3.3.1 Spectroscopic and Binding Behavior of Free PAR and Nickel-Bound PAR

The properties of PAR make it a generally useful indicator for determining micromolar concentrations of transition metal cations (M^{n+}) in aqueous samples (McCall & Fierke, 2000). PAR is a tridentate ligand known to coordinate metal ions in M^{n+} to PAR stoichiometries of 1:1 and 1:2, as illustrated in Fig. 2. The resulting M^{n+}/PAR complexes typically display good water solubility at neutral to alkaline pH and high molar extinction coefficients ($\varepsilon_{max} \sim 10^4 M^{-1}$ cm^{-1}). The pH dependence of the spectral properties of free PAR and M^{2+}/PAR, including Co^{2+}, Ni^{2+}, and Cu^{2+}-coordinated PAR, has been reported previously (Kocyła, Pomorski, & Krężel, 2015). PAR has three acid–base groups with aqueous pK_a(1), pK_a(2), and pK_a(3) values of 2.9, 5.4, and 12.1, respectively, as displayed in Fig. 2. The mole fraction of the monoprotonated species is >0.95 for the pH 7.0–10.5

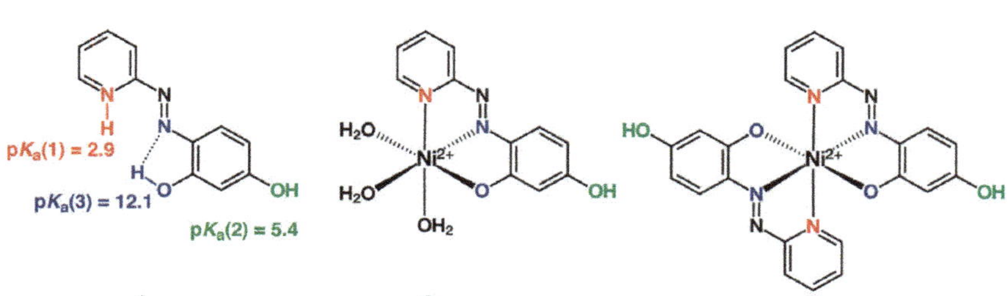

1 2 3

Fig. 2 Chemical structures of (**1**) fully protonated PAR in its free form, (**2**) monocoordinated $[Ni(H_2O)_3(PAR)]^{2+}$, and (**3**) double-coordinated $[Ni(PAR)_2]^{2+}$. Free PAR has three acid–base groups and their pK_a values are also shown. An ε_{413} of $32{,}900 \pm 500 M^{-1}$ cm^{-1} was determined for PAR in milliQ water (Section 3.3.1).

range (Kocyła et al., 2015) and this state exhibits a characteristic absorption maximum at 413 nm, as shown in Fig. 3A (black spectrum). We determined a ε_{413} of 32,900 ± 500 M^{-1} cm^{-1} for PAR in milliQ water. This value was obtained by using a 100 μ*M* PAR stock made as described in Section 3.1.1 and preparing a concentration series ranging from 2 to 100 μ*M* PAR via large-volume serial dilutions. Cuvettes with pathlengths of 0.2, 0.5, and 1.0 cm were used such that the optical density (OD) did not exceed unity. An ε_{413} was determined from the linear dependence in the OD_{413}/pathlength values vs the PAR concentration. Upon binding to M^{2+}, the $pK_a(3)$ group deprotonates and the M^{2+}/PAR system exhibits a significantly red-shifted spectrum with an absorption maximum in the 490–510 nm range depending on the specific M^{2+} ion (Gómez, Estela, Cerdà, & Blanco, 1992; Kocyła et al., 2015; McCall & Fierke, 2000). Spectral changes induced by the Ni^{2+} to PAR complexation are shown in Fig. 3A (colored spectra).

$[Ni(PAR)_2]^{2+}$ formation is governed by the following equilibria:

$$Ni^{2+} + PAR \leftrightarrows [Ni(PAR)]^{2+} \quad K_1 \tag{1}$$

$$[Ni(PAR)]^{2+} + PAR \leftrightarrows [Ni(PAR)_2]^{2+} \quad K_2 \tag{2}$$

Recently, Krężel and coworkers presented M^{2+}/PAR titration curves obtained at multiple points across a pH 7.0–9.9 range (see figure 4 in Kocyła et al., 2015). Mn^{2+}, Cu^{2+}, Zn^{2+}, Pd^{2+}, Cd^{2+}, and Hg^{2+} to PAR complexation reactions were characterized by titration curves with significant curvature and pH dependence. This behavior is consistent with an increasing presence of the monocoordinated $[M(PAR)]^{2+}$ species as the M^{2+}:PAR ratio increases and approaches a 1:2 value. In contrast, Ni^{2+} to PAR and Co^{2+} to PAR complexation gave rise to linear and pH-insensitive titration curves. Thus, free PAR and double-coordinated $[Ni(PAR)_2]^{2+}$ are the dominant species when PAR is in excess to Ni^{2+} and the $[Ni(PAR)_2]^{2+}$ complex is highly stable at neutral to alkaline pH (dissociation constant $< 7 \times 10^{-13}$ M^2, Kocyła et al., 2015).

3.3.2 Determining Effective Molar Extinction Coefficients and the Upper Ni^{2+} Detection Limit

Fig. 3A shows a series of spectra measured for samples containing 50 μ*M* PAR and 0–50 μ*M* Ni^{2+} in 25 m*M* NaH_2PO_4, 150 m*M* NaCl buffer at pH 8.0. The OD at the absorption maxima for free PAR (413 nm) and nickel-bound PAR (495 nm) is plotted as a function of the Ni^{2+} concentration in Fig. 3C. The difference spectra displayed in Fig. 3B were obtained by

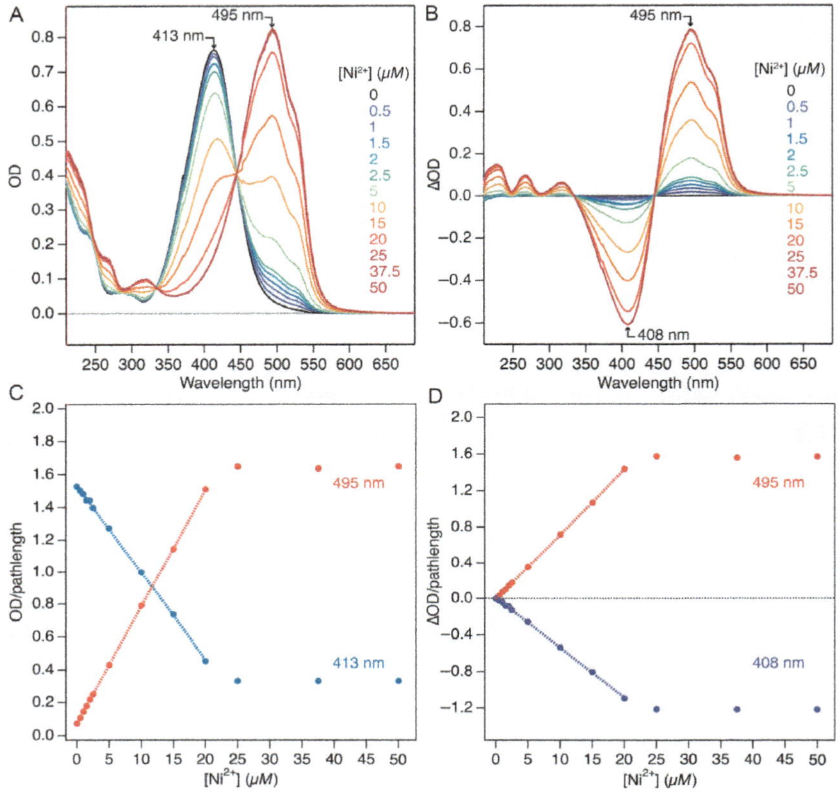

Fig. 3 (A) Absorption spectra following a Ni^{2+}/PAR titration series with samples containing 25 m*M* NaH_2PO_4, 150 m*M* NaCl, pH 8.0, 50 μ*M* PAR, and 0–50 μ*M* Ni^{2+}. The spectra of free PAR and nickel-bound PAR display a peak maximum at 413 and 495 nm, respectively. The titration series was made by mixing equal volumes (750 μL) of a 100 μ*M* PAR (in milliQ water) stock and 0–100 μ*M* Ni^{2+} (in Buffer A) solutions. (B) The Ni^{2+}/PAR titration series displayed as a set of difference spectra obtained by subtracting the free PAR (0 μ*M* Ni^{2+}) spectrum. The difference spectra of free and nickel-bound PAR display a peak maximum at 408 and 495 nm, respectively. (C) Changes in the 413 nm (*blue*) and 495 nm (*red*) absorbance as a function of the Ni^{2+} concentration. The linear regions display a slope of $-53{,}678 \pm 408$ and $71{,}739 \pm 90$ for the 413 and 495 nm data, respectively. (D) Changes in the 408 nm (*purple*) and 495 nm (*red*) difference absorbance as a function of the Ni^{2+} concentration. The linear regions display a slope of $-54{,}829 \pm 415$ and $71{,}798 \pm 123$ for the 408 and 495 nm data, respectively. The spectra were recorded on an Evolution 300 UV–Vis double-beam spectrophotometer equipped with a Smart Peltier thermostatted cell holder using the following settings: blank 25 m*M* NaH_2PO_4, 150 m*M* NaCl, pH 8.0, pathlength 0.5 cm, sample volume 600 μL, spectral range 200–700 nm, bandwidth 2 nm, step size 1 nm, scan rate 120 nm min^{-1}, and temperature 25°C. Data were processed using Igor Pro 6.37.

subtracting the spectrum recorded for the 0 μ*M* Ni^{2+} sample from the spectra collected from the 0–50 μ*M* Ni^{2+} samples. Fig. 3D displays the resulting difference OD (ΔOD) values, reflecting the depletion of free PAR at 408 nm and the formation of nickel-bound PAR at 495 nm, as a function of the Ni^{2+} concentration. The titration plots exhibit a clear linear dependence in the 0–20 μ*M* Ni^{2+} range. The slopes of the displayed linear fits correspond to the effective molar extinction coefficients (ε^{eff}, Fig. 3C) and the effective difference molar extinction coefficients ($\Delta\varepsilon^{eff}$, Fig. 3D). The ε^{eff} and $\Delta\varepsilon^{eff}$ values capture changes in the PAR absorbance due to formation of $[Ni(PAR)_n]^{2+}$ ($n=1$, 2) as the Ni^{2+} content increases. We found $\varepsilon^{eff}_{413} = -55{,}000 \pm 400\,M^{-1}\,cm^{-1}$ and $\varepsilon^{eff}_{495} = 72{,}000 \pm 200\,M^{-1}\,cm^{-1}$ for the 25 m*M* NaH_2PO_4, 150 m*M* NaCl, pH 8.0 buffer system used here. The difference spectra analysis produced consistent numbers with $\Delta\varepsilon^{eff}_{408} = -54{,}000 \pm 400\,M^{-1}\,cm^{-1}$ and $\Delta\varepsilon^{eff}_{495} = 72{,}000 \pm 200\,M^{-1}\,cm^{-1}$. Effective ε and $\Delta\varepsilon$ values for other aqueous Ni^{2+}/PAR systems have been reported previously (ε^{eff}_{494} 72,200 $M^{-1}\,cm^{-1}$ for a 50 m*M* NH_3 solution, Gómez et al., 1992; ε^{eff}_{494} 61,100 ± 400 $M^{-1}\,cm^{-1}$ for 50 m*M* HEPES, 100 m*M* NaCl, pH 7.4, Kocyła et al., 2015).

The linear region of the titration data shown in Fig. 3 also defines an upper detection limit of 20 μ*M* Ni^{2+} when the concentration of PAR is 50 μ*M*. The lower detection limit for our method will be discussed in Section 4.1. In general, as long as the Ni^{2+}:PAR ratio is ≤4:10, ε^{eff} can be used to determine the Ni^{2+} concentration without the risk of underestimating the amount of Ni^{2+} in a given sample. The ε^{eff} and $\Delta\varepsilon^{eff}$ values listed above were determined for 25 m*M* NaH_2PO_4, 150 m*M* NaCl, pH 8.0. The experimental data set shown in Fig. 3A and the analysis illustrated in Fig. 3B–D can be reproduced using other buffer conditions, if required for a specific protein of interest. A key advantage of using effective ε and $\Delta\varepsilon$ values is that one need not delineate the individual concentrations of $[Ni(PAR)]^{2+}$ and $[Ni(PAR)_2]^{2+}$ for an accurate determination of the Ni^{2+} concentration.

3.4 Description of the Spectroscopic Assay to Quantify Possible Ni^{2+} Contaminations

The quantitative colorimetric assay involves a spectroscopic comparison of two solutions: a control containing 50 μ*M* free PAR and a sample containing 50 μ*M* PAR dissolved in the protein solution of interest. Spectra of the two samples are recorded and a difference spectrum is obtained by subtracting the

PAR spectrum from the protein-PAR spectrum. Using Beer–Lambert's law and the effective $\Delta\varepsilon_{495}$ obtained in Section 3.3.2, the concentration of nickel-bound PAR can be calculated. Samples prepared for the qualitative protocol can be utilized as is for the quantitative protocol described in Section 3.5.

Different solvent conditions (buffer, pH, polarity, ionic strength) can influence the spectral properties of PAR and may cause shifts in peak maxima by a few nm. A 1–2 nm variance is not unusual for different spectrometers. Thus, make sure to use values at the observed peak maximum for the (protein-PAR − PAR) difference spectra to make the Ni^{2+} concentration calculations. Using a 100 μ*M* PAR stock solution sets an upper threshold of 20 μ*M* for the Ni^{2+} content to be reliably measured. The final pH should be ≥7.0 for the PAR control and the protein-PAR samples studied.

3.5 Protocol for the Spectroscopic Assay to Quantitate Possible Ni^{2+} Contaminations

3.5.1 Prepare the PAR Stock Solution

Prepare 100 μ*M* PAR in milliQ water or in a pH 7–10 buffer (see Section 3.1.1).

3.5.2 Prepare the Free PAR Control

Titrate 1:1 volumetric ratios of the 100 μ*M* PAR stock and the buffer used for the protein sample into a translucent vial. The final pH should be ≥7.0. The control should be a vibrant yellow in color.

3.5.3 Prepare the Protein-PAR Sample

Titrate 1:1 volumetric ratios of the 100 μ*M* PAR stock and the protein sample solution into a translucent vial. Gently mix sample and let stand for 5–10 min or until color does not change. The final pH should be ≥7.0. If the protein-PAR sample is coral in color (see Fig. 1 for color scale), dilute the protein sample solution as described in Section 3.2.5.

3.5.4 Prepare the Spectroscopic Blank

Titrate 1:1 volumetric ratios of the solution used to prepare the 100 μ*M* PAR stock and the buffer used for the protein sample.

3.5.5 Record UV–Vis Spectra and Obtain a Difference Spectrum

Measure the spectra of the PAR control and the protein-PAR samples from 200 to 700 nm using 0.5 cm cuvettes. If using a double-beam spectrometer (recommended), remember to use the blank sample prepared in Section 3.5.2. Subtract the PAR control spectrum from protein-PAR spectrum and note the difference in absorbance at 495 nm (ΔAbs_{495}).

3.5.6 Use Beer–Lambert's Law to Calculate the Concentration of Ni^{2+}

$$\left[Ni^{2+}\right] = \frac{\Delta Abs_{495}}{\ell \times \varepsilon_{495}^{eff}} \times D = \frac{\Delta Abs_{495}}{0.5 \times 72000} \times 2 \quad (3)$$

where ℓ is the cuvette pathlength (in cm), ε_{495}^{eff} is the effective difference extinction coefficient at 495 nm ($cm^{-1}\,M^{-1}$), ΔAbs_{495} is the difference in absorption at 495 nm (OD), defined as $\Delta Abs_{495} = Abs_{495}^{[Ni(PAR)_n]2+} - Abs_{495}^{PAR}$, and D is the dilution factor. Use $D=2$ if the screened protein material/sample was only diluted in the 1:1 volumetric titration of Section 3.5.3. If the screened protein material/sample underwent additional dilutions, correct D accordingly.

3.5.7 Tip for Obtaining Good Quality Difference Spectra

The use of a double-beam spectrophotometer is highly recommended. It is important that the buffer composition in the blank, the PAR control, and the protein-PAR sample is (near) identical. The difference spectra should have a baseline centered around zero (e.g., see isosbestic point at 445 nm and the 600–700 nm region in Fig. 3B or Fig. 4). Difference spectra not centered on the x-axis indicate that good quality baseline and/or PAR and protein-PAR spectra were not obtained.

3.5.8 Interpreting Difference Spectra

A negative intensity at ca. 410 nm indicates the depletion of free PAR, while the positive intensity at 495 nm indicates the formation of $[Ni(PAR)_n]^{2+}$ ($n=1, 2$). A flat line for difference spectra would indicate equal concentrations of free PAR in the control and in the protein-PAR sample and, consequently, no Ni^{2+} in the latter. Fig. 4A and B illustrate these effects, as described in Section 4.1.

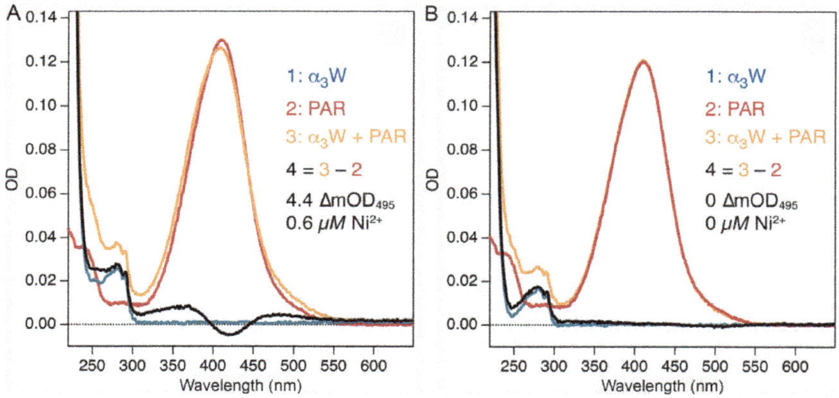

Fig. 4 Screening for Ni^{2+} in precious material and small volumes. The following protocol can be used if material is scarce: Mix 50 μL protein solution, 50 μL 200 μ*M* PAR in Buffer A, and 150 μL Buffer A (or some other pH ≥ 7.0 buffer). Prepare the blank, PAR control, and protein-PAR samples so that the final buffer composition is the same in the three solutions. A 250 μL sample can be measured accurately in a 0.1 cm pathlength cuvette (nominal volume 400 μL) but confirm that the sample when placed in the cell holder covers the light beam of the spectrophotometer with good margin. The displayed spectra were recorded with sample volumes as low as 210 μL. (A) A ^{15}N-labeled α_3W preparation (*blue* spectrum) was screened for Ni^{2+} by subtracting the PAR control spectrum (*red*; 40 μ*M* PAR) from the corresponding α_3W-PAR spectrum (*orange*; 50 μ*M* α_3W + 40 μ*M* PAR). The difference spectrum (*black*) displays a ΔAbs_{495} of 4.4 mOD, which corresponds to 0.6 μ*M* Ni^{2+}. This sets the Ni^{2+} content to 2.5 μ*M* in the 50 μL, 200 μ*M* α_3W aliquot removed for testing. (B) The ^{15}N-labeled α_3W material was treated with EDTA as described in Section 4.1 (*blue* spectrum) and then screened for Ni^{2+}. The EDTA-treated α_3W sample gave rise to identical control (*red*; 35 μ*M* PAR) and α_3W-PAR (*orange*; 35 μ*M* α_3W + 35 μ*M* PAR) spectra and no Ni^{2+} was detected. Spectral settings as in Fig. 3.

4. TWO EXAMPLES OF USEFUL APPLICATIONS FOR THE DEVELOPED COLORIMETRIC METHOD

4.1 Checking NMR Protein Material for Ni^{2+}

Protein NMR samples are often precious and the general guidelines presented above can be modified to preserve material. The following protocol to check and prepare NMR samples works well for the α_3X family of Ni^{2+}/IMAC-purified model proteins developed by the Tommos group (Berry, Martínez-Rivera, & Tommos, 2012; Glover et al., 2014; Martínez-Rivera et al., 2011; Ravichandran et al., 2017; Tommos, Valentine, Martínez-Rivera, Liang, & Moorman, 2013): Place 2.5 mL 50 m*M* sodium acetate-d_3, 50 m*M* NaCl, pH 5.7 buffer in a disposable UV cuvette and titrate freeze-dried α_3X powder to a protein concentration of about 200 μ*M*. Remove 50 μL for Ni^{2+} analysis. Add EDTA to a final concentration of 220 μ*M*, let the sample sit for 15 min, and buffer exchange using a Sephadex G-25 PD10 column equilibrated with 3 m*M* sodium acetate-d_3, 3 m*M* NaCl, pH 5.7 buffer. Remove 50 μL for Ni^{2+} analysis. Freeze–dry the buffer-exchanged sample and dissolve in 500 μL H_2O and/or D_2O containing 75 μ*M* DSS. Check and adjust pH.

Fig. 4 shows a Ni^{2+} screen of ^{15}N-labeled α_3W prepared using this protocol. For the spectroscopic assay, it is important that the blank contains the same final buffer composition as the PAR and protein-PAR samples to ensure good baseline recording and spectral subtraction results. If done carefully and using a double-beam spectrometer, a ΔAbs_{495} of a few mOD can be detected. This is illustrated in Fig. 4. A Ni^{2+} content of 2.5 μ*M* was determined for the 50 μL, 200 μ*M* α_3W aliquot removed prior to the EDTA step (Fig. 4A). No Ni^{2+} could be detected for the 50 μL, 130 μ*M* α_3W aliquot removed after the EDTA step (Fig. 4B). More experimental details are provided in Fig. 4 legend. Notably, when screening multiple batches of purified α_3X proteins, we consistently observed the Ni^{2+} content to vary by one order of magnitude even when using the same type of Ni^{2+}/IMAC resin and set of stock solutions. We estimate that the lower detection limit is about 0.3–0.5 μ*M* Ni^{2+} for the spectroscopic assay. Practically, the lower limit will vary somewhat depending on the quality of the spectrophotometer, the Ni^{2+} to target protein stoichiometry, protein availability, and/or cuvette pathlength used.

4.2 NMR Spectral Line Broadening Due to Low-Level Ni^{2+} Contaminations

Fig. 5A and B presents ^{15}N-HSQC spectra of α_3W samples prepared from the same material that was used for the Ni^{2+} screen. The NMR samples were made excluding (Fig. 5A and C) and including (Fig. 5B and D) an EDTA treatment step. A Ni^{2+} contamination of ~12 μ*M*, or one metal ion per 80 protein molecules, results in a clearly observable line broadening of both water and protein resonances. The positive and negative contour levels are identical for the blue and the red spectra and emphasize the more prominent background water streaks (colored black) in the blue spectrum representing the Ni^{2+}-containing sample. Additionally, the average peak height is lower and the peak intensities more uneven in the spectrum from the Ni^{2+}-containing sample (Fig. 5C) as compared to EDTA-treated sample (Fig. 5D).

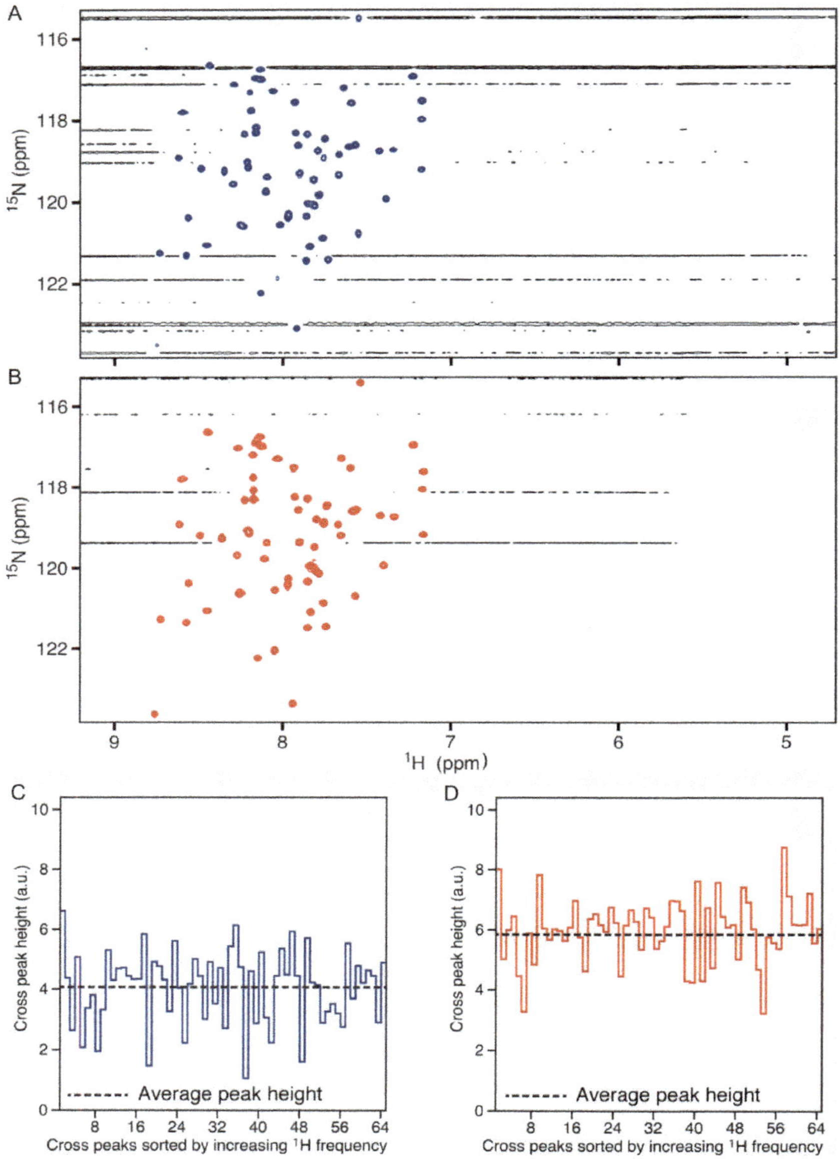

Fig. 5 ^{15}N-HSQC spectra of (A) a 950 μ*M* α_3W sample containing ~12 μ*M* Ni^{2+} and (B) a 800 μ*M* α_3W sample with no detectable Ni^{2+}. Panels (C) and (D) display the peak heights for the protein resonances in panels (A) (*blue*) and (B) (*red*), respectively. The spectra were collected, processed, and displayed using identical parameters. Experimental conditions: Bruker Avance 500 MHz spectrometer equipped with a cryoprobe, ^{1}H dimension spectral width of 10 ppm and 512 complex points, ^{15}N dimension spectral width of 8.8 ppm and 160 complex points, buffer conditions about 23 m*M* sodium acetate-d_3 and 23 m*M* NaCl, 75 μ*M* DSS, pH 5.5 ± 0.2, and temperature 30°C. The spectra were processed using Felix95 (Accelrys Inc., San Diego, CA), analyzed with NMRFAM-SPARKY (Lee, Tonelli, & Markley, 2015), and displayed using Adobe Illustrator (Adobe Creative Cloud).

4.3 Measuring Ni^{2+} Release From Three Common IMAC Resins

The α_3X proteins are purified using two Ni^{2+}/IMAC steps. The first step isolates the expressed thioredoxin-His_6-thrombinsite-α_3X fusion and the second step separates the thioredoxin-His_6 part from α_3X following thrombin digestion. The α_3X-containing flow-through fraction from the second Ni^{2+} column is further purified by reversed-phase HPLC using a semi-preparative C18 column and a linear 30%–60% acetonitrile/water/0.1% trifluoroacetic acid gradient. Despite the final HPLC step, Ni^{2+} copurifies with the α_3X proteins as described above. To refine the α_3X purification protocol and minimize Ni^{2+} copurification, we investigated Ni^{2+} leakage from three IMAC products: His•Bind resin (Novagen; chelator chemistry IDA, stationary matrix agarose, binding capacity 8 mg protein/mL resin), Ni^{2+}-preloaded His60 Ni Superflow resin (Clontech Laboratories; chelator chemistry IDA, stationary matrix Superflow 6 agarose beads, binding capacity ≤60 mg protein/mL), and Ni^{2+}-preloaded cOmplete His-Tag Purification resin (Roche; chelator chemistry proprietary; stationary matrix Sepharose-CL 6B, binding capacity ≥40 mg protein/mL resin). 10 mL of new resin from each brand was prepared, equilibrated, washed, and eluted with an imidazole gradient following the manufacturer's instructions. Fractions were continuously collected during the equilibration, wash, and elution steps and their Ni^{2+} content measured using the quantitative colorimetric methods described in Section 3.5. The results are presented in Fig. 6 and more experimental details are provided in the legend.

There are major differences in Ni^{2+} leakage between the His•Bind and His60 resins and the cOmplete resin. Ni^{2+} concentrations on the tens to hundreds of μM range were observed across the equilibration, wash, and elute fractions from the His•Bind resin. The Ni^{2+} leakage was lower for the His60 resin with concentrations on the ~6 μM to hundreds of μM range. The cOmplete resin releases very little with no detectable Ni^{2+} during the equilibration step and ~2 μM at the end of the 0–250 mM imidazole elution step. Based on this information, the α_3X protein purification protocol was modified to include His•Bind and cOmplete resins for the first and second IMAC steps, respectively. The former is easily charged and stripped and this procedure can be repeated many times without a notable change in the effective binding capacity. This provides the advantage that multiple members of the α_3X protein family, which typically differ by only one or two residues in their primary sequence, can be purified using a single column

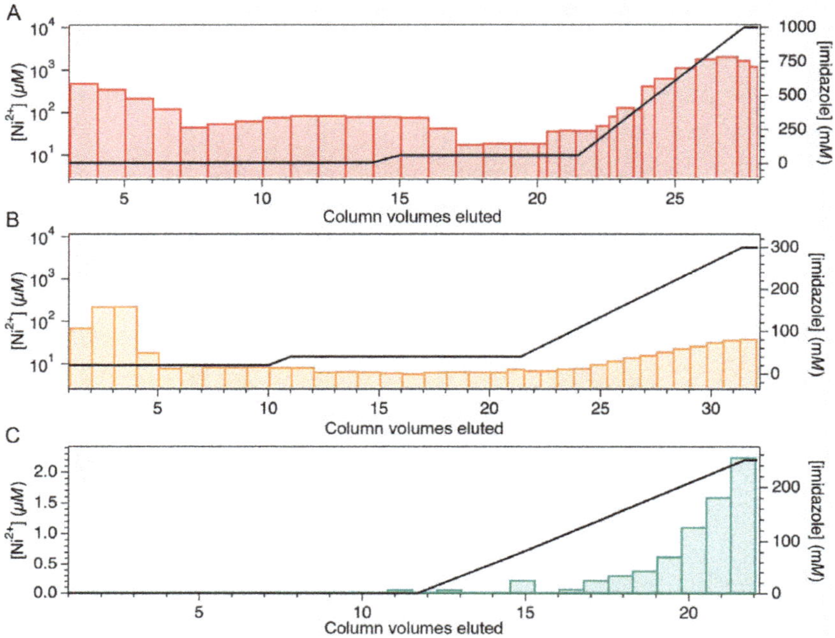

Fig. 6 Ni^{2+} leakage from three commonly used IMAC resins. (A) A column containing 10 mL of pristine His•Bind resin was rinsed with 3 column volumes (CV) of milliQ water, charged with 5 CV of 50 m*M* $NiSO_4$, and rinsed with 5 CV of binding buffer (20 m*M* Tris–HCl, 500 m*M* NaCl, 5 m*M* imidazole, pH 7.9). The column was then (*a*) equilibrated with an additional 10 CV of binding buffer, (*b*) washed with 6 CV of wash buffer (20 m*M* Tris–HCl, 500 m*M* NaCl, 60 m*M* imidazole, pH 7.9), and (*c*) eluted using a 60–1000 m*M* imidazole in 20 m*M* Tris–HCl, 500 m*M* NaCl, pH 7.9, gradient. Panel (A) displays the elution profile of steps *a* to *c*. (B) A column containing 10 mL of pristine Ni^{2+} precharged His60 Ni Superflow resin was rinsed with 10 CV of equilibration buffer (50 m*M* NaH_2PO_4, 300 m*M* NaCl, 20 m*M* imidazole, pH 7.4). The column was then (*a*) washed with an additional 10 CV of equilibration buffer, (*b*) washed with 10 CV of wash buffer (50 m*M* NaH_2PO_4, 300 m*M* NaCl, 40 m*M* imidazole, pH 7.4), and (*c*) eluted using a 40–300 m*M* imidazole in 50 m*M* NaH_2PO_4, 300 m*M* NaCl, pH 7.4, gradient. Panel (B) displays the elution profile of steps *a* to *c*. (C) A column containing 10 mL of pristine Ni^{2+} precharged cOmplete His-Tag Purification resin was rinsed with 10 CV of equilibration buffer (50 m*M* NaH_2PO_4, 300 m*M* NaCl, pH 8.0). The column was then (*a*) washed with an additional 10 CV of equilibration buffer and (*b*) eluted using a 0–250 m*M* imidazole in 50 m*M* NaH_2PO_4, 300 m*M* NaCl, pH 8.0, gradient. Panel (C) displays the elution profile of steps *a* and *b*.

without the risk of cross contamination. Introducing the Ni^{2+} precharged cOmplete resin for the second IMAC step dramatically lowered the Ni^{2+} content of the final HPLC-purified material to the point where no obvious NMR spectral perturbations could be observed.

5. CONCLUSION

Protein production involving expression of His-tagged proteins and/or IMAC-based purification strategies presents the likelihood of contamination by the Mn^{+} species used to charge the resin. Ni^{2+}, and other commonly used transition metals including Co^{2+} and Cu^{2+}, are paramagnetic and their presence in protein NMR samples may introduce unwanted spectral perturbations. The colorimetric assays presented here provide a simple and quick means to determine the presence of Ni^{2+} in low μM concentrations.

ACKNOWLEDGMENTS

This work was supported by the National Institutes of Health Grant GM079190 (C.T.) and the Swedish Research Council Grant 2017-04992 (S.D.G.). Valuable discussions with Drs. Josh Wand and Kathy Valentine, Ms. Li Liang, and Mr. Delfin Buyco are gratefully acknowledged.

REFERENCES

Berry, B. W., Martínez-Rivera, M. C., & Tommos, C. (2012). Reversible voltammograms and a Pourbaix diagram for a protein tyrosine radical. *Proceedings of the National Academy of Sciences of the United States of America*, *109*, 9739–9743.

Dai, Q. H., Tommos, C., Fuentes, E. J., Blomberg, M. R. A., Dutton, P. L., & Wand, A. J. (2002). Structure of a de novo designed protein model of radical enzymes. *Journal of the American Chemical Society*, *124*, 10952–10953.

Glover, S. D., Jorge, C., Liang, L., Valentine, K. G., Hammarström, L., & Tommos, C. (2014). Photochemical tyrosine oxidation in the structurally well-defined α_3Y protein: Proton-coupled electron transfer and a long-lived tyrosine radical. *Journal of the American Chemical Society*, *136*, 14039–14051.

Gómez, E., Estela, J. M., Cerdà, V., & Blanco, M. (1992). Simultaneous spectrophotometric determination of metal ions with 4-(pyridyl-2-azo)resorcinol (PAR). *Fresenius' Journal of Analytical Chemistry*, *342*, 318–321.

Hearon, J. Z. (1948). The configuration of coboltodihistidine and oxy-bis(coboltodihistidine). *Journal of the National Cancer Institute*, *9*, 1–11.

Hochuli, E., Bannwarth, W., Döbeli, H., Gentz, R., & Stüber, D. (1988). Genetic approach to facilitate purification of recombinant proteins with a novel metal chelate adsorbent. *Biotechnology*, *6*, 1321–1325.

Hochuli, E., Döbeli, H., & Schacher, A. (1987). New metal chelate adsorbent selective for proteins and peptides containing neighbouring histidine residues. *Journal of Chromatography*, *411*, 177–184.

Kocyła, A., Pomorski, A., & Krężel, A. (2015). Molar absorption coefficients and stability constants of metal complexes of 4-(2-pyridylazo)resorcinol (PAR): Revisiting common chelating probe for the study of metalloproteins. *Journal of Inorganic Biochemistry*, *152*, 82–92.

Lee, W., Tonelli, M., & Markley, J. L. (2015). NMRFAM-SPARKY: Enhanced software for biomolecular NMR spectroscopy. *Bioinformatics*, *31*, 1325–1327.

Martínez-Rivera, M. C., Berry, B. W., Valentine, K. G., Westerlund, K., Hay, S., & Tommos, C. (2011). Electrochemical and structural properties of a protein system designed to generate tyrosine Pourbaix diagrams. *Journal of the American Chemical Society*, *133*, 17786–17795.

McCall, K. A., & Fierke, C. A. (2000). Colorimetric and fluorimetric assays to quantitate micromolar concentrations of transition metals. *Analytical Biochemistry*, *284*, 307–315.

Porath, J. (1992). Immobilized metal ion affinity chromatography. *Protein Expression and Purification*, *3*, 263–281.

Porath, J., Carlsson, J., Olsson, I., & Belfrage, G. (1975). Metal chelate affinity chromatography, a new approach to protein fractionation. *Nature*, *258*, 598–599.

Ravichandran, K. R., Zong, A. B., Taguchi, A. T., Nocera, D. G., Stubbe, J., & Tommos, C. (2017). Formal reduction potentials of difluorotyrosine and trifluorotyrosine protein residues: Defining the thermodynamics of multistep radical transfer. *Journal of the American Chemical Society*, *139*, 2994–3004.

Smith, M. C., Furman, T. C., Ingolia, T. D., & Pidgeon, C. (1988). Chelating peptide-immobilized metal ion affinity chromatography. A new concept in affinity chromatography for recombinant proteins. *Journal of Biological Chemistry*, *263*, 7211–7215.

Smith, M. C., Furman, T. C., & Pidgeon, C. (1987). Immobilized iminodiacetic acid metal peptide complexes. Identification of chelating peptide purification handles for recombinant proteins. *Inorganic Chemistry*, *26*, 1965–1969.

Sulkowski, E. (1985). Purification of proteins by IMAC. *Trends in Biotechnology*, *3*, 1–7.

Tommos, C., Valentine, K. G., Martínez-Rivera, M. C., Liang, L., & Moorman, V. R. (2013). Reversible phenol oxidation-reduction in the structurally well-defined 2-mercaptophenol-α_3C protein. *Biochemistry*, *52*, 1409–1418.

FURTHER READING

Corsini, A., Yih, I. M., Fernando, Q., & Freiser, H. (1962). Potentiometric investigation of the metal complexes of 1-(2-pyridylazo)-2-naphthol and 4-(2-pyridylazo)resorcinol. *Analytical Chemistry*, *34*, 1090–1093.

Stanley, R. W., & Cheney, G. E. (1966). Metal complexes of 4-(2'-thiazolylazo)-resorcinol, a comparative study with 4-(2'-pyridylazo)-resorcinol. *Talanta*, *13*, 1619–1629.

CHAPTER FIVE

Toolkit for NMR Studies of Methyl-Labeled Proteins

Paolo Rossi*,[1], Yoan R. Monneau[†], Youlin Xia*, Yojiro Ishida[‡], Charalampos G. Kalodimos*,[1]

*Department of Structural Biology, St Jude Children's Research Hospital, Memphis, TN, United States
[†]University Grenoble Alpes, CEA, CNRS, IBS, Grenoble, France
[‡]CABM, Rutgers University, Piscataway, NJ, United States
[1]Corresponding authors: e-mail address: paolo.rossi@stjude.org; babis.kalodimos@stjude.org

Contents

Abstract

Selective methyl labeling is an extremely powerful approach to study the structure, dynamics, and explore mechanistic insights of large biomolecules by solution NMR. Methyls are relatively insensitive to chemical exchange-induced depolarization and provide superior probes of supramolecular interactions and allostery in such systems. In this chapter, we describe our systematic approach and contributions in the areas of sample preparation, data collection, and data analysis that streamline the application of methyl labeling in solution NMR studies of large proteins. We focus our effort on the initial and often onerous task of methyl resonance assignment and here we detail our approaches to simplify the process. We produce new methyl labeling combinations using *Escherichia coli* auxotrophs, increase speed, sensitivity, and resolution of NOESY experiments by employing 3D SOFAST–NOESY, and assign methyl resonances from raw data with spectral simulation tools and(or) automatically with minimal expert supervision using the MAGIC algorithm.

Methods in Enzymology, Volume 614
ISSN 0076-6879
https://doi.org/10.1016/bs.mie.2018.08.036

1. INTRODUCTION

Multidimensional nuclear magnetic resonance (NMR) has emerged as an important tool to study structure, dynamics, and allostery in biomolecules (Huang & Kalodimos, 2017; Kiteyski-LeBlanc et al., 2018; Rosenzweig & Kay, 2014; Saio, Kawagoe, Ishimori, & Kalodimos, 2018; Wiesnerl & Sprangers, 2015). The structure of weak complexes can be determined and their dynamic behavior examined by monitoring spin-relaxation rates from picosecond to second timescale with atomic accuracy (Palmer, Kroenke, & Loria, 2001). Using backbone amide $^{1}H–^{15}N$ as reporters, events ranging from enzyme catalysis to molecular recognition are captured at atomic level in small to medium size systems (Palmer, 2016). Significant efforts have gone into making NMR of larger systems possible first with the discovery of the TROSY effect in amides (Igumenova & Palmer, 2006; Pervushin, Riek, Wider, & Wuthrich, 1997) and more recently with "methyl-TROSY" spectroscopy (Kay, 2011; Ollerenshaw, Tugarinov, & Kay, 2003). Methyls are relatively immune to chemical exchange and relaxation phenomena that can severely impact amides above ~50 kDa and that has expanded the scope of NMR to biomolecules that were previously the exclusive domain of crystallography. This is achieved by producing samples containing isolated $^{13}C–^{1}H_3$ spin systems in a uniformly deuterated and relatively "NMR-silent" background and by acquiring $^{1}H–^{13}C$ heteronuclear multiple quantum coherence (HMQC)-based data (Tugarinov & Kay, 2004). In HMQC-based multidimensional experiments, the intensity is weighted by the slow relaxing multiple-quantum (MQ) transition in the $^{1}H–^{13}C$ methyl manifold giving better sensitivity and resolution. Lowering proton density also reduces spin diffusion phenomena and enables the observation of more reliable methyl–methyl distances by nuclear Overhauser (NOE) experiments (Gardner & Kay, 1998), empowering NMR in the study of difficult systems (Huang, Rossi, Saio, & Kalodimos, 2016). Knowledge of the spatial relationships of methyl groups provides the tertiary structure-defining connectivity framework for use in sparse restraints structure determination or modeling (Lange et al., 2012). The six methyl-containing residues—alanine, isoleucine, leucine, methionine, threonine, and valine—can be selectively and(or) stereospecifically $^{13}C–^{1}H_3$-labeled using well-established protocols (Kerfah, Plevin, Sounier, Gans, & Boisbouvier, 2015). These combined technical advances have allowed unprecedented atomic level understanding in system that had remained impervious to structural characterization by crystallography (Saio, Guan, Rossi, Economou, & Kalodimos, 2014). In spite of the advances in the field, aspects of methyl assignment and reagent cost still pose a challenge to routine implementation in structural biology laboratories (Alphonse & Ghose, 2018; Gorman, Debashish, O'Rourke, & Boehr, 2018).

A schematic approach to methyl NMR spectroscopy of biomolecules is outlined in Fig. 1. Herein, we address specific steps relating to labeling, data collection, and data analysis protocols that are currently in use in our laboratory to assign methyl-labeled proteins. Methyl labeling conducted using commercially available alpha-keto acid precursors allows for highly specific and scramble-free labeling of isoleucine- (ketobutyric acid), leucine- and valine- (ketoisovaleric acid), methyls, and their isotopomers. This combination of

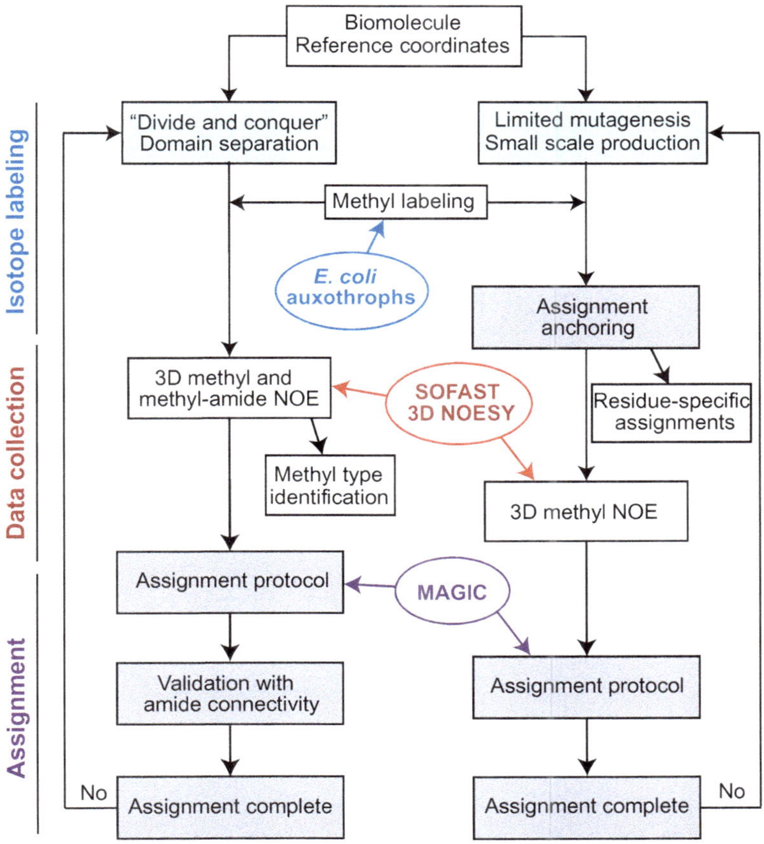

Fig. 1 The workflow to study proteins using methyl-labeling technology.

methyls provides the starter set of core-defining methyls groups information (Tugarinov, Choy, Orekhov, & Kay, 2005). Additionally, [^{1}H,^{13}C] methyl labeling of alanine (Ayala, Sounier, Use, Gans, & Boisbouvier, 2009; Isaacson et al., 2007), methionine (Gelis et al., 2007), isoleucine-γ2 (Ruschak, Velyvis, & Kay, 2010), and threonine (Velyvis, Ruschak, & Kay, 2012) has been reported to date. A representative spectrum for a 28 kDa protein with all methyl types labeled included is shown in Fig. 2.

Overcoming resonance overlap, particularly in the Leu and Val spectral region, is key in successful application of methyl labeling to biological problems. Alpha keto-isovaleric acid is the precursor responsible for simultaneous labeling of Leu and Val methyls. Leu and Val methyl can be labeled as $\delta1/\delta2\text{-}^{12}CD_3/^{13}CH_3$-Leu, $\gamma1/\gamma2\text{-}^{12}CD_3/^{13}CH_3$-Val racemate (monomethyl) or uniformly $\delta1/\delta2\text{-}^{13}CH_3/^{13}CH_3$-Leu, $\gamma1/\gamma2\text{-}^{13}CH_3/^{13}CH_3$-Val

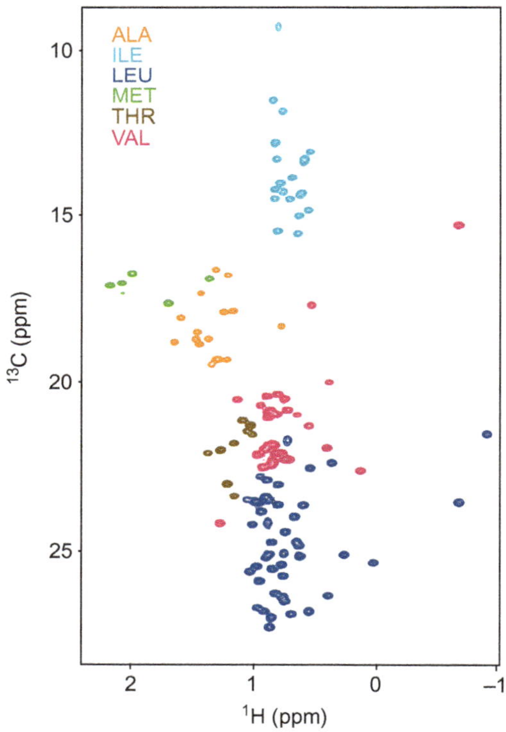

Fig. 2 2D SOFAST–HMQC of Hsp90-N domain. The peaks are *color coded* as to the actual assigned methyl type. Barring a few exceptions, methyl types cluster in distinct regions of the spectrum. Leu and Val are produced from the same keto-acid precursor and are also poorly dispersed.

(dimethyl). Monomethyl labeling is better-suited for larger systems because of the lower overall proton density, while dimethyl gives 50% stronger Leu/Val signal and allows for straightforward pairing of intraresidue methyls and methyl type discrimination. Stereospecific labeling (either *pro-R* or *pro-S*) was developed to reduce the overlap in the Leu/Val region by using commercially available acetolactate precursors (Gans et al., 2010). To further simplify Leu/Val spectral region, Val-only methyl labeling is possible by suppressing Leu resonances via addition of *U*-^{2}H Leu (Mas, Crublet, Hamelin, Gans, & Boisbouvier, 2013). However, the method yields residual Leu cross-peaks that complicate the analysis even after following strict expression protocol (Monneau et al., 2016). In addition, Leu-only methyl labeling can be achieved using recently developed new precursor (Lichtenecker, Coudevylle, Konrat, & Schmid, 2013) for mono- or dimethyl labeling purposes. Stereospecific Leu-only methyl labeling was not possible until recently and prior to the development of a series of auxotrophic *Escherichia coli* cells in our laboratory. The auxotrophs enable clean Leu- or Val-only labeling as mono-, di-, or stereospecific methyl positions (Monneau et al., 2016). Section 2.1 of this manuscript is dedicated to detailing the approach.

Following sample preparation, assignment of methyl resonances is carried out. In principle, assignment could be conducted by brute force by producing single mutant variants of each methyl-containing residue. However, this approach in large systems is highly impractical and its efficiency is questionable. A far better approach consists in matching an experimentally derived NOE connectivity network to the distance network in the 3D model using two complementary strategies: (i) "divide and conquer" (Gelis et al., 2007; Pickford & Campbell, 2004; Sprangers & Kay, 2007) and (ii) limited mutagenesis (Amero et al., 2011). The former is applied to multidomain proteins by assigning each domain separately and then transferring the assignments to the complete protein; the latter is used when the protein cannot be separated into domains or if ambiguity remains after "divide and conquer" strategy. In the limited mutagenesis approach, conservative mutations of methyl-containing amino acids are performed, such as Ile to Leu, and the assignment is obtained by identifying the cross-peaks absent from the 2D HMQC spectrum as a result of mutation. Small-scale expression can be used here since concentrations as low as 5–10 μ*M* can provide the answer by 2D HMQC. In addition, cost is mitigated by recycling ^2H_2O in preparation for qualitative NMR experiments (see below). Using a combinatorial approach, it is possible to minimize the number of samples/methyls ratio (Alphonse & Ghose, 2018). The assignments from mutagenesis serve to bootstrap the subsequent NOESY-based assignment. A drawback of the limited mutagenesis approach is that even the most conservative mutations can have significant structure- or spectrum-altering consequences that are often difficult to deconvolute.

In the "divide and conquer" approach individual domains are typically of size suitable for backbone assignment by TROSY-based triple resonance spectroscopy (Salzmann, Pervushin, Wider, Senn, & Wuthrich, 1998) and other more traditional 3D TOCSY- and COSY-based experiments (Bax et al., 1990; Kay, Xu, Singer, Muhandiram, & Formankay, 1993). Provided that the model has the same geometry as the query protein, NOESY-based experiments can have the discriminating power necessary to complete the assignments in medium size or oligomer systems (Rossi et al., 2015). The ^{13}C-edited NOESY experiment is used to establish the connectivity network. Variants of the "methyl-NOESY" experiment such as 3D CCH- and HCH or 4D HCCH–HMQC–NOESY–HMQC are possible. Amides are used to provide rapid and independent validation of the methyl-only NOE network by heteronuclear NOESY experiments such as 3D ^{15}N–HMQC–NOESY–^{13}C–HMQC (NCH-) (Kay, Clore, Bax, & Gronenborn, 1990) and its counterpart (CNH-) version in addition to the traditional 3D ^{13}C-edited NOESY–HMQC (Ikura, Kay, Tschudin, & Bax, 1990) and the optional ^{15}N-edited NOESY–TROSY (Zhu, Kong, & Sze, 1999).

If amides assignments are available, the Ile, Leu, and Val methyl types can be assigned directly by out-and-back 3D experiments starting from the methyl proton in which the H_MC_M-CγCβCα and H_MC_M-CβCα are obtained for Ile/Leu and Val, respectively (Tugarinov & Kay, 2003). Given Leu/Val overlap, out-and-back experiments provide a straightforward means for distinguishing between Leu and Val methyl type simply based on the Cβ chemical shift. The experiment requires racemic ($^{13}CH_3$/$^{12}CD_3$) methyl labeling and [U-^{2}H,^{13}C] on the sidechain of Leu and Val and due to the multiple J-coupling transfers involved, the experiments are affected by transverse relaxation and may not scale well to the largest systems.

In parallel to hardware-derived improvements in sensitivity and resolution in the forms of ultra-high magnetic fields and cryoprobes, there has been an intense effort in the NMR community to improve the experiment design with the introduction of fast pulsing NMR (Schanda, Kupce, & Brutscher, 2005) and nonuniform sampling (Rovnyak, Hoch, Stern, & Wagner, 2004). We chose to improve sampling and reduce data collection time in NOESY experiments by introducing fast pulsing schemes to multidimensional NOESY (Rossi, Xia, Khanra, Veglia, & Kalodimos, 2016). We produced a suite of

NOESY experiments based on the SOFAST–HMQC that is ideally suited to methyl TROSY implementation and to flexible and rapid data collection. SOFAST–HMQC editing allows for short (~0.2 s) recovery delay and sensitivity approaching the traditional HMQC experiment acquired with 5 × (1 s) the recovery delay. In Section 2.2 of this chapter, we will detail the implementation of these experiments in the context of methyl spectroscopy of high molecular weight systems and highlight the subset of spectra we currently employ for methyl assignments.

The methyl assignment task of matching patterns is very well-suited to computational approaches. A useful methyl assignment algorithm must (i) implement ways to alleviate the combinatorial problem and run in a reasonable amount of time and (ii) to produce accurate assignment with minimal expert supervision starting from raw NOESY peak list. Earlier, stochastic Monte Carlo swapping algorithms were developed (Chao et al., 2014; Xu & Matthews, 2013) and more recently, it was realized that graph theory could be utilized to achieve the goal (Pritisanac et al., 2017). A direct way to alleviate the combinatorial problem is to provide methyl type information as input. This can be achieved very simply and unambiguously with two or three small-scale protein expressions with alternative methyl labeling. It is also useful to establish the Leu/Val intraresidue pairings by performing a short mixing time NOESY on a dimethyl Leu/Val-labeled sample. Combining such information with any available results from limited mutagenesis can simplify the assignment problem considerably. We have developed Methyl Assignment by Graphing Inference Construct (MAGIC), an exhaustive search algorithm that uses raw NMR data and achieves accurate methyl assignment with minimal expert supervision (Monneau et al., 2017). Our algorithm is designed around the NOE network density concept to bypass the manual network-definition step, and use raw data and residue type and pairing information when available. MAGIC manages the combinatorial problem by proceeding in a hierarchical fashion, identifying *local*, high-density and high-confidence NOE networks and prioritizing their assignment via a *local* exhaustive search protocol. In this chapter, Section 2.3 will describe the algorithm in its current (version 2.0) implementation.

2. METHODS

2.1 Val- or Leu-Only Selective Methyl Labeling by Auxotrophic *E. coli* Strains

Resonance overlap presents a serious challenge when scaling up to very large systems. Even in the well-resolved spectrum of a 30 kDa protein in Fig. 2, partial crowding in the leucine/valine region can emerge. Overlap hampers the ability to follow molecular events at atomic scale and introduce significant technical challenges to assignment. Thus, strategies to decrease spectral complexity in Leu/Val area are highly desirable. Precursors for stereospecific labeling of Leu/Val methyls were introduced (Gans et al., 2010). In addition, Leu- or Val-only labeling was also adressed either by using a new precursor for Leu labeling (Lichtenecker et al., 2013) or by the addition of deuterated leucine to suppress Leu labeling and thereby achieving a Val-only labeling (Mas et al., 2013). However, the former does not enable stereospecific labeling and the latter can lead to residual leucine incorporation that result in spurious Leu peaks. In order to achieve a pure leucine or valine labeling and to also enable residue-specific and stereospecific labeling of leucine, we developed a series of *E. coli* auxotrophic strains (Monneau et al., 2016). Those strains are either auxotroph for Leu or Val, triggering the isotopes routing to the desired residue type (Fig. 3). Table 1 lists the available *E. coli* strains and their naming is given according to their phenotype (within parenthesis) and genotype. Strains are freely available upon request. The biosynthetic pathways of branched-chain residues are displayed in Fig. 3A, highlightening the different enzymatic steps targeted for gene deletions. An example of the corresponding spectra for those achieved labeling schemes is also displayed in Fig. 3B. Simultaneous incorporation of Leu/Val with Met, Ala, and Thr is possible, with the exception of Ile. The auxotrophic strains need 5–10 times less precursors to achieve complete labeling and therefore are then well-suited for mutagenesis-based assignment that may require production of numerous mutants for 2D screening and for unambiguous methyl type determination. Hereafter, we describe the detailed procedure to use our auxotrophic strains.

2.1.1 The Protocols

Before using our auxotrophic cells, they need to be made competent for transformation:

1. Spread on plate (LB/agar +50 mg/L kanamycin) the cells from frozen vials and incubate overnight at 37°C
2. Inoculate 3 mL of LB/kanamycin (50 mg/L) as preculture overnight at 37°C and transfer the cells culture to the desire volume for cells production
3. Follow a standard chemically competent cell protocols

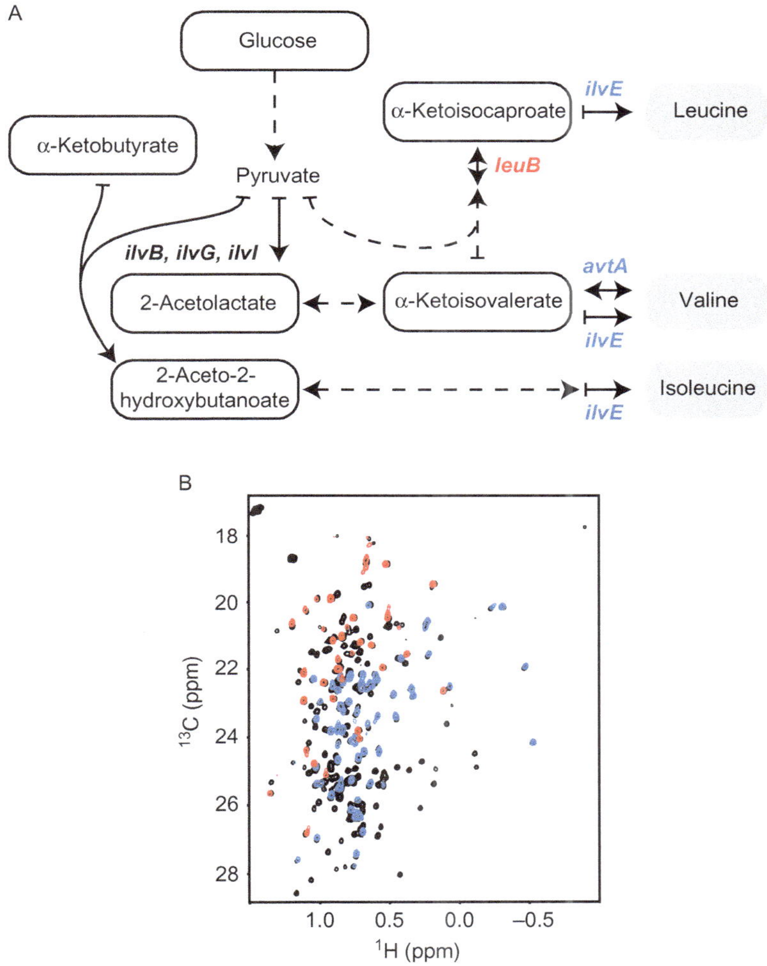

Fig. 3 The isotope routing thanks by auxotrophic strains and the resulting expected spectrum. (A) Simplified pathways for Ile, Leu, and Val biosynthesis are outlined starting from glucose, as the main carbon source, according to the Kyoto Encyclopedia of Genes and Genomes (http://www. genome.jp/kegg/). The commercially available isotope-labeled amino acid precursors (α-ketobutyrate, α-ketoisovalerate, α-ketoisocaproate, 2-aceto-2-hydroxybutanoate, and 2-acetolactate) are also encircled. Each *solid arrow* is indicative of one enzymatic reaction. *Dashed arrows* indicate multistep reactions and *double arrows* represent reversible enzymatic reaction. The deleted genes are shown for the AcL- (*black*), Leu- (*red*), Val- (*blue*) strain phenotypes. (B) The simplification of methyl spectra achieved using our auxotrophic strains is depicted by the overlapping of ^{13}C-HMQC spectra recorded on MSG samples either expressed with (*black*) BL21 (AcL-) and 50 mg/L of α-^{13}C-ketoisovalerate, containing Leu-δ1/δ2 and Val-γ1/γ2 cross-peaks, or (*red*) BL21(AcL-/Leu-) with 40 mg/L of ^{13}C-acetolactate, containing only Val-γ2 cross-peaks, or (*green*) BL21(AcL-/Val-) with 40 mg/L of ^{13}C-acetolactate, containing only Leu-δ2 cross-peaks.

Table 1 *E. coli* Auxotrophs Strains for Methyl Labeling

Strains (Phenotypes) Compatible Labeling[a]	Genotypes	Targeted Labeling Schemes	Precursors
BL21(Leu-) AIMT	Δ*leuB*	$\gamma1/\gamma2$-$^{12}CD_3/^{13}CH_3$-Val	120 mg/L Isovalerate monomethyl
		$\gamma1/\gamma2$-$^{13}CH_3/^{13}CH_3$-Val	120 mg/L Isovalerate dimethyl
		$\gamma1$ or $\gamma2$-$^{12}CD_3/^{13}CH_3$-Val	300 mg/L Acetolactate (either proS or proR)
BL21(AcL-) AMT	Δ*hisB*, Δ*trpC*, Δ*ilvB*, Δ*ilvG*, Δ*ilvI*	$\delta1/\delta2$-$^{12}CD_3/^{13}CH_3$-Leu $\gamma1/\gamma2$-$^{12}CD_3/^{13}CH_3$-Val	50 mg/L Isovalerate monomethyl
		$\delta1/\delta2$-$^{13}CH_3/^{13}CH_3$-Leu $\gamma1/\gamma2$-$^{13}CH_3/^{13}CH_3$-Val	50 mg/L Isovalerate dimethyl
		$\delta1$ or $\delta2$-$^{12}CD_3/^{13}CH_3$-Leu $\gamma1$ or $\gamma2$-$^{12}CD_3/^{13}CH_3$-Val	40 mg/L Acetolactate (either proS or proR)
BL21(AcL-/Val-) AMT	Δ*hisB*, Δ*trpC*, Δ*ilvB*, Δ*ilvG*, Δ*ilvI*, Δ*ilvE*, Δ*avtA*	$\delta1/\delta2$-$^{12}CD_3/^{13}CH_3$-Leu	50 mg/L Isovalerate monomethyl
		$\delta1/\delta2$-$^{13}CH_3/^{13}CH_3$-Leu	50 mg/L Isovalerate dimethyl
		$\delta1$ or $\delta2$-$^{12}CD_3/^{13}CH_3$-Leu	40 mg/L Acetolactate (either proS or proR)
BL21(AcL-/Leu-) AMT	Δ*hisB*, Δ*trpC*, Δ*ilvB*, Δ*ilvG*, Δ*ilvI*, Δ*leuB*	$\gamma1/\gamma2$-$^{12}CD_3/^{13}CH_3$-Val	50 mg/L Isovalerate monomethyl
		$\gamma1/\gamma2$-$^{13}CH_3/^{13}CH_3$-Val	50 mg/L Isovalerate dimethyl
		$\gamma1$ or $\gamma2$-$^{12}CD_3/^{13}CH_3$-Val	40 mg/L Acetolactate (either proS or proR)

[a] A, I, M, and T are for methyl-labeling of either Alanine, Isoleucine, Methionine, and Threonine, respectively.

Then, the user can transformed the cells with their favorite plasmid. Basically, the production of protein using our auxotrophic strain is performed in two steps:

- The growing step, in which cells exponentielly grow with deuterated amino acids (Val, Leu, Ile, His, and Trp) and glucose in a standard M9-D_2O medium
- The induction step, where the cells are grown in same condition as for the growing step except that the amino acids to isotopically label are replaced by the appropriate precursors.

The following protocol is set up for 1 L production of deuterated protein with either stereospecific labeling of both Leu and Val, or stereospecific labeling of Leu-only or Val-only.

1. Prepare both M9-D_2O media according to Table 2 to be used with BL21 (AcL-), BL21(AcL-/Val-), and BL21(AcL-/Leu-).
2. The cells are incubated in the growth medium up to $OD_{600} \sim 0.8$ prior to gentle pelleting, washed twice with M9-D_2O (10 mL) and resuspended in the induction medium that contains all needed amino acids and isotopically labeled precursors. The precursors can be either ^{13}C-proS-acetolactate or ^{13}C-α-ketoisovalerate, according to the desired specific labeling (see Table 1). Regarding the dilution factor, the cell concentration roughly corresponds to $OD_{600} \sim 0.2$. For stereospecific labeling, 20 mg/L of acetolactate are added just after resuspention of the cells and 20 mg/L along with IPTG. For mono- or dimethyl labeling, 50 mg/L of the relevant ^{13}C-α-ketoisovalerate are added right after resuspention of the cells.

Table 2 Base Growth and Induction Media

Growth Medium	**Induction Medium**
M9-D_2O (250 mL)	M9-D_2O (1000 mL)
Antibiotic	Antibiotic
[^{2}H]-celtone base powder (0.1 g/L)	D_8-glucose (2 g/L)
D_8-glucose (2 g/L)	D_{10}-Leu (15 mg/L) or precursor (Leu labeling)
D_{10}-Leu (15 mg/L)	D_{10}-Ile (15 mg/L)
D_{10}-Ile (15 mg/L)	D_8-Val (15 mg/L) or precursor (Val labeling)
D_8-Val (15 mg/L)	Trp (15 mg/L)
Trp (15 mg/L)	His (15 mg/L)
His (15 mg/L)	

3. After 1.5 h incubation at 37°C ($OD_{600} \sim 0.4$), the production can be induced by IPTG and incubation conditions set at a temperature and for a period previously optimized by the user using a standard BL21 strain.
4. After harvesting the cells by centrifugation, the residual D_2O can be recycled by distillation using a rotavapor at 42°C and 35 mbar. Codistilled metabolites are removed adding 10 g/L activated charcoal (Sigma cat. C9157) that is agitated and allowed to stand until filtered with 0.22 μm membrane (Autofil PES) immediately prior to use. The D_2O purity can be assessed by 1D NMR against a standard 1% H_2O in 100% D_2O.

2.2 Rapid NOESY Data Collection With SOFAST–NOESY

Improving data quality while reducing data collection time motivated us to implement SOFAST–HMQC (Schanda et al., 2005) as the frequency-editing scheme in multidimensional NOESY experiments (Rossi et al., 2016). The experiments are very well-suited to methyl-labeled biomolecules that is characterized by a ^{1}H spectrum separated into distinct regions for methyls, amides, and/or aromatics. Selective T_1 relaxation (or L-optimized effect) (Pervushin, Vogeli, & Eletsky, 2002) and Ernst angle optimization (Schanda et al., 2005) together with improved water suppression combine to produce excellent data quality in which the sensitivity can be enhanced by over twofold in favorable instances. Fig. 4 gives a conceptual view of the experiments and the flexibility afforded by isolating proton regions with selective pulses vs the traditional broadband approach using hard pulses.

Two basic variants are the SOFAST–NOESY–HMQC and SOFAST–HMQC–NOESY–HMQC. Each HMQC can be either ^{15}N- or ^{13}C-based, the latter can be focused on either the aromatic or methyl region that are well separated along both the ^{1}H and ^{13}C chemical shift dimension. Figs. 5–7 exemplify the standard data collection pulse sequence (parameters described in (Rossi et al., 2016)) and data quality results in the case of a 200 kDa oligomer that was separated into domains for assignment in a divide-and-conquer approach. The methyl and amides cross-peaks are observed and separated along both the ^{15}N and ^{13}C (Figs. 5 and 6) and the protons dimensions (Fig. 7). Knowledge of amide assignment greatly facilitates the manual task, insuring high assignment accuracy. The details for the observed sensitivity improvements are given elsewhere (Rossi et al., 2016). Briefly, amides and methyl moieties can experience enhanced longitudinal (T_1)

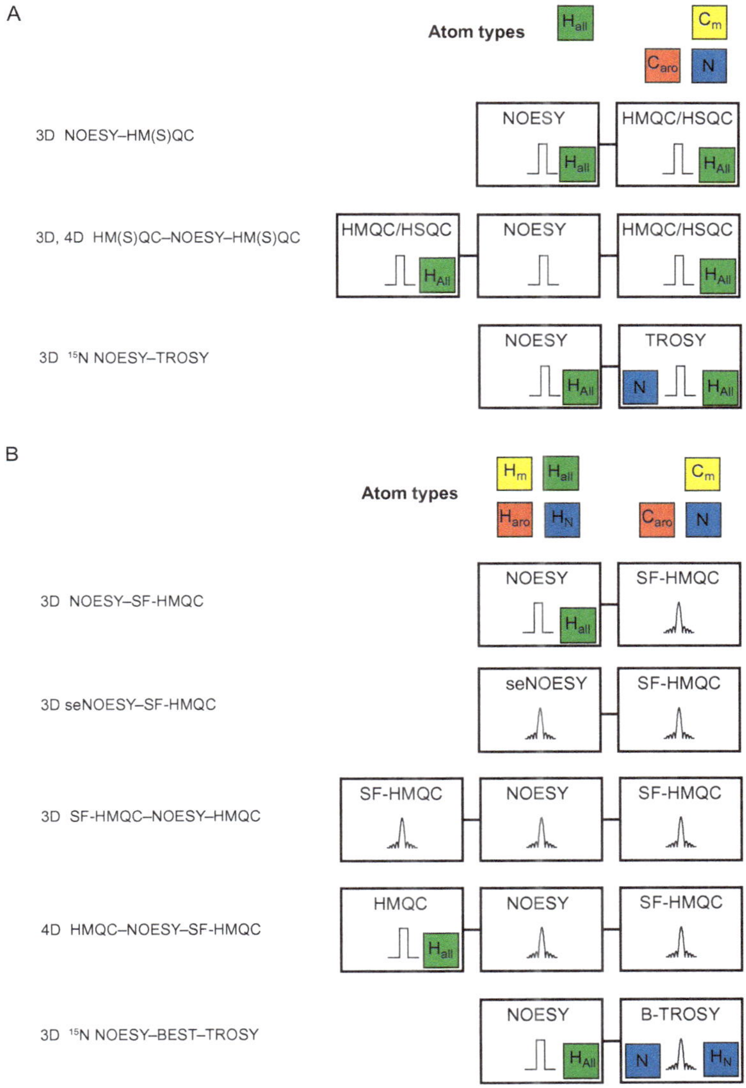

Fig. 4 Conceptual representation of traditional (A) vs SOFAST-based (B) multi-dimensional NOESY experiments. Each *rectangle* represents a module for frequency editing or for NOESY mixing/evolution and *square boxes* are colored based on the selective excitation chosen. Combinations of modules and frequency selection produce the complete 3D or 4D experiment. Frequency-selective pulse modules make additional experiments possible such as diagonal-free aromatic-methyl or amide–methyl NOESY. In contrast, the traditional experiment is limited to all-proton excitation with hard pulses (H_{all}). Selective NOESY (seNOESY) indicates that mixing and evolution periods use selective ^{1}H pulses.

relaxation. Under these conditions, amides experience ~50% while methyls ~25% relaxation enhancement. Since the signal intensity is related to T_1 relaxation via the expression: $I(0)(1-e^{-d1/T_1})$ a reduction in longitudinal relaxation produces an increase in sensitivity as a function of recovery delay ($d1$) with greater benefit at short $d1$ (~0.2 s). Provided that the sequence starts and ends on the same ^{1}H moiety, further ~10%–15% enhancement is obtained by altering the angle of the flip-down pulse after NOE mixing period (Ernst angle). In the case of heteronuclear editing, for example, in an amide to methyl experiment (Fig. 5) signal enhancement is observed due to differential T_1 relaxation. Under these conditions, the mixing time (τ) adds to the total relaxation time for the spins that are pulsed only prior to the NOE mixing period and left undisturbed until the subsequent scan. The expressions for magnetization go from $M_0[1-e^{-(d1+aq)/T_1}]$ in a traditional experiment to $M_0[1-e^{-(\tau+d1+aq)/T_1}]$ in heteronuclear SOFAST–NOESY. The effect accounts for a ~40% sensitivity enhancement vs traditional heteronuclear experiments. All the above effects combined result in up to ~2.5-fold sensitivity enhancement for the SOFAST–NOESY vs traditional experiments.

For ^{15}N frequency editing, we favor the simpler HMQC experiment that gives superior raw sensitivity particularly in the 3D heteronuclear NOESY experiments (Fig. 5), the drawback of this approach is that the amide protons are sensitive to ^{15}N decoupling schemes and give broader lines (Rossi et al., 2016). For the traditional ^{15}N-edited NOESY the BEST–TROSY sequence can be utilized for editing to produce the 3D ^{15}N-edited NOESY–BEST–TROSY experiment and produce sharper cross-peaks due to the amide TROSY effect (Pervushin et al., 1997). The experiment is not specific to methyl-labeled samples and could be of general use in biomolecular NMR and in favorable cases provides reasonable sensitivity improvement over the traditional hard pulse-based NOESY (Fig. 8).

2.2.1 SOFAST–NOESY Protocol

The standard experimental parameters for 2D and 3D SOFAST-based data collection are summarized in Table 1 including optional 3D ^{15}N-edited NOESY. A traditional ^{15}N TROSY is recorded for matching the backbone (Salzmann et al., 1998) to the methyl-labeled datasets. All pulse sequences and python scripts are compatible with Bruker Topspin 2.1–4.0 and can be downloaded at: https://github.com/NMRsoftware/SOFAST.

The pulse library and scripts use the standard Bruker "prosol" relation tables for pulse width calibration. Bruker standard channel configuration

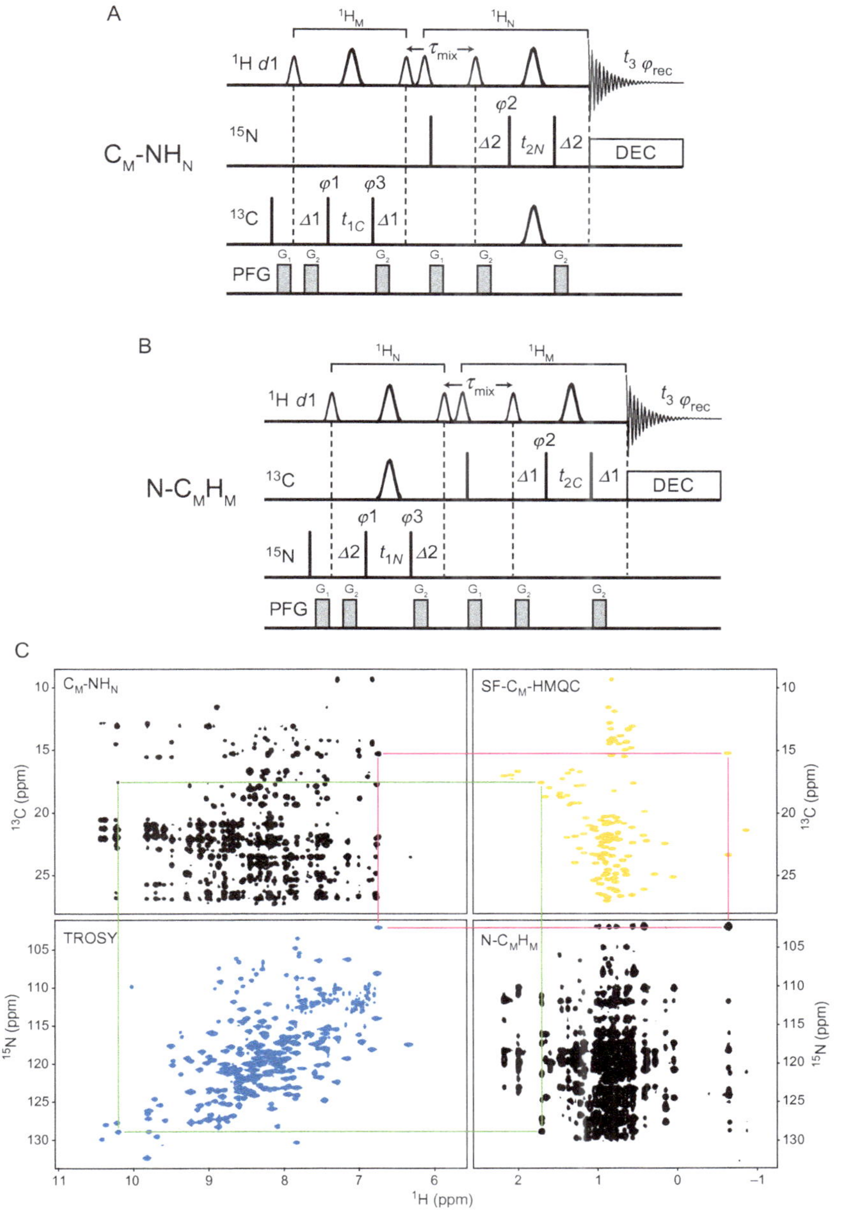

Fig. 5 Symmetric SOFAST-based heteronuclear ^{15}N, ^{13}C 3D HMQC–NOESY–HMQC (X-YH-type) pulse sequences (A) and (B), respectively. Application to Hsp90 N-domain is shown in panel (C). In *black* are the full 3D projections from (A) and (B) experiments aligned to their respective 2D ^{15}N TROSY and ^{13}C SOFAST–HMQC (methyl-TROSY). Examples of assignments validation via their symmetric amide–methyl patterns are *color coded* to Met (*green*) and Val (*magenta*) methyls as in Fig. 2.

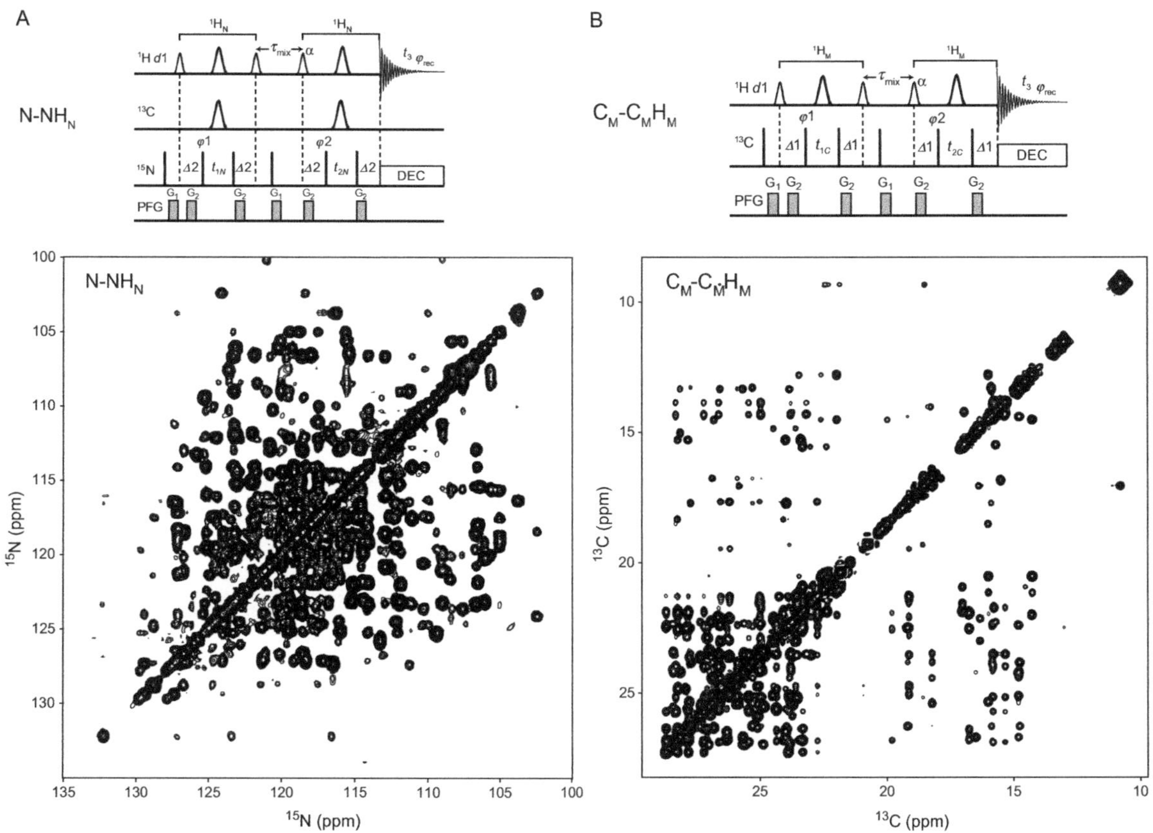

Fig. 6 Symmetric SOFAST-based homonuclear ^{15}N and ^{13}C 3D HMQC–NOESY–HMQC (X-XH-type) pulse sequences (A) and (B) *top panels*, respectively. On the *bottom panels* are the corresponding projections from full 3D data along the heavy atom dimensions. Experiment in (B) is the SOFAST implementation of the 3D methyl-NOESY.

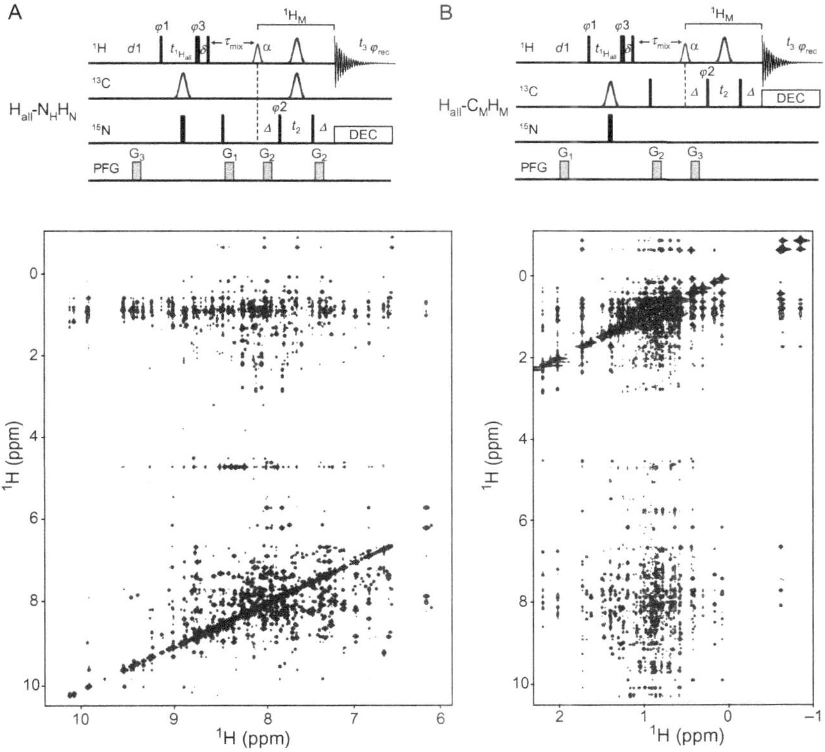

Fig. 7 SOFAST-based ^{15}N- and ^{13}C-edited 3D NOESY–HMQC pulse sequences (*top panel*) with hard pulse NOESY module for all proton resonances and SOFAST–HMQC editing (A) and (B), respectively. *Bottom panel* are projections of 3D data for Hsp90 N-domain.

is used. Channel F1, F2, and F3 in edasp are for ^{1}H, ^{13}C, and ^{15}N, respectively. Installation on the spectrometer is done as follows:

1 Using Topspin command prompt create two parameter files HC2D_N (a) and HNN3D (b):

Parameter set (a):

type in command: rpar HSQCGPPH

type in command: edasp; click arrow under F3; choose ^{15}N; click Default at bottom; and click "Save and Close"

type in command: o3p 118

type in command: wpar HC2D_N

Parameter set (b):

type in command: rpar HNCAGPWG3D

type in command: eda; find out parameter NUC1 and change ^{13}C to ^{15}N for F1 dimension

type in command: wpar HNN3D

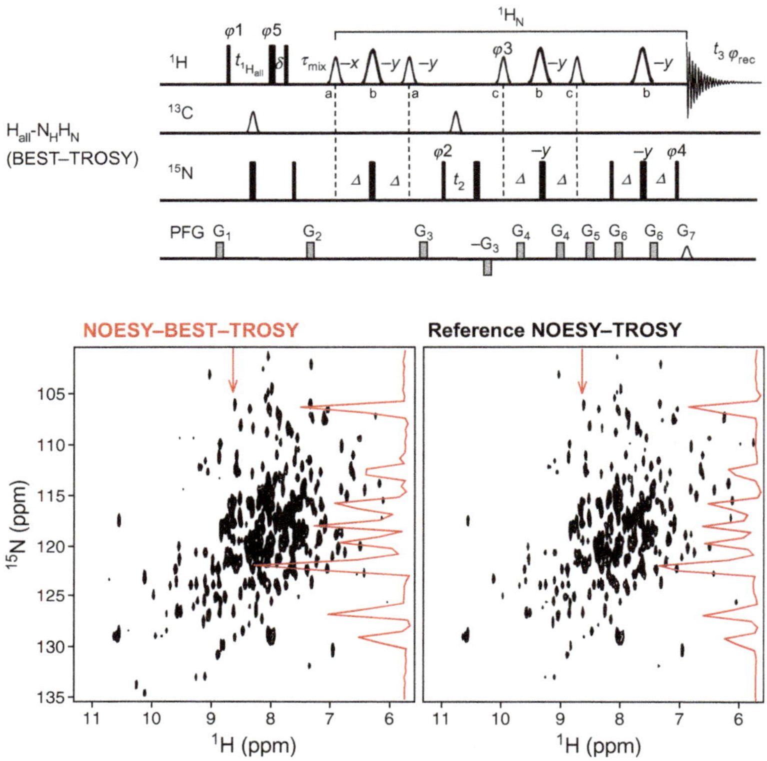

Fig. 8 3D ^{15}N-NOESY-BEST_TROSY pulse sequence (*top panel*) for frequency labeling $H_{All}(F_1)$-$N(F_2)H_N(F_3)$. The 90° Pc9_4 shaped pulse (Kupce & Freeman, 1994) is 1.69 ms long, the 180° Reburp is 1.15 ms long, and the 90° Eburp2 pulse is 1.08 ms long (Geen & Freeman, 1991). The offsets of all ^{1}H shaped pulses are 3230 Hz (at 8.5 ppm). The *narrow and wide bars* represent 90° and 180° hard pulse. The shaped pulse on ^{13}C channel represents a 500 ms long 180° smoothed Chirp (Hwang, van Zijl, & Garwood, 1997). The spectral centers of $^1H(F_1)$, $^{15}N(F_2)$, $^1H(F_3)$ dimensions are at 4.7, 118.0, and 4.7 ppm, respectively. The delays are $d1=0.2$ s, $\Delta=2.6$ ms, $\delta=6\,\mu$s (initial t_1 value), $t_{mix}=0.3$ s. The phase cycling is $\phi1=(x, x, -x, -x)$, $\phi2=(x, -x)$, $\phi3=y$, $\phi4=y$, $\phi5=4(x), 4(y)$, $\phi_{rec}=(x, -x, -x, x, -x, x, x, -x)$. The duration of G1, G2, G3, G5, and G7 is 1 ms, and the duration of G4 and G6 is 0.3 (ms). The quadrature detections in t_1 dimension is acquired via States-TPPI of $\phi1$. The gradient strengths of G1, G2, G4, and G6 are 15, 15, 5, and 5 G/cm, respectively. Acquisition mode in t_1 dimension uses echo-AntiEcho: setting $\phi3=y$, $\phi4=y$, and (G3:G5:G7) = (1.0000:0.6667:1.0000) selects Echo, whereas setting $\phi3=-y$, $\phi4=-y$, and (G3:G5:G7) = (0.8750:1.0000:0.6595) selects anti-Echo.

2 Shaped pulses:
Bruker standard shaped pulses Pc9_4_90.1000 and Reburp.1000 and two new shaped pulses, consine- and sine-modulated Pc9_4_90.1000 shaped pulses are automatically generated and calibrated by the script sofast.py.

The pulse width is determined by the excitation bandwidth (4.5–5.0 ppm), and the modulation frequency is 3.7 ppm. For detail, please see the subroutine myshape in sofast.py.

3 In the 'py' folder edit the path in sofast.py script to adhere to directory structure on the local spectrometer:

Find the line:

execfile("/opt/MNMR/py/mymodule.py"),

Replace the path/opt/MNMR/py with your correct path:

<TOPSPIN HOME>/exp/stan/nmr/py/user

Copy all pulse sequences in folder pp to <TOPSPIN HOME>/exp/stan/nmr/lists/pp/user

Copy all python scripts in folder py to <TOPSPIN HOME>/exp/stan/nmr/py/user

Copy the bi_garp_2pl.2 in folder cpd to <TOPSPIN HOME>/exp/stan/nmr/lists/cpd

4 Special notes:

Water suppression maybe improved by increasing the gpz1 or the last pair of PFGs, or by adjusting the ^{1}H excitation frequency and bandwidth (cnst1 and cnst2) to keep distance from the water resonance.

The experiments use garp for both ^{15}N- and ^{13}C-decoupling. There are instances where results maybe improved and artifacts removed using bi_garp_2pl for ^{15}N-decoupling (van Ingen, Vuister, & Tessari, 2002).

The python script is started by typing the "sofast" command in Topspin command prompt and selecting the desired experiment type. Default parameters for ILVM-type methyls are used by the python script as in Table 3. Adjustments in SW and offset maybe required for specific cases.

All SOFAST–NOESY experiments are carried out with the following common parameters: ^{15}N $^{1}J_{NH}=95$ Hz, ^{13}C$_{M}$ $^{1}J_{CH}=125$ Hz corresponding to 5.2, and 4.0 ms value for delay $\Delta=1/(2\times{}^{1}J_{XH})$, recycle delay of 0.2 s, mixing time 0.3 s, sufficient number of increments in the indirect dimensions are used to approach optimal sampling in the indirect dimension (Rovnyak et al., 2004). Sweep width and center frequency position of the polychromatic ^{1}H shaped pulses (Pc4_9), REBURP, and EBURP are directly calibrated in the pulse program based on the value of ^{1}H 90-degree square pulse. SOFAST–NOESY experiments were developed on Bruker spectrometers. All spectra examples depicted herein were collected on Bruker AVANCE NEO 700 and 850 MHz equipped with 5-mm TCI cryoprobes. All examples experiments (Figs. 5–8) are recorded at 30°C with the above parameters, spectra are processed with NMRPipe (Delaglio et al., 1995) and analyzed with NMRFAM-Sparky (Lee et al., 2016).

Table 3 Standard Experiment and Parameter Sets for SOFAST–NOESY Data Collection

Experiment Name	Python Setup Entry No.	Observe Atom Type	Frequency SW/Offset (ppm)[a]			ns[b]	Complex Points		
			^{13}C	^{15}N	^{1}H dir(ind.)		^{13}C	^{15}N	^{1}H dir(ind.)
2D HMQC	2	C_M-H_M			14.0	8	256		2k
2D HMQC	1	NH_N		36/118	14.0	8		256	2k
3D HMQC–NOESY–HMQC	6	C_M-C_MH_M	22/17		14.0	16	140		2k
3D HMQC–NOESY–HMQC	8	N-C_MH_M	22/17	36/118	14.0	16	140	100	2k
3D HMQC–NOESY–HMQC	9	C_M-NH_N	22/17		14.0	16	140	100	2k
3D NOESY–HMQC	14	H_{all}-C_MH_M	22/17		14.0(12)/4.7	8	140		2k(440)
3D NOESY–HMQC[c]	19	H_{all}-NH_N		36/118	14.0(12)/4.7	8	100		2k(440)
3D NOESY-BEST_TROSY[c]	5	H_{all}-NH_N		36/118	14.0(12)/4.7	8	100		2k(440)

[a]The parameters listed here are default from the python setup scripts. The cnst1 and cnst2 constants allow to set the ^{1}H offset and sweep width (SW) of the selective pulses, typically the bandwidth is 4.5–5.0 ppm and the offset position is ~8.5 ppm (amides), ~0.9 ppm for methyls, and 7.5 ppm for aromatics, if present. In addition, cnst3 sets the flip angle value as appropriate (for instance in X-XH-type experiments) while cnst4 and cnst5 set the $^{13}C_M$ and ^{15}N $^{1}J_{CH}$, respectively.

[b]Number of scans and experiment time are for reference only and may change based on sample characteristics and spectrometer frequency. The parameters here are reasonable for a 700 MHz ^{1}H frequency spectrometer.

[c]Optional.

2.2.2 Expert-Driven Methyl Type Assignment/Validation Protocol

Our amide-based methyl assignment protocol builds on well-established expert-driven methodologies (Gardner, Zhang, Gehring, & Kay, 1998; Tugarinov & Kay, 2003) and makes use of new python modules (Monneau et al., 2017) that function within NMRFAM-Sparky. The Sparky interface is user-friendly, well maintained and integration of new modules is straightforward. Amides resonances are assigned PINE server (Bahrami, Assadi, Markley, & Eghbalnia, 2009) with peak lists from triple-resonance. Results are then verified and extended manually using 3D ^{15}N-edited SOFAST–NOESY–HMQC or ^{15}N–NOESY–TROSY. For efficient and accurate methyl assignment, it is important to establish methyl type, in the case of Ile–Leu–Val–Met (ILVM) labeling the task is simple as the Ile and Met are very well separated from each other and from the Leu/Val group. Inclusion of Met ^{13}Cε is extremely useful in making the methyl–methyl network more robust and unambiguous due to the distinct ^{1}H and ^{13}C chemical shifts. The NOE pattern recognition is most effectively conducted by guided elimination simulating the amide region using the coordinates and the pdb2noe.py script (see Section 2.3). Ileδ1 is assigned by identifying the Cδ1 position in the 3D C_M-NH_N after navigating to each Ile amide strip (Fig. 9A). The Met is assigned by examining the pattern in the methyl-NOESY looking for distinctive patterns to Ileδ1. Given the low number of Met residues, simply guessing the assignment and swapping until the amide matching in 3D ^{13}C-edited SOFAST–NOESY–HMQC (H_{all}-C_MH_M) is achieved may work well as shown for an example Met in Fig. 9B. Leu and Val are assigned first by pairing intraresidue methyl using a short mixing time (~40 ms) methyl-NOESY (3D C_M-C_MH_M) in a sample with both δ1/δ2-$^{13}CH_3$/$^{13}CH_3$-Leu and γ1/γ2-$^{13}CH_3$/$^{13}CH_3$-Val. Such spectrum should be featureless except for the geminal cross-peak pairs. Pairs can be further analyzed by examining their patterns in the full 300 ms mixing time methyl-NOESY spectrum. Each intraresidue pair should display very similar patterns due to proximity to the same set of neighbors. Throughout the procedure, simulating the 3D H_{all}-C_MH_M spectrum provides rapid visual confirmation and validation. In short sidechains such as Ala, Thr, and Val, the intraresidue amide generally produces the largest NOE cross-peak as shown for a Val in Fig. 9C. Conversely, Leu has higher degree of ambiguity because the longer sidechain is such that the intraresidue amide may not have the strongest cross-peak in any of the NOESY spectra, consequently Leu may require more expert manual analysis. We now prefer to delay Leu assignment until after MAGIC automated assignment step is complete.

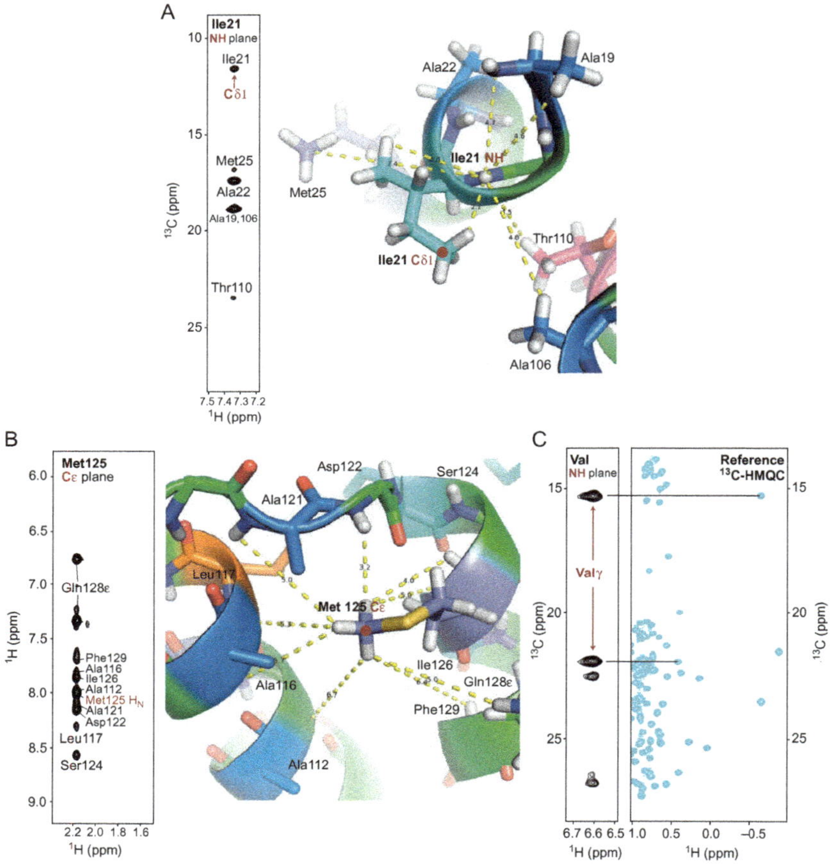

Fig. 9 Divide and conquer strategy and the presence of amide assignment simplifies the assignment and validation using the "pdb2noe" python program and a coordinate file. Here simulation of 3D ^{13}C–SFHMQC–NOESY–^{15}N–SF-HMQC (A) and 3D ^{13}C-edited NOESY–SF-HMQC (B) are conducted. Examples NOE strips show the correctly assigned Met Cε and Ile Cδ1 producing peak patterns that match the structure patterns. Short side chains such as Ala, Thr, and Val have strong intraresidue amide–methyl cross-peak that facilitates their direct assignment as in the example for a very characteristically shifted Val in Hsp90 HMQC spectrum.

Ala Hβ is assigned from backbone Cβ value and confirmed by identifying the corresponding amide NOE in the 3D ^{13}C-edited SOFAST–NOESY–HMQC (H_{all}-$C_M H_M$) strip. Slight systematic discrepancies in the Cβ value maybe present due to isotope effect when analyzing ^{2}H- to ^{1}H-bound Cβ from triply-labeled vs methyl-labeled, respectively. Thr γ2 are assigned in the same manner. However, we now prefer to postpone the Ala and Thr assignment in all but the simplest cases and produce a separate sample with all labeled methyls for that purpose. Ala and Thr generally have low number of NOE connectivity and their inclusion together with the ILVM residues complicates automated assignment.

A partial version of the strategy outlined above is followed in order to obtain the methyl type for the automated assignment and prepare for the run using MAGIC (see Section 2.3). The step is limited to rapid type identification and no interresidue cross-peak analysis or assignment, are needed. In most cases of ILMV methyl labeling the I- and M-types can be established visually and the short mixing time NOESY is helpful for complete Leu/Val pairing.

2.3 Methyl Assignment by Graphing Inference Construct (MAGIC)

Methyl assignment algorithms are based on matching two networks, a methyl–methyl proximity network from 3D coordinate file and an experimental peak–peak NOE-based connectivity network (Chao et al., 2014; Monneau et al., 2017; Pritisanac et al., 2017; Xu & Matthews, 2013). In the MAGIC algorithm, we use exhaustive search for network matching and our optimizer routine ranks the matches and provides a statistical measure of assignment ambiguity. Importantly, we include a front end that handles raw NOESY data and integrates with the visualization software for easy evaluation of the results. Exhaustive search insures that, as long as the 3D model is relevant to the tested sample in solution, the "true" assignment be represented in the output by one or more possible assignments. The NOESY data only affect the statistics in that lower overlap and higher completeness data increase the ranking of the accurate assignment among all the possibilities. The only major limitation of exhaustive search procedure is the combinatorial problem that grows with N! (N, the number of labeled methyls).

To overcome the limitation, we leverage the local nature of the NOE effect and define a set of *local* distance network (or peak clusters) constructed from automatic analysis of raw NOESY peak lists that are matched to the methyl–methyl proximity network. The impact of false-positive peak–peak connections is moderated by a confidence score that ranks all peak–peak connections (Monneau et al., 2017) (Fig. 10A). The $N \times N$ peak connectivity matrix $\boldsymbol{P}$ (see Fig. 10A and B), where N is the number of peaks, is automatically constructed from the NOESY peak list and incorporates the confidence scores. This matrix provides a map of the local peak network around each peak. The robustness of the peak cluster definition is further

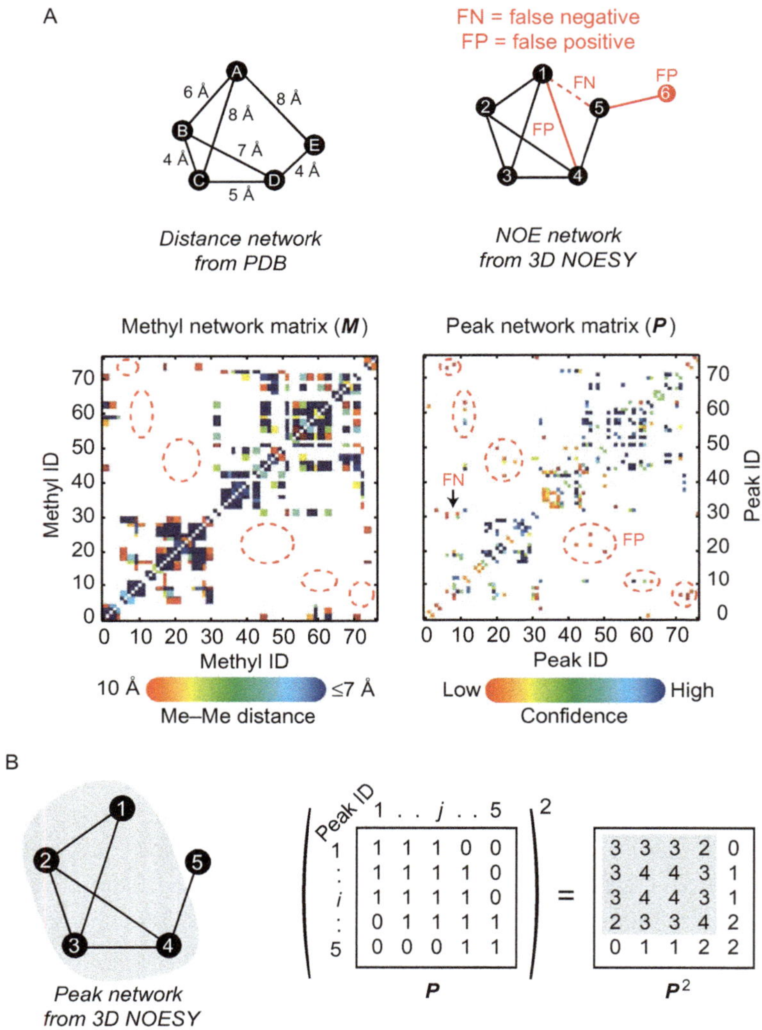

Fig. 10 The peak–peak connection map. (A) Methyl assignment algorithms are based on matching two networks, a methyl–methyl proximity network from 3D coordinate file and an experimental peak–peak NOE-based connectivity network. The structure-based methyl network matrix (***M***) and the experimental 3D NOESY peak network matrix (***P***), permuted according to the "true" assignment, are displayed with the *red to blue* color scale from 0 to 1 for the ***M*** matrix elements and from 0 to 4 for the ***P*** matrix elements. The ***P*** matrix starts from randomly distributed and converges during calculation to resemble ***M*** at the end of the assignment calculation. The *dotted red circles* highlight discrepancies between the ***M*** and ***P*** that can be schematically represented by the simple networks above. ***M*** shows much higher density and represents the true connection pattern, while the ***P*** matrix shows several false negatives (FN) and false positive (FP).

improved by defining an NOE "density" score matrix as the square of the NOE connection matrix (see Fig. 10B). The elements of the $\boldsymbol{P}^2$ matrix increase in value when more methyl–methyl NOEs are present and interconnected, indicating high density, high confidence connectivity. The benefit in considering $\boldsymbol{P}^2$ instead of $\boldsymbol{P}$ is to lessen the impact of false-negative peak–peak connection and to allow MAGIC to very accurately define peaks cluster around each peak.

The software is built with a hierarchical and iterative assignment procedure based on the confidence score function that prioritizes assignment of high-confidence, dense local NOE networks (Fig. 11). Once a cluster of peaks matches the criteria (e.g., enough cluster size or low assignment ambiguity), its assignments are merged with global assignments. At each step, the top scoring assignments are selected. MAGIC does not need methyl type and geminal pairing information to run, however, such information is key to accuracy and speed. Running with methyl type is highly recommended. In our test cases, accuracy was lowered when methyl type was not included.

FN connections are those present in the structure but not in the data (due to dynamics or low s/n data or incomplete peak selection), and FP connections are those present in the data but not actually represented in $\boldsymbol{M}$. The latter category indicates noise, incorrectly selected peaks, overlapping or discrepancies between the structure and the data (for example, crystal vs solution structure). The MAGIC algorithm deals with FP connections via the confidence score matrix (see in $\boldsymbol{P}$, connections inside *red circles* have low confident scores). (B) Given the simple network (*on the left*) composed of five peaks and their connections, which are depicted by vertices and edges, respectively, the peak–peak connection matrix, $\boldsymbol{P}$, is built with all the non-zero values for peak–peak interactions (here all set to one for simplicity instead of the confidence scores calculated by MAGIC) and the ith-row/column corresponding to the ith-peak. The square of $\boldsymbol{P}$ is the peak network density matrix. Higher numbers indicate denser networks. The $\boldsymbol{P}^2$ elements that describe the association between the 2nd-peak and 3rd-peak have high values because both peaks are involved in the same network sharing interaction with the 1st- and the 4th-peaks. Note that the values describing the networking between the 1st- and 4th-peaks, which are not directly connected, are not null because both are connected to the 2nd- and the 3rd-peak. The 4×4 submatrix (*highlighted in gray*) is the dense portion of this simple network. Conversely, the $\boldsymbol{P}^2$ row corresponding to the 5th-peak contains only low values because it is not involved in a high-density network. The NOE density threshold, used to define the peak clusters, is the value above which one an element of $\boldsymbol{P}^2$ is consider for defining a local network. For instance, a NOE density threshold set to three will enable the definition of a cluster of peaks around the 2nd-peak including the 1st-, the 2nd-, the 3rd-, and the 4th-peak of our simple network example. In this manner, decreasing the NOE density threshold value will cause the peak clusters to expand. Moreover, $\boldsymbol{P}^2$ matrix helps moderate the impact of FN connections in peak cluster definition as long as the involved peaks share other connections within their network.

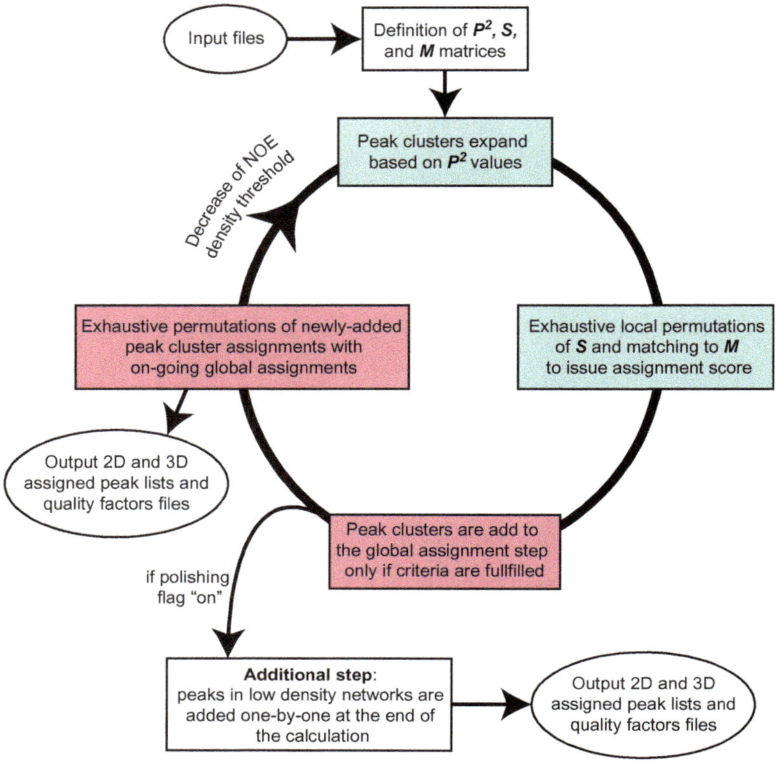

Fig. 11 The simplified MAGIC workchart. From the input data, MAGIC defines the three essentials matrices: the $\boldsymbol{P^2}$ matrix representing the confidence and the density of the peak–peak networking; the $\boldsymbol{M}$ matrix, defined according to the methyl–methyl proximity networking; and the score matrix ($\boldsymbol{S}$). Each score matrix element is the product of the confidence score and the relative intensity of the NOE from which the connection has been defined. $\boldsymbol{P^2}$ is about the definition of peak clusters, while $\boldsymbol{S}$ and $\boldsymbol{M}$ are about the assignment score calculation. The schematic of the MAGIC algorithm is then subdivided into two main trunks of local (*cyan*) and global assignment (*purple*) and consists in an iterative procedure, starting from the highly dense and confident NOE networks (i.e., the procedure starts with the highest NOE density value of $\boldsymbol{P^2}$) to the less ones at the end of the iteration. The local assignment involves the isolated peak cluster growing with decreasing NOE density threshold (see Fig. 10B for NOE density threshold definition), and the subsequent exhaustive search round within each peak cluster, while global assignment merges peak cluster assignments, in the order of decreasing NOE density threshold, when the cluster size and precision criteria are fulfilled. The score calculation for each assignment is the Hadamard product of the reordered $\boldsymbol{S}$ matrix by the $\boldsymbol{M}$ matrix. For each value of the NOE density threshold, the best current assignment is produced in the form of 2D HMQC and 3D NOESY peaklists. At the end of the calculation, only if the polishing flag is set to "on," isolated peaks or very low confidence peak networks are individually added to their best matching clusters.

Hereafter, detailed protocol is described to use MAGIC. Briefly, MAGIC handles NOESY peak lists, 2D HMQC peak list, and PDB coordinate file and outputs the assigned NOESY and HMQC peak lists along with a ".pml" file containing instruction for PyMol to map all individual scores (NOESY assignment completeness for each 2D peak and confidence score for each peak–peak connection) into the PDB structure. The user should be prepared to complete an initial run, refine the NOESY peak lists, verifying the atom type, and adding geminal pairings and rerun the program to improve the results.

2.3.1 System Requirement and the Input File Formats

MAGIC can be downloaded from https://github.com/NMRsoftware/MAGIC. The program is written in Python language, using several add-on modules. We recommend the installation of the Anaconda Python distribution (https://www.anaconda.com/download/) since it meets all module requirements and the most recent NMRFAM-Sparky distribution (https://nmrfam.wisc.edu/nmrfam-sparky-distribution/) to easily satisfy MAGIC's input/output needs (Lee et al., 2016). The Python modules required by MAGIC program are *types*, *re*, *numpy*, *math*, *datetime*, *sys*, *os*, *multiprocessing*, *cPickle*, *shutil*, *resource*, *psutil*, and *shuffle*. The following files are required to run MAGIC:

2D HMQC peak list
3D or 4D [^{13}C,^{1}H]-HMQC–NOESY–[^{13}C,^{1}H]-HMQC peak list
PDB file
fasta file (with the actual sequence of the sample construct)
Magic.py
generate.py
start.txt (the launcher file)

MAGIC needs all files to be properly formatted. All peak lists are in sparky format without any header. The format of the 2D peak list has to be as following:

1	13.058	0.786
2	20.855	0.749
3	20.875	0.886
4	21.449	1.075
5	10.332	0.328
6	23.844	0.769

with the first column defining the name (or index) of each peak, and the second and the third columns, the ^{13}C and ^{1}H chemical shifts, respectively. The 2D peak list has to be first reformatted by the *generate.py* script before running MAGIC (see below). Any information at the right of the proton frequency column is discarded during the reformatting process. In the case the peak names are already correctly indexed or named from prior manual assignment, a flag enables the user to avoid peak renaming. The format of the NOESY peak list (for instance the CCH–NOESY peak list) is based on sparky format and two header lines:

13C;13C;1H				
0.1;0.1;0.01				
?-?-?	10.027	13.679	0.692	577300
?-?-?	10.029	14.241	0.862	1602024
?-?-?	10.032	14.434	0.728	1652766
?-?-?	11.602	16.755	2.021	349449
?-?-?	11.624	23.538	1.086	1078639

with the first column defining the name of each NOE peak, the second, the third, and the fourth columns, the ^{13}C chemical shifts of the NOE dimension, the ^{13}C and ^{1}H chemical shifts of the editing plane, respectively. The header, to be added by the user, defines the nuclei type for each columns and the allowed matching tolerances. The last column is the cross-peak intensity, which is used by MAGIC to calculate the weight of each NOE relative to the overall intensity in each resolved strip. Diagonal (autocorrelation peaks) should be purged from the NOESY peak lists prior to running. MAGIC applies a filter in order to remove the diagonal, but careful peak list preparation that purges any diagonal and noise peaks and includes as many real peaks as possible is recommended. MAGIC now can handle the 3D or 4D version of the [^{13}C,^{1}H]-HMQC–NOESY–[^{13}C,^{1}H]-HMQC pulse scheme.

The coordinate file is in PDB format (https://www.pdb.org). While proton coordinates that are normally absent in crystal structures are not required to run MAGIC, they are required for subsequent amide-based validation and can be added using MolProbity (Chen et al., 2010) webserver (http://molprobity.biochem.duke.edu/). The validation is conducted using *pdb2noe.py* within NMRFAM-Sparky under the "Extensions/Assignment" menu or keyboard shortcut "SN."

The FASTA file shall contain the actual primary sequence of the sample construct in one-letter code without header. This file is needed because the PDB file does not necessary contain all the methyl-containing residues included in the sample construct.

2.3.2 The Protocol

1: Create a directory with all input files and navigate to the directory in a shell terminal.

2: Convert the 2D HMQC peak list using the script *generate.py* (type the command 'python generate.py –h' for further information about how to call the script). This script will generate two files including the list of all methyl-labeled residues according to your labeling scheme and the formatted 2D HMQC peak list. The methyl type is suggested for each 2D peak according to the chemical shift distribution extracted from the BMRB database. If the 3D NOESY peak list using a short mixing time is provided, the script will also suggest ambiguous geminal pairings for each Leu/Val peak in the case of double-methyl labeling. The converted 2D HMQC peak list should be checked and edited by the user and we emphasize that the methyl type and geminal pairing information is used as-is by MAGIC. The following command line example is used to successfully run the script:

```
> python generate.py [2D peak list file name] [construct sequence file name, fasta format] [labeling, e.g., AILMTV] [starting sequence number, e.g., 48] [rename peaks: y or n] {optional: [CCH short mixing time peak list]}.
e.g. > python generate.py hmqc.list seq.fasta AILMV 48 y
```

The format of the new 2D HMQC peak list is as follows:

1	13.058	0.786	I	
2	20.855	0.749	ALV	;3;6
3	20.875	0.886	ALV	;2
4	21.449	1.075	ALV	;6
I345CD1-HD1	10.332	0.328	-	
6	23.844	0.769	VL	;2

with the first column, the arbitrary generated 2D peak names (optional); the second and the third columns, both ^{13}C and ^{1}H frequencies, respectively; the fourth column, the proposed methyl type; and the last column, the geminal pairing. In the presence of partial assignments, known from mutagenesis or otherwise, the user can replace the methyl-type information by "-" and format the peak names according to sparky-labeling requirement (e.g., I345CD1-HD1).

The generate.py script also converts the FASTA sequence file to the 'seq.auto' file that contains methyl-only residues numbered according to the structure in the following format:

M447
I452
I457
V459
I463
V471

3: Once the new 2D HMQC peak list has been reviewed for accuracy, the "start.txt" file needs to be properly set. This file contains all the information needed by MAGIC to run, such as file pathways or labeling scheme. The format of this file is as following:

```
############ Input ###########
##2d HMQC peak list path:
./hmqc.list
##3d NOESY peak list path:
./cch_Tm300ms.list
##PDB path:
./construct.pdb
##SEQ path:
seq.auto
##labeling:
I,CD1;L,CD1,CD2;V,CG1,CG2;A;M;T
############ parameters ###########
##stereospecific=2;double methyl=1;mono-methyl=0:
1
##score threshold factor:
1
##Distance threshold:
7 10
##Final one-by-one step (off or on):
on
```

with line 3, 5, 7, and 9, the pathways of the four mandatory files to run MAGIC; line 11, the labeling scheme (please strictly follow the format and do not use semi-codon at the end of this line); line 14, the Leu/Val code according to the labeling scheme; line 16, the score factor (usual values are from 0.1 to 2), with 1 the standard value, lower or higher values than 1 makes the exclusion of assignments more or less stringent, respectively; line 18, the lower and higher distance thresholds to define methyl network (Monneau et al., 2017), the default values are "7 10"; line 20, enable ("on") or disable ("off") the last polishing step which consist in adding unassigned peak to the global assignment one-by-one. Once the "start.txt" file has been properly edited, the calculation can be launched using the following command line:

```
> python Magic.py start.txt
```

Then, MAGIC generates a time-stamped folder, in which are three folders: (i) the input folder containing all input files, for the record of calculation conditions; (ii) the output folder, continuously edited, containing the current assigned 2D and 3D peak lists along with the mapping file to visualize into PDB structure the quality factors; (iii) the run folder, which contains all intermediate files used by the software. The contents of the "run" directory can be removed at the end of the calculation.

2.3.3 The Results

Inside the output folder, the 2D and 3D NOESY peak lists are updated all along the run. The formatting of the assigned 2D peak list, that follows the Sparky format, is detailed below:

A106CB-HB	18.708	1.392	A	4.0	0.7	{'A106CB': 290.531, 'A22CB': 290.012}
I99CD1-HD1	14.912	0.584	I	9.0	0.9	{'I99CD1': 290.531}
L138CD1-HD1	24.224	1.03	L	12.63	0.4	{'L138CD': 290.531}
L138CD2-HD2	26.736	0.992	L	15.39	0.8	{'L138CD': 290.531}
T89CG2-HG2	23.297	1.182	T	17.74	1.0	{'T89CG2': 290.531}
V131CG1-HD1	22.504	0.952	V	12.5	0.67	{'V131CG': 290.531}
V131CG2-HD2	17.723	0.56	V	18.0	0.67	{'V131CG': 290.531}

with the highest score assignments on the first column; the second and the third columns, the ^{13}C and ^{1}H frequencies, respectively; the fourth column, the methyl types defined by the user (or by the script *generate.py*); the fifth column, the sum over all confidence scores of all connections involving this assigned peak; the sixth column, the NOE completeness (rate of assigned NOEs within the NOE-strip related to the 2D peak); the seventh column, all possible assignments and their related scores. Low completeness may signify a wrong assignment, an unassigned nearby methyl, or overlapping 2D peaks instead at those frequencies. The case of low completeness along with high total confidence score (as for the third line) should be investigated by the user as a possible problem area with NOEs assignment.

The assigned 3D NOESY peak list in the same Sparky-compatible format is as detailed below:

I213CD1-I44CD1-HD1	14.245	13.683	0.687	3.333	$d=4.3$	1
I75CD1-I44CD1-HD1	14.224	13.683	0.687	4.5	$d=4.0$	1
L215CD-I44CD1-HD1	26.226	13.683	0.687	1.25	$d=9.4$	1
L40CD-I44CD1-HD1	25.323	13.683	0.687	9.0	$d=4.2$	1
L40CD-I44CD1-HD1	24.427	13.683	0.687	7.0	$d=4.2$	0

with the first column, the best-scored assignment; the second, the third, and the fourth columns, the ^{13}C chemical shifts of the NOE dimension, the ^{13}C and ^{1}H chemical shifts of the editing plane, respectively; the fifth column, the connection confidence score; the sixth column, the actual distance in Å between the NOE-producing methyls; the seventh column, a code to notify the user about whether this connection has been used for peak clustering ("1") or not ("0"). The third NOE assignment is an example of connection the user has to verify, first of all the methyl–methyl distance is large (9.4 Å) and the related confidence score is low (1.25). *Note*: in Sparky visualizing columns beyond the frequencies columns requires the user to enable "note" display.

3. SUMMARY

Harnessing the power of methyl labeling by NMR involves the essential step of methyl resonance assignment. The task presents an interesting technical challenge to which we have contributed from the standpoints of sample preparation (*E. coli* auxotrophic strains), experimental NMR development (SOFAST–NOESY), and algorithm design (MAGIC) that we have outlined in this chapter. Our approach to methyl assignment is proven on a variety of biomolecules ranging from ~30 to 100 kDa. This comprehensive recipe for methyl assignment may prove useful to structural biology groups interested in developing an integrative approach to biological systems that incorporates solution NMR.

REFERENCES

Alphonse, S., & Ghose, R. (2018). Methyl NMR spectroscopy: Measurement of dynamics in viral RNA-directed RNA polymerases. *Methods*, *148*, 100–114. https://doi.org/10.1016/j.ymeth.2018.05.021.

Amero, C., Dura, M. A., Noirclerc-Savoye, M., Perollier, A., Gallet, B., Plevin, M. J., et al. (2011). A systematic mutagenesis-driven strategy for site-resolved NMR studies of supramolecular assemblies. *Journal of Biomolecular NMR*, *50*(3), 229–236. https://doi.org/10.1007/s10858-011-9513-5.

Ayala, I., Sounier, R., Use, N., Gans, P., & Boisbouvier, J. (2009). An efficient protocol for the complete incorporation of methyl-protonated alanine in perdeuterated protein. *Journal of Biomolecular NMR*, *43*(2), 111–119. https://doi.org/10.1007/s10858-008-9294-7.

Bahrami, A., Assadi, A. H., Markley, J. L., & Eghbalnia, H. R. (2009). Probabilistic interaction network of evidence algorithm and its application to complete labeling of peak lists from protein NMR spectroscopy. *PLoS Computational Biology*, *5*(3), e1000307https://doi.org/10.1371/journal.pcbi.1000307.

Bax, A., Clore, G. M., Driscoll, P. C., Gronenborn, A. M., Ikura, M., & Kay, L. E. (1990). Practical aspects of proton carbon carbon proton 3-dimensional correlation spectroscopy of C-13-labeled proteins. *Journal of Magnetic Resonance*, *87*(3), 620–627. https://doi.org/10.1016/0022-2364(90)90320-9.

Chao, F. A., Kim, J. G., Xia, Y. L., Milligan, M., Rowe, N., & Veglia, G. (2014). FLAMEnGO 2.0: An enhanced fuzzy logic algorithm for structure-based assignment of methyl group resonances. *Journal of Magnetic Resonance*, *245*, 17–23. https://doi.org/10.1016/j.jmr.2014.04.012.

Chen, V. B., Arendall, W. B., Headd, J. J., Keedy, D. A., Immormino, R. M., Kapral, G. J., et al. (2010). MolProbity: All-atom structure validation for macromolecular crystallography. *Acta Crystallographica. Section D, Biological Crystallography*, *66*, 12–21. https://doi.org/10.1107/S0907444909042073.

Delaglio, F., Grzesiek, S., Vuister, G. W., Zhu, G., Pfeifer, J., & Bax, A. (1995). NMRpipe—A multidimensional spectral processing system based on UNIX pipes. *Journal of Biomolecular NMR*, *6*(3), 277–293. https://doi.org/10.1007/Bf00197809.

Gans, P., Hamelin, O., Sounier, R., Ayala, I., Dura, M. A., Amero, C. D., et al. (2010). Stereospecific isotopic labeling of methyl groups for NMR spectroscopic studies of high-molecular-weight proteins. *Angewandte Chemie-International Edition*, *49*(11), 1958–1962. https://doi.org/10.1002/anie.200905660.

Gardner, K. H., & Kay, L. E. (1998). The use of H-2, C-13, N-15 multidimensional NMR to study the structure and dynamics of proteins. *Annual Review of Biophysics and Biomolecular Structure*, *27*, 357–406. https://doi.org/10.1146/annurev.biophys.27.1.357.

Gardner, K. H., Zhang, X. C., Gehring, K., & Kay, L. E. (1998). Solution NMR studies of a 42 KDa *Escherichia coli* maltose binding protein beta-cyclodextrin complex: Chemical shift assignments and analysis. *Journal of the American Chemical Society*, *120*(45), 11738–11748. https://doi.org/10.1021/ja982019w.

Geen, H., & Freeman, R. (1991). Band-selective radiofrequency pulses. *Journal of Magnetic Resonance*, *93*(1), 93–141. https://doi.org/10.1016/0022-2364(91)90034-Q.

Gelis, I., Bonvin, A. M. J. J., Keramisanou, D., Koukaki, M., Gouridis, G., Karamanou, S., et al. (2007). Structural basis for signal-sequence recognition by the translocase motor SecA as determined by NMR. *Cell*, *131*(4), 756–769. https://doi.org/10.1016/j.cell.2007.09.039.

Gorman, S. D., Debashish, S., O'Rourke, K. F., & Boehr, D. D. (2018). Assigning methyl resonances for protein solution-state NMR studies. *Methods*, *148*, 88–99. https://doi.org/10.1016/j.ymeth.2018.06.010.

Huang, C. D., & Kalodimos, C. G. (2017). Structures of large protein complexes determined by nuclear magnetic resonance spectroscopy. *Annual Review of Biophysics*, *46*, 317–336. https://doi.org/10.1146/annurev-biophys-070816-033701.

Huang, C. D., Rossi, P., Saio, T., & Kalodimos, C. G. (2016). Structural basis for the antifolding activity of a molecular chaperone. *Nature*, *537*(7619), 202–206. https://doi.org/10.1038/nature18965.

Hwang, T. L., van Zijl, P. C. M., & Garwood, M. (1997). Broadband adiabatic refocusing without phase distortion. *Journal of Magnetic Resonance*, *124*(1), 250–254. https://doi.org/10.1006/jmre.1996.1049.

Igumenova, T. I., & Palmer, A. G. (2006). Off-resonance TROSY-selected R-1p experiment with improved sensitivity for medium- and high-molecular-weight proteins. *Journal of the American Chemical Society*, *128*(25), 8110–8111. https://doi.org/10.1021/ja061692f.

Ikura, M., Kay, L. E., Tschudin, R., & Bax, A. (1990). 3-Dimensional NOESY-HMQC spectroscopy of a C-13-labeled protein. *Journal of Magnetic Resonance*, *86*(1), 204–209. https://doi.org/10.1016/0022-2364(90)90227-Z.

Isaacson, R. L., Simpson, P. J., Liu, M., Cota, E., Zhang, X., Freemont, P., et al. (2007). A new labeling method for methyl transverse relaxation-optimized spectroscopy NMR spectra of alanine residues. *Journal of the American Chemical Society*, *129*(50), 15428–15429. https://doi.org/10.1021/ja0761784.

Kay, L. E. (2011). Solution NMR spectroscopy of supra-molecular systems, why bother? A methyl-TROSY view. *Journal of Magnetic Resonance*, *210*(2), 159–170. https://doi.org/10.1016/j.jmr.2011.03.008.

Kay, L. E., Clore, G. M., Bax, A., & Gronenborn, A. M. (1990). 4-Dimensional heteronuclear triple-resonance NMR-spectroscopy of interleukin-1-beta in solution. *Science*, *249*(4967), 411–414. https://doi.org/10.1126/science.2377896.

Kay, L. E., Xu, G. Y., Singer, A. U., Muhandiram, D. R., & Formankay, J. D. (1993). A gradient-enhanced HCCH TOCSY experiment for recording side-chain H-1 and C-13 correlations in H_2O samples of proteins. *Journal of Magnetic Resonance Series B*, *101*(3), 333–337. https://doi.org/10.1006/jmrb.1993.1053.

Kerfah, R., Plevin, M. J., Sounier, R., Gans, P., & Boisbouvier, J. (2015). Methyl-specific isotopic labeling: A molecular tool box for solution NMR studies of large proteins. *Current Opinion in Structural Biology*, *32*, 113–122. https://doi.org/10.1016/j.sbi.2015.03.009.

Kiteyski-LeBlanc, J. L., Yuwen, T., Dyer, P. N., Rudolph, J., Luger, K., & Kay, L. E. (2018). Investigating the dynamics of destabilized nucleosomes using methyl-TROSY NMR. *Journal of the American Chemical Society*, *140*(14), 4774–4777. https://doi.org/10.1021/jacs.8b00931.

Kupce, E., & Freeman, R. (1994). Wide-band excitation with polychromatic pulses. *Journal of Magnetic Resonance Series A*, *108*(2), 268–273. https://doi.org/10.1006/jmra.1994.1123.

Lange, O. F., Rossi, P., Sgourakis, N. G., Song, Y. F., Lee, H. W., Aramini, J. M., et al. (2012). Determination of solution structures of proteins up to 40 kDa using CS-Rosetta with sparse NMR data from deuterated samples. *Proceedings of the National Academy of Sciences of the United States of America*, *109*(27), 10873–10878. https://doi.org/10.1073/pnas.1203013109.

Lee, W., Cornilescu, G., Dashti, H., Eghbalnia, H. R., Tonelli, M., Westler, W. M., et al. (2016). Integrative NMR for biomolecular research. *Journal of Biomolecular NMR*, *64*(4), 307–332. https://doi.org/10.1007/s10858-016-0029-x.

Lichtenecker, R. J., Coudevylle, N., Konrat, R., & Schmid, W. (2013). Selective isotope labelling of leucine residues by using alpha-ketoacid precursor compounds. *Chembiochem*, *14*(7), 818–821. https://doi.org/10.1002/cbic.201200737.

Mas, G., Crublet, E., Hamelin, O., Gans, P., & Boisbouvier, J. (2013). Specific labeling and assignment strategies of valine methyl groups for NMR studies of high molecular weight proteins. *Journal of Biomolecular NMR*, *57*(3), 251–262. https://doi.org/10.1007/s10858-013-9785-z.

Monneau, Y. R., Ishida, Y., Rossi, P., Saio, T., Tzeng, S. R., Inouye, M., et al. (2016). Exploiting *E-coli* auxotrophs for leucine, valine, and threonine specific methyl labeling of large proteins for NMR applications. *Journal of Biomolecular NMR*, *65*(2), 99–108. https://doi.org/10.1007/s10858-016-0041-1.

Monneau, Y. R., Rossi, P., Bhaumik, A., Huang, C. D., Jiang, Y. J., Saleh, T., et al. (2017). Automatic methyl assignment in large proteins by the MAGIC algorithm. *Journal of Biomolecular NMR*, *69*(4), 215–227. https://doi.org/10.1007/s10858-017-0149-y.

Ollerenshaw, J. E., Tugarinov, V., & Kay, L. E. (2003). Methyl TROSY: Explanation and experimental verification. *Magnetic Resonance in Chemistry*, *41*(10), 843–852. https://doi.org/10.1002/mrc.1256.

Palmer, A. G. (2016). A dynamic look backward and forward. *Journal of Magnetic Resonance*, *266*, 73–80. https://doi.org/10.1016/j.jmr.2016.01.018.

Palmer, A. G., Kroenke, C. D., & Loria, J. P. (2001). Nuclear magnetic resonance methods for quantifying microsecond-to-millisecond motions in biological macromolecules. *Methods in Enzymology*, *339*, 204–238. https://doi.org/10.1016/S0076-6879(01)39315-1.

Pervushin, K., Riek, R., Wider, G., & Wuthrich, K. (1997). Attenuated T-2 relaxation by mutual cancellation of dipole-dipole coupling and chemical shift anisotropy indicates an avenue to NMR structures of very large biological macromolecules in solution. *Proceedings of the National Academy of Sciences of the United States of America*, *94*(23), 12366–12371. https://doi.org/10.1073/pnas.94.23.12366.

Pervushin, K., Vogeli, B., & Eletsky, A. (2002). Longitudinal H-1 relaxation optimization in TROSY NMR spectroscopy. *Journal of the American Chemical Society*, *124*(43), 12898–12902. https://doi.org/10.1021/ja027149q.

Pickford, A. R., & Campbell, I. D. (2004). NMR studies of modular protein structures and their interactions. *Chemical Reviews*, *104*(8), 3557–3565. https://doi.org/10.1021/cr0304018.

Pritisanac, I., Degiacomi, M. T., Alderson, T. R., Carneiro, M. G., Eiso, A. B., Siegal, G., et al. (2017). Automatic assignment of methyl-NMR spectra of supramolecular machines using graph theory. *Journal of the American Chemical Society*, *139*(28), 9523–9533. https://doi.org/10.1021/jacs.6b11358.

Rosenzweig, R., & Kay, L. E. (2014). Bringing dynamic molecular machines into focus by methyl-TROSY NMR. *Annual Review of Biochemistry*, *83*, 291–315. https://doi.org/10.1146/annurev-biochem-060713-035829.

Rossi, P., Shi, L., Liu, G. H., Barbieri, C. M., Lee, H. W., Grant, T. D., et al. (2015). A hybrid NMR/SAXS-based approach for discriminating oligomeric protein interfaces using Rosetta. *Proteins-Structure Function and Bioinformatics*, *83*(2), 309–317. https://doi.org/10.1002/prot.24719.

Rossi, P., Xia, Y. L., Khanra, N., Veglia, G., & Kalodimos, C. G. (2016). N-15 and C-13-SOFAST-HMQC editing enhances 3D-NOESY sensitivity in highly deuterated, selectively [H-1,C-13]-labeled proteins. *Journal of Biomolecular NMR*, *66*(4), 259–271. https://doi.org/10.1007/s10858-016-0074-5.

Rovnyak, D., Hoch, J. C., Stern, A. S., & Wagner, G. (2004). Resolution and sensitivity of high field nuclear magnetic resonance spectroscopy. *Journal of Biomolecular NMR*, *30*(1), 1–10. https://doi.org/10.1023/B:Jnmr.0000042946.04002.19.

Ruschak, A. M., Velyvis, A., & Kay, L. E. (2010). A simple strategy for C-13,H-1 labeling at the Ile-gamma 2 methyl position in highly deuterated proteins. *Journal of Biomolecular NMR*, *48*(3), 129–135. https://doi.org/10.1007/s10858-010-9449-1.

Saio, T., Guan, X., Rossi, P., Economou, A., & Kalodimos, C. G. (2014). Structural basis for protein antiaggregation activity of the trigger factor chaperone. *Science*, *344*(6184), 1250494. https://doi.org/10.1126/science.1250494. ARTN.

Saio, T., Kawagoe, S., Ishimori, K., & Kalodimos, C. G. (2018). Oligomerization of a molecular chaperone modulates its activity. *Elife* 7, https://doi.org/10.7554/eLife.35731.001. ARTN 35731.

Salzmann, M., Pervushin, K., Wider, G., Senn, H., & Wuthrich, K. (1998). TROSY in triple-resonance experiments: New perspectives for sequential NMR assignment of large proteins. *Proceedings of the National Academy of Sciences of the United States of America*, *95*(23), 13585–13590. https://doi.org/10.1073/pnas.95.23.13585.

Schanda, P., Kupce, E., & Brutscher, B. (2005). SOFAST-HMQC experiments for recording two-dimensional heteronuclear correlation spectra of proteins within a few seconds. *Journal of Biomolecular NMR*, *33*(4), 199–211. https://doi.org/10.1007/s10858-005-4425-x.

Sprangers, R., & Kay, L. E. (2007). Quantitative dynamics and binding studies of the 20S proteasome by NMR. *Nature*, *445*(7128), 618–622. https://doi.org/10.1038/nature05512.

Tugarinov, V., Choy, W. Y., Orekhov, V. Y., & Kay, L. E. (2005). Solution NMR-derived global fold of a monomeric 82-kDa enzyme. *Proceedings of the National Academy of Sciences of the United States of America*, *102*(3), 622–627. https://doi.org/10.1073/pnas.0407792102.

Tugarinov, V., & Kay, L. E. (2003). Ile, Leu, and Val methyl assignments of the 723-residue malate synthase G using a new labeling strategy and novel NMR methods. *Journal of the American Chemical Society*, *125*(45), 13868–13878. https://doi.org/10.1021/ja030345s.

Tugarinov, V., & Kay, L. E. (2004). An isotope labeling strategy for methyl TROSY spectroscopy. *Journal of Biomolecular NMR*, *28*(2), 165–172. https://doi.org/10.1023/B:JNMR.0000013824.93994.1f.

van Ingen, H., Vuister, G. W., & Tessari, M. (2002). A two-dimensional artifact from asynchronous decoupling. *Journal of Magnetic Resonance*, *156*(2), 258–261. https://doi.org/10.1006/jmre.2002.2564.

Velyvis, A., Ruschak, A. M., & Kay, L. E. (2012). An economical method for production of H-2, (CH3)-C-13-threonine for solution NMR studies of large protein complexes: Application to the 670 kDa proteasome. *PLoS One*, 7(9), e43725. https://doi.org/10.1371/journal.pone.0043725.

Wiesnerl, S., & Sprangers, R. (2015). Methyl groups as NMR probes for biomolecular interactions. *Current Opinion in Structural Biology*, *35*, 60–67. https://doi.org/10.1016/j.sbi.2015.08.010.

Xu, Y. Q., & Matthews, S. (2013). MAP-XSII: An improved program for the automatic assignment of methyl resonances in large proteins. *Journal of Biomolecular NMR*, *55*(2), 179–187. https://doi.org/10.1007/s10858-012-9700-z.

Zhu, G., Kong, X. M., & Sze, K. H. (1999). Gradient and sensitivity enhancement of 2D TROSY with water flip-back, 3D NOESY-TROSY and TOCSY-TROSY experiments. *Journal of Biomolecular NMR*, *13*(1), 77–81. https://doi.org/10.1023/A:1008398227519.

CHAPTER SIX

Synthetic Biology-Based Solution NMR Studies on Membrane Proteins in Lipid Environments

Erik Henrich, Frank Löhr, Julija Mezhyrova, Aisha Laguerre, Frank Bernhard, Volker Dötsch[1]

Institute of Biophysical Chemistry, Centre for Biomolecular Magnetic Resonance, J.W. Goethe-University, Frankfurt am Main, Germany

[1]Corresponding author: e-mail address: vdoetsch@em.uni-frankfurt.de

Contents

Methods in Enzymology, Volume 614
ISSN 0076-6879
https://doi.org/10.1016/bs.mie.2018.08.019

Abstract

Although membrane proteins are in the focus of biochemical research for many decades the general knowledge of this important class is far behind soluble proteins. Despite several recent technical developments, the most challenging feature still is the generation of high-quality samples in environments suitable for the selected application. Reconstitution of membrane proteins into lipid bilayers will generate the most native-like environment and is therefore commonly desired. However, it poses tremendous problems to solution-state NMR analysis due to the dramatic increase in particle size resulting in high rotational correlation times. Nevertheless, a few promising strategies for the solution NMR analysis of membrane inserted proteins are emerging and will be discussed in this chapter. We focus on the generation of membrane protein samples in nanodisc membranes by cell-free systems and will describe the characteristic advantages of that platform in providing tailored protein expression and folding environments. We indicate frequent problems that have to be overcome in cell-free synthesis, nanodisc preparation, and customization for samples dedicated for solution-state NMR. Detailed instructions for sample preparation are given, and solution NMR approaches suitable for membrane proteins in bilayers are compiled. We further discuss the current strategies applied.for signal detection from such difficult samples and describe the type of information that can be extracted from the various experiments. In summary, a comprehensive guideline for the analysis of membrane proteins in native-like membrane environments by solution-state NMR techniques will be provided.

1. INTRODUCTION

Within the last 20 years, techniques to characterize membrane proteins improved dramatically and especially examples of their structural investigation increased nearly exponentially (membrane proteins of known structure database: http://blanco.biomol.uci.edu/mpstruc/). The majority of known structures have been determined by X-ray crystallography and solution NMR spectroscopy, but with recent developments in detector technology the determination of high-resolution structures (Kühlbrandt, 2014) by cryo-EM is catching up. The drastic growth of structural data can be mainly attributed to improvements in measurement techniques and methodological advancements regarding protein production, purification, and sample preparation. High-end protein engineering (Serrano-Vega, Magnani, Shibata, & Tate, 2008), stabilization by small molecules or antibody fragments (Kim et al., 2015; Rasmussen et al., 2011, 2007), in meso crystallization (Caffrey, 2015), or lipid nanodiscs (Bayburt & Sligar, 2003) are only some of these developments. A general tendency is to focus on natural-like membrane environments for membrane protein research as more and more studies focus on the role of lipids for membrane protein structure and stability (Barrera, Zhou, & Robinson, 2013; Gupta et al., 2017; Henrich et al., 2017; Saliba, Vonkova, & Gavin, 2015). In particular, the use of nanolipid particles as solubilization environment for membrane proteins in structural studies has set new standards (Denisov & Sligar, 2016; Efremov, Gatsogiannis, & Raunser, 2017; Nikolaev et al., 2017). This is not surprising as detergents can have a dramatic influence on the overall structural integrity or dynamics of membrane proteins (Dehez, Schanda, King, Kunji, & Chipot, 2017). Interestingly, the most preferred detergents for membrane solubilization and protein purification like *n*-dodecyl β-D-maltoside (He, Wang, & Yan, 2014) preserve a high amount of lipids still attached to the extracted membrane protein (Ilgü et al., 2014).

In this respect sample preparation techniques that avoid detergent contacts and allow subsequent analysis within natural membranes or membrane mimetics are desirable. Here, the cotranslational insertion of membrane proteins into defined membranes by synthetic approaches offers a remarkable advantage as any detergent contact is avoided during the direct folding of the nascent protein chain into a provided lipid bilayer (Harris et al., 2017). Nanodiscs are extremely well suited for this production process via cell-free expression systems (Roos et al., 2014). The extraordinary features and physicochemical properties of nanodiscs streamline protein purification and enable a sophisticated analysis by many state-of-the-art techniques (Denisov & Sligar, 2017). Control of the bilayer composition allows information about the effect of specific lipids (Dawaliby et al., 2016; Henrich et al., 2016; Rues, Dötsch, & Bernhard, 2016) or about the influence of general bilayer characteristics on membrane protein structure or stability (McClary, Sumida, Scian, Paço, & Atkins, 2016) to be obtained. In solution NMR spectroscopy, short-chain detergents have almost exclusively been used to provide a hydrophobic environment as they produce reasonable small particles that allow sufficiently fast tumbling. Bicelles were the first bilayer containing environment that gave high-resolution spectra of membrane proteins (Morrison et al., 2012). While they are widely used in solution NMR (Dürr, Gildenberg, & Ramamoorthy, 2012), their classical preparation still requires treatment of the solubilized membrane protein with detergent (Table 1). Nanolipoprotein particles only slowly found their entrance into the NMR field (Glück et al., 2009) since the first nanodiscs with a diameter of ~10 nm still were of unfavorable size (Denisov, Grinkova, Lazarides, & Sligar, 2004). Several recent improvements in nanodisc design (Hagn et al., 2013; Nasr et al., 2017) and sample preparation (Laguerre et al., 2016) initiated a broader usage of nanodiscs for solution NMR (Table 1).

Table 1 Strategies for Preparing Solution NMR Samples in Bilayer Environments

	Conventional		**Cell-Free**	
	Bicelles	**Nanodiscs**	**Nanodiscs**	**Bicelles**
Workflow	Cellular protein expression		Cell-free protein expression	
	Membrane solubilization		—	
	Purification		Purification	
	Reconstitution		—	
	—		—	Reverse q-titration
Preparation[a]	3–4 days	3–4 days	<24 h	<24 h
Detergent solubilization[b]	Yes	Yes	No	No
Detergent screen[c]	Membrane solubilization and purification		No	Reverse q-titration
Lipid selection	Synthetic and short-chain lipids		Any	Any
Particle size	~3–7 nm defined by q-ratio	~8–15 nm defined by MSP type	~8–15 nm defined by MSP type	Variable, defined by q-ratio
Label scrambling	Reduced by auxotrophic strains		Reduced by inhibitors	
Protocols and	Piai, Fu, Dev, and Chou (2017)	Hagn, Etzkorn, Raschle, and Wagner (2013)	Roos et al. (2014)	Laguerre et al. (2016)
Applications[d]	Morrison et al. (2012)	Bibow et al. (2014); Kofuku et al. (2014)	Lyukmanova et al. (2012)	

[a]Time estimates include all steps listed in the workflow section.
[b]Workflow requires complete solubilization of the membrane protein in detergent micelles.
[c]Workflow steps requiring detergent screening procedure.
[d]Selected references.

In this chapter, we describe the technical details of a complete pipeline from cell-free membrane protein production to NMR analysis in membrane environment. This involves the setup of customized cell-free protein production (Schwarz et al., 2007) allowing specialized labeling strategies (Hein, Löhr, Schwarz, & Dötsch, 2017; Laguerre et al., 2016; Löhr et al., 2012; Löhr, Tumulka, Bock, Abele, & Dötsch, 2015), the cotranslational membrane protein insertion into defined nanodiscs (Henrich et al., 2016; Rues et al., 2016), and their transition into small bicelles for solution NMR measurements (Laguerre et al., 2016). The presented workflow will serve as a guideline to alleviate solution NMR characterizations of membrane proteins in native-like environments and potentially critical steps will be indicated.

2. BASIC CELL-FREE EXPRESSION SYSTEM

Expression systems based on cellular lysates have continuously been optimized for approximately the past 20 years. Systems utilizing lysates from different organisms including prokaryotic as well as eukaryotic sources (Kigawa et al., 2004; Madin, Sawasaki, Ogasawara, & Endo, 2000) are described and to some extent commercially available. Commercial systems are widely used and enable the quick access to cell-free technology, but they restrict options to optimize production and quality of the synthesized proteins. Individually prepared systems offer a more diverse variety of opportunities to adjust the expression conditions, as all supplemented components are known and under control (Schwarz et al., 2007). In addition, recent proteomics studies identified the residual protein components of *Escherichia coli* cell-free lysates obtained by standard preparation procedures (Foshag et al., 2018; Hurst et al., 2017). For economic reasons, tunable and most efficient protein synthesis, systems based on *E. coli* (Henrich, Hein, Dötsch, & Bernhard, 2015) or wheat germ (Harbers, 2014) cell lysates are thus currently recommended. *E. coli* is preferred for the on-site preparation of cell-free lysates as the preparation protocol and the quality of the lysate source are better controllable and more reliable.

2.1 *E. coli* Cell-Free Lysate Preparation

Various strains can be used as lysate source and the most common are *E. coli* A19 and BL21. Special applications may require engineered strains having, for example, eliminated release factors for the efficient incorporation of non-natural amino acids (Hong et al., 2014; Peuker et al., 2016). A standard S30 lysate preparation procedure out of a 10-L fermenter is described, which is able to provide 60–100 mL of cell-free lysate yielding 0.3–1.5 g of expressed protein (Foshag et al., 2018; Schwarz et al., 2007), but it should be noted that a variety of protocol modifications exist (Kigawa et al., 2004; Kim et al., 2006; Kwon & Jewett, 2015; Shrestha, Holland, & Bundy, 2012; Yang, Patel, Wong, & Swartz, 2012; Zubay, 1973):

1. Strike out *E. coli* A19 cells from a glycerol stock on a fresh agar plate and incubate the plate for 6–8 h. Inoculate two flasks containing 150 mL LB medium (10 g tryptone, 10 g NaCl, 5 g yeast extract/L) with the freshly grown cells and incubate them overnight at 37°C under vigorous shaking. The cultures usually grow to an OD_{600} of 6–7 overnight.
2. Inoculate 10 L of culture medium (16 g tryptone, 5 g NaCl, 10 g yeast extract/L) supplemented with 100 m*M* glucose and potassium phosphate buffer (22 m*M* KH_2PO_4, 40 m*M* K_2HPO_4) 1:66–100 with the preculture. At this stage, we recommend performing the cultivation in a fermenter to achieve higher cell densities and obtain a better yields and a higher quality of extract. It is also possible to use conventional culture flasks, but in our hands the lysate performance is reduced drastically. Cultivate the cells at 37°C under high oxygen saturation and stir until they reach mid-log phase (~4–5 OD_{600}) before cooling them down to ~18–20°C. The cooling step should not exceed 30 min. A growth curve should be recorded beforehand to assess the mid-log phase period in a given fermenter setup.
3. Cells are harvested by centrifugation (6700 × *g*, 20 min, 4°C) and the pellet is washed three times with 300 mL ice-cold buffer S30-A (10 m*M* Tris–acetate, pH 8.2, 14 m*M* $Mg(oAc)_2$, 60 m*M* KCl, 6 m*M* β-me) with 10 min centrifugation (8000 × *g*, 4°C) between the washes.
4. Pellet is resuspended in 110% (v/w) ice-cold buffer S30-B (10 m*M* Tris–acetate, pH 8.2, 14 m*M* $Mg(oAc)_2$, 60 m*M* KCl, 1 m*M* DTT, 1 × complete protease inhibitor) before cell breakage by a French Press device or similar automated cell breakage systems with standard settings for *E. coli*. The presented protocol recommends the cell breakage by mechanical force, but in general also sonication seems feasible (Shrestha et al., 2012).
5. Lysate is centrifuged at 30,000 × *g* for 30 min at 4°C. The supernatant is transferred to fresh tubes and the centrifugation step is repeated.
6. The supernatant is stepwise adjusted to 400 m*M* NaCl (5 *M* stock solution) before 45 min incubation at 42°C, which leads to a turbid solution.
7. The lysate is filled in dialysis bags (12–14 kDa MWCO) and dialyzed overnight at 4°C against 5 L buffer S30-C (10 m*M* Tris–acetate, pH 8.2, 14 m*M* $Mg(oAc)_2$, 60 m*M* KoAc, 0.5 m*M* DTT) with one buffer exchange after 3–4 h.
8. At the next day the lysate is centrifuged again (30,000 × *g*, 30 min, 4°C), split into aliquots and shock frozen with liquid nitrogen. The lysate can be stored at −80°C till usage and also can be shipped in lyophilized condition.

It should be noted that this procedure can be adapted to achieve desired lysate modifications, for example, to enhance the concentration of chaperones by an additional heat-shock step during the cultivation (Foshag et al., 2018). In this case, it should always be considered that any modification in lysate production may affect the production efficiency in subsequent cell-free protein production reactions (Foshag et al., 2018). By using appropriate blends, the advantages of lysates, e.g., giving high expression yields or improved folding by increased chaperone concentrations, can be combined (Rues, Dong, Dötsch, & Bernhard, 2018).

2.2 T7 RNA-Polymerase Preparation

The fast and efficient T7 RNA-polymerase is commonly employed for transcription within a coupled transcription and translation cell-free system (Schwarz et al., 2007). Since *E. coli* lysates still contain the endogenous RNA-polymerase (Foshag et al., 2018), cell-free reactions can alternatively be carried out with standard *E. coli* promotors and without the addition of T7 RNA-polymerase (Shin & Noireaux, 2010), but a significant drop in expression yield may occur. The following section illustrates the preparation of T7 RNA-polymerase by overexpression in *E. coli* cells (Schwarz et al., 2007). The yield of T7 RNA-polymerase out of 1 L *E. coli* culture is usually sufficient to supplement 0.5–1 L of cell-free lysate.

1. Inoculate an overnight culture (37°C, 200 rpm shaking) of LB medium (e.g., 150 mL) supplemented with 100 μg/mL ampicillin with BL21 (DE3) Star cells carrying the plasmid pAR1219 from a glycerol stock freshly grown on an agar plate.
2. Inoculate TB medium (12 g tryptone, 24 g yeast extract, 4 mL glycerol/L), 1–2 L is usually sufficient for a T7 RNA-polymerase batch, 1:100 (~0.1 OD_{600}) with the fresh overnight culture and incubate the culture at 37°C and 200 rpm shaking. When the OD_{600} reaches 0.6–0.8, induce T7 expression by the addition of 1 m*M* IPTG and incubate for further 5 h under the same conditions. Harvest the cells by centrifugation (6500 × *g*, 15 min, 4°C).
3. Resuspend the cells in 30 mL buffer (30 m*M* Tris–HCl, pH 8.0, 50 m*M* NaCl, 10 m*M* EDTA, 10 m*M* β-me, 5% (v/v) glycerol, 1 × complete protease inhibitors)/L of expression and sonicate eight times for 1 min with 30 s resting periods. Alternatively a French press can be used. Lysate is centrifuged (20,000 × *g*, 30 min, 4°C) and the supernatant is collected.

4. DNA is removed by the stepwise addition of 20% (w/v) streptomycin to a final concentration of 2%. Ensure incubation on ice and sufficient mixing after each step. DNA precipitate is removed by centrifugation (20,000 × *g*, 30 min, 4°C).
5. The supernatant is purified by anion exchange chromatography. The lysate is loaded on a Q-Sepharose column (75 mL bed volume for 1–2 L of expression culture), which was equilibrated with buffer A (30 m*M* Tris–HCl, pH 8.0, 50 m*M* NaCl, 1 m*M* EDTA, 10 m*M* β-me, 5% glycerol). After loading, the column is washed with buffer A until a stable baseline is reached. A gradient of 50–500 m*M* in 90 min (flow rate ~4 mL/min if allowed by pressure limit) is used for elution. Collect 2–3 mL fractions and analyze them by SDS-PAGE as elution profile might vary between runs. To save time the usage of gels enabling direct detection under UV is recommended. Combine fractions carrying high T7 RNA-polymerase concentrations (prominent band between 90 and 100 kDa). The T7 RNA-polymerase is not pure but sufficiently enriched after this step. Discard fractions with low concentrations and a high degree of impurities.
6. Dialyze collected fractions against 5 L buffer (10 m*M* K_2HPO_4, pH 8.0, 100 m*M* NaCl, 0.5 m*M* EDTA, 1 m*M* DTT, 5% glycerin) and afterward adjust the solution to 10% glycerol.
7. Concentrate the protein solution in Centriprep devices (50 kDa MWCO) by centrifugation (1500 × *g* for 10 min, 4°C) to a concentration of 2–4 mg/mL total protein. Adjust the concentrated samples to 50% glycerol and shock freeze in liquid nitrogen before storage at −80°C.
8. To determine the optimal amount of T7 RNA-polymerase for supplementation to cell-free reactions, perform a test cell-free reaction with increasing amounts of T7 polymerase. At a certain concentration a plateau is reached and the lowest T7 concentration of the plateau should be used for future reactions.

It is possible to skip this procedure by using commercially available T7 RNA-polymerase, but a concentration screen for obtaining the most efficient protein expression should be performed as well.

3. NANODISC PREPARATION

Nanodiscs are nanolipid particles that consist of a disc-shaped lipid bilayer, which is encircled by an ApoA1 derivative, the so-called membrane scaffold protein (MSP) (Fig. 1; Denisov & Sligar, 2017). The MSP protects the hydrophobic lipid tails and thus keeps the disc soluble.

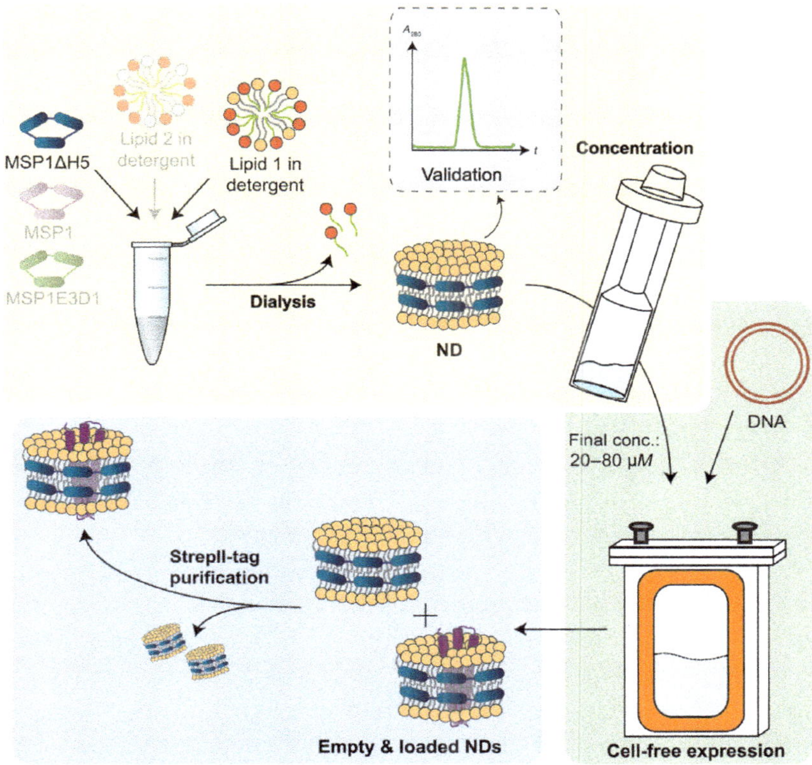

Fig. 1 General L-CF workflow including nanodisc formation (*yellow*), cell-free production (*green*) as well as sample purification (*blue*). Pictograms illustrate molecules or devices and workflow steps are indicated.

The reconstitution of "empty" nanodiscs (i.e., nonmembrane protein containing) as they are provided in cell-free reactions (Katzen et al., 2008; Roos et al., 2012) has been well established for more than 10 years (Bayburt, Grinkova, & Sligar, 2002; Denisov et al., 2004). Over the years engineered nanodiscs were designed for particular tasks, as, for example, nanodiscs with a smaller diameter that are more suitable for solution NMR experiments (Hagn et al., 2013). Particular comments and guidelines for successful nanodisc formation are given.

3.1 Preparation of Lipid Stock Solutions

For lipid stock solutions dedicated for nanodisc reconstitutions 50 m*M* is a reasonable concentration (Roos et al., 2014). The sufficient solubilization of the lipids is crucial for the successful reconstitution of a homogeneous population of nanodiscs. As nanodiscs are formed upon detergent removal, detergents with a relatively high critical micellar concentration (cmc) are suitable as they can be easily removed from the mixture. Commonly the bile acid salt sodium cholate is used (Bayburt et al., 2002). In general, it is also possible to use other detergents, but detergents with a very low cmc can become problematic. Most lipids can be completely solubilized by 100 m*M* sodium cholate in water, but increased detergent concentrations should be considered if the solution still stays turbid. In this respect especially lipid mixtures extracted from natural sources can become problematic. Lipid stocks can be stored at −20°C.

3.2 MSP Preparation

The MSP can be produced by conventional expression in *E. coli* cells and purified via the N-terminal His_6-tag (Denisov et al., 2004). Thereby the same protocol is used for all different variants of MSP:

1. Freshly transform the selected MSP expression plasmid (Addgene) in BL21 (DE3) star cells, streak them out on an agar plate containing kanamycin and incubate the plate overnight at 37°C. Inoculate a 600-mL preculture of LB medium supplemented with kanamycin (30 μg/mL) and incubate overnight (37°C, 200 rpm shaking).
2. Inoculate LB medium (8 × 600 mL) supplemented with kanamycin (30 μg/mL) and glucose (0.5% (v/v)) with 50 mL of the grown overnight culture and incubate at 37°C and 200 rpm shaking until OD_{600} reaches 1–1.2. Induce MSP expression with 1 m*M* IPTG with additional incubation of 1 h at 37°C and 4 h at 28°C at 200 rpm shaking. Harvest the cells by centrifugation (6500 × *g*, 10 min, 4°C) and transfer the pellets to a falcon tube. Pellets can be stored at −20°C for few months or used directly for purification.
3. Resuspend cell pellet (~15–25 g) in 45 mL MSP-buffer (40 m*M* Tris–HCl, pH 8.0, 300 m*M* NaCl). Add 5 mL of a 10% Triton X-100 solution and gently mix the solution. Cell breakage is performed by sonication with 3 × 60 s and 3 × 45 s cycles with always 60 s resting time. Within the resting time ensure proper mixing. Centrifuge the suspension at 30,000 × *g* for 20 min at 4°C, collect the supernatant, and pass it through 0.45 μ*M* filters.
4. The filtered lysate is loaded on a His-trap column (bed volume ~20–25 mL) equilibrated with MSP-buffer supplemented with 1% Triton X-100. Afterward the column is washed with each five column

volumes of MSP-buffer with 1% Triton X-100, MSP-buffer at pH 8.9 containing additional 50 m*M* cholic acid, MSP-buffer, and MSP-buffer with addition of 50 m*M* imidazole. The protein is eluted by five column volumes of MSP-buffer containing additional 300 m*M* imidazole.

5. MSP-containing fractions are pooled and adjusted to 10% glycerol (v/v) before dialysis overnight against 5 L of MSP-buffer supplemented with 10% glycerol (v/v) with one buffer exchange after 3–4 h.

For some experiments it can be beneficial to label the MSP with heavy isotopes (Li, Kijac, Sligar, & Rienstra, 2006; Peetz et al., 2017). For labeling with ^{15}N and ^{13}C isotopes, LB overnight cultures are harvested by centrifugation (7000 × *g*, 10 min, 4°C) and resuspended in minimal media (2 g/L ^{13}C-glucose, 1 g/L ^{15}N-NH_4Cl, 0.2 g/L $FeCl_3$, 0.67 g/L $ZnCl_2$, 1 m*M* $MgSO_4$, 1 m*M* $CaCl_2$, 0.01 g/L thiamine, 0.1 g/L ampicillin, 3 g/L KH_2PO_4, 12.8 g/L Na_2HPO_4, 0.5 g/L NaCl). Expression is induced after 2 h at 37°C and 200 rpm shaking and the cultures are incubated for additional 3 h before harvesting (Peetz et al., 2017).

3.3 Nanodisc Reconstitution

For the formation of nanodiscs the MSP is first mixed with detergent-solubilized lipids (Fig. 1). It is crucial to consider previously defined MSP to lipid ratios for the formation of homogeneous discs (Marty, Wilcox, Klein, & Sligar, 2013; Roos et al., 2012). Due to the dilution of the lipid stocks it should be verified that the final detergent concentration is not reduced below its cmc. In that case either additional sodium cholate or *n*-dodecylphosphocholine (DPC) could be added to reach concentrations above cmc (Roos et al., 2012) or higher concentrated MSP stocks may be used. If the appropriate ratio of the selected lipid/MSP types is not known, a corresponding prescreen on an analytical scale (~100 μL samples) should be carried out. The samples are incubated with gentle shaking for about 1 h at temperatures above the lipid transition temperature, preferentially room temperature. Nanodisc formation is then induced by detergent removal with biobeads or by dialysis. For dialysis, a ~250 × volume (i.e., 5 L for 20 mL of mixture) of DF-buffer (40 m*M* Tris–HCl, pH 8.0, 100 m*M* NaCl) is recommended with incubation for 3 days and one buffer exchange each day. To separate the nanodiscs from soluble aggregates or aggregated lipids, the solution is centrifuged (22,000 × *g*, 20 min, 4°C) and the supernatant can be concentrated with DF-buffer equilibrated Centriprep tubes (10 kDa MWCO) by multiple centrifugation steps (20 min, 2000 × *g*, 4°C) to a final concentration of 0.5–1 mM (Fig. 1). The concentrated solutions can be shock frozen by liquid nitrogen and stored at −80°C until usage.

3.4 Nanodisc Quality Control

The homogeneity of nanodisc stocks and the MSP:lipid ratio screens is analyzed by size-exclusion chromatography: 50 μL of a 20–50 μM nanodisc solution is separated with an analytical size-exclusion column (e.g., Superdex 200 increase) in DF-buffer (Roos et al., 2012). Size and shape of prepared nanodiscs may be further evaluated by negative stain cryo-EM (Henrich et al., 2017). To prevent aggregation or extensive stacking on the grids, the nanodisc concentration should be kept ~20 μM.

4. CELL-FREE INSERTION OF MEMBRANE PROTEINS INTO NANODISCS AND QUALITY OPTIMIZATION

In general, cell-free production approaches can be carried out in three major expression modes (Schwarz et al., 2007). The P-CF mode is performed without any supplied hydrophobic environment, but the resulting membrane protein precipitates can readily be solubilized in detergents. In the D-CF mode, detergents are supplied to the reaction leading to an immediate solubilization of the expressed protein. L-CF expressions contain bilayer structures such as liposomes, bicelles, or nanodiscs for membrane protein solubilization. Due to its simplicity and efficiency P-CF expression is usually the first choice for the production of a protein in a cell-free system, whereas the L-CF mode offers the opportunity to synthesize solubilized membrane proteins without any detergent contacts. Depending on the nature of the synthesized membrane protein, liposomes, bicelles, or nanodiscs might have different efficiencies for its proper cotranslational insertion and folding.

4.1 L-CF Expression

To setup a cell-free reaction, a variety of compounds responsible for energy supply (e.g., acetyl phosphate), a suitable chemical environment (e.g., Mg^{2+}, HEPES-buffer), precursors (e.g., amino acids), and preservation agents (e.g., protease inhibitors) are mixed together in a feeding mixture (FM), ensuring supply with nutrients as well as dilution of toxic byproducts. The FM is separated by a membrane from the reaction mixture (RM) holding the components responsible for transcription (e.g., T7 RNA-polymerase) and the protein production machinery (e.g., cell lysate). We highly recommend the usage of such a continuous exchange setup with an RM to FM ratio of 1:15–20 to achieve an efficient production of material required for NMR experiments. The RM volume should be adjusted to the total amount of desired protein in the NMR sample and is usually between 1 and 10 mL. In Table 2 all ingredients with stock solution and final concentrations for the expression of an unlabeled protein are displayed and the required volumes for a 1-mL reaction with a 1:20 ratio of reaction to FM and a Mg^{2+} concentration of 18 m*M* are indicated. For preparative scale expressions in

Table 2 Compounds and Concentrations for a 1-mL Cell-Free Reaction

			Volume (μL)	
Compound	**Stock**	**Final**	**FM**	**RM**
RCWMDE[a]	16.67 m*M* each	1 m*M*	1200	60
20 Amino acids[b]	25 m*M* each	0.55 m*M* each	440	22
Acetyl phosphate	1000 m*M*	20 m*M*	400	20
Phosphoenolpyruvate	1000 m*M*	20 m*M*	400	20
NTP-mix[c]	75 ×	1 ×	266.7	13.3
DTT	500 m*M*	2 m*M*	80	4
Folinic acid	10 mg/mL	0.1 mg/mL	200	10
Complete cocktail[d]	50 ×	1 ×	400	20
HEPES-KOH (pH 8.0)/EDTA	2.5 *M*/0.5 *M*	100 m*M*/20 m*M*	800	40
$Mg(OAc)_2$	1000 m*M*	13.1 m*M*	262	13.1
KOAc	10,000 m*M*	110 m*M*	220	11
PEG8000	40%	2%	1000	50
Sodium azide	10%	0.05%	100	5
Addition to FM				
20 Amino acids[e]	25 m*M* each	0.5 m*M* each	400	
Buffer S30-C[f]	—	—	7000	
H_2O	—	—	6831	
Addition to RM				
[NDs (DMPG)][f]	750 μ*M*	100 μ*M*	133.3	
S30-Extract	100%	35%	350	

Continued

Table 2 Compounds and Concentrations for a 1-mL Cell-Free Reaction—cont'd

Compound	Stock	Final	Volume (μL) FM	RM
Plasmid DNA	1 mg/mL	0.015 mg/mL	15	
RNAse inhibitor[g]	40 U/μL	0.30 U/μL	7.5	
T7-RNA polymerase[h]	3.63 mg/mL	0.03 mg/mL	5.5	
tRNA *E. coli*	40 mg/mL	0.50 mg/mL	12.5	
Pyruvate kinase	10 mg/mL	0.04 mg/mL	4	
H_2O	—	—	183.7	
Total volume			20,000	1000

[a]Is not added in protein-labeling reactions.
[b]For protein-labeling reactions adjust stock concentration to 30 mg/mL (total amino acid concentration) and final concentration to 1.5 mg/mL (total amino acid concentration) giving a final concentration of ~0.5 m*M* (each amino acid) in the RM.
[c]ATP, GTP, CTP, and UTP in the concentration of 360, 240, 240, and 240 m*M*, respectively.
[d]Protease inhibitor cocktail (one tablet in 1 mL H_2O, Roche Diagnostics).
[e]For labeled expressions adjust stock concentration to 30 mg/mL (total amino acid concentration) and final concentration to 0.5 mg/mL (total amino acid concentration). Together with the portion of amino acids from the general mix, which is used for the FM, this leads to a final concentration of ~0.6 m*M* (each amino acid) in the FM.
[f]Optional compound for L-CF expression with NDs, final concentration depends on target protein expression and its membrane insertion efficiency.
[g]Ribolock RNase inhibitor (Thermo Scientific).
[h]The stock and the final concentration of the T7-RNA polymerase varies from batch to batch and has to be determined in pilot experiments.
Adapted from Henrich, E., Dötsch, V., & Bernhard, F. (2015). Screening for lipid requirements of membrane proteins by combining cell-free expression with nanodiscs. *Methods in Enzymology, 556*, 351–369.

the mL range, the utilization of Slyde-A-lyzer devices with an MWCO of 10 kDa (Thermo Scientific) as container for the RM is very convenient. These are placed in special plastic containers that hold the FM (Schwarz et al., 2007). For screening purposes, smaller containers holding ~55 μL of RM are sealed with a dialysis membrane with an MWCO of 12–14 kDa fixed with a Teflon ring and then placed into standard 24-well cell culture plates holding 825 μL of FM. The basic steps for setting up preparative cell-free reactions are indicated here:

1. Stock solutions should be prepared in advance and stored at −20°C prior to usage. For salts and buffers 50 mL stocks and for amino acids 5–10 mL are suitable, whereas the other stock solutions may not exceed 2 mL due to lower stability of the compounds.
2. A premix of all low molecular weight components (Table 2, first section), which are present in the reaction as well as the FM, is prepared.

It is important to ensure proper mixing (at this step best performed by vortexing) before removing aliquots from the stock solutions (some stocks remain as suspensions) as well as after completing the premix.

3. The FM is completed by the addition of amino acids, buffer S30-C and H_2O (Table 2, second section). Mixing is performed by vortexing and the homogeneous FM is transferred to the reaction container.
4. The RM is completed by addition of the high molecular weight compounds (Table 2, third section). The completed RM is carefully mixed with a pipette. The membranes of the Slyde-A-lyzer are equilibrated from the outside with FM and then the RM is transferred into the Slyde-A-lyzer with a syringe. Excess air should be removed from the Slyde-A-lyzer to facilitate proper exchange of FM and RM.
5. Finally, the reaction container is closed and incubated under gentle shaking at 30°C for 12–20 h. After incubation, the RM is centrifuged at 16,000–20,000 × g for 10 min at 4°C to separate precipitates from the soluble fraction.

For protein-labeling experiments, the final amino acid concentration in the RM and FM should be adjusted to ~0.5–1 mM and ~0.6–1.5 mM, respectively, for each amino acid (Klammt et al., 2004; Löhr et al., 2015). Depending on the desired labeling, customized stock solutions, containing a fully labeled amino acid mix or a mixture of unlabeled and labeled amino acids, are required for specific experiments. For simplification, the amino acid stock solution can be prepared with a total amino acid concentration of 30 mg/mL giving 1.5 mg/mL of each amino acid. The final concentration in the RM and FM is in total 1.5 mg/mL and the extra shot for the FM is 0.5 mg/mL. Further detailed protocols for the setup of standard cell-free reactions including compound concentrations as well as dimensions of customized reaction containers have been described before (Henrich, Dötsch, & Bernhard, 2015; Klammt et al., 2004; Reckel et al., 2010; Schwarz et al., 2007).

For membrane protein solubilization the preformed nanodiscs are supplemented to the RM (Fig. 1 and Table 2) only. The suitable concentration as well as the type of nanodiscs (i.e., selected MSP type and lipid composition of the membrane) should be assessed in pilot experiments by considering the following points:

1. The MSP type should be selected according to the size of the membrane-integrated domains of the target protein to ensure an adequate size of the resulting nanodisc membrane. For solution-state NMR applications a rather short MSP type like MSP1D1ΔH5 is recommended as it forms

nanodiscs of only ~8 nm. This size is usually sufficient to accommodate most membrane proteins and even the formation of large assemblies such as the pentameric complex of the seven-transmembrane helix containing proton pump proteorhodopsin with a total molecular mass of ~135 kDa can be achieved (Peetz et al., 2017). On the other side, the limited membrane space can also hamper the insertion of specific membrane proteins or complexes. Thus, the expression of the target into different nanodisc types should be tested to evaluate the formation of soluble membrane protein/nanodisc complexes.

2. The composition of the bilayer should be customized to the protein of interest. Some proteins require a specific lipid environment for their functional fold and activity (Henrich et al., 2016). If a lipid dependency is not known, the insertion and subsequent analysis of the membrane protein into nanodiscs assembled with different lipids is recommended as an initial screen.
3. The nanodisc concentration should be adjusted to the target concentration in the reaction. Furthermore, individual membrane proteins strongly differ in their efficiency of nanodisc insertion (Roos et al., 2012). Screening different nanodisc concentrations (we recommend a range of 10–100 μ*M*) for an optimal solubilization should be considered and the lowest nanodisc concentration that achieves saturated solubilization is recommended for subsequent expression experiments. If multimeric proteins are analyzed, it should be taken into account that the nanodisc concentration can have an influence on the oligomeric state formation (Peetz et al., 2017). The oligomeric assembly of a protein in nanodiscs might be analyzed, e.g., by native mass spectrometry (Henrich et al., 2017).

Samples dedicated for NMR experiments require the supplementation of isotope-labeled amino acids. A fully labeled sample can be generated by specific mixtures containing all 20 amino acids with the desired isotopic labeling scheme (^{15}N, ^{13}C, and ^{2}H). Uniform labeling, however, can usually easily be achieved expression in *E. coli* in media containing ^{13}C-glucose or ^{15}N-ammonium chloride. The strength of cell-free expression is the ability to achieve specific labeling patterns due to control over the added amino acids and reduced metabolic scrambling (Reckel et al., 2010). Sophisticated labeling schemes facilitate the assignment of soluble as well as of membrane proteins (Hein et al., 2017; Löhr et al., 2012, 2015) and are described in detail in a recent review (Hoffmann, Löhr, Laguerre, Bernhard, & Dötsch, 2018). Moreover, residual amino acid scrambling can be addressed by the addition of distinct enzyme inhibitors (LaGuerre, Löhr, Bernhard, & Dötsch, 2015).

4.2 Customizing Cell-Free Expression Conditions

Since proteins fulfill numerous tasks in different environments, they exhibit a huge variety in structure and function. This diversity generates different demands on their surrounding environment or their interaction with other molecules. As protein folding depends on environmental factors, the open nature of cell-free systems is often a valuable benefit for the optimization of sample conditions (Fig. 2). For example, the addition of specific inhibitors (Laguerre et al., 2016) or cofactors (Waberer et al., 2017) can significantly enhance the general folding and stability of produced proteins. Furthermore, the correct formation of disulfide bridges, crucial for particular proteins, relies on redox conditions that in cell-free systems can be nicely adjusted to individual folding requirements (Rues et al., 2018). Increased concentrations of chaperones and other stabilizers can have a positive effect on protein folding (Foshag et al., 2018). While many diverse additives could be added to cell-free reactions, several issues need to be considered:

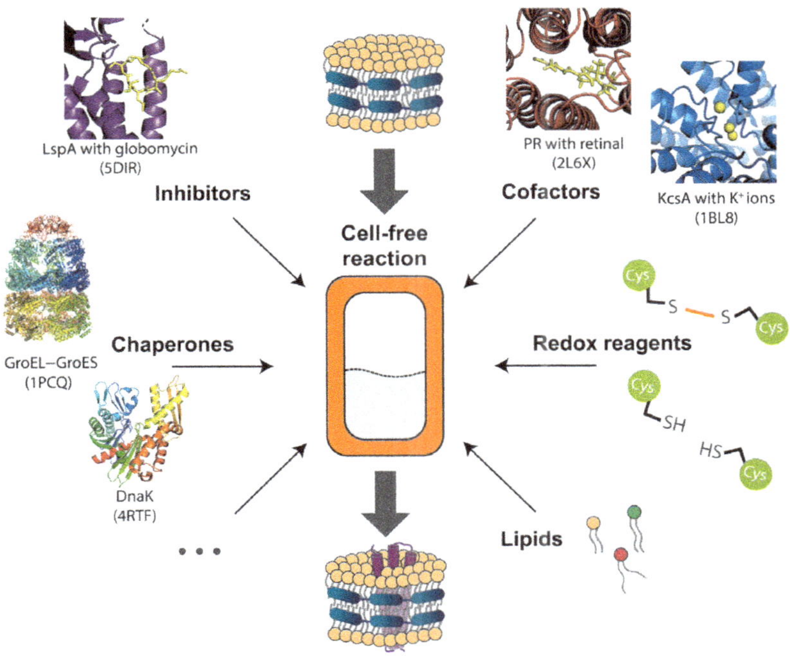

Fig. 2 Customizing cell-free expression conditions for improved sample quality. Protein structures are displayed with the corresponding PDB code. Cell-free device, lipids, nanodiscs, and redox systems are illustrated with pictograms.

1. The additive should be prepared in high stock concentrations as the available volume in the reaction is limited.
2. The pH of the additive should be similar to the pH in the reaction. Furthermore, the additive should not contain organic solvents. Exceptions can be ethanol or DMSO below 2% final concentration.
3. Potential negative effects of any new additive on the system should be tested in advance. Test expression screens of a reporter protein like GFP or luciferase, which enable a quick readout and covering different concentration ranges of the additive, are recommended.
4. It should be carefully considered if the additive can interfere with certain substances from the system. For example, Zn^{2+} supplementation requires the substitution of DTT with another reducing agent (e.g., TCEP; Waberer et al., 2017). In those cases the system has to be adapted before the additive tests.

4.3 Sample Purification

For NMR experiments, the purification of the target protein is necessary even in the reduced cell-free lysate proteome. In general, it is feasible to do NMR experiments directly from the RM of cell-free systems (Guignard, Ozawa, Pursglove, Otting, & Dixon, 2002). However, there is a high possibility that the resolution of the resulting spectra is decreased due to the viscosity of the RM and additional signals, which can arise due to conversion of labeled amino acids into other metabolites. To achieve suitable purity of membrane protein/nanodisc complexes, a one-step affinity purification by utilizing, e.g., His_6- or StrepII-tags is sufficient. As nanodiscs have a His_6-tag, using a different purification tag at the inserted membrane protein would remove empty discs (Fig. 1). Using StrepII-tagged membrane proteins is straightforward, while removing the His_6-tag from the nanodiscs by cleavage with TEV-protease would be an alternative option. The removal of empty discs from the sample significantly improves the resulting NMR spectra. In case of residual soluble aggregates, a final size-exclusion chromatography step can be added.

Suitable buffers for NMR experiments should have lower pH values to minimize signal loss due to amide protein exchange and low salt concentrations to ameliorate dielectric loss effects on the probe (e.g., 20 m*M* sodium acetate pH 4.0; 25 m*M* Bis–Tris pH 6.5; 20 m*M* potassium phosphate pH 7.2 (Laguerre et al., 2016; Peetz et al., 2017)). Exchange to such buffers can be done upon size-exclusion chromatography, by dialysis or during sample concentration by centrifugation in Amicon tubes with an MWCO of

10 kDa. Here, the concentration gradient within the tube should be considered. To prevent too high concentrations of the protein at the bottom of the tube, short centrifugation times (e.g., 5 min) should be applied, followed by removal of the filtrate and mixing of the concentrated solution. Buffer exchange during concentration can be achieved by multiple dilutions of the concentrated sample with the new buffer and subsequent concentration. At this state it can be highly beneficial to first screen various buffers with gradually lower pH for their suitability.

5. BASIC NMR SAMPLE PROCESSING AND ANALYSIS

Structure determination of membrane proteins by solution NMR is a general challenge and most so far studies were performed with proteomicelles (Liang & Tamm, 2016). While detergent micelles frequently represent a stabilizing environment similar to membranes (Hagn et al., 2013), a variety of membrane proteins strictly require lipid contacts (Henrich et al., 2016). Moreover, detergents suitable for solution-state NMR may drastically alter protein dynamics (Dehez et al., 2017; Kurauskas et al., 2018). The previous chapters focused on the workflow of sample preparation based on cell-free expression in the presence of supplied nanodiscs, with the option of subsequent bicelle formation by reverse q-titration as outlined in the next paragraph. The experiments described in Sections 5–7 can be performed with either bicelles or nanodiscs and are independent of the employed sample preparation procedure. Conventional cell-based expression systems might thus be considered if a membrane protein is sufficiently synthesized, can efficiently be labeled with stable isotopes, and tolerates detergent extraction from the membrane. An overview on the different sample production strategies is given in Table 1.

5.1 Environment and Interactions

Structure, function, and stability of membrane proteins are drastically modulated by their hydrophobic environment, making the identification of a suitable solvation agent a crucial task. While in the past usually a variety of different detergents were only screened for giving good spectral resolution, the increasing variety of solvation agents now includes also other membrane mimetics or amphipols (Etzkorn et al., 2013). Any hydrophobic environment must be evaluated for its spectral resolution (Etzkorn et al., 2013; Frey, Lakomek, Riek, & Bibow, 2017), but effects on structural state and functional behavior should be assessed as well by orthogonal studies like activity assays (Etzkorn et al., 2013; Laguerre et al., 2016). For the spectral assessment, ^{15}N HSQC experiments are commonly performed to define the general number of signals and their distribution (Fig. 3A). This type of experiments is highly advantageous as it can be performed with even quite low sample concentrations in the order of 100 μ*M*. Unfavorable line broadening of membrane protein samples can be addressed by applying transverse relaxation-optimized spectroscopy (TROSY)-type experiments (Pervushin, Riek, Wider, & Wüthrich, 1997). For the acceleration of the data collection the band-selective excitation short-transient (BEST) pulse sequence should be applied as well (Schanda, Van Melckebeke, & Brutscher, 2006).

Classical chemical shift studies will identify protein/ligand interactions upon ligand titration. The spectral comparison of apo and holo protein can reveal information about binding sites, protein dynamics, and conformational changes (Fig. 3B). If the chemical shift dispersion of the spectra is sufficient, these experiments can be performed on fully labeled samples (Laguerre et al., 2016). Using nanodiscs can involve the problem of signal overlap due to the size of the particle. In these cases, a smaller bicelle environment, that still comprises a lipid bilayer, can be generated by reverse *q*-titration (Laguerre et al., 2016). This is achieved by the stepwise titration of detergents like DPC or 1,2-diheptanoyl-*sn*-glycero-3-phosphocholine (DH^7PC) into the nanodisc sample, which breaks up the nanodisc particle, leaving particles with a smaller size and thus with faster tumbling. The progress of signal improvement and the lipid/detergent concentrations can be monitored by TROSY and 1D spectra, respectively. To prevent a substantial dilution of the sample, detergent stock concentrations should be around 10% (w/v). For elevated molecular weight proteins, which usually display an even more severe signal overlap, amino acid-type-selective labeling, potentially also focusing on methyl groups by methionine (Kofuku et al., 2012; Nygaard et al., 2013) or valine (Isogai et al., 2016), represents another possibility to simplify the analysis. Extremely large complexes like GPCRs incorporated into nanodiscs will require in addition the production of deuterated samples (Kofuku et al., 2014).

5.2 ^{31}P NMR

Like nanodiscs, bicelles are small disc-like membrane patches that in contrast to the former are not stabilized by an amphipathic protein, but by detergent molecules. They are typically formed by short-chain lipids such as

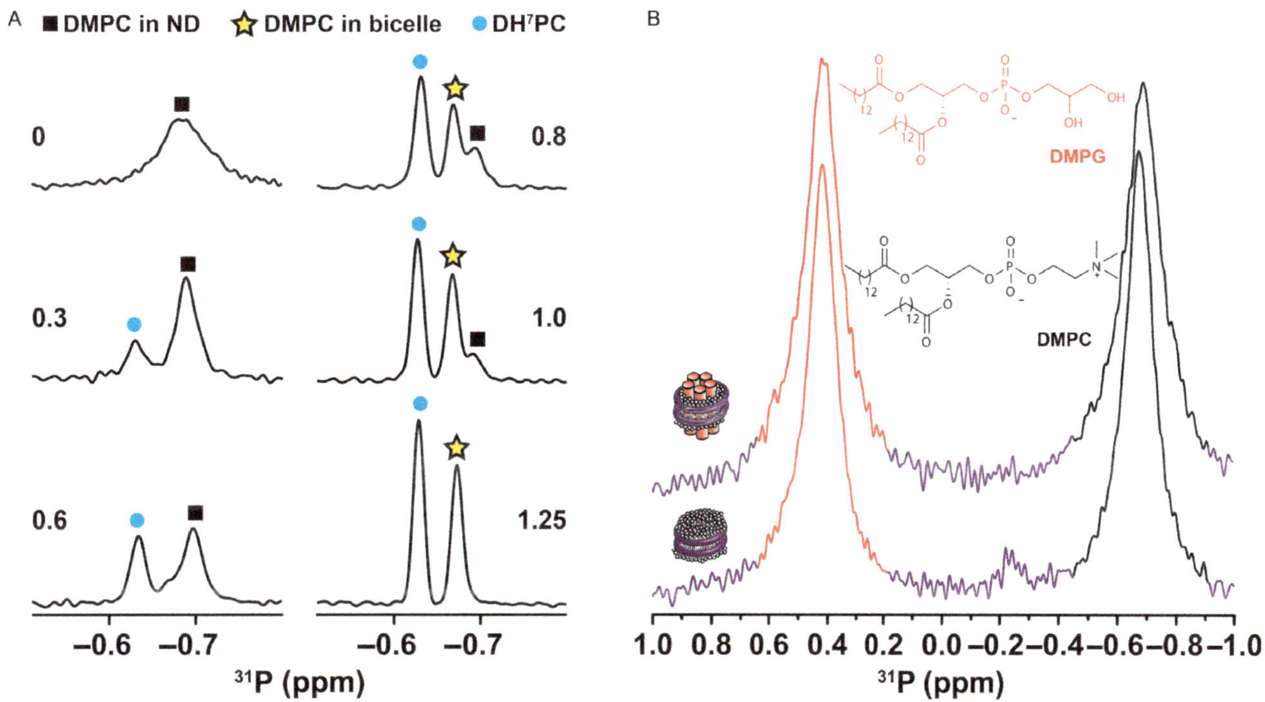

Fig. 3 $^{15}N,^{1}H$-TROSY of LspA in different conditions. The influence of different hydrophobic environments on LspA is demonstrated in (A). In (B) the stabilization of LspA by the inhibitor globomycin is shown. The color-codes of the spectra are indicated in the legends. *Adapted from Laguerre, A., Löhr, F., Henrich, E., Hoffmann, B., Abdul-Manan, N., Connolly, P. J., et al. (2016). From nanodiscs to isotropic bicelles: A procedure for solution nuclear magnetic resonance studies of detergent-sensitive integral membrane proteins.* Structure, *24(10), 1830–1841.*

1,2-dimyristoyl-*sn*-glycero-3-phosphocholine (DMPC) or 1,2-dimyristoyl-*sn*-glycero-3-phospho-(1-rac-glycerol) (DMPG) with DH^7PC, DPC or 3-[(cholamidopropyl)dimethylammonio]-1-propanesulfonate (CHAPS) ensuring the solubilization. In first experiments, the size and composition of a desired bicelle are adjusted by defining the optimal ratio between lipids and detergents, the so-called *q*-value (1):

$$q = \frac{[\text{lipid}]}{[\text{detergent}]} \tag{1}$$

In the case of bicelle generation from membrane protein/nanodisc complexes, here referred as "reverse *q*-titration," we use the reciprocal *q*-value (2):

$$q^{-1} = \frac{[\text{detergent}]}{[\text{lipid}]} \tag{2}$$

At known lipid concentrations, the *q*-value can easily be adjusted by supplying appropriate detergent concentrations. However, absolute concentrations are often unknown and then NMR analysis offers an easy tool for the determination of the *q*-value by detecting the naturally NMR active phosphorus nucleus (^{31}P) of lipids in 1D-^{31}P experiments. Similarly, phosphate-containing detergents like DH^7PC or DPC can be detected (Fig. 4A). By calculating the integrals of NMR signals arising from either lipid or detergent the *q*-value can be defined. Due to the large chemical shift anisotropy (CSA) contribution to transverse relaxation, best resolution is often obtained with low field spectrometers. In case of phosphate-free detergents such as cholesterol or CHAPS, the NMR quantification, e.g., by detection of ^{13}C nuclei, can be problematic due to the possible overlap with protein resonances. Peak overlap might also prevent the distinction between DPC or DH^7PC and particular lipids. Here, a calculation of the added amounts is indispensable. Interestingly, in the case of reverse *q*-titration the transition of nanodiscs into bicelles can be followed and estimated by a correlated peak splitting and downshift of the lipid signals (Laguerre et al., 2016). By quantifying the lipid amount inside nanodiscs before and after the cotranslational insertion of a membrane protein, the release of lipids from the bilayer and the membrane area occupied by the inserted protein can be estimated (Peetz et al., 2017). The NMR quantification of ^{31}P-containing lipids can therefore serve as a valuable future tool to characterize lipid mixtures, membrane protein dimensions, or general changes of nanoparticle topologies (Fig. 4B).

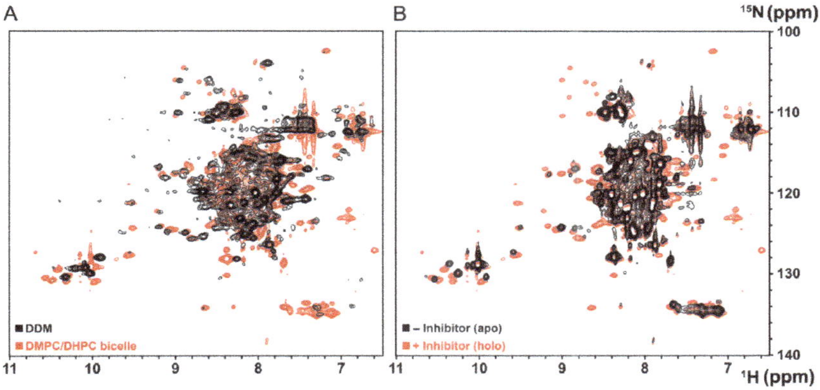

Fig. 4 1D-^{31}P NMR of nanodiscs. 1D-NMR spectra of the reverse *q*-titration with the detergent DH^{7}PC of nanodiscs assembled with MSP1D1ΔH5 and DMPC are shown in (A). Peaks of different species are highlighted with symbols (see legend) and *q*-ratios are displayed. (B) 1D-NMR spectra of nanodiscs formed by MSP1D1ΔH5 with an equimolar mixture of DMPG and DMPC before (*lower curve*) and after the insertion of proteorhodopsin (*upper curve*). The chemical structures of the lipids are displayed and the corresponding peaks are highlighted in *matched colors*. Pictograms illustrate the analyzed particles. Spectra (A, B) were recorded at 202 MHz and a sample temperature of 318 K on a Bruker DRX500 spectrometer equipped with a broadband inverse probe. *Panel (A): Adapted from Laguerre, A., Löhr, F., Henrich, E., Hoffmann, B., Abdul-Manan, N., Connolly, P. J., et al. (2016). From nanodiscs to isotropic bicelles: A procedure for solution nuclear magnetic resonance studies of detergent-sensitive integral membrane proteins.* Structure, 24*(10), 1830–1841.*

5.3 TRACT

The definition of particle size is highly desirable as significant variations exist among membrane proteins due to attached hydrophobic molecules (Ilgü et al., 2014). Therefore, TRACT (TROSY for rotational correlation times) measurements giving a fast approximation of rotational correlation times represent a perfect tool (Lee, Hilty, Wider, & Wüthrich, 2006). It should, however, be noted that the true correlation time of membrane protein can be severely underestimated if the entire amide region is chosen for integration in 1D spectra, because already a small number of mobile residues or side-chain signals may contribute a large portion to the total integrals. In favorable cases, e.g., with β-barrel proteins or α-helical proteins featuring well-dispersed amide signals, suitable regions may be selected easily.

Alternatively, selectively labeled samples may be used for 1D TRACT analysis, allowing it to obtain residue-specific τ_c values from resolved signals (Etzkorn et al., 2013). Although more time consuming, 2D TRACT versions are in general advantageous in this respect as they allow single signals or groups of signals from structured regions of the protein to be selected for integration in a straightforward manner using uniformly labeled samples (Laguerre et al., 2016). Here, differential observations for single amino acids can be made that provide single residue mobility information allowing for estimations regarding membrane partitioning or involvement in specific protein motions.

6. BACKBONE ASSIGNMENT OF MEMBRANE PROTEINS

The unfortunate coincidence of inherently low chemical shift dispersion and slow rotational reorientation of membrane proteins—in particular in lipid-like environments—renders their resonance assignment by solution NMR techniques challenging (Kim, Howell, Van Horn, Jeon, & Sanders, 2009; Maslennikov & Choe, 2013; Sanders & Sönnichsen, 2006; Viegas, Viennet, & Etzkorn, 2016). Routinely used 3D experiments for backbone assignment rely on matching ^{13}C chemical shifts observed in intra- and interresidual correlations with amide ^{1}H and ^{15}N resonances. This procedure is severely compromised by spectral crowding in the NH plane as well as ^{13}C chemical shift degeneracy caused by the prevalence of one secondary structure type and the high abundance of certain amino acid types in transmembrane regions (Reckel et al., 2008). Long rotational correlation times entail fast transverse relaxation which translates into broad resonances and inefficient magnetization transfer. In addition, exchange between different environments and interconverting conformations at a timescale in the order of the frequency difference of the corresponding resonances may lead to excessive line broadening.

Although the effect of high molecular mass of samples is alleviated to some extent by the use of TROSY, perdeuteration at nonexchangeable sites, and high measurement temperatures, the performance of crucial experiments, such as HNCACB (Wittekind & Mueller, 1993), HN(CO)CACB (Yamazaki, Lee, Arrowsmith, Muhandiram, & Kay, 1994), or HN(CA)CO (Clubb, Thanabal, & Wagner, 1992) that involve ^{13}C–^{13}C transfer steps, is severely hampered for ordered residues in membrane proteins (Etzkorn et al., 2013; Laguerre et al., 2016; Löhr et al., 2012, 2015; Schubert, Kolbe, Kessler, Oesterhelt, & Schmieder, 2002). On the other hand, exclusively relying on C^{α} connectivities resulting from the more sensitive HNCA (Ikura, Kay, & Bax, 1990) and HN(CO)CA (Bax & Ikura, 1991) experiments is usually insufficient and requires additional information to resolve ambiguities. In principle such information can be obtained from amino-acid-type-selective triple-resonance experiments (Dötsch, Oswald, & Wagner, 1996; Feuerstein, Plevin, Willbold, & Brutscher, 2012; Pantoja-Uceda & Santoro, 2008; Schubert, Smalla, Schmieder, & Oschkinat, 1999) which are, however, unsuitable for large proteins due to their multistep coherence transfer or editing elements. Alternatively, amino acid-type-specific isotope labeling (Griffey, Redfield, Loomis, & Dahlquist, 1985; LeMaster & Richards, 1985; Muchmore, McIntosh, Russell, Anderson, & Dahlquist, 1989; Senn, Eugster, Otting, Suter, & Wüthrich, 1987) provides valuable assignment information in a very sensitive manner and leads to substantial spectral simplifications. The simultaneous use of both, ^{15}N- and 1-^{13}C (carbonyl)-labeled amino acid types, results in sequence-specific backbone NH assignments for unique amino acid pairs by exploiting the scalar ^{13}C–^{15}N coupling across the peptide bond (Griffey, Redfield, McIntosh, Oas, & Dahlquist, 1986; Guignard et al., 2002; Ikura, Krinks, Torchia, & Bax, 1990; Kainosho & Tsuji, 1982; Takahashi et al., 1991; Tate et al., 1992; Weigelt, van Dongen, Uppenberg, Schultz, & Wikström, 2002; Westler et al., 1988; Yabuki et al., 1998). Since the latter interaction is relatively strong and can be detected in a straightforward manner, this assignment strategy has been applied for larger proteins (Kato et al., 1989; Rule, Tjandra, Simplaceanu, & Ho, 1993). On the downside, the use of these strategies to identify all residue types or even all amino acid pairs occurring in a protein would require preparation of a prohibitively large number of labeled samples. The same goal can, however, be achieved with considerably higher efficiency using combinatorial methods (Hefke et al., 2011; Parker, Aulton-Jones, Hounslow, & Craven, 2004; Shortle, 1994; Staunton, Schlinkert, Zanetti, Colebrook, & Campbell, 2006; Wu et al., 2006).

Combinatorial amino acid-selective labeling (see Hoffmann et al., 2018; Jaipuria, Krishnarjuna, Mondal, Dubey, & Atreya, 2012 for recent reviews) employs a limited number of samples, each containing different subsets of labeled amino acids rather than a single labeling type per sample. In an early application of dual-selective combinatorial ^{15}N- and 1-^{13}C-labeling to a detergent-solubilized membrane protein (Trbovic et al., 2005), ^{15}N HSQC and 2D HN(CO) spectra recorded on a set of three samples provided amino-acid-type information and sequence-specific assignments that served as "anchor points" to facilitate the classical 3D-based approach, finally resulting in 85% overall backbone assignment. As mentioned earlier, some important 3D triple-resonance experiments fail for large protein/lipid complexes. It is therefore desirable to increase the information content of a combinatorial labeling scheme such that finally only short gaps in the protein sequence have to be filled that can be assigned with the most sensitive 3D experiments recorded on a uniformly labeled sample. Three options are conceivable to achieve this goal: (1) increasing the number of samples, (2) use of multiple ^{15}N and ^{13}C enrichment levels, and (3) introducing more isotopomerically distinct dipeptide species. The first approach has been successfully employed to assign a number of proteins containing two to four transmembrane helices with the help of up to eight dual-selectively labeled samples (Maslennikov et al., 2010). The second method allows the distinction of all 19 nonproline amino acid types with as few as 3 samples and requires evaluation of signal intensities to decode enrichment levels (Kasai, Koshiba, Yokoyama, & Kigawa, 2015). Taking into account the large dynamic range of signal intensities often observed in membrane protein spectra, the latter issue may be critical at some instances. Implementation of the third option can be achieved by labeling amide nitrogens, carbonyl carbons, and α-carbons individually as well as simultaneously in different amino acid types of the same sample (Löhr et al., 2012, 2015). In this combinatorial triple-selective labeling strategy, sample preparation efforts are reduced at the expense of a larger variety of 2D ^{1}H–^{15}N correlation experiments that need to be carried out per sample in order to distinguish more amino-acid-type pairs based on their distinct labeling patterns. Nevertheless, the required set of spectra can usually be acquired in a reasonable amount of time especially when applying time-shared pulse sequences (Hein et al., 2017; Löhr et al., 2014). It should however be noted that despite the higher sensitivity of the latter experiments compared to 3D triple-resonance experiments comprising ^{13}C evolution times, application to slowly tumbling protein complexes still requires the use of deuterated amino acids (Laguerre et al., 2016).

7. NMR STRUCTURAL APPROACHES FOR MEMBRANE PROTEINS IN LIPID BILAYERS

7.1 Distance Restraints

The most important source of structural information of biomolecules in solution NMR is the nuclear Overhauser effect (NOE). For slowly tumbling proteins the collection of NOE-based distance constraints is to a large extent confined to methyl groups. Together with aromatic rings these are involved in forming the hydrophobic cores, and in membrane proteins methyl-containing amino acids are highly abundant in transmembrane regions. Being located at the end of the side chains methyl groups feature increased local mobility compared to other side-chain or backbone positions, thus diminishing linewidth limitations. In addition, fast rotation around the carbon–carbon bond makes the methyl protons chemically equivalent, resulting in threefold enhanced signal intensities. Spectroscopically, significant line narrowing can be achieved by virtue of cancellation of intramethyl ^{1}H–^{1}H and ^{1}H–^{13}C dipolar relaxation mechanisms (Ollerenshaw, Tugarinov, & Kay, 2003; Tugarinov, Hwang, Ollerenshaw, & Kay, 2003) when using of HMQC-type pulse sequences (methyl-TROSY). The effect is most pronounced when the influence of external dipolar relaxation sources is minimized such as in side chains containing a single-protonated, ^{13}C-labeled methyl group, whereas all other sites are deuterated but not ^{13}C labeled (Tugarinov & Kay, 2004). At the same time, such a labeling scheme eliminates signal broadening due to passive scalar couplings. Most importantly, the absence of $^{1}J_{CC}$ couplings allows for enormous resolution gains without reverting to low-sensitivity constant-time methods. Protein sample production involves either incorporation of methyl-labeled amino acids (Miyanoiri et al., 2013) in *E. coli* cellular or cell-free expression systems or, more commonly, α-ketoacid precursor-based cellular expression using deuterated minimal media (Gardner & Kay, 1997; Goto, Gardner, Mueller, Willis, & Kay, 1999; Hagn & Wagner, 2015; Lichtenecker, Coudevylle, Konrat, & Schmid, 2013; Tugarinov, Kanelis, & Kay, 2006). High label costs and insufficient precursor conversion can be addressed by a recently developed new strategy based on the methyl side-chain labeling of proteins by cell-free synthesis using amino acid precursors (Lazarova et al., 2018). Special *E. coli* lysates enriched with the branched chain aminotransferase IlvE showed almost complete conversion of the labeled precursors 2-ketoisovaleric acid (KIV) or 4-methyl-2-oxovalerate (MOV) to methyl-labeled L-Val or L-Leu. Besides a tremendous reduction in label costs, additional advantages are the options to exclusively label either L-Val or L-Leu methyl residues, the lack of any precursor scrambling and to preserve high production efficiencies of even complex membrane proteins.

Taking into account the above considerations the most useful setup for the measurement of NOEs in membrane protein/lipid complexes is the application of HMQC–NOESY–HMQC-type pulse sequences (Clore, Kay, Bax, & Gronenborn, 1991; Frenkiel, Bauer, Carr, Birdsall, & Feeney, 1990; Kay, Clore, Bax, & Gronenborn, 1990; Vuister et al., 1993; Zuiderweg, Petros, Fesik, & Olejniczak, 1991) to selectively ^{15}N- and ^{1}H/^{13}C-methyl-labeled samples in a ^{2}H/^{12}C background—a combination which provides the highest achievable sensitivity and resolution. The purpose of ^{15}N labeling is the sequence-specific assignment of methyl resonances by virtue of intraresidual amide-to-methyl NOE contacts detected in $^{15}N,^{13}C$-separated HMQC–NOESY–HMQC variants. Here, it should be noted that for slowly tumbling molecules the [^{1}H–^{15}N]-HMQC module takes advantage of relaxation interference between ^{1}H CSA and ^{1}H–^{15}N dipolar mechanisms ("CRINEPT–HMQC–[^{1}H]-TROSY" (Riek, Wider, Pervushin, & Wüthrich, 1999)) allowing for the use of magnetization transfer periods that are considerably shorter than the nominal $1/(2 * {}^{1}J_{NH})$ duration, thus minimizing relaxation losses. In a second set of experiments, $^{13}C,^{13}C$-separated HMQC–NOESY–HMQC spectra are acquired in order to obtain interresidue methyl–methyl distance information. Although more time consuming, it may be advisable to record 3D (two experiments incorporating either ^{1}H or ^{13}C evolution times before NOE mixing) as well as 4D (with both evolution times) versions. The 3D versions are more sensitive and can be recorded with the conventional, full sampling method. Therefore, all NOE cross-peaks that can potentially be detected for a given system should be present in the 3D spectra. However, being spread along a single indirect dimension, only a small fraction of the donor methyl resonances can usually be assigned without prior structural information. Owing to the dispersion along two indirect dimensions, vastly removing overlap, and the fact that donor ^{1}H and ^{13}C chemical shifts are directly correlated, their four-dimensional counterparts can in principle provide NOE assignments in an unambiguous manner. On the other hand, lower sensitivity and possibly artifacts arising from restricted phase cycling or sampling (see later) may raise doubts about the reliability of an observed cross-peak. In such cases, it is helpful to verify the putative NOE by the presence of cross-peaks at the corresponding coordinates in the 3D spectra.

In order to exploit the full potential of resolution optimized methyl-selective labeling schemes, long acquisition times must be employed in all indirect dimensions of an NOESY experiment. In the case of 4D experiments, this would require unmanageable long periods of measurement time (in the order of months to years). Therefore, nonuniform sampling schemes (Holland, Bostock, Gladden, & Nietlispach, 2011; Mobli & Hoch, 2014; Orekhov, Ibraghimov, & Billeter, 2003; Sun, Frueh, Selenko, Hoch, & Wagner, 2005; Zawadzka-Kazimierczuk, Kazimierczuk, & Koźmiński, 2010), where only a few percent of the full time-domain data grid is actually sampled, are now routinely used to obtain high-resolution 4D NOESY spectra (Hiller, Ibraghimov, Wagner, & Orekhov, 2009; Luan, Jaravine, Yee, Arrowsmith, & Orekhov, 2005; Orekhov & Jaravine, 2011; Reckel et al., 2011; Stanek et al., 2013; Werner-Allen, Coggins, & Zhou, 2010).

Moderate sensitivity gains can sometimes be realized by sampling more data points at short evolution time values, where intensities are comparatively high, and fewer points are present at later stages of the interferograms (Barna & Laue, 1987; Rovnyak, Hoch, Stern, & Wagner, 2004; Schmieder, Stern, Wagner, & Hoch, 1993).

Significant time savings and/or sensitivity improvements in HMQC-type experiments can also be accomplished by the use of band-selective proton pulses in combination with small flip angles (SOFAST-HMQC), as demonstrated for amide ^{1}H–^{15}N (Schanda & Brutscher, 2005) as well as for methyl ^{1}H–^{13}C correlations (Amero et al., 2009). Likewise, HMQC–NOESY–HMQC experiments can be implemented to benefit from the underlying longitudinal relaxation enhancement (Rossi, Xia, Khanra, Veglia, & Kalodimos, 2016). In our experience application of such pulse sequences to membrane proteins often suffer from extensive t_1-noise in the aliphatic region, especially if nondeuterated lipids or detergents have been employed for NMR sample preparation. Such artifacts can however be avoided by the use of a gradient-echo/antiecho coherence selection scheme (Davis, Laue, & Keeler, 1991). Although this modification degrades sensitivity by a theoretical factor of $2^{1/2}$, it is finally the signal-to-artifact ratio that is more relevant for obtaining reliable NOE information than the signal-to-(white) noise ratio.

Besides typical distance restraints generated by NOEs the introduction of spin probes, usually by coupling of paramagnetic spin labels to specific cysteine mutants, can deliver additional information. These paramagnetic relaxation enhancement (PRE) experiments can be complementary to NOE information, provide distance information in the range of 15–24 Å (Liang, Bushweller, & Tamm, 2006), and have been used for structural determinations of membrane proteins from various types (Hagn et al., 2013; Reckel et al., 2011). The efficient incorporation should be checked by MS or EPR spectroscopy to guarantee uniform labeling. Unlabeled protein would decrease the overall PRE effect and thus lead to inaccurate distance restraints. For the visualization of the PRE effect [^{15}N,^{1}H]-TROSY–HSQC spectra of diamagnetic and paramagnetic labeled proteins are recorded and compared. Signals of residues affected by the paramagnetic probe are reduced compared to diamagnetic conditions and the distance of the residue toward the probe can be back calculated from the magnitude of signal reduction. To exclude intermolecular relaxation these experiments should be repeated at different concentrations (Liang et al., 2006). With respect to α-helical membrane proteins one spin-labeled mutant for every transmembrane helix should be generated to obtain enough distance restraints from PRE data for structure calculations (Gottstein, Reckel, Dötsch, & Güntert, 2012).

7.2 Transmembrane Partitioning

Another interesting application of the PRE effect is the assessment of the solvent PRE for single residues of the membrane protein that is inserted into a bicelle or a nanodisc (Piai et al., 2017). [^{1}H,^{15}N]-TROSY–HSQC spectra are recorded with increasing concentration (e.g., from 0 to 20 mM in the case of Gd-DOTA) of a paramagnetic agent and the reduction of the peak signals is compared. If the paramagnetic agent is water-soluble and membrane impenetrable (e.g., Gd-DOTA) the signal reduction gradually decreases from the solvent-accessible parts of the membrane protein to the membrane center. Thereby, the membrane protein can be localized with respect to the center of the membrane (Piai et al., 2017) and it can provide information about membrane thickness and the membrane partitioning of a multidomain protein (Chen et al., 2018). A further prerequisite is that the bicelles or nanodiscs are of a sufficient size to prevent lateral contributions from the surrounding paramagnetic field to the relaxation (Piai et al., 2017), but most nanodiscs (Frey et al., 2017; Hagn et al., 2013; Kofuku et al., 2014) or bicelles (Laguerre et al., 2016; Morrison et al., 2012) offer this feature. In a similar fashion using spin-labeled lipids, this procedure can also be used to characterize the amino acids located within a micelle or other hydrophobic environment (Fig. 5; Sobhanifar et al., 2010).

7.3 Residual Dipolar Couplings

In general, it is beneficial to gain as much information as possible for structure calculations and therefore residual dipolar couplings could deliver additional restraints. Therefore, coupling constants under isotropic and anisotropic conditions [^{15}N^{1}H]-TROSY–HSQC and [^{15}N^{1}H]-COCAINE (Lee, Vögeli, & Pervushin, 2005) experiments are measured under unaligned and aligned conditions. As an alignment medium the bacteriophage Pf1 can be used since controls revealed no significant unspecific interaction between the discs and the phage (Bibow et al., 2014). To exclude possible interactions with a membrane protein/nanodisc or bicelle complex, the correlation times before and after addition of Pf1 should be checked by TRACT measurements (Bibow et al., 2014). If the signal overlap becomes severe, RDC measurements by HNCO experiments could be performed (Permi, Rosevear, & Annila, 2000).

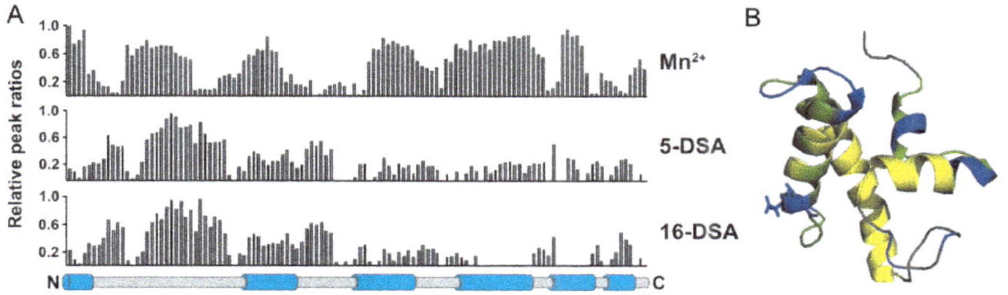

Fig. 5 Determination of protein surface accessibility by the application of paramagnetic reagents. Peak height ratios of resonances demonstrate the broadening effect by the supplementation of different paramagnetic reagents (A). Mn^{2+} indicates surface accessibility and DSA reagents identify micelle insertion. Due to the different position of the paramagnetic group at DSA, these reagents give also information about the penetration depth. The schematic representation at the bottom demonstrates the structural arrangement of CTF with helices highlighted in *blue*. In (B) the information about surface accessibility is mapped onto the CTF structure where *blue* indicates surface accessibility, *green* partially accessible regions, and *yellow* nonaccessible residues. *Adapted from Sobhanifar, S., Schneider, B., Lohr, F., Gottstein, D., Ikeya, T., Mlynarczyk, K., et al. (2010). Structural investigation of the C-terminal catalytic fragment of presenilin 1.* Proceedings of the National Academy of Sciences of the United States of America, 107*(21), 9644–9649.*

7.4 Lipid-NOEs

As lipids play a crucial role in connection with membrane proteins the elucidation of the lipid contact is of high interest (Saliba et al., 2015), especially when considering that detergent contact can drastically influence functionality and dynamics. For a first estimation if the protein is in direct contact with lipids or detergent molecules in the desired condition [^{1}H,^{15}N]-TROSY–HSQCs of the membrane protein in different environments can be compared. For example, if the resonances of a sample in bicelles are more similar to a spectrum in nanodiscs rather then detergent, it can be assumed that the membrane protein is surrounded by lipids (Frey et al., 2017).

Another more specific methodology is the detection of specific protein–lipid NOEs. Two samples are required one with conventional, protonated lipids and one with deuterated lipids. With a 3D NOESY–[^{15}N,^{1}H]-TROSY experiment, signals are visible in the sample with protonated lipids and should disappear in the sample with deuterated lipid (Fig. 6A; Laguerre et al., 2016). Within lipid mixtures or bicelles the differentiation between the lipid and detergent species could be achieved by multiple experiments with varying deuteration of the individual compounds. Thus, the presence of lipid molecules next to the membrane protein, often called the annual lipid belt,

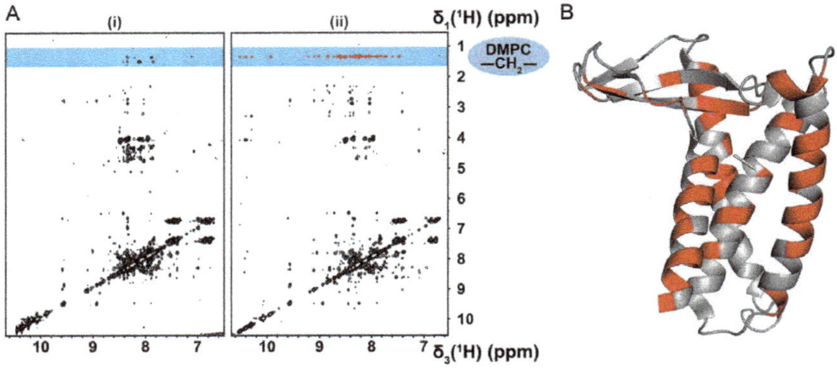

Fig. 6 Lipid-NOE measurements of LspA in bicelles after reverse q-titration. (A) NOESY–^{15}N–^{1}H-TROSY spectra of LspA. The *blue section* highlights the resonance area of the specific lipid NOEs. In (i) deuterated lipid and protonated detergent were used and in (ii) the lipid is protonated and the detergent deuterated. The residues that show NOEs to lipids are displayed in *red* on the crystal structure of LspA (PDB: 5DIR) (B). *Panel A, adapted from Laguerre, A., Löhr, F., Henrich, E., Hoffmann, B., Abdul-Manan, N., Connolly, P. J., et al. (2016). From nanodiscs to isotropic bicelles: A procedure for solution nuclear magnetic resonance studies of detergent-sensitive integral membrane proteins.* Structure, *24(10), 1830–1841.*

can be detected and lipid contact sites can be mapped (Fig. 6B; Laguerre et al., 2016). Here, also information about the membrane partitioning can be extracted by assigning the residues presenting lipid NOEs (Chen et al., 2018). This technique can also be used for verifying specific interactions of the membrane protein with a certain lipid on a certain position (Zhao, Wang, Run, OuYang, & Chou, 2016). Since the chemical environment of methylene groups in acyl chain moieties of different lipids or detergents is quite similar, multiple samples with differential deuteration of the individual components are required as well.

8. OUTLOOK

With perspectives of EM resolution revolution or in meso crystallization the role for NMR in structure determination of membrane proteins becomes smaller. On the other hand, solution NMR is the method of choice to address questions of protein dynamics in solution (Liang & Tamm, 2016; Palmer, 2004), which is problematic with static techniques. Taken into account that the role of lipids for membrane proteins with respect to complex integrity (Gupta et al., 2017; Henrich et al., 2017), modulation of dynamics (Martens et al., 2016), or regulation (Cong, Liu, Liu, Liang, & Laganowsky, 2017) gets into the focus of current research, it is beneficial that bilayer environments like nanodiscs or bicelles become compatible with solution-state NMR. By detecting the overall characteristics in bilayers or distinct interactions with lipids by, for example, protein–lipid NOEs, NMR provides manifold opportunities for the analysis of membrane proteins. Here, the combination with other techniques such as native mass spectrometry (Barrera & Robinson, 2011) could be highly beneficial for the identification of the interplay with specific lipids. As lipid mixtures get more and more popular the differentiation between different lipid species could be achieved by directed delipidation strategies (Bechara et al., 2015).

The problematic increase in sample particle size when using bilayer environments for solution NMR will be addressed in future by continuously probing new hydrophobic environments. Already for a few years nanodiscs based on chemical polymers (SMALPs) are evaluated especially for solid-state NMR (Bersch, Dörr, Hessel, Killian, & Schanda, 2017; Orwick-Rydmark et al., 2012), but due to their low pH and salt tolerance as well as their still large size not applicable to solution NMR so far. However, the continuous modification of the polymers probably will enable their use in the future in solution NMR as well (Fiori, Jiang, Altenberg, & Liang, 2017; Ravula, Ramadugu, Di Mauro, & Ramamoorthy, 2017). For enhanced spectral resolution of membrane proteins inserted into nanodiscs the Wagner lab recently proposed an approach to covalently circulate the MSP scaffold protein, thus limiting the size variation of resulting nanodiscs (Nasr et al., 2017). Alternative nanodisc-like environments having a scaffold not based on Apo-A1 derivatives but on the lipid-binding protein saposin emerged recently (Frauenfeld et al., 2016). The advantage of such salipro particles is the increased flexibility in adjusting their size according to the dimension of the inserted membrane protein, while the required amount of lipids to solubilize the membrane protein is kept at a minimum. This reduces the overall size of the particle and their entry into solution-state NMR experiments is currently emerging (Chien et al., 2017).

Further improvements are expected in methods and costs for stable isotope labeling. Cell-free expression is already an efficient platform for the production of specifically labeled samples also due to the reduced scrambling (LaGuerre et al., 2015). With the recently gained information on the lysate proteome (Foshag et al., 2018), it is now possible to directly use and optimize residual metabolic pathways for the generation of specifically labeled amino acids (Lazarova et al., 2018). Also the efficient incorporation of nonnatural amino acids (Jones et al., 2010) with NMR active isotopes (Manglik et al., 2015) becomes feasible with cell-free expression approaches (Martin et al., 2018; Peuker et al., 2016).

ACKNOWLEDGMENTS

This work was funded by the Collaborative Research Center (SFB) 807 of the German Research Foundation (DFG), the Center for Biomolecular Magnetic Resonance (BMRZ) by the state of Hessen, and the Cluster of Excellence in Frankfurt (Macromolecular Complexes).

REFERENCES

Amero, C., Schanda, P., Durá, M. A., Ayala, I., Marion, D., Franzetti, B., et al. (2009). Fast two-dimensional NMR spectroscopy of high molecular weight protein assemblies. *Journal of the American Chemical Society*, *131*(10), 3448–3449.

Barna, J. C. J., & Laue, E. D. (1987). Conventional and exponential sampling for 2D NMR experiments with application to a 2D NMR spectrum of a protein. *Journal of Magnetic Resonance (1969)*, *75*(2), 384–389.

Barrera, N. P., & Robinson, C. V. (2011). Advances in the mass spectrometry of membrane proteins: From individual proteins to intact complexes. *Annual Review of Biochemistry*, *80*, 247–271.

Barrera, N. P., Zhou, M., & Robinson, C. V. (2013). The role of lipids in defining membrane protein interactions: Insights from mass spectrometry. *Trends in Cell Biology*, *23*(1), 1–8.

Bax, A., & Ikura, M. (1991). An efficient 3D NMR technique for correlating the proton and 15N backbone amide resonances with the alpha-carbon of the preceding residue in uniformly 15N/13C enriched proteins. *Journal of Biomolecular NMR*, *1*(1), 99–104.

Bayburt, T. H., Grinkova, Y. V., & Sligar, S. G. (2002). Self-assembly of discoidal phospholipid bilayer nanoparticles with membrane scaffold proteins. *Nano Letters*, *2*(8), 853–856.

Bayburt, T. H., & Sligar, S. G. (2003). Self-assembly of single integral membrane proteins into soluble nanoscale phospholipid bilayers. *Protein Science*, *12*(11), 2476–2481.

Bechara, C., Nöll, A., Morgner, N., Degiacomi, M. T., Tampé, R., & Robinson, C. V. (2015). A subset of annular lipids is linked to the flippase activity of an ABC transporter. *Nature Chemistry*, *7*(3), 255–262.

Bersch, B., Dörr, J. M., Hessel, A., Killian, J. A., & Schanda, P. (2017). Proton-detected solid-state NMR spectroscopy of a zinc diffusion facilitator protein in native nanodiscs. *Angewandte Chemie (International Ed. in English)*, *56*(9), 2508–2512.

Bibow, S., Carneiro, M. G., Sabo, T. M., Schwiegk, C., Becker, S., Riek, R., et al. (2014). Measuring membrane protein bond orientations in nanodiscs via residual dipolar couplings. *Protein Science*, *23*(7), 851–856.

Caffrey, M. (2015). A comprehensive review of the lipid cubic phase or in meso method for crystallizing membrane and soluble proteins and complexes. *Acta Crystallographica. Section F, Structural Biology Communications*, *71*(Pt. 1), 3–18.

Chen, W., Dev, J., Mezhyrova, J., Pan, L., Piai, A., & Chou, J. J. (2018). The unusual transmembrane partition of the hexameric channel of the hepatitis C virus. *Structure*, *26*(4), 627–634.e624.

Chien, C. H., Helfinger, L. R., Bostock, M. J., Solt, A., Tan, Y. L., & Nietlispach, D. (2017). An adaptable phospholipid membrane mimetic system for solution NMR studies of membrane proteins. *Journal of the American Chemical Society*, *139*(42), 14829–14832.

Clore, G. M., Kay, L. E., Bax, A., & Gronenborn, A. M. (1991). Four-dimensional 13C/13C-edited nuclear Overhauser enhancement spectroscopy of a protein in solution: Application to interleukin 1 beta. *Biochemistry*, *30*(1), 12–18.

Clubb, A. T., Thanabal, V., & Wagner, G. (1992). A constant-time three-dimensional triple-resonance pulse scheme to correlate intraresidue ^{1}H, ^{15}N, ^{13}C' chemical shifts in ^{15}N-^{13}C-labeled proteins. *Journal of Magnetic Resonance*, *97*, 213–217.

Cong, X., Liu, Y., Liu, W., Liang, X., & Laganowsky, A. (2017). Allosteric modulation of protein-protein interactions by individual lipid binding events. *Nature Communications*, *8*(1), 2203.

Davis, A. L., Laue, E. D., & Keeler, J. (1991). Absorption-mode two-dimensional NMR spectra recorded using pulsed field gradients. *Journal of Magnetic Resonance (1969)*, *94*(3), 637–644.

Dawaliby, R., Trubbia, C., Delporte, C., Masureel, M., Van Antwerpen, P., Kobilka, B. K., et al. (2016). Allosteric regulation of G protein-coupled receptor activity by phospholipids. *Nature Chemical Biology*, *12*(1), 35–39.

Dehez, F., Schanda, P., King, M. S., Kunji, E. R. S., & Chipot, C. (2017). Mitochondrial ADP/ATP carrier in dodecylphosphocholine binds cardiolipins with non-native affinity. *Biophysical Journal*, *113*(11), 2311–2315.

Denisov, I. G., Grinkova, Y. V., Lazarides, A. A., & Sligar, S. G. (2004). Directed self-assembly of monodisperse phospholipid bilayer nanodiscs with controlled size. *Journal of the American Chemical Society*, *126*(11), 3477–3487.

Denisov, I. G., & Sligar, S. G. (2016). Nanodiscs for structural and functional studies of membrane proteins. *Nature Structural & Molecular Biology*, *23*(6), 481–486.

Denisov, I. G., & Sligar, S. G. (2017). Nanodiscs in membrane biochemistry and biophysics. *Chemical Reviews*, *117*(6), 4669–4713.

Dötsch, V., Oswald, R. E., & Wagner, G. (1996). Amino-acid-type-selective triple-resonance experiments. *Journal of Magnetic Resonance. Series B*, *110*(1), 107–111.

Dürr, U. H., Gildenberg, M., & Ramamoorthy, A. (2012). The magic of bicelles lights up membrane protein structure. *Chemical Reviews*, *112*(11), 6054–6074.

Efremov, R. G., Gatsogiannis, C., & Raunser, S. (2017). Lipid nanodiscs as a tool for high-resolution structure determination of membrane proteins by single-particle cryo-EM. *Methods in Enzymology*, *594*, 1–30.

Etzkorn, M., Raschle, T., Hagn, F., Gelev, V., Rice, A. J., Walz, T., et al. (2013). Cell-free expressed bacteriorhodopsin in different soluble membrane mimetics: Biophysical properties and NMR accessibility. *Structure*, *21*(3), 394–401.

Feuerstein, S., Plevin, M. J., Willbold, D., & Brutscher, B. (2012). iHADAMAC: A complementary tool for sequential resonance assignment of globular and highly disordered proteins. *Journal of Magnetic Resonance*, *214*(1), 329–334.

Fiori, M. C., Jiang, Y., Altenberg, G. A., & Liang, H. (2017). Polymer-encased nanodiscs with improved buffer compatibility. *Scientific Reports*, 7(1), 7432.

Foshag, D., Henrich, E., Hiller, E., Schäfer, M., Kerger, C., Burger-Kentischer, A., et al. (2018). The E. coli S30 lysate proteome: A prototype for cell-free protein production. *New Biotechnology*, *40*(Pt. B), 245–260.

Frauenfeld, J., Löving, R., Armache, J. P., Sonnen, A. F., Guettou, F., Moberg, P., et al. (2016). A saposin-lipoprotein nanoparticle system for membrane proteins. *Nature Methods*, *13*(4), 345–351.

Frenkiel, T., Bauer, C., Carr, M. D., Birdsall, B., & Feeney, J. (1990). HMQC-NOESY-HMQC, a three-dimensional NMR experiment which allows detection of nuclear overhauser effects between protons with overlapping signals. *Journal of Magnetic Resonance (1969)*, *90*(2), 420–425.

Frey, L., Lakomek, N. A., Riek, R., & Bibow, S. (2017). Micelles, bicelles, and nanodiscs: Comparing the impact of membrane mimetics on membrane protein backbone dynamics. *Angewandte Chemie (International Ed. in English)*, *56*(1), 380–383.

Gardner, K. H., & Kay, L. E. (1997). Production and incorporation of 15N, 13C, 2H (1H-δ1 methyl) isoleucine into proteins for multidimensional NMR studies. *Journal of the American Chemical Society, 119*(32), 7599–7600.

Glück, J. M., Wittlich, M., Feuerstein, S., Hoffmann, S., Willbold, D., & Koenig, B. W. (2009). Integral membrane proteins in nanodiscs can be studied by solution NMR spectroscopy. *Journal of the American Chemical Society, 131*(34), 12060–12061.

Goto, N. K., Gardner, K. H., Mueller, G. A., Willis, R. C., & Kay, L. E. (1999). A robust and cost-effective method for the production of Val, Leu, Ile (delta 1) methyl-protonated 15N-, 13C-, 2H-labeled proteins. *Journal of Biomolecular NMR, 13*(4), 369–374.

Gottstein, D., Reckel, S., Dötsch, V., & Güntert, P. (2012). Requirements on paramagnetic relaxation enhancement data for membrane protein structure determination by NMR. *Structure, 20*(6), 1019–1027.

Griffey, R. H., Redfield, A. G., Loomis, R. E., & Dahlquist, F. W. (1985). Nuclear magnetic resonance observation and dynamics of specific amide protons in T4 lysozyme. *Biochemistry, 24*(4), 817–822.

Griffey, R. H., Redfield, A. G., McIntosh, L. P., Oas, T. G., & Dahlquist, F. W. (1986). Assignment of proton amide resonances of T4 lysozyme by carbon-13 and nitrogen-15 multiple isotopic labeling. *Journal of the American Chemical Society, 108*(21), 6816–6817.

Guignard, L., Ozawa, K., Pursglove, S. E., Otting, G., & Dixon, N. E. (2002). NMR analysis of in vitro-synthesized proteins without purification: A high-throughput approach. *FEBS Letters, 524*(1–3), 159–162.

Gupta, K., Donlan, J. A. C., Hopper, J. T. S., Uzdavinys, P., Landreh, M., Struwe, W. B., et al. (2017). The role of interfacial lipids in stabilizing membrane protein oligomers. *Nature, 541*(7637), 421–424.

Hagn, F., Etzkorn, M., Raschle, T., & Wagner, G. (2013). Optimized phospholipid bilayer nanodiscs facilitate high-resolution structure determination of membrane proteins. *Journal of the American Chemical Society, 135*(5), 1919–1925.

Hagn, F., & Wagner, G. (2015). Structure refinement and membrane positioning of selectively labeled OmpX in phospholipid nanodiscs. *Journal of Biomolecular NMR, 61*(3–4), 249–260.

Harbers, M. (2014). Wheat germ systems for cell-free protein expression. *FEBS Letters, 588*(17), 2762–2773.

Harris, N. J., Reading, E., Ataka, K., Grzegorzewski, L., Charalambous, K., Liu, X., et al. (2017). Structure formation during translocon-unassisted co-translational membrane protein folding. *Scientific Reports*, 7(1), 8021.

He, Y., Wang, K., & Yan, N. (2014). The recombinant expression systems for structure determination of eukaryotic membrane proteins. *Protein & Cell, 5*(9), 658–672.

Hefke, F., Bagaria, A., Reckel, S., Ullrich, S. J., Dötsch, V., Glaubitz, C., et al. (2011). Optimization of amino acid type-specific 13C and 15N labeling for the backbone assignment of membrane proteins by solution- and solid-state NMR with the UPLABEL algorithm. *Journal of Biomolecular NMR, 49*(2), 75–84.

Hein, C., Löhr, F., Schwarz, D., & Dötsch, V. (2017). Acceleration of protein backbone NMR assignment by combinatorial labeling: Application to a small molecule binding study. *Biopolymers, 107*(5), e23013.

Henrich, E., Dötsch, V., & Bernhard, F. (2015). Screening for lipid requirements of membrane proteins by combining cell-free expression with nanodiscs. *Methods in Enzymology, 556*, 351–369.

Henrich, E., Hein, C., Dötsch, V., & Bernhard, F. (2015). Membrane protein production in Escherichia coli cell-free lysates. *FEBS Letters, 589*(15), 1713–1722.

Henrich, E., Ma, Y., Engels, I., Münch, D., Otten, C., Schneider, T., et al. (2016). Lipid requirements for the enzymatic activity of MraY translocases and in vitro reconstitution of the lipid II synthesis pathway. *The Journal of Biological Chemistry, 291*(5), 2535–2546.

Henrich, E., Peetz, O., Hein, C., Laguerre, A., Hoffmann, B., Hoffmann, J., et al. (2017). Analyzing native membrane protein assembly in nanodiscs by combined non-covalent mass spectrometry and synthetic biology. *eLife*, *6*(6), e20954.

Hiller, S., Ibraghimov, I., Wagner, G., & Orekhov, V. Y. (2009). Coupled decomposition of four-dimensional NOESY spectra. *Journal of the American Chemical Society*, *131*(36), 12970–12978.

Hoffmann, B., Löhr, F., Laguerre, A., Bernhard, F., & Dötsch, V. (2018). Protein labeling strategies for liquid-state NMR spectroscopy using cell-free synthesis. *Progress in Nuclear Magnetic Resonance Spectroscopy*, *105*, 1–22.

Holland, D. J., Bostock, M. J., Gladden, L. F., & Nietlispach, D. (2011). Fast multidimensional NMR spectroscopy using compressed sensing. *Angewandte Chemie (International Ed. in English)*, *50*(29), 6548–6551.

Hong, S. H., Ntai, I., Haimovich, A. D., Kelleher, N. L., Isaacs, F. J., & Jewett, M. C. (2014). Cell-free protein synthesis from a release factor 1 deficient Escherichia coli activates efficient and multiple site-specific nonstandard amino acid incorporation. *ACS Synthetic Biology*, *3*(6), 398–409.

Hurst, G. B., Asano, K. G., Doktycz, C. J., Consoli, E. J., Doktycz, W. L., Foster, C. M., et al. (2017). Proteomics-based tools for evaluation of cell-free protein synthesis. *Analytical Chemistry*, *89*(21), 11443–11451.

Ikura, M., Kay, L. E., & Bax, A. (1990). A novel approach for sequential assignment of 1H, 13C, and 15N spectra of proteins: Heteronuclear triple-resonance three-dimensional NMR spectroscopy. Application to calmodulin. *Biochemistry*, *29*(19), 4659–4667.

Ikura, M., Krinks, M., Torchia, D. A., & Bax, A. (1990). An efficient NMR approach for obtaining sequence-specific resonance assignments of larger proteins based on multiple isotopic labeling. *FEBS Letters*, *266*(1–2), 155–158.

Ilgü, H., Jeckelmann, J. M., Gachet, M. S., Boggavarapu, R., Ucurum, Z., Gertsch, J., et al. (2014). Variation of the detergent-binding capacity and phospholipid content of membrane proteins when purified in different detergents. *Biophysical Journal*, *106*(8), 1660–1670.

Isogai, S., Deupi, X., Opitz, C., Heydenreich, F. M., Tsai, C. J., Brueckner, F., et al. (2016). Backbone NMR reveals allosteric signal transduction networks in the β1-adrenergic receptor. *Nature*, *530*(7589), 237–241.

Jaipuria, G., Krishnarjuna, B., Mondal, S., Dubey, A., & Atreya, H. S. (2012). Amino acid selective labeling and unlabeling for protein resonance assignments. *Advances in Experimental Medicine and Biology*, *992*, 95–118.

Jones, D. H., Cellitti, S. E., Hao, X., Zhang, Q., Jahnz, M., Summerer, D., et al. (2010). Site-specific labeling of proteins with NMR-active unnatural amino acids. *Journal of Biomolecular NMR*, *46*(1), 89–100.

Kainosho, M., & Tsuji, T. (1982). Assignment of the three methionyl carbonyl carbon resonances in Streptomyces subtilisin inhibitor by a carbon-13 and nitrogen-15 double-labeling technique. A new strategy for structural studies of proteins in solution. *Biochemistry*, *21*(24), 6273–6279.

Kasai, T., Koshiba, S., Yokoyama, J., & Kigawa, T. (2015). Stable isotope labeling strategy based on coding theory. *Journal of Biomolecular NMR*, *63*(2), 213–221.

Kato, K., Matsunaga, C., Nishimura, Y., Waelchli, M., Kainosho, M., & Arata, Y. (1989). Application of 13C nuclear magnetic resonance spectroscopy to molecular structural analyses of antibody molecules. *Journal of Biochemistry*, *105*(6), 867–869.

Katzen, F., Fletcher, J. E., Yang, J. P., Kang, D., Peterson, T. C., Cappuccio, J. A., et al. (2008). Insertion of membrane proteins into discoidal membranes using a cell-free protein expression approach. *Journal of Proteome Research*, 7(8), 3535–3542.

Kay, L. E., Clore, G. M., Bax, A., & Gronenborn, A. M. (1990). Four-dimensional heteronuclear triple-resonance NMR spectroscopy of interleukin-1 beta in solution. *Science*, *249*(4967), 411–414.

Kigawa, T., Yabuki, T., Matsuda, N., Matsuda, T., Nakajima, R., Tanaka, A., et al. (2004). Preparation of Escherichia coli cell extract for highly productive cell-free protein expression. *Journal of Structural and Functional Genomics*, *5*(1–2), 63–68.

Kim, H. J., Howell, S. C., Van Horn, W. D., Jeon, Y. H., & Sanders, C. R. (2009). Recent advances in the application of solution NMR spectroscopy to multi-span integral membrane proteins. *Progress in Nuclear Magnetic Resonance Spectroscopy*, *55*(4), 335–360.

Kim, T. W., Keum, J. W., Oh, I. S., Choi, C. Y., Park, C. G., & Kim, D. M. (2006). Simple procedures for the construction of a robust and cost-effective cell-free protein synthesis system. *Journal of Biotechnology*, *126*(4), 554–561.

Kim, J., Wu, S., Tomasiak, T. M., Mergel, C., Winter, M. B., Stiller, S. B., et al. (2015). Subnanometre-resolution electron cryomicroscopy structure of a heterodimeric ABC exporter. *Nature*, *517*(7534), 396–400.

Klammt, C., Löhr, F., Schäfer, B., Haase, W., Dötsch, V., Rüterjans, H., et al. (2004). High level cell-free expression and specific labeling of integral membrane proteins. *European Journal of Biochemistry*, *271*(3), 568–580.

Kofuku, Y., Ueda, T., Okude, J., Shiraishi, Y., Kondo, K., Maeda, M., et al. (2012). Efficacy of the β_2-adrenergic receptor is determined by conformational equilibrium in the transmembrane region. *Nature Communications*, *3*, 1045.

Kofuku, Y., Ueda, T., Okude, J., Shiraishi, Y., Kondo, K., Mizumura, T., et al. (2014). Functional dynamics of deuterated β2-adrenergic receptor in lipid bilayers revealed by NMR spectroscopy. *Angewandte Chemie (International Ed. in English)*, *53*(49), 13376–13379.

Kühlbrandt, W. (2014). The resolution revolution. *Science*, *343*(6178), 1443–1444.

Kurauskas, V., Hessel, A., Ma, P., Lunetti, P., Weinhäupl, K., Imbert, L., et al. (2018). How detergent impacts membrane proteins: Atomic-level views of mitochondrial carriers in dodecylphosphocholine. *Journal of Physical Chemistry Letters*, *9*(5), 933–938.

Kwon, Y. C., & Jewett, M. C. (2015). High-throughput preparation methods of crude extract for robust cell-free protein synthesis. *Scientific Reports*, *5*, 8663.

LaGuerre, A., Löhr, F., Bernhard, F., & Dötsch, V. (2015). Labeling of membrane proteins by cell-free expression. *Methods in Enzymology*, *565*, 367–388.

Laguerre, A., Löhr, F., Henrich, E., Hoffmann, B., Abdul-Manan, N., Connolly, P. J., et al. (2016). From nanodiscs to isotropic bicelles: A procedure for solution nuclear magnetic resonance studies of detergent-sensitive integral membrane proteins. *Structure*, *24*(10), 1830–1841.

Lazarova, M., Lőhr, F., Rues, R. B., Kleebach, R., Dötsch, V., & Bernhard, F. (2018). Precursor-based selective methyl labelling of cell-free synthesized proteins. *ACS Chemical Biology*, *13*(8), 2170–2178. https://doi.org/10.1021/acschembio.8b00338.

Lee, D., Hilty, C., Wider, G., & Wüthrich, K. (2006). Effective rotational correlation times of proteins from NMR relaxation interference. *Journal of Magnetic Resonance*, *178*(1), 72–76.

Lee, D., Vögeli, B., & Pervushin, K. (2005). Detection of C′,Calpha correlations in proteins using a new time- and sensitivity-optimal experiment. *Journal of Biomolecular NMR*, *31*(4), 273–278.

LeMaster, D. M., & Richards, F. M. (1985). 1H-15N heteronuclear NMR studies of Escherichia coli thioredoxin in samples isotopically labeled by residue type. *Biochemistry*, *24*(25), 7263–7268.

Li, Y., Kijac, A. Z., Sligar, S. G., & Rienstra, C. M. (2006). Structural analysis of nanoscale self-assembled discoidal lipid bilayers by solid-state NMR spectroscopy. *Biophysical Journal*, *91*(10), 3819–3828.

Liang, B., Bushweller, J. H., & Tamm, L. K. (2006). Site-directed parallel spin-labeling and paramagnetic relaxation enhancement in structure determination of membrane proteins by solution NMR spectroscopy. *Journal of the American Chemical Society*, *128*(13), 4389–4397.

Liang, B., & Tamm, L. K. (2016). NMR as a tool to investigate the structure, dynamics and function of membrane proteins. *Nature Structural & Molecular Biology*, *23*(6), 468–474.

Lichtenecker, R. J., Coudevylle, N., Konrat, R., & Schmid, W. (2013). Selective isotope labelling of leucine residues by using α-ketoacid precursor compounds. *Chembiochem*, *14*(7), 818–821.

Löhr, F., Laguerre, A., Bock, C., Reckel, S., Connolly, P. J., Abdul-Manan, N., et al. (2014). Time-shared experiments for efficient assignment of triple-selectively labeled proteins. *Journal of Magnetic Resonance*, *248*, 81–95.

Löhr, F., Reckel, S., Karbyshev, M., Connolly, P. J., Abdul-Manan, N., Bernhard, F., et al. (2012). Combinatorial triple-selective labeling as a tool to assist membrane protein backbone resonance assignment. *Journal of Biomolecular NMR*, *52*(3), 197–210.

Löhr, F., Tumulka, F., Bock, C., Abele, R., & Dötsch, V. (2015). An extended combinatorial 15N, 13Cα, and 13C' labeling approach to protein backbone resonance assignment. *Journal of Biomolecular NMR*, *62*(3), 263–279.

Luan, T., Jaravine, V., Yee, A., Arrowsmith, C. H., & Orekhov, V. Y. (2005). Optimization of resolution and sensitivity of 4D NOESY using multi-dimensional decomposition. *Journal of Biomolecular NMR*, *33*(1), 1–14.

Lyukmanova, E. N., Shenkarev, Z. O., Khabibullina, N. F., Kopeina, G. S., Shulepko, M. A., Paramonov, A. S., et al. (2012). Lipid-protein nanodiscs for cell-free production of integral membrane proteins in a soluble and folded state: Comparison with detergent micelles, bicelles and liposomes. *Biochimica et Biophysica Acta*, *1818*(3), 349–358.

Madin, K., Sawasaki, T., Ogasawara, T., & Endo, Y. (2000). A highly efficient and robust cell-free protein synthesis system prepared from wheat embryos: Plants apparently contain a suicide system directed at ribosomes. *Proceedings of the National Academy of Sciences of the United States of America*, *97*(2), 559–564.

Manglik, A., Kim, T. H., Masureel, M., Altenbach, C., Yang, Z., Hilger, D., et al. (2015). Structural insights into the dynamic process of β2-adrenergic receptor signaling. *Cell*, *161*(5), 1101–1111.

Martens, C., Stein, R. A., Masureel, M., Roth, A., Mishra, S., Dawaliby, R., et al. (2016). Lipids modulate the conformational dynamics of a secondary multidrug transporter. *Nature Structural & Molecular Biology*, *23*(8), 744–751.

Martin, R. W., Des Soye, B. J., Kwon, Y. C., Kay, J., Davis, R. G., Thomas, P. M., et al. (2018). Cell-free protein synthesis from genomically recoded bacteria enables multisite incorporation of noncanonical amino acids. *Nature Communications*, *9*(1), 1203.

Marty, M. T., Wilcox, K. C., Klein, W. L., & Sligar, S. G. (2013). Nanodisc-solubilized membrane protein library reflects the membrane proteome. *Analytical and Bioanalytical Chemistry*, *405*(12), 4009–4016.

Maslennikov, I., & Choe, S. (2013). Advances in NMR structures of integral membrane proteins. *Current Opinion in Structural Biology*, *23*(4), 555–562.

Maslennikov, I., Klammt, C., Hwang, E., Kefala, G., Okamura, M., Esquivies, L., et al. (2010). Membrane domain structures of three classes of histidine kinase receptors by cell-free expression and rapid NMR analysis. *Proceedings of the National Academy of Sciences of the United States of America*, *107*(24), 10902–10907.

McClary, W. D., Sumida, J. P., Scian, M., Paço, L., & Atkins, W. M. (2016). Membrane fluidity modulates thermal stability and ligand binding of cytochrome P4503A4 in lipid nanodiscs. *Biochemistry*, *55*(45), 6258–6268.

Miyanoiri, Y., Takeda, M., Okuma, K., Ono, A. M., Terauchi, T., & Kainosho, M. (2013). Differential isotope-labeling for Leu and Val residues in a protein by E. coli cellular expression using stereo-specifically methyl labeled amino acids. *Journal of Biomolecular NMR*, *57*(3), 237–249.

Mobli, M., & Hoch, J. C. (2014). Nonuniform sampling and non-Fourier signal processing methods in multidimensional NMR. *Progress in Nuclear Magnetic Resonance Spectroscopy*, *83*, 21–41.

Morrison, E. A., DeKoster, G. T., Dutta, S., Vafabakhsh, R., Clarkson, M. W., Bahl, A., et al. (2012). Antiparallel EmrE exports drugs by exchanging between asymmetric structures. *Nature*, *481*(7379), 45–50.

Muchmore, D. C., McIntosh, L. P., Russell, C. B., Anderson, D. E., & Dahlquist, F. W. (1989). Expression and nitrogen-15 labeling of proteins for proton and nitrogen-15 nuclear magnetic resonance. *Methods in Enzymology*, *177*, 44–73.

Nasr, M. L., Baptista, D., Strauss, M., Sun, Z. J., Grigoriu, S., Huser, S., et al. (2017). Covalently circularized nanodiscs for studying membrane proteins and viral entry. *Nature Methods*, *14*(1), 49–52.

Nikolaev, M., Round, E., Gushchin, I., Polovinkin, V., Balandin, T., Kuzmichev, P., et al. (2017). Integral membrane proteins can be crystallized directly from nanodiscs. *Crystal Growth & Design*, *17*(3), 945–948.

Nygaard, R., Zou, Y., Dror, R. O., Mildorf, T. J., Arlow, D. H., Manglik, A., et al. (2013). The dynamic process of β(2)-adrenergic receptor activation. *Cell*, *152*(3), 532–542.

Ollerenshaw, J. E., Tugarinov, V., & Kay, L. E. (2003). Methyl TROSY: Explanation and experimental verification. *Magnetic Resonance in Chemistry*, *41*(10), 843–852.

Orekhov, V. Y., Ibraghimov, I., & Billeter, M. (2003). Optimizing resolution in multidimensional NMR by three-way decomposition. *Journal of Biomolecular NMR*, *27*(2), 165–173.

Orekhov, V. Y., & Jaravine, V. A. (2011). Analysis of non-uniformly sampled spectra with multi-dimensional decomposition. *Progress in Nuclear Magnetic Resonance Spectroscopy*, *59*(3), 271–292.

Orwick-Rydmark, M., Lovett, J. E., Graziadei, A., Lindholm, L., Hicks, M. R., & Watts, A. (2012). Detergent-free incorporation of a seven-transmembrane receptor protein into nanosized bilayer Lipodisq particles for functional and biophysical studies. *Nano Letters*, *12*(9), 4687–4692.

Palmer, A. G. (2004). NMR characterization of the dynamics of biomacromolecules. *Chemical Reviews*, *104*(8), 3623–3640.

Pantoja-Uceda, D., & Santoro, J. (2008). Amino acid type identification in NMR spectra of proteins via beta- and gamma-carbon edited experiments. *Journal of Magnetic Resonance*, *195*(2), 187–195.

Parker, M. J., Aulton-Jones, M., Hounslow, A. M., & Craven, C. J. (2004). A combinatorial selective labeling method for the assignment of backbone amide NMR resonances. *Journal of the American Chemical Society*, *126*(16), 5020–5021.

Peetz, O., Henrich, E., Laguerre, A., Löhr, F., Hein, C., Dötsch, V., et al. (2017). Insights into cotranslational membrane protein insertion by combined LILBID-mass spectrometry and NMR spectroscopy. *Analytical Chemistry*, *89*(22), 12314–12318.

Permi, P., Rosevear, P. R., & Annila, A. (2000). A set of HNCO-based experiments for measurement of residual dipolar couplings in 15N, 13C, (2H)-labeled proteins. *Journal of Biomolecular NMR*, *17*(1), 43–54.

Pervushin, K., Riek, R., Wider, G., & Wüthrich, K. (1997). Attenuated T2 relaxation by mutual cancellation of dipole-dipole coupling and chemical shift anisotropy indicates an avenue to NMR structures of very large biological macromolecules in solution. *Proceedings of the National Academy of Sciences of the United States of America*, *94*(23), 12366–12371.

Peuker, S., Andersson, H., Gustavsson, E., Maiti, K. S., Kania, R., Karim, A., et al. (2016). Efficient isotope editing of proteins for site-directed vibrational spectroscopy. *Journal of the American Chemical Society*, *138*(7), 2312–2318.

Piai, A., Fu, Q., Dev, J., & Chou, J. J. (2017). Optimal bicelle size q for solution NMR studies of the protein transmembrane partition. *Chemistry*, *23*(6), 1361–1367.

Rasmussen, S. G., Choi, H. J., Fung, J. J., Pardon, E., Casarosa, P., Chae, P. S., et al. (2011). Structure of a nanobody-stabilized active state of the beta(2) adrenoceptor. *Nature*, *469*(7329), 175–180.

Rasmussen, S. G., Choi, H. J., Rosenbaum, D. M., Kobilka, T. S., Thian, F. S., Edwards, P. C., et al. (2007). Crystal structure of the human beta2 adrenergic G-protein-coupled receptor. *Nature*, *450*(7168), 383–387.

Ravula, T., Ramadugu, S. K., Di Mauro, G., & Ramamoorthy, A. (2017). Bioinspired, size-tunable self-assembly of polymer-lipid bilayer nanodiscs. *Angewandte Chemie (International Ed. in English)*, *56*(38), 11466–11470.

Reckel, S., Gottstein, D., Stehle, J., Löhr, F., Verhoefen, M. K., Takeda, M., et al. (2011). Solution NMR structure of proteorhodopsin. *Angewandte Chemie (International Ed. in English)*, *50*(50), 11942–11946.

Reckel, S., Sobhanifar, S., Durst, F., Lohr, F., Shirokov, V. A., Dötsch, V., et al. (2010). Strategies for the cell-free expression of membrane proteins. *Methods in Molecular Biology*, *607*, 187–212.

Reckel, S., Sobhanifar, S., Schneider, B., Junge, F., Schwarz, D., Durst, F., et al. (2008). Transmembrane segment enhanced labeling as a tool for the backbone assignment of alpha-helical membrane proteins. *Proceedings of the National Academy of Sciences of the United States of America*, *105*(24), 8262–8267.

Riek, R., Wider, G., Pervushin, K., & Wüthrich, K. (1999). Polarization transfer by cross-correlated relaxation in solution NMR with very large molecules. *Proceedings of the National Academy of Sciences of the United States of America*, *96*(9), 4918–4923.

Roos, C., Kai, L., Haberstock, S., Proverbio, D., Ghoshdastider, U., Ma, Y., et al. (2014). High-level cell-free production of membrane proteins with nanodiscs. *Methods in Molecular Biology*, *1118*, 109–130.

Roos, C., Zocher, M., Müller, D., Münch, D., Schneider, T., Sahl, H. G., et al. (2012). Characterization of co-translationally formed nanodisc complexes with small multidrug transporters, proteorhodopsin and with the E. coli MraY translocase. *Biochimica et Biophysica Acta*, *1818*(12), 3098–3106.

Rossi, P., Xia, Y., Khanra, N., Veglia, G., & Kalodimos, C. G. (2016). ^{15}N and ^{13}C-SOFAST-HMQC editing enhances 3D-NOESY sensitivity in highly deuterated, selectively [^{1}H,^{13}C]-labeled proteins. *Journal of Biomolecular NMR*, *66*(4), 259–271.

Rovnyak, D., Hoch, J. C., Stern, A. S., & Wagner, G. (2004). Resolution and sensitivity of high field nuclear magnetic resonance spectroscopy. *Journal of Biomolecular NMR*, *30*(1), 1–10.

Rues, R. B., Dong, F., Dötsch, V., & Bernhard, F. (2018). Systematic optimization of cell-free synthesized human endothelin B receptor folding. *Methods*, *147*, 73–83. https://doi.org/10.1016/j.ymeth.2018.01.012.

Rues, R. B., Dötsch, V., & Bernhard, F. (2016). Co-translational formation and pharmacological characterization of beta1-adrenergic receptor/nanodisc complexes with different lipid environments. *Biochimica et Biophysica Acta*, *1858*(6), 1306–1316.

Rule, G. S., Tjandra, N., Simplaceanu, V., & Ho, C. (1993). Assignment strategies for ^{15}N-^{1}H correlated spectra of large proteins in solution. *Journal of Magnetic Resonance. Series B*, *102*, 126–128.

Saliba, A. E., Vonkova, I., & Gavin, A. C. (2015). The systematic analysis of protein-lipid interactions comes of age. *Nature Reviews. Molecular Cell Biology*, *16*(12), 753–761.

Sanders, C. R., & Sönnichsen, F. (2006). Solution NMR of membrane proteins: Practice and challenges. *Magnetic Resonance in Chemistry: MRC*, *44*, S24–S40. Spec. No.

Schanda, P., & Brutscher, B. (2005). Very fast two-dimensional NMR spectroscopy for real-time investigation of dynamic events in proteins on the time scale of seconds. *Journal of the American Chemical Society*, *127*(22), 8014–8015.

Schanda, P., Van Melckebeke, H., & Brutscher, B. (2006). Speeding up three-dimensional protein NMR experiments to a few minutes. *Journal of the American Chemical Society, 128*(28), 9042–9043.

Schmieder, P., Stern, A. S., Wagner, G., & Hoch, J. C. (1993). Application of nonlinear sampling schemes to COSY-type spectra. *Journal of Biomolecular NMR, 3*(5), 569–576.

Schubert, M., Kolbe, M., Kessler, B., Oesterhelt, D., & Schmieder, P. (2002). Heteronuclear multidimensional NMR spectroscopy of solubilized membrane proteins: Resonance assignment of native bacteriorhodopsin. *Chembiochem, 3*(10), 1019–1023.

Schubert, M., Smalla, M., Schmieder, P., & Oschkinat, H. (1999). MUSIC in triple-resonance experiments: Amino acid type-selective (1)H-(15)N correlations. *Journal of Magnetic Resonance, 141*(1), 34–43.

Schwarz, D., Junge, F., Durst, F., Frölich, N., Schneider, B., Reckel, S., et al. (2007). Preparative scale expression of membrane proteins in Escherichia coli-based continuous exchange cell-free systems. *Nature Protocols, 2*(11), 2945–2957.

Senn, H., Eugster, A., Otting, G., Suter, F., & Wüthrich, K. (1987). 15N-labeled P22 c2 repressor for nuclear magnetic resonance studies of protein-DNA interactions. *European Biophysics Journal, 14*(5), 301–306.

Serrano-Vega, M. J., Magnani, F., Shibata, Y., & Tate, C. G. (2008). Conformational thermostabilization of the beta1-adrenergic receptor in a detergent-resistant form. *Proceedings of the National Academy of Sciences of the United States of America, 105*(3), 877–882.

Shin, J., & Noireaux, V. (2010). Efficient cell-free expression with the endogenous E. Coli RNA polymerase and sigma factor 70. *Journal of Biological Engineering, 4*(1), 8.

Shortle, D. (1994). Assignment of amino acid type in ^{1}H-^{15}N correlation spectra by labeling with ^{14}N-amino acids. *Journal of Magnetic Resonance. Series B, 105*, 88–90.

Shrestha, P., Holland, T. M., & Bundy, B. C. (2012). Streamlined extract preparation for Escherichia coli-based cell-free protein synthesis by sonication or bead vortex mixing. *BioTechniques, 53*, 163–174.

Sobhanifar, S., Schneider, B., Lohr, F., Gottstein, D., Ikeya, T., Mlynarczyk, K., et al. (2010). Structural investigation of the C-terminal catalytic fragment of presenilin 1. *Proceedings of the National Academy of Sciences of the United States of America, 107*(21), 9644–9649.

Stanek, J., Nowakowski, M., Saxena, S., Ruszczyńska-Bartnik, K., Ejchart, A., & Koź miński, W. (2013). Selective diagonal-free (13)C, (13)C-edited aliphatic-aromatic NOESY experiment with non-uniform sampling. *Journal of Biomolecular NMR, 56*(3), 217–226.

Staunton, D., Schlinkert, R., Zanetti, G., Colebrook, S. A., & Campbell, I. D. (2006). Cell-free expression and selective isotope labelling in protein NMR. *Magnetic Resonance in Chemistry: MRC, 44*, S2–S9. Spec. No.

Sun, Z. Y., Frueh, D. P., Selenko, P., Hoch, J. C., & Wagner, G. (2005). Fast assignment of 15N-HSQC peaks using high-resolution 3D HNcocaNH experiments with non-uniform sampling. *Journal of Biomolecular NMR, 33*(1), 43–50.

Takahashi, H., Odaka, A., Kawaminami, S., Matsunaga, C., Kato, K., Shimada, I., et al. (1991). Multinuclear NMR study of the structure of the Fv fragment of anti-dansyl mouse IgG2a antibody. *Biochemistry, 30*(26), 6611–6619.

Tate, S., Kikumoto, Y., Ichikawa, S., Kaneko, M., Masui, Y., Kamogashira, T., et al. (1992). Stable isotope aided nuclear magnetic resonance study to investigate the receptor-binding site of human interleukin 1 beta. *Biochemistry, 31*(8), 2435–2442.

Trbovic, N., Klammt, C., Koglin, A., Löhr, F., Bernhard, F., & Dötsch, V. (2005). Efficient strategy for the rapid backbone assignment of membrane proteins. *Journal of the American Chemical Society, 127*(39), 13504–13505.

Tugarinov, V., Hwang, P. M., Ollerenshaw, J. E., & Kay, L. E. (2003). Cross-correlated relaxation enhanced 1H[bond]13C NMR spectroscopy of methyl groups in very high molecular weight proteins and protein complexes. *Journal of the American Chemical Society, 125*(34), 10420–10428.

Tugarinov, V., Kanelis, V., & Kay, L. E. (2006). Isotope labeling strategies for the study of high-molecular-weight proteins by solution NMR spectroscopy. *Nature Protocols*, *1*(2), 749–754.

Tugarinov, V., & Kay, L. E. (2004). An isotope labeling strategy for methyl TROSY spectroscopy. *Journal of Biomolecular NMR*, *28*(2), 165–172.

Viegas, A., Viennet, T., & Etzkorn, M. (2016). The power, pitfalls and potential of the nanodisc system for NMR-based studies. *Biological Chemistry*, *397*(12), 1335–1354.

Vuister, G. W., Clore, G. M., Gronenborn, A. M., Powers, R., Garrett, D. S., Tschudin, R., et al. (1993). Increased resolution and improved spectral quality in four-dimensional 13C/13C-separated HMQC-NOESY-HMQC spectra using pulsed field gradients. *Journal of Magnetic Resonance. Series B*, *101*(2), 210–213.

Waberer, L., Henrich, E., Peetz, O., Morgner, N., Dötsch, V., Bernhard, F., et al. (2017). The synaptic vesicle protein SV31 assembles into a dimer and transports Zn(2). *Journal of Neurochemistry*, *140*(2), 280–293.

Weigelt, J., van Dongen, M., Uppenberg, J., Schultz, J., & Wikström, M. (2002). Site-selective screening by NMR spectroscopy with labeled amino acid pairs. *Journal of the American Chemical Society*, *124*(11), 2446–2447.

Werner-Allen, J. W., Coggins, B. E., & Zhou, P. (2010). Fast acquisition of high resolution 4-D amide-amide NOESY with diagonal suppression, sparse sampling and FFT-CLEAN. *Journal of Magnetic Resonance*, *204*(1), 173–178.

Westler, W. M., Stockman, B. J., Hosoya, Y., Miyake, Y., Kainosho, M., & Markley, J. L. (1988). Correlation of carbon-13 and nitrogen-15 chemical shifts in selectively and uniformly labeled proteins by heteronuclear two-dimensional NMR spectroscopy. *Journal of the American Chemical Society*, *110*(18), 6256–6258.

Wittekind, M., & Mueller, L. (1993). HNCACB, a high-sensitivity 3D NMR experiment to correlate amide-proton and nitrogen resonances with the α- and β-carbon resonances in proteins. *Journal of Magnetic Resonance. Series B*, *101*, 201–205.

Wu, P. S., Ozawa, K., Jergic, S., Su, X. C., Dixon, N. E., & Otting, G. (2006). Amino-acid type identification in 15N-HSQC spectra by combinatorial selective 15N-labelling. *Journal of Biomolecular NMR*, *34*(1), 13–21.

Yabuki, T., Kigawa, T., Dohmae, N., Takio, K., Terada, T., Ito, Y., et al. (1998). Dual amino acid-selective and site-directed stable-isotope labeling of the human c-Ha-Ras protein by cell-free synthesis. *Journal of Biomolecular NMR*, *11*(3), 295–306.

Yamazaki, T., Lee, W., Arrowsmith, C. H., Muhandiram, D. R., & Kay, L. E. (1994). A suite of triple resonance NMR experiments for the backbone assignment of 15N, 13C, 2H labeled proteins with high sensitivity. *Journal of the American Chemical Society*, *116*(26), 11655–11666.

Yang, W. C., Patel, K. G., Wong, H. E., & Swartz, J. R. (2012). Simplifying and streamlining Escherichia coli-based cell-free protein synthesis. *Biotechnology Progress*, *28*(2), 413–420.

Zawadzka-Kazimierczuk, A., Kazimierczuk, K., & Koźmiński, W. (2010). A set of 4D NMR experiments of enhanced resolution for easy resonance assignment in proteins. *Journal of Magnetic Resonance*, *202*(1), 109–116.

Zhao, L., Wang, S., Run, C., OuYang, B., & Chou, J. J. (2016). Specific lipid binding of membrane proteins in detergent micelles characterized by NMR and molecular dynamics. *Biochemistry*, *55*(38), 5317–5320.

Zubay, G. (1973). In vitro synthesis of protein in microbial systems. *Annual Review of Genetics*, *7*, 267–287.

Zuiderweg, E. R. P., Petros, A. M., Fesik, S. W., & Olejniczak, E. T. (1991). Four-dimensional [^{13}C, ^{1}H, ^{13}C, ^{1}H] HMQC-NOE-HMQC NMR spectroscopy: Resolving tertiary NOW distance restraints in spectra of larger proteins. *Journal of the American Chemical Society*, *113*, 370–372.

CHAPTER SEVEN

Preparation of Phosphorylated Proteins for NMR Spectroscopy

Ganesan Senthil Kumar, Rebecca Page, Wolfgang Peti[1]
Department of Chemistry and Biochemistry, University of Arizona, Tucson, AZ, United States
[1]Corresponding author: e-mail address: wolfgangpeti@email.arizona.edu

Contents

Abstract

Phosphorylation is a ubiquitous posttranslational modification that is essential for the regulation of many cellular processes. The human genome consists of more than 200,000 phosphorylation sites, whose phosphorylation is tightly controlled by $\geq$500 kinases and $\sim$200 phosphatases. Given the large number of phosphorylation sites and the key role phosphorylation plays in regulating cellular processes, it is essential to characterize the impact of phosphorylation on substrate structure, dynamics, and function. However, a major challenge is the large-scale production of phosphorylated proteins in vitro for these structural, functional, and dynamic studies. Here, we describe an efficient protocol used routinely in our laboratory for the production of phosphorylated proteins. We also describe the methods used for identifying, characterizing, and separating the resulting phosphorylated proteins for subsequent studies.

Methods in Enzymology, Volume 614
ISSN 0076-6879
https://doi.org/10.1016/bs.mie.2018.07.004

1. INTRODUCTION

Reversible protein phosphorylation affects an estimated ~70% of all eukaryotic cellular proteins (Olsen et al., 2010). As a consequence, phosphorylation is an essential regulatory mechanism that affects almost every aspect of cellular life, including protein synthesis, cell division, signal transduction, development, metabolism, gene expression, and aging, among many others (Fig. 1A) (Humphrey, James, & Mann, 2015; Hunter & Karin, 1992). The level of phosphorylation is maintained by the concerted action of kinases, enzymes that phosphorylate substrates by transferring the γ-phosphatase group from ATP to specific amino acids, and phosphatases, enzymes that remove the phosphate groups.

Substrates are primarily phosphorylated on Serine, Threonine, Tyrosine, Histidine, and Arginine residues (Fig. 1B) (Cohen, 2000; Fuhs & Hunter, 2017; Rossetto, Avvakumov, & Côté, 2012; Sawicka & Seiser, 2014). This phosphorylation often alters the substrate in one or more ways, such as (i) changing the substrate's conformation, (ii) altering its ability to bind other proteins, or (iii) changing its intrinsic dynamics. Thus, these changes typically change enzyme activity and/or protein:protein interactions and, in turn, signaling outcomes. Because of this, many signaling events function as switches that turn signaling cascades "on" or "off." Thus, a thorough understanding of how protein kinases and phosphatases orchestrate phosphorylation events is essential for understanding cellular signaling (Day, Sosale, & Lazzara, 2016).

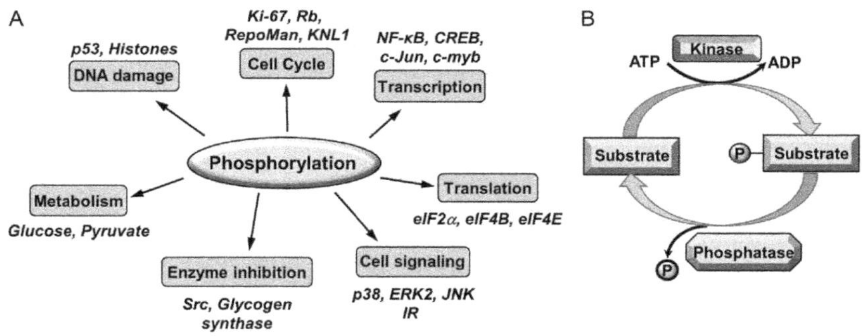

Fig. 1 (A) Phosphorylation is essential in many key cellular processes including signaling, transcription, and the cell cycle, among many others. (B) Phosphorylation involves the transfer of phosphate group from ATP to substrates. Kinases catalyze the phosphate transfer from ATP to substrates, whereas phosphatases catalyze the removal of phosphate group. The reciprocal actions of kinases and phosphatases maintain the level of phosphorylation in the cell.

Both folded and intrinsically unstructured/disordered proteins (IDPs) are regulated by phosphorylation. However, even in folded proteins, most phosphorylation sites are primarily located in disordered regions (i.e., loops) (Iakoucheva et al., 2004). For instance, in many ser/thr kinases, phosphorylation occurs at residues that are located in a disordered activation loop. Phosphorylation of these residues leads to the activation of the kinase (Canagarajah, Khokhlatchev, Cobb, & Goldsmith, 1997; Johnson, Noble, & Owen, 1996; Kishimoto, Matsumoto, & Ninomiya-Tsuji, 2000). Although less common, phosphorylation also occurs at the N-terminus of α-helices, especially at residue positions 1 and 2. Examples of phosphorylation on α-helices include the KIX–pKID complex (Radhakrishnan et al., 1997), sigma factor SpoIIAA (Seavers et al., 2001), ovalbumin (Stein, Leslie, Finch, & Carrell, 1991), and the leucine zipper motif of the bzip protein. In the last example, phosphorylation destabilized the α-helix, which reduced its affinity for DNA (Szilák, Moitra, Krylov, & Vinson, 1997).

In some cases, phosphorylation alters the ensemble of conformations adopted by an IDP in solution. For example, phosphorylation of IDPs can either increase or decrease partially populated secondary structure elements (Johnson & Lewis, 2001). In very rare cases, it can also lead to a stable, folded structure. For example, phosphorylation of the eukaryotic translation initiation factor 4E-binding protein 2 (4E-BP2) caused it to adopt a single stable fold. This phosphorylation-based folding reduced the affinity of 4E-BP2 for eIF4E and suppressed the cap-dependent initiation of translation (Bah et al., 2015). However, phosphorylation can also have the opposite effect. Namely, it was recently shown that phosphorylation of the C-terminal domain of Cdc-37 caused it to unfold and, in turn, promote the phosphorylation of Hsp90, which altered the chaperone function of Hsp90 (Bachman et al., 2018).

In addition to altering protein structure, phosphorylation can also induce changes in protein dynamics. For example, Xiao et al. showed that phosphorylation of the MAPK ERK2 leads to global changes in ERK2 conformational exchange dynamics, resulting in ERK2 activation (Xiao et al., 2014). Similarly, we have also shown that phosphorylation of the MAPK p38α alters both its fast and intermediate time-scale dynamics, ultimately leading to p38 activation (Kumar et al., 2018).

To investigate the role of phosphorylation on protein structure and dynamics, one must isolate the protein of interest and then phosphorylate it on the desired residues. There are two major challenges encountered when phosphorylating substrates in vitro by kinases: (i) the availability of functional kinases for catalyzing the phosphorylation reactions and (ii) optimizing the phosphorylation conditions in order to obtain sufficient quantities of the phosphorylated products for biophysical and molecular studies. It is also important to develop protocols that allow for the separation of phosphorylated from nonphosphorylated substrates in order to obtain samples that are uniformly and fully phosphorylated.

In our laboratory, we have successfully used multiple strategies for the expression and purification of kinases using both eukaryotic and prokaryotic expression systems (Peti & Page, 2007). Here, we describe the protocol used for the in vitro phosphorylation of two substrates, p38 and Ki-67, using two of these kinases, active MKK6 and Aurora kinase B, respectively (Kumar et al., 2018, 2016). p38 is a folded 40.3 kDa protein kinase, whereas Ki-67 is a 5.2 kDa IDP; hence these proteins exemplify the use of our protocol when phosphorylating two very different types of substrates. We also describe the methods used routinely in our laboratory for the purification and characterization of uniformly phosphorylated substrates that can then be studied using NMR spectroscopy.

2. PHOSPHORYLATION

In order to perform in vitro phosphorylation of substrates one needs to: (i) identify the endogenous phosphorylation site(s), (ii) identify the kinase(s) responsible for the phosphorylation, and (iii) produce or purchase the enzymatically active kinases in order to carry out the reaction in vitro. Various databases can be readily used to identify endogenous phosphorylation sites, including PhosphoSite (https://www.phosphosite.org) (Hornbeck et al., 2015), PhosIDA (http://141.61.102.18/phosida/index.aspx) (Gnad, Gunawardena, & Mann, 2011), Phospho.ELM (http://phospho.elm.eu.org/) (Dinkel et al., 2011), and Phospho3D (http://www.phospho3d.org/) (Zanzoni et al., 2011), among others. An added benefit of using these databases is that they commonly also identify the corresponding kinase(s) necessary for a particular phosphorylation event. Next the active kinase must be obtained. Activated kinases have been successfully produced in *Escherichia coli* (i.e., MKK6-EE/DD, pp-ERK2, CDKs, PKA), but more often they are expressed in Sf9 insect cells (insulin receptor kinase, Abl, Src, Lck, BRAF) (Canagarajah et al., 1997; Hubbard, Wei, Ellis, & Hendrickson, 1994; Min et al., 2009; Nagar et al., 2003; Wilson et al., 1996; Xu, Harrison, & Eck, 1997; Yamaguchi & Hendrickson, 1996).

More recently, suspension culture mammalian expression systems (Expi293) have been used to express active eukaryotic kinases at medium to high yields (2–30 mg/L of cell culture); this has become our main choice for producing active kinases. Lastly, a large number of kinases are also commercially available from a broad variety of vendors. It is important to note that, in some cases, kinases require phosphorylation in order to become fully active (p38, ERK2, MKK6, among others), i.e., the kinases are also *substrates* of other kinases. In these cases, activated kinase variants can be expressed and purified (i.e., the MKK6-DD/EE mutant, which is fully active can be expressed and purified from *E. coli* [DD/EE refers to a double Asp/Glu point mutation in the activation loop of MKK6 that leads to the activation of MKK6 (Khokhlatchev et al., 1997; Nielsen & Schwalbe, 2011)]). With the active kinase(s) in hand, it is necessary to optimize the substrate phosphorylation to create a uniformly and quantitatively phosphorylated product for structural and functional studies. The strategies most commonly used to obtain these samples, which in many cases are similar to those needed to obtain activated kinases, are outlined in detail later.

2.1 Coexpression of Substrates With the Activated Kinases

Substrates can be phosphorylated by coexpressing the activated kinase with the substrate. For instance, as stated earlier, the MAPK p38 (which is, in this case, the substrate) can be coexpressed with its upstream kinase MKK6-EE/DD in order to achieve full phosphorylation and activation (*pp*-p38). There are two commonly used strategies to achieve robust coexpression. First, p38 and MKK6-EE/DD can be coexpressed from a single vector that is designed for the coexpression of two target genes (i.e., pET Duet; this vector includes two multiple cloning sites with a T7 promoter/lac operator and a ribosome-binding site at the 5′-end). Second, the substrate and active kinase can be subcloned into different plasmids with different selection genes (i.e., Kan and Amp) and then coexpressed. Careful construct design, i.e., by engineering different affinity tags on each protein, allows for rapid purification of only the activated kinase.

2.2 In Vitro Phosphorylation of Substrates With Purified Activated Kinases

A frequently used strategy for phosphorylating substrates is in vitro phosphorylation. First, the substrate and the active kinase must be purified to homogeneity using standard methods. The substrate is then incubated with

Table 1 Summary of Reagents Necessary for the Preparation of Phosphorylated p38 by $MKK6_{EE}$

Reagents (Stock Solution)	Final Conc.	Vol. (μL)
Water	—	Fill to 15 mL
0.2 *M* HEPES 7.5 + 5.0 m*M* EDTA	20 m*M* 0.5 m*M*	1500
1.0 *M* DTT	2 m*M*	30
0.1 *M* $MgCl_2$	20 m*M*	3000
13.8 μ*M* MKK6[a]	28.0 μg (0.05 μ*M*)	54.5
0.16 m*M* p38α[a]	1.2 mg (2 μ*M*)	186.0
ATP	4.8 m*M*	36.5 mg

[a]p38 and active $MKK6_{EE}$ were purified to homogeneity and were stored at −20°C. Thawed prior to use.

the appropriate active kinase in the presence of ATP and Mg^{2+} (a systematic protocol for the in vitro phosphorylation of a variety of substrates is described in Section 4; see Table 1). This protocol generally requires optimization, as reaction kinetics differ for each kinase and substrate combination. Steps that are often optimized include: (i) the incubation time, (ii) the incubation temperature, and (iii) ATP/Mg^{2+} concentrations. Typically, phosphorylation reactions are complete in 3–5 h, but variations are possible and depend on multiple variables. These include the activity and stability of the kinase, access to phosphorylation sites on substrates, among other factors. The optimal reaction temperature for the phosphor-transfer reaction is between 20°C and 37°C and is highly dependent on the stabilities of the substrate and kinase. Thus, using common methods for measuring protein stability (circular dichroism or differential scanning fluorimetry to obtain protein melting temperatures, T_m) prior to reaction setup is imperative. Enzymes are most stable ~30°C below their T_ms. Finally, incubation of the phosphorylation reaction at higher temperatures and/or for longer times can result in the nonproductive hydrolysis of ATP to ADP, decreasing the efficiency of the phosphorylation reaction (ADP can function as a kinase inhibitor at high concentrations). This can be overcome by adding fresh ATP during the course of the incubation, often allowing for a more efficient completion of the phosphorylation transfer reaction.

We have used in vitro phosphorylation to obtain uniformly phosphorylated samples of dually phosphorylated MAPK p38, *pp*-p38, and the well-known proliferation marker and PP1 regulatory protein Ki-67, *p*-Ki-67, for structural studies. Additional examples of phosphorylated proteins that have been subsequently used for NMR-based structural studies include but are not limited to pKID (by Protein Kinase A) (Radhakrishnan et al., 1997), ERK2 (by MAP kinase 1) (Xiao et al., 2014), PTEN (by CK2 kinase) (Cordier, Chaffotte, & Wolff, 2015), RepoMan (by Aurora kinase B) (Kumar et al., 2016), DARPP32 (by Protein Kinase A) (Dancheck, Nairn, & Peti, 2008), and NHE-1 (by ERK2) (Hendus-Altenburger et al., 2016), among many others.

3. METHODS TO MONITOR PHOSPHORYLATION

Multiple techniques are used to detect phosphorylation. One of the oldest is SDS-PAGE. In this case, differences in the charge between the phosphorylated and nonphosphorylated protein lead to different migration times. However, these migration differences are often small, especially when detecting only small changes in phosphorylation (1–2 residues). In these cases, the normally small changes observed in SDS-PAGE migration can be enhanced by using Phos-tag-PAGE (Kinoshita, Kinoshita-Kikuta, & Koike, 2009). This is a modified version of SDS-PAGE in which a phosphate-binding tag (Mg^{2+}-Phos-tag, which itself is linked to the polyacrylamide in the gel) binds to phosphomonoester dianions (i.e., the phosphate groups on ser/thr/tyr residues). This interaction greatly slows the migration of the phosphorylated samples, leading to a readily detectable separation of the nonphosphorylated and phosphorylated species. Selective stains, such as Pro-Q® Diamond phosphoprotein gel stain (ThermoFisher), can also be used with SDS-PAGE for the rapid detection of phosphorylated proteins. These approaches do not require a high degree of optimization; however, they are not specific for individual proteins and/or phosphorylation sites. Phospho-specific antibodies are a powerful tool to detect whether particular protein is phosphorylated at a particular residue. Hundreds of phospho-specific antibodies are commercially available; however, due to their specificity, their use is often limited to a single target protein. Furthermore, it is necessary to rigorously test the specificity of any used antibody, as they can sometimes lack the specificity advertised.

Clearly, none of these techniques can be used for novel phosphorylation sites, nor are they useful for a quantitative analysis of the extent of phosphorylation. Finally, none of these methods provide any molecular details about which residues are phosphorylated (the only exception is if a site-specific antibody is already available). Thus, mass spectrometry and NMR spectroscopy have emerged as most reliable and efficient methods for the identification and characterization of distinct phosphorylation sites.

3.1 Mass Spectrometry

Mass spectrometry (MS) is readily used to identify phosphorylated proteins/peptides as the stability of the modification is not technique limiting (Aebersold & Goodlett, 2001; Witze, Old, Resing, & Ahn, 2007). Phosphorylation of an amino acid is readily detectable as it increases an amino acid mass by 80 Da (MW of the HPO_3). Thus, increases in a protein's molecular weight of 80, 160, or 240 Da corresponds to one, two, or three phosphorylation events, respectively. Commonly, a mass spectrum shows a single peak upon complete phosphorylation of a protein. However, incomplete phosphorylation reactions will lead to MS data with multiple peaks, i.e., both the phosphorylated and nonphosphorylated protein species will be detected. It is also possible to characterize the level of phosphorylation from the relative intensities of the peaks, assuming both phosphorylated and nonphosphorylated proteins ionize similarly. Thus, MS can be used for the optimization of in vitro reaction conditions that ultimately allow for complete phosphorylation to be achieved.

Mass shifts of fragment ions in tandem mass spectrometry (MS/MS) can be used to identify the amino acid that is phosphorylated (i.e., the phosphorylation site). However, the use of MS/MS for identifying phosphorylation sites is only possible for peptides. Thus, for proteins that are larger >3 kDa, chemical digestion (i.e., using proteases such as trypsin and chymotrypsin) converts the protein under investigation into smaller peptide-sized fragments that allow for high quality MS/MS spectra to be recorded (bottom-up proteomics) (Aebersold & Mann, 2003). As an example, we phosphorylated p38 using an activated variant of its upstream kinase, MKK6-DD/EE. MKK6-DD/EE phosphorylates p38 on residues Thr180 and Tyr182 (*pp*-p38). Using ESI-MS, we detected an increase in 160 Da in the *pp*-p38 sample which was accompanied by the loss of the peak corresponding to nonphosphorylated p38. This data confirmed that both residues were phosphorylated (Fig. 2A). On the other hand, Ki-67 has a single phosphorylation site, as was evident from a mass increase of only 80 Da (Fig. 2B).

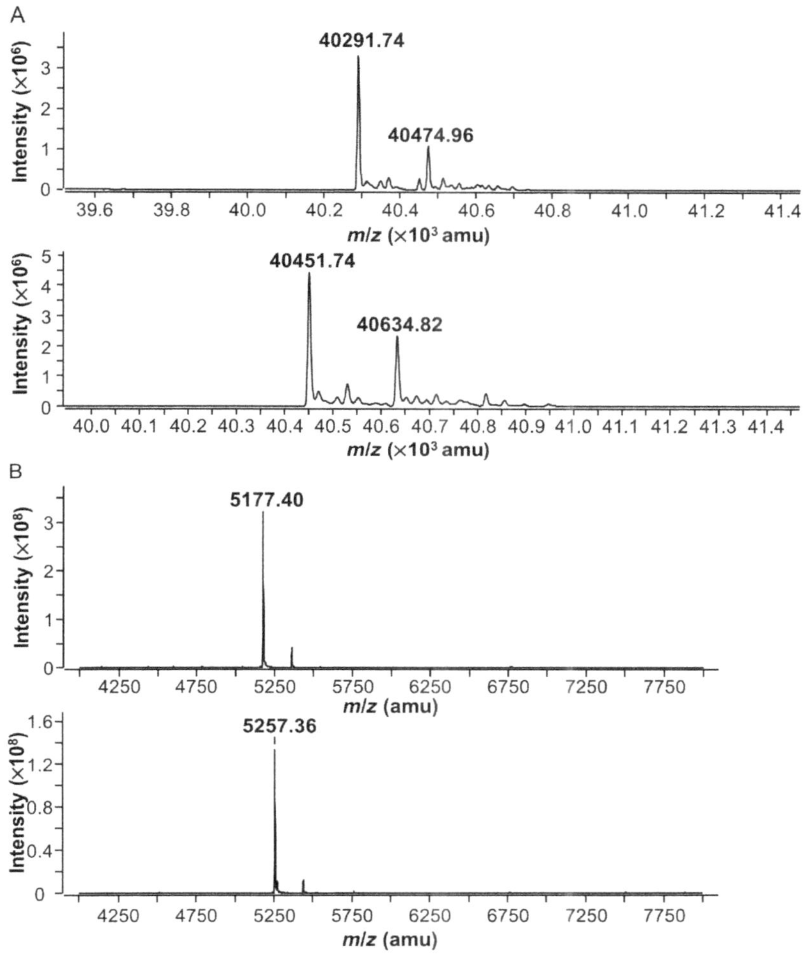

Fig. 2 Identification of phosphorylation using mass spectrometry. (A) ESI-MS spectrum of phosphorylated (*bottom panel*) and nonphosphorylated (*top panel*) p38. Phosphorylated p38 shows a 160 Da increase in mass indicating the phosphorylation of two amino acids. (B) MALDI-TOF spectrum of nonphosphorylated (*top panel*) and phosphorylated (*bottom panel*) Ki-67 shows an 80 Da difference in mass indicating a single phosphorylation site (S507) in Ki-67.

3.2 NMR Spectroscopy

NMR spectroscopy has emerged as a complementary technique for studying posttranslational modifications (Theillet et al., 2012). Chemical shifts, an important observable in NMR spectroscopy, are directly dependent on the local chemical environment. The incorporation of highly electronegative groups, such as a phosphate ion, onto an amino acid will result in a significantly changed chemical shift (most commonly a downfield shift). This change in chemical shift is particularly strong for ser and thr residues; the chemical shift changes for tyr phosphorylation are less dramatic but also often lead to changes in the chemical shifts of neighboring residues. Thus, both are readily detected. Hence, NMR spectroscopy is an exceptional tool for identifying phosphorylation sites by monitoring the chemical shifts of ser/thr/tyr residues. These changes are detected by comparing the 2D [^{1}H,^{15}N] HSQC (Heteronuclear Single Quantum Coherence) spectra of ^{15}N-labeled phosphorylated and nonphosphorylated proteins. Specifically, upon phosphorylation, the (^{1}H,^{15}N) cross peak that corresponds to the nonphosphorylated ser/thr residue will disappear and reappear at a different, now downfield-shifted, position in the 2D [^{1}H,^{15}N] HSQC spectrum. While the ^{1}H,^{15}N chemical shifts are valuable and easy to access reporters, other chemical shifts, including side chain C^{β} chemical shifts of phosphorylated ser and thr residues, also show chemical shift changes due to closest neighbor effects of the phosphate group.

Importantly, the total number of new cross peaks that appear in the spectrum upon phosphorylation directly report on the number of phosphorylated residues (exceptions would be a slow exchange process that can happen upon phosphorylation). In addition, NMR spectroscopy can also be used to derive the kinetics, the order as well as possible interdependency of phosphorylation. This is readily achieved by acquiring a series of 2D [^{1}H,^{15}N] HSQC/TROSY spectra over time and monitoring the appearing and disappearing intensities of the phosphorylated and unphosphorylated peaks, respectively. The rate constant of phosphorylation is determined by fitting the amplitudes of the peaks to a single exponential function (Cordier et al., 2012).

Comparing the 2D [^{1}H,^{15}N] TROSY spectrum of the ^{2}H,^{15}N-labeled MAP kinase p38 before and after phosphorylation shows two peaks that experience large chemical shift changes (Fig. 3A, indicated by arrow; small chemical shifts are commonly seen for residues neighboring the phosphorylated residues). As the sequence-specific backbone assignment of p38 is known (Francis et al., 2011), it is readily possible to identify the residues that undergo phosphorylation. Indeed, the cross peaks corresponding to Thr180 and Tyr182 disappear upon phosphorylation and reappear at a different position. As anticipated, small additional chemical shift changes are observed

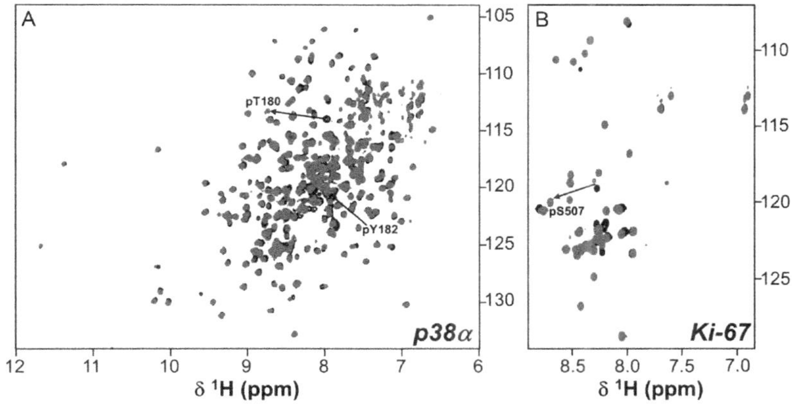

Fig. 3 Characterization of phosphorylation using NMR spectroscopy. (A) 2D [^{1}H,^{15}N] TROSY spectrum of p38 before (*black*) and after (*gray*) phosphorylation by MKK6$_{EE}$. Note that the peaks corresponding to phosphorylated residues T180 and Y182 disappear and reappear at a changed chemical shift position upon phosphorylation (indicated by *arrows*). (B) 2D [^{1}H,^{15}N] HSQC spectrum of Aurora kinase B phosphorylated (*gray*) and nonphosphorylated (*black*) Ki-67$_{496-536}$. Residue S507 is phosphorylated and the change in the peak position is indicated by *black arrow*. *Panel (A): Modified with permission from Kumar, G. S., Clarkson, M. W., Kunze, M. B. A., Granata, D., Wand, A. J., Lindorff-Larsen, K., et al.(2018). Dynamic activation and regulation of the mitogen-activated protein kinase p38.* Proceedings of the National Academy of Sciences of the United States of America, 115, *4655–4660. https://doi.org/10.1073/pnas.1721441115. Panel (B): Modified with permission from Kumar, G. S., Gokhan, E., De Munter, S., Bollen, M., Vagnarelli, P., Peti, W., et al. (2016). The Ki-67 and RepoMan mitotic phosphatases assemble via an identical, yet novel mechanism.* eLife, 5. *https://doi.org/10.7554/eLife.16539.*

upon phosphorylation, which are all localized to the residues around Thr180 and Tyr182 (Kumar et al., 2018). This shows that p38 phosphorylation only affects the local environment and that there are no phosphorylation-induced structural changes in the protein. Similarly, in the 2D [^{1}H,^{15}N] HSQC spectrum of ^{15}N-labeled Ki-67, a cross peak corresponding to Ser507 disappears and reappears at a downfield-shifted position upon phosphorylation with Aurora B kinase (Fig. 3B, indicated by arrow).

4. PROBLEMS, PITFALLS, AND ALTERNATIVE ROUTES

A key defining factor for substrate phosphorylation is the specificity of its cognate kinase (Miller & Turk, 2018; Ubersax & Ferrell, 2007; Zhu, Liu, & Shaw, 2005). Kinase specificity is determined by: (i) the primary sequence around the phosphorylation site, (ii) cellular substrate recruitment, and (iii) the cellular localization of the substrate. However, the high specificity observed in vivo is not always recapitulated in vitro. This is because, in vitro, it is possible (and observed) that residues that are not normally specific substrates of the kinase are inadvertently phosphorylated. This is due, in large part, to the high concentrations of activated kinase and substrate. The most effective way to overcome this problem is to mutate the nonspecifically phosphorylated residue to an Ala (Ser/Thr) or Phe (Tyr). This will prevent the undesired phosphorylation.

The other challenge can be the production of the active kinase. Often, kinases become active only after being phosphorylated by another kinase or by binding to a specific binding partner. For instance, MAPK signaling pathways involve the sequential activation of multiple kinases, ultimately leading to the phosphorylation of pathway-specific substrates (Peti & Page, 2013). MAPK p38 is phosphorylated by its upstream kinases MKK3 or MKK6. However, MKK6 is only active after it is phosphorylated by its upstream kinase, MEKK2/3. One strategy for overcoming this issue is to generate phosphomimetic mutations (Glu or Asp) in the upstream kinase at the phosphorylated residues in order to artificially induce its activation. We have successfully used this approach for phosphorylating p38. In this case, the residues that are normally phosphorylated to make MKK6 active are mutated to Glu. This results in an active MKK6 variant, MKK6-DD/EE. This variant can be then used to phosphorylate p38 in vitro or in vivo (via coexpression). Another example includes the in vitro phosphorylation of ERK2 by an artificially activated variant of MAP kinase kinase 1 (MEK1), where the residues that must be phosphorylated in order to render MEK1 active are mutated to Asp. It is important to know that phosphomimetic mutations do not always result in an activated kinase; thus it is important to test that these variants indeed lead to an active kinase.

Finally, many kinases are also often activated by a direct interaction with binding and/or scaffolding proteins. For example, cyclin-dependent kinases (CDKs) form stable complexes with cyclins, which are necessary both for CDK activation and to confer substrate specificity (Schulman, Lindstrom, & Harlow, 1998). Another example is the complex formed between INCENP and Aurora kinase B (AuKB) (Santaguida, Tighe, D'Alise, Taylor, & Musacchio, 2010). In these cases, it is necessary to either coexpress the kinase with its binding partner or provide the binding partner sequentially, in order to achieve activity.

5. PROTOCOL FOR IN VITRO PHOSPHORYLATION

This protocol assumes that purified substrate and kinase are available.

5.1 Equipment

GE Healthcare AKTA Purifier/Pure or BioRad NGC protein purification system
BioRad Standard SDS-PAGE equipment
BioRad Precast SDS-PAGE gels (4%–20% Mini-PROTEAN TGX Stain-Free Precast SDS-PAGE Gels)
Water Bath
Anion-Exchange column Mono Q (5/50 GL, GE Healthcare, 1 CV = 1.0 mL)
Amicon/Sartorius Ultra-15 Concentrator (10 or 3 kDa MWCO)

5.2 Buffers and Reagents

All the reagents were purchased from Fisher, Sigma, or Gold Bio Com unless specified
0.2 *M* HEPES pH 7.5 containing 5 m*M* EDTA
1 *M* DTT
0.1 *M* $MgCl_2$
ATP powder (Roche)
Mono Q buffers—buffer A (20 m*M* Tris pH 7.6 75 m*M* NaCl, 0.5 m*M* TCEP) and buffer B (20 m*M* Tris pH 7.6, 400 m*M* NaCl, 0.5 m*M* TCEP)

5.3 Procedure

1. Prepare stock solutions of 0.2 *M* HEPES pH 7.5, containing 5 m*M* EDTA, 1 *M* DTT, 0.1 *M* $MgCl_2$.
2. Mix HEPES, DTT, $MgCl_2$, and water as shown in Table 1. The ingredients shown in Table 1 are for a total reaction volume of 15 mL. For large-scale production, perform reactions in duplicates or multiplets.
3. Add ATP powder to the reaction to a final concentration of 4.8 m*M* (this concentration of ATP is optimized for p38α/MKK6-DD/EE and Ki-67/AuKB reactions).
4. Adjust the pH to 7.5.

5. Add the substrate protein that is to be phosphorylated (e.g., p38/Ki-67) and the appropriate active kinase (e.g., MKK6-DD/EE/Aurora kinase B, respectively).
6. Optimize the incubation time and temperature for a particular concentration of ATP and kinase–substrate pair. For example, incubate the reaction mixture at two different temperatures (i.e., 25°C and 30°C), withdraw samples at different time intervals (0, 1, 2, 3, 4, and 5 h), and then analyze for the formation of phosphorylated substrate using mass spectrometry. For p38–MKK6-DD/EE and Ki-67-Aurora kinase B, phosphorylation was complete in 5 h at 27°C.
7. Once the in vitro phosphorylation reaction is complete, concentrate and exchange the phosphorylated substrate into a buffer suitable for Mono Q chromatography using an Amicon/Sartorius Centrifugal Filter Device (10 or 3 kDa MWCO); this step will also remove ATP and ADP.
8. Purify the substrate using Mono Q chromatography; Buffer A (no/low salt, i.e., 0/0.05 *M* NaCl) and Buffer B (high salt, i.e., 1 *M* NaCl) by running a gradient of 0%–100% B over a period of 60–120 min at a flow rate of 0.5–1 mL/min.
9. Pool the fractions containing the eluted substrate. If multiple peaks are observed during the elution, there are likely differentially phosphorylated species (this is often observed when the reaction is not 100% complete).
10. Concentrate and exchange the uniformly phosphorylated protein into the optimal buffer for NMR spectroscopy using an Amicon/Sartorius Centrifugal Filter Device (10 or 3 kDa MWCO) or similar methods.
11. Confirm phosphorylation by MS or NMR spectroscopy.

5.4 Notes

1. In the case of Aurora kinase B, the presence of a GST-tag did not inhibit its activity.
2. The choice of buffer and pH depends on the stability of the substrate and kinase. In general, a buffer (Tris/HEPES/phosphate) in the pH range of 6.5–8 is optimal for phosphorylation.
3. The optimal temperature for phosphorylation reactions can vary between 10°C and 30°C, depending on the stability of the protein. Note that at lower temperatures, the phosphorylation reaction will require longer incubation times (up to 12–16 h, or even longer).
4. Adjusting the pH after the addition of ATP powder is essential, as the addition of ATP alters the pH of the buffer solution.

5. The amount of ATP, protein, and the incubation time for the phosphorylation reaction needs to be fine-tuned depending on the number of phosphorylation sites.
6. In our experience, final ATP concentrations of 5–10 m*M* in the reaction mixture work well for single to triple phosphorylation sites, respectively. However, this is clearly kinase–substrate dependent.
7. Anion exchange chromatography removes the nonphosphorylated substrate (which elutes earlier than the phosphorylated substrate) as well as the kinase (which may or may not bind to anion exchange column depending on the pH/pI used).
8. Use of Mono Q buffer A with slightly higher salt concentrations (75 m*M*) often helps to overcome the nonspecific binding of the kinase to the column and provides a quick way to perform cleanup.
9. Alternatively, for IDPs, heat purification at 65–95°C will separate the kinase from the phosphorylated IDP substrate, as only IDPs will stay in solution.
10. The use of size-exclusion chromatography after Mono Q/heat purification is often a necessary polishing step.
11. In some cases, the solubility of the kinases can be lower in buffers used for phosphorylation (Table 1). In these cases, it is often helpful to add low levels of nonionic detergents (~0.01%) like NP40 or Triton X-100 in the buffers.
12. The addition of $MnCl_2$, as opposed to $MgCl_2$, can help boost the activity of some kinases, such as integrin-linked kinases. However, several kinases like lipid kinases show lower activity when Mn^{2+} is present as the divalent cation. When Mn^{2+} is present, use only TCEP as the reducing agent.
13. Real-time phosphorylation reaction monitoring by NMR spectroscopy will often show T_1 noise artifact signals from ATP (due to the high concentration used) that can be identified in the 2D [^1H,^{15}N] HSQC spectrum at 6.2, 8.2, and 8.5 ppm (Kanelis, Hudson, Thibodeau, Thomas, & Forman-Kay, 2010).

6. SUMMARY AND CONCLUSIONS

Here, we describe a general protocol for the large-scale production of phosphorylated proteins for structural and dynamic studies. These methods have worked efficiently for the production of multiple proteins in many laboratories including ours. Monitoring phosphorylation using mass spectrometry and solution NMR spectroscopy is readily achieved and is critical in order to know when the reaction is complete. Although this protocol is often successful and reliable, it almost always requires fine-tuning, including identifying optimal incubation times, incubation temperatures, and ATP concentrations for a particular kinase–substrate pair. It is also important to optimize the purification protocols for each individual protein under investigation, especially when removing the phosphorylated substrate from the kinase and nonphosphorylated substrate after the in vitro reaction.

ACKNOWLEDGMENTS

This work was supported by the American Diabetes Association Pathway to Stop Diabetes Grant 1-14-ACN-31 to W.P. and the National Institute of Health Grants R01GM100910 and R01NS091336 to W.P. and R01GM098482 to R.P.

REFERENCES

Aebersold, R., & Goodlett, D. R. (2001). Mass spectrometry in proteomics. *Chemical Reviews*, *101*, 269–295. 11712248.

Aebersold, R., & Mann, M. (2003). Mass spectrometry-based proteomics. *Nature*, *422*, 198–207. https://doi.org/10.1038/nature01511.

Bachman, A. B., Keramisanou, D., Xu, W., Beebe, K., Moses, M. A., Vasantha Kumar, M. V., et al. (2018). Phosphorylation induced cochaperone unfolding promotes kinase recruitment and client class-specific Hsp90 phosphorylation. *Nature Communications*, *9*, 265. https://doi.org/10.1038/s41467-017-02711-w.

Bah, A., Vernon, R. M., Siddiqui, Z., Krzeminski, M., Muhandiram, R., Zhao, C., et al. (2015). Folding of an intrinsically disordered protein by phosphorylation as a regulatory switch. *Nature*, *519*, 106–109. https://doi.org/10.1038/nature13999.

Canagarajah, B. J., Khokhlatchev, A., Cobb, M. H., & Goldsmith, E. J. (1997). Activation mechanism of the MAP kinase ERK2 by dual phosphorylation. *Cell*, *90*, 859–869. 9298898.

Cohen, P. (2000). The regulation of protein function by multisite phosphorylation—A 25 year update. *Trends in Biochemical Sciences*, *25*, 596–601. 11116185.

Cordier, F., Chaffotte, A., Terrien, E., Préhaud, C., Theillet, F.-X., Delepierre, M., et al. (2012). Ordered phosphorylation events in two independent cascades of the PTEN C-tail revealed by NMR. *Journal of the American Chemical Society*, *134*, 20533–20543. https://doi.org/10.1021/ja310214g.

Cordier, F., Chaffotte, A., & Wolff, N. (2015). Quantitative and dynamic analysis of PTEN phosphorylation by NMR. *Methods*, *77–78*, 82–91. https://doi.org/10.1016/j.ymeth.2014.10.007.

Dancheck, B., Nairn, A. C., & Peti, W. (2008). Detailed structural characterization of unbound protein phosphatase 1 inhibitors. *Biochemistry*, *47*, 12346–12356. https://doi.org/10.1021/bi801308y.

Day, E. K., Sosale, N. G., & Lazzara, M. J. (2016). Cell signaling regulation by protein phosphorylation: A multivariate, heterogeneous, and context-dependent process. *Current Opinion in Biotechnology*, *40*, 185–192. https://doi.org/10.1016/j.copbio.2016.06.005.

Dinkel, H., Chica, C., Via, A., Gould, C. M., Jensen, L. J., Gibson, T. J., et al. (2011). Phospho.ELM: A database of phosphorylation sites-update 2011. *Nucleic Acids Research*, *39*, D261–D267. https://doi.org/10.1093/nar/gkq1104.

Francis, D. M., Różycki, B., Koveal, D., Hummer, G., Page, R., & Peti, W. (2011). Structural basis of p38α regulation by hematopoietic tyrosine phosphatase. *Nature Chemical Biology*, 7, 916–924. https://doi.org/10.1038/nchembio.707.

Fuhs, S. R., & Hunter, T. (2017). pHisphorylation: The emergence of histidine phosphorylation as a reversible regulatory modification. *Current Opinion in Cell Biology*, *45*, 8–16. https://doi.org/10.1016/j.ceb.2016.12.010.

Gnad, F., Gunawardena, J., & Mann, M. (2011). PHOSIDA 2011: The posttranslational modification database. *Nucleic Acids Research*, *39*, D253–D260. https://doi.org/10.1093/nar/gkq1159.

Hendus-Altenburger, R., Pedraz-Cuesta, E., Olesen, C. W., Papaleo, E., Schnell, J. A., Hopper, J. T. S., et al. (2016). The human Na(+)/H(+) exchanger 1 is a membrane scaffold protein for extracellular signal-regulated kinase 2. *BMC Biology*, *14*, 31. https://doi.org/10.1186/s12915-016-0252-7.

Hornbeck, P. V., Zhang, B., Murray, B., Kornhauser, J. M., Latham, V., & Skrzypek, E. (2015). PhosphoSitePlus, 2014: Mutations, PTMs and recalibrations. *Nucleic Acids Research*, *43*, D512–D520. https://doi.org/10.1093/nar/gku1267.

Hubbard, S. R., Wei, L., Ellis, L., & Hendrickson, W. A. (1994). Crystal structure of the tyrosine kinase domain of the human insulin receptor. *Nature*, *372*, 746–754. https://doi.org/10.1038/372746a0.

Humphrey, S. J., James, D. E., & Mann, M. (2015). Protein phosphorylation: A major switch mechanism for metabolic regulation. *Trends in Endocrinology and Metabolism*, *26*, 676–687. https://doi.org/10.1016/j.tem.2015.09.013.

Hunter, T., & Karin, M. (1992). The regulation of transcription by phosphorylation. *Cell*, *70*, 375–387. 1643656.

Iakoucheva, L. M., Radivojac, P., Brown, C. J., O'Connor, T. R., Sikes, J. G., Obradovic, Z., et al. (2004). The importance of intrinsic disorder for protein phosphorylation. *Nucleic Acids Research*, *32*, 1037–1049. https://doi.org/10.1093/nar/gkh253.

Johnson, L. N., & Lewis, R. J. (2001). Structural basis for control by phosphorylation. *Chemical Reviews*, *101*, 2209–2242. 11749371.

Johnson, L. N., Noble, M. E., & Owen, D. J. (1996). Active and inactive protein kinases: Structural basis for regulation. *Cell*, *85*, 149–158. 8612268.

Kanelis, V., Hudson, R. P., Thibodeau, P. H., Thomas, P. J., & Forman-Kay, J. D. (2010). NMR evidence for differential phosphorylation-dependent interactions in WT and DeltaF508 CFTR. *The EMBO Journal*, *29*, 263–277. https://doi.org/10.1038/emboj.2009.329.

Khokhlatchev, A., Xu, S., English, J., Wu, P., Schaefer, E., & Cobb, M. H. (1997). Reconstitution of mitogen-activated protein kinase phosphorylation cascades in bacteria. Efficient synthesis of active protein kinases. *The Journal of Biological Chemistry*, *272*, 11057–11062. 9110999.

Kinoshita, E., Kinoshita-Kikuta, E., & Koike, T. (2009). Separation and detection of large phosphoproteins using Phos-tag SDS-PAGE. *Nature Protocols*, *4*, 1513–1521. https://doi.org/10.1038/nprot.2009.154.

Kishimoto, K., Matsumoto, K., & Ninomiya-Tsuji, J. (2000). TAK1 mitogen-activated protein kinase kinase kinase is activated by autophosphorylation within its activation loop. *The Journal of Biological Chemistry*, *275*, 7359–7364. 10702308.

Kumar, G. S., Clarkson, M. W., Kunze, M. B. A., Granata, D., Wand, A. J., Lindorff-Larsen, K., et al. (2018). Dynamic activation and regulation of the mitogen-activated protein kinase p38. *Proceedings of the National Academy of Sciences of the United States of America*, *115*, 4655–4660. https://doi.org/10.1073/pnas.1721441115.

Kumar, G. S., Gokhan, E., De Munter, S., Bollen, M., Vagnarelli, P., Peti, W., et al. (2016). The Ki-67 and RepoMan mitotic phosphatases assemble via an identical, yet novel mechanism. *eLife*, *5*, (Aug 30, pii: e16539). https://doi.org/10.7554/eLife.16539.

Miller, C. J., & Turk, B. E. (2018). Homing in: Mechanisms of substrate targeting by protein kinases. *Trends in Biochemical Sciences*, *43*, 380–394. https://doi.org/10.1016/j.tibs.2018.02.009.

Min, X., Akella, R., He, H., Humphreys, J. M., Tsutakawa, S. E., Lee, S.-J., et al. (2009). The structure of the MAP2K MEK6 reveals an autoinhibitory dimer. *Structure*, *17*, 96–104. https://doi.org/10.1016/j.str.2008.11.007.

Nagar, B., Hantschel, O., Young, M. A., Scheffzek, K., Veach, D., Bornmann, W., et al. (2003). Structural basis for the autoinhibition of c-Abl tyrosine kinase. *Cell*, *112*, 859–871. 12654251.

Nielsen, G., & Schwalbe, H. (2011). NMR spectroscopic investigations of the activated p38α mitogen-activated protein kinase. *Chembiochem*, *12*, 2599–2607. https://doi.org/10.1002/cbic.201100527.

Olsen, J. V., Vermeulen, M., Santamaria, A., Kumar, C., Miller, M. L., Jensen, L. J., et al. (2010). Quantitative phosphoproteomics reveals widespread full phosphorylation site occupancy during mitosis. *Science Signaling*, *3*, ra3. https://doi.org/10.1126/scisignal.2000475.

Peti, W., & Page, R. (2007). Strategies to maximize heterologous protein expression in Escherichia coli with minimal cost. *Protein Expression and Purification*, *51*, 1–10. https://doi.org/10.1016/j.pep.2006.06.024.

Peti, W., & Page, R. (2013). Molecular basis of MAP kinase regulation. *Protein Science*, *22*, 1698–1710. https://doi.org/10.1002/pro.2374.

Radhakrishnan, I., Pérez-Alvarado, G. C., Parker, D., Dyson, H. J., Montminy, M. R., & Wright, P. E. (1997). Solution structure of the KIX domain of CBP bound to the transactivation domain of CREB: A model for activator:Coactivator interactions. *Cell*, *91*, 741–752. 9412984.

Rossetto, D., Avvakumov, N., & Côté, J. (2012). Histone phosphorylation: A chromatin modification involved in diverse nuclear events. *Epigenetics*, 7, 1098–1108. https://doi.org/10.4161/epi.21975.

Santaguida, S., Tighe, A., D'Alise, A. M., Taylor, S. S., & Musacchio, A. (2010). Dissecting the role of MPS1 in chromosome biorientation and the spindle checkpoint through the small molecule inhibitor reversine. *The Journal of Cell Biology*, *190*, 73–87. https://doi.org/10.1083/jcb.201001036.

Sawicka, A., & Seiser, C. (2014). Sensing core histone phosphorylation—A matter of perfect timing. *Biochimica et Biophysica Acta*, *1839*, 711–718. https://doi.org/10.1016/j.bbagrm.2014.04.013.

Schulman, B. A., Lindstrom, D. L., & Harlow, E. (1998). Substrate recruitment to cyclin-dependent kinase 2 by a multipurpose docking site on cyclin A. *Proceedings of the National Academy of Sciences of the United States of America*, *95*, 10453–10458. 9724724.

Seavers, P. R., Lewis, R. J., Brannigan, J. A., Verschueren, K. H., Murshudov, G. N., & Wilkinson, A. J. (2001). Structure of the Bacillus cell fate determinant SpoIIAA in phosphorylated and unphosphorylated forms. *Structure*, *9*, 605–614. 11470435.

Stein, P. E., Leslie, A. G., Finch, J. T., & Carrell, R. W. (1991). Crystal structure of uncleaved ovalbumin at 1.95 A resolution. *Journal of Molecular Biology*, *221*, 941–959. 1942038.

Szilák, L., Moitra, J., Krylov, D., & Vinson, C. (1997). Phosphorylation destabilizes alpha-helices. *Nature Structural Biology*, *4*, 112–114. 9033589.

Theillet, F.-X., Smet-Nocca, C., Liokatis, S., Thongwichian, R., Kosten, J., Yoon, M.-K., et al. (2012). Cell signaling, post-translational protein modifications and NMR spectroscopy. *Journal of Biomolecular NMR*, *54*, 217–236. https://doi.org/10.1007/s10858-012-9674-x.

Ubersax, J. A., & Ferrell, J. E. (2007). Mechanisms of specificity in protein phosphorylation. *Nature Reviews. Molecular Cell Biology*, *8*, 530–541. https://doi.org/10.1038/nrm2203.

Wilson, K. P., Fitzgibbon, M. J., Caron, P. R., Griffith, J. P., Chen, W., McCaffrey, P. G., et al. (1996). Crystal structure of p38 mitogen-activated protein kinase. *The Journal of Biological Chemistry*, *271*, 27696–27700. 8910361.

Witze, E. S., Old, W. M., Resing, K. A., & Ahn, N. G. (2007). Mapping protein post-translational modifications with mass spectrometry. *Nature Methods*, *4*, 798–806. https://doi.org/10.1038/nmeth1100.

Xiao, Y., Lee, T., Latham, M. P., Warner, L. R., Tanimoto, A., Pardi, A., et al. (2014). Phosphorylation releases constraints to domain motion in ERK2. *Proceedings of the National Academy of Sciences of the United States of America*, *111*, 2506–2511. https://doi.org/10.1073/pnas.1318899111.

Xu, W., Harrison, S. C., & Eck, M. J. (1997). Three-dimensional structure of the tyrosine kinase c-Src. *Nature*, *385*, 595–602. https://doi.org/10.1038/385595a0.

Yamaguchi, H., & Hendrickson, W. A. (1996). Structural basis for activation of human lymphocyte kinase Lck upon tyrosine phosphorylation. *Nature*, *384*, 484–489. https://doi.org/10.1038/384484a0.

Zanzoni, A., Carbajo, D., Diella, F., Gherardini, P. F., Tramontano, A., Helmer-Citterich, M., et al. (2011). Phospho3D 2.0: An enhanced database of three-dimensional structures of phosphorylation sites. *Nucleic Acids Research*, *39*, D268–D271. https://doi.org/10.1093/nar/gkq936.

Zhu, G., Liu, Y., & Shaw, S. (2005). Protein kinase specificity. A strategic collaboration between kinase peptide specificity and substrate recruitment. *Cell Cycle*, *4*, 52–56. https://doi.org/10.4161/cc.4.1.1353.

CHAPTER EIGHT

Synthesis and NMR Characterization of the Prenylated Peptide, a-Factor

Taysir K. Bader, Todd M. Rappe, Gianlugi Veglia, Mark D. Distefano[1]
University of Minnesota, Twin Cities, Minneapolis, MN, United States
[1]Corresponding author: e-mail address: diste001@umn.edu

Contents

Abstract

Protein and peptide prenylation is an essential biological process involved in many signal transduction pathways. Hence, it plays a critical role in establishing many major human ailments, including Alzheimer's disease, amyotrophic lateral sclerosis (ALS), malaria, and Ras-related cancers. Yeast mating pheromone a-factor is a small dodecameric peptide that undergoes prenylation and subsequent processing in a manner identical to larger proteins. Due to its small size in addition to its well-characterized behavior in yeast, a-factor is an attractive model system to study the prenylation pathway. Traditionally, chemical synthesis and characterization of a-factor have been challenging, which has limited its use in prenylation studies. In this chapter, a robust method for the synthesis of a-factor is presented along with a description of the characterization of the peptide using MALDI and NMR. Finally, complete assignments of resonances from

Methods in Enzymology, Volume 614
ISSN 0076-6879
https://doi.org/10.1016/bs.mie.2018.09.025

the isoprenoid moiety and a-factor from COSY, TOCSY, HSQC, and long-range HMBC NMR spectra are presented. This methodology should be useful for the synthesis and characterization of other mature prenylated peptides and proteins.

1. INTRODUCTION

Protein and peptide prenylation is an essential biological process involved in many signal transduction pathways (Kassai & Fukada, 2011). It involves the covalent attachment of an isoprenoid moiety to the cysteine residue of a C-terminal CaaX box sequence via a thioester linkage (Casey & Seabra, 1996). The attached isoprenoid unit can be either a 15-carbon farnesyl group or a 20-carbon geranylgeranyl group, and facilitates anchoring in the lipid membrane (Kassai & Fukada, 2011). Prenylated proteins and changes in the levels thereof are at the heart of many significant human ailments including malaria (Chakrabarti et al., 2002; Suazo, Schaber, Palsuledesai, Odom John, & Distefano, 2016), Alzheimer's disease (Eckert et al., 2009), amyotrophic lateral sclerosis (ALS) (Li et al., 2016), as well as many different cancers (Ochocki & Distefano, 2013). One noteworthy lipidated protein is the oncogenic Ras protein, which is involved in more than 30% of all human cancers. As such, understanding and inhibiting the prenylation pathway have been a prime research target (Ochocki & Distefano, 2013). The prenylation of the Ras protein (as well as other prenylated proteins) with an isoprenoid group typically involves a three-step pathway and three distinct enzymatic activities. It begins in the cytosol where protein farnesyltransferase (PFTase) recognizes the CaaX sequence and attaches a farnesyl group to the cysteine's thiol. This leads to anchoring in the lipid bilayer of the endoplasmic reticulum, where a membrane-attached protease proteolytically cleaves off the "aaX" sequence. Finally, the cysteine's newly exposed C-terminal carboxylate becomes methylated by a membrane-bound carboxymethyl transferase to yield the fully modified protein (Diaz-Rodriguez & Distefano, 2017).

Significant strides in understanding the prenylation pathway have emerged as a result of studying the *Saccharomyces cerevisiae* mating pheromone a-factor (Sinensky, 2000). a-Factor is a hydrophobic dodecapeptide that is processed through the prenylation pathway (Caldwell, Naider, & Becker, 1995). The small size of a-factor makes it an attractive system to probe various aspects of the prenylation pathway. However, availability of a-factor has traditionally been limited due to its significant hydrophobicity as well as having an esterified C-terminus, which complicates its synthesis using standard Fmoc-based solid phase peptide synthesis (SPPS). Earlier strategies for a-factor synthesis involved solution state fragment condensation or HF-based SPPS methods (Diaz-Rodriguez & Distefano, 2017). Additionally, while in Vitro enzymatic farnesylation of CaaX substrates has been described previously (Hosokawa et al., 2007; Palsuledesai et al., 2016; Rowell, Kowalczyk, Lewis, & Garcia, 1997), we were not able to find evidence of such a reaction applied to a-factor, likely due to the hydrophobic nature of the peptides. Our group has developed more streamlined methods for the synthesis of a-factor, first using hydrazine-containing resin followed by oxidative resin cleavage (Mullen et al., 2011), and then more recently by side-chain anchoring to trityl chloride resin (Diaz-Rodriguez, Ganusova, Rappe, Becker, & Distefano, 2015; Diaz-Rodriguez, Mullen, Ganusova, Becker, & Distefano, 2012). The later method showed greater versatility and marked increases in yield with minimal epimerization observed after prolonged exposure to basic conditions mimicking extended rounds of SPPS coupling deprotection, and hence that procedure is described here. In brief, Fmoc-Cys-OMe is synthesized and coupled to a trityl-chloride resin via its thiol side-chain. The peptide is next extended via standard Fmoc chemistry and is then cleaved from resin and side-chain-deprotected using acidic conditions. The resulting peptide is subsequently purified and farnesylated in solution, followed by HPLC purification to homogeneity. This methodology has been employed to synthesize a-factor analogs and incorporate farnesylated peptides with C-terminal methyl esters into whole proteins using native chemical ligation (Zhang et al., 2017).

Once the final peptide is obtained and purified, it can be further analyzed using mass spectrometry and NMR. Here, MALDI was used for mass spectrometric analysis, as it resulted in better ionization of the peptide and simpler spectra relative to ESI. In fact, MALDI typically produces singly charged species and more predictable adducts (Nadler et al., 2017). Next, NMR was used to study the chemical structure of a-factor at the atomic level. Previous NMR studies showed that a-factor adopts a predominantly disordered conformation in DMSO, which is largely unaffected by farnesylation (Gounarides, Broido, Xue, Becker, & Naider, 1991). Therefore, the NMR experiments presented here focus on the characterization of the a-factor peptide using J coupling experiments to validate the chemical structure of a-factor and its analogues. (Diaz-Rodriguez & Distefano, 2017; Diaz-Rodriguez et al., 2012; Vervacke et al., 2014). The experiments and the methods described here can be used in structural studies of other prenylated peptides of biological importance.

2. SYNTHESIS OF A-FACTOR

2.1 Fmoc-Cys-OMe Synthesis

The first step in the synthesis of a-factor is to obtain Fmoc-protected cysteine with a free side-chain thiol and C-terminal methyl ester. This can be achieved through an acid catalyzed reaction of commercially available Fmoc-L-Cys hydrate with methanol, as outlined in Scheme 1. This reaction has also been employed to produce other alkyl esters of a-factor (Diaz-Rodriguez et al., 2015).

2.1.1 Equipment

- TLC plates
- Shortwave UV light
- Standard synthesis labware
- Rotary evaporator

2.1.2 Reagents

- Fmoc-L-cysteine hydrate
- Reagent grade methanol
- Concentrated hydrochloric acid
- Acetone
- 50:50 Hexanes/ethyl acetate

2.1.3 Procedure

1. Dissolve 361 mg (1.00 mmol) Fmoc-L-cysteine hydrate in 10.0 mL of methanol. The solution should be clear and colorless. Add 6–10 drops of concentrated HCl and stir the solution overnight with a magnetic stir bar.
2. Within 24 h, a white paste suspended in a slurry should be observed. Gently shake to homogenize the solution then draw one drop, dilute in acetone then spot on a TLC plate, and develop using 50:50

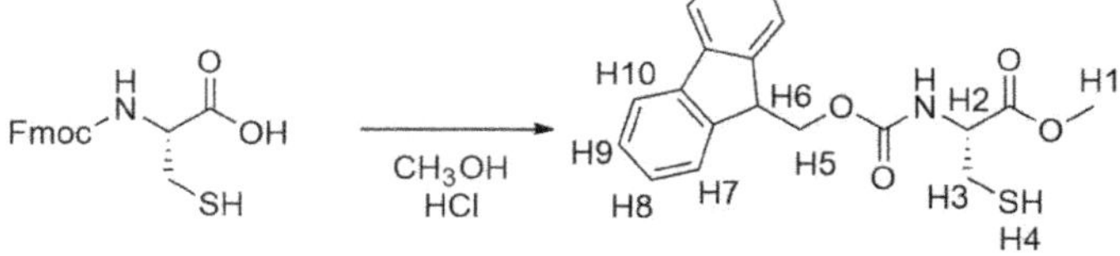

Scheme 1 Synthesis of Fmoc-L-CysOMe.

hexane/ethyl acetate. Check for complete conversion of the starting materials under shortwave UV light. The R_f of Fmoc-L-CysOMe is 0.9, while the corresponding value for Fmoc-L-Cys·H_2O is 0.3.

3. Proceed to the next step regardless of the extent of completion of the reaction. Dilute the solution with acetone until a clear, colorless solution is observed. Transfer to a preweighed round-bottom flask, then remove solvent in vacuo using a rotary evaporator. The heating bath temperature should not exceed 30°C, since at elevated temperatures the product can phase separate as an oil instead of a white precipitate powder. Once the solvent is evaporated, remove from the heat and put the precipitate under vacuum for several hours to remove any residual solvent.
4. If the initial reaction did not proceed to completion, repeat steps 1 and 2 to convert any remaining starting material; otherwise proceed to step 5.
5. Confirm the identity of the product by dissolving a small amount (~10 mg) in 0.7 mL $CDCl_3$ and then acquire ^1H 1D NMR spectrum. Full assignment of the molecule resonances is shown in Table 1. Additionally, completion of the reaction can be further monitored by comparing the ^1H spectrum of the product to that of the starting material; no starting material should be observed. Furthermore, if any remaining

Table 1 ^1H NMR Assignment of Fmoc-L-CysOMe in $CDCl_3$ was Carried Out With TMS as an Internal Standard

^1H Chemical Shift δ (ppm)	Integration	Multiplicity	J_{H-H} (Hz)	Assignment (From Scheme 1)
1.39	1	t	8.65	H^4
3.03	2	d	7.87	H^3
3.83	3	s	—	H^1
4.26	1	t	6.85	H^6
4.44	2	d	6.30	H^5
4.67	1	t	3.71	H^2
5.71	1	b	—	NH
7.35	2	t	7.20	H^8
7.44	2	t	7.48	H^9
7.64	2	d	7.48	H^7
7.8	2	d	7.6	H^{10}

solvent peaks are observed in the ^{1}H NMR spectrum, the flask should be subjected to vacuum for a few more hours to completely remove it.

6. Weigh the flask after no solvent remains and calculate the final yield. A 99% yield is typical for this protocol.

2.1.4 Notes

1. The final product should be stored at 2–8°C under nitrogen prior to use.
2. The reaction can be scaled up or down as needed. It was successfully carried out under the same conditions on a 0.1, 0.2, 0.4, 4, and 10 mmol scales. 15 Drops HCl were used when carried out on 4 and 10 mmol scales.

2.2 Loading of Trityl Chloride Resin

Once Fmoc-L-CysOMe is obtained, it can be side-chain anchored to trityl chloride resin as shown in Scheme 2. Trityl chloride resin is chosen for two reasons; it offers high yields (Diaz-Rodriguez et al., 2015), and little cysteine epimerization is observed during loading (Barany, Han, Hargittai, Liu, & Varkey, 2003).

2.2.1 Equipment

- Rotator
- Fritted polypropylene syringe
- Syringe stopcock

2.2.2 Reagents

- Trityl chloride resin
- Diisopropylethylamine
- Synthesized Fmoc-L-CysOMe

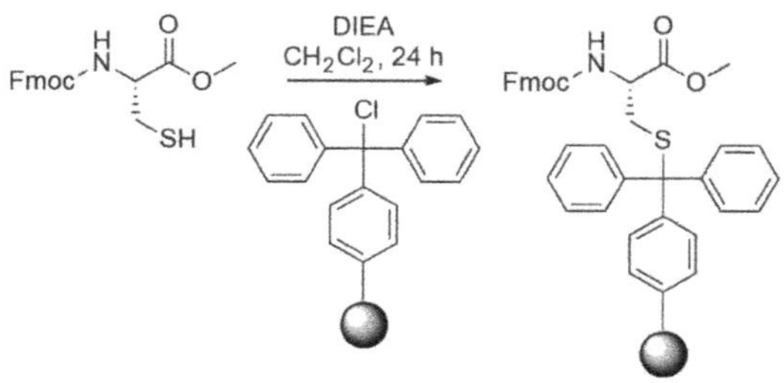

Scheme 2 Loading of Fmoc-Cys-OCH_3 onto trityl chloride resin.

2.2.3 Procedure

1. Load 0.2 mmol resin into fritted syringe, then wash with 5 mL DMF × 3 times, followed by a wash with 5 mL CH_2Cl_2 × 3 times. For the CH_2Cl_2 washes, place syringe on rotator for 5 min each time to swell it. Drain the resin after each wash.
2. Dissolve 286 mg (4.00 *M* equiv.) of Fmoc-L-CysOMe in 3 mL CH_2Cl_2 and add 140 μL (4.00 *M* equiv.) of diisopropylethylamine to the solution. Add this solution to the resin and leave overnight on rotating mixer.
3. The next day, add 200 μL methanol to the syringe containing the reaction and place on the rotating mixer for 15 min. This will cap any unreacted sites, resulting in fewer side reactions. Do not leave the reaction mixture on the rotating mixer for extended periods as the methanol can displace the Fmoc-L-CysOMe.
4. Drain the solution and then wash with 5 mL DMF × 3 times followed by 5 mL CH_2Cl_2 × 3 times. Dry resin by pulling vacuum through the syringe for at least 1 h. Store the resin at 2–8°C under nitrogen.

2.2.4 Notes

1. Amount of resin used will vary from batch to batch depending on the resin substitution.
2. Unmodified trityl chloride resin should be stored at 2–8°C under nitrogen. Failure to do so can lead to hydrolysis, which will lead to lower coupling efficiencies. If hydrolysis occurs, the resin can be regenerated using HCl, acetyl chloride, or $SOCl_2$ (García-Martín, Bayo, Cruz, Bohling, & Albericio, 2006)
3. Loadings of Fmoc-L-CysOMe in the range of 0.7–1.2 mmol/g are common. While it is possible to obtain even higher loadings (up to 1.8 mmol/g) with extended coupling times, this procedure is undesired as it leads to decreased yields during subsequent peptide synthesis and side-product formation due to aggregation of the peptide on resin surface.

2.3 Quantification of Loading Using Fmoc-Assay

Once the resin loaded with Fmoc-Cys-OMe is obtained, it is important to quantify the loading efficiency to be able to proceed with SPPS, as it is the basis for all subsequent calculations such as amount of resin to use as well as amount of amino acids and coupling reagent. The most prevalent method relies on a cuvette-based quantitative Fmoc assay where the Fmoc group on the loaded amino acid is deprotected using piperidine, and the resulting pyridine-Fmoc adduct is quantified using a known molar absorptivity constant (ε) value (Chan & White, 2000; Eissler et al., 2017). This method, however, was found to give inconsistent results. To address that issue, a more accurate standard-curve-based assay was developed and used here.

2.3.1 Equipment

- UV/vis plate reader
- Standard 96-well plate
- Hamilton syringes
- 25 mL Volumetric flasks
- 20 mL Glass vials
- Eppendorf tubes

2.3.2 Reagents

- Fmoc-OSu
- Absolute ethanol
- 20% Piperidine in DMF

2.3.3 Procedure

1. In triplicate, weigh ~10 mg resin aliquots in glass vials and record the exact weight.
2. Using a Hamilton syringe, add 0.5 mL 20% piperidine in DMF to each vial and incubate for 30 min to deprotect Fmoc groups from Fmoc-L-CysOMe on resin. Make sure most of the resin beads are at the bottom of the vial and in contact with the solution. In the meantime, prepare a stock of 1 m*M* Fmoc-OSu in ethanol, then dilute it in Eppendorf tubes or glass vials as described in Table 2 to make a series of standards and a standard blank. Also make a sample blank consisting of 0.5 mL 20% piperidine in DMF diluted to 25 mL in ethanol using a volumetric flask.
3. Use ethanol to dilute the deprotected samples and transfer them to 25 mL volumetric flasks. Adjust the volume to 25 mL with more ethanol.
4. In triplicate, transfer 150 µL of each sample, sample blank, and standard into a 96-well plate. Read absorbance at 301 nm.
5. Construct a standard curve by first subtracting standard H (standard blank) from all other standard absorbance values, then plotting average absorbance measurements against Fmoc-OSu concentration.
6. Subtract the average sample blank measurement from each average sample measurement and then calculate the Fmoc concertation using the best line fit to the standard.

Table 2 Dilutions of Fmoc-OSu in Ethanol Used for Fmoc Quantitative Assay

Standard	Amount of Standard to Add (μL)	Source	Amount of Ethanol to Add (μL)	Final Concentration (m*M*)	Absorbance Range
A	300	1 m*M* Stock	0	1	1.820–1.875
B	200	1 m*M* Stock	100	0.75	1.300–1.391
C	300	1 m*M* Stock	300	0.5	0.999–1.029
D	300	C	300	0.25	0.626–0.689
E	300	D	300	0.125	0.443–0.468
F	300	E	300	0.063	0.344–0.353
G	300	F	300	0.031	0.276–0.295
H	0	—	300	0	0.204–0.220

7. Calculate loading for each sample according the following equation before averaging:

$$\text{Resin loading (mmol/g)} = \frac{\text{Fmoc concetration m}M \times 0.025\,\text{L}}{\text{Amount of resin cleavege (g)}}.$$

2.3.4 Notes

1. While it is possible to make a larger amount of standard stock and store it for later use, more reproducible results are obtained if this standard is made freshly before use.

2.4 SPPS of a-Factor Precursor

With the loaded trityl resin in hand, the rest of the peptide (YIIKGVFWDPAC-OMe) can be extended using standard Fmoc-chemistry as seen in Scheme 3. A potential consideration in this process is epimerization of the C-terminal cysteine due to repeated exposure to piperidine during the synthesis (Han, Albericio, & Barany, 1997). However, using the side-chain anchoring methodology described here, negligible amount of epimerization was observed (Diaz-Rodriguez et al., 2015). The synthesis was carried out using an automated peptide synthesizer (PS3, Protein Technologies Inc., Memphis, TN). Sufficiently high yields

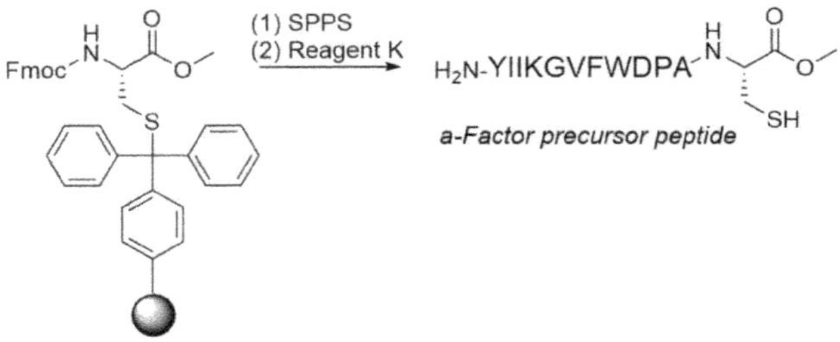

Scheme 3 Synthesis of a-factor precursor peptide via standard solid phase peptide synthesis starting with Fmoc-Cys-OCH_3 anchored to trityl resin.

were obtained using the conditions described (>95% for crude peptide). Based on these results, different types of automated synthesizers or manual synthesis should also give similar results.

2.4.1 Equipment

- Peptide synthesizer
- Fritted propylene syringe and syringe stopcock
- Rotating mixer
- Centrifuge

2.4.2 Reagents

- Fmoc-L-CysOMe-loaded trityl resin
- Coupling agent. *O*-(1H-6-Chlorobenzotriazole-1-yl)-1,1,3,3-tetramethyluronium hexafluorophosphate (HCTU) was used
- Standard Fmoc-protected amino acids with acid labile side-chain protecting groups
- Activator base (20% *N*-methylmorpholine in DMF used here)
- 20% Piperidine in DMF
- DMF
- Dry ice
- Acetone

2.4.3 Procedure

1. Carry out synthesis on a 0.1-mmol scale. Swell the resin in DMF for 10 min. Extend the peptide using single couplings with 4.0 equiv. of activator and amino acids in activator base. Each coupling step should be allowed to proceed for at least 30 min. Deprotect amino acid side chains during the synthesis by treatment with 20% piperidine in DMF for 10 min twice. Repeat for the final deprotection. Formation of the free

amine can then be observed using a ninhydrin test (Kaiser, Colescott, Bossinger, & Cook, 1970).

2. Once free amine formation is confirmed, wash resin with 5 mL DMF × times, followed by $CH_2Cl_2 \times 3$ times. Make 10 mL Reagent K (8.25 mL TFA, 500 mg phenol, 500 μL H_2O, 500 μL anisole, 250 μL 1,2 ethanedithiol) and sparge nitrogen through solution for 10 min to remove oxygen from it. Add to resin and place on rotary mixer for 2 h to cleave peptide.
3. After 2 h, fill two 50-mL falcon tubes with 30 mL diethyl ether each, then split the cleavage cocktail into both tubes. Wash resin with 5 more mL of TFA, then split into falcon tubes again. Dilute to the top with more ether, then cool tubes using acetone/dry ice bath. Spin down at 4000 RPM for 5 min to pellet peptide. Decant ether, refill tubes with fresh diethyl ether, and vortex tubes to resuspend pellet in solution. Once the pellet is resuspended, centrifuge and repeat the wash one more time for a total of three precipitation steps.
4. Use ~5 methanol to transfer peptide to preweighed glass vial, then dilute with equal amount H_2O. Evaporate methanol and remaining ether using a gentle stream of nitrogen until ~5 mL of solution remains. This could take several hours. Flash-freeze using liquid nitrogen and then place on a lyophilizer overnight.
5. The correct product can be confirmed using MALDI (Section 2.4) after solubilization as described in Section 2.5.3, step 6.

2.4.4 Notes

1. The synthesis was also performed using a microwave-assisted synthesizer (CEM Liberty/Discover microwave synthesizer). In this case double couplings were used. The coupling reagent was HBTU and the activator base was 35% v/v DIEA in NMM. The synthesis with the CEM microwave synthesizer gave similar results to the PS3 synthesizer. It is also worth mentioning that very little epimerization of the C-terminal cysteine residue was observed as determined by NMR integration (<5%).
2. Reagent K is extremely caustic, and so all handling should be done inside a well-ventilated fume hood. Appropriate safety precautions should be taken into consideration as well when handling of TFA, as it is highly corrosive.
3. It is possible to shorten step 4 in length by omitting the methanol transfer and lyophilizing the peptide directly in the falcon tubes after the last ether precipitation. This, however, could lead to inaccuracy when

measuring the mass of the crude material. To acomplish this, begin by preweighing the falcon tubes before step 3. After decanting the ether layer at the end of step 3, remove any remaining ether using a gentle stream of nitrogen applied for at least 30 min. Once the ether has evaporated completely, suspend the peptide in 5 mL H_2O (solution will be cloudy), flash-freeze, and place the suspension on a lyophilizer.

4. Lyophilizing the peptide suspended in water as opposed to simply removing the solvent in vacuo or using N_2 stream results in a powder rather than a residue, which helps with subsequent handling.

2.5 HPLC Purification of Unfarnesylated a-Factor

The crude a-factor precursor peptide has low solubility in aqueous solutions and many organic solvents. This can be due, in part, to adventitious disulfide bond formation. Additionally, the solubilized peptide tends to precipitate out remarkably fast, likely due to nonsolubilized material acting as nucleation sites. A method has been developed here to solubilize the peptide completely and reduce any disulfide bonds using DTT. One issue that was encountered was that the peptide requires acidic conditions to dissolve, but DTT is more reactive for disulfide reduction under basic conditions. This issue was overcome by employing stepwise addition of organic solvents to a $NaHCO_3$ buffered aqueous DTT solution with extended sonication in between, followed by the addition of a large excess of TFA.

2.5.1 Equipment

- HPLC instrument
- Semi-prep RP C-18 HPLC column (9.4 × 250 mm column used)
- Analytical RP C-18 HPLC column (4.6 × 250 mm column used)
- Sonicator
- Lyophilizer

2.5.2 Reagents

- HPLC grade water with 0.1% TFA (buffer A)
- HPLC grade CH_3CN with 0.1% TFA (buffer B)
- HPLC grade water buffered to pH 8.0 with $NaHCO_3$
- HPLC grade CH_3CN
- TFA
- HPLC grade methanol
- 0.5 *M* DTT aqueous solution

2.5.3 Procedure

1. Equilibrate the column with buffer A containing 1% buffer B. This should be done for at least 45 min to avoid that the peptide eluting prematurely from the column due to the large amount of TFA utilized for its solubilization. If an analytical column is used, a flow rate of 1 mL/min is appropriate. For a semi-prep column, it is advisable to use a flow rate of 3 mL/min.
2. To a 10 mg powder of crude peptide add 1 mL of HPLC grade water buffered to pH 8.0 with $NaHCO_3$ and 30 μL 0.5 *M* DTT solution, then sonicate for 15 min.
3. Add 1 mL HPLC grade CH_3CN and sonicate for 15 min.
4. Add 1 mL HPLC grade methanol and sonicate for 15 min.
5. Add TFA until solution becomes clear. Anywhere from 0.1 to 1 mL can be expected depending on the amount of peptide used as well as the impurities present. Some particulates may still be floating in solution. The least amount of TFA required should be used for solubilization as it interferes with the purification where it will prevent the peptide from binding the column resin.
6. Centrifuge at 4000 RPM for 5 min. This should be done immediately, as a-factor can precipitate out of solution if particulates are present. If this occurs, an additional sonication step and more TFA can be used to resolubilize peptide. A small pellet may appear at the bottom.
7. Transfer the supernatant to a new falcon tube being careful not to disturb the pellet. Centrifuge again to ensure no particulates were transferred over. Dilute at least twofold with 50:50 buffer A/buffer B solution. This is critical as the excess amount of TFA used can cause the peptide to elute prematurely.
8. Inject the desired amount into HPLC injection loop. For the analytical column, 20 μL peptide solution diluted to 100 μL in 50:50 buffer with a 200-μL injection loop gave the best results. For the semi-prep column, up to 4 mL of solution can be used with a 5-mL injection loop.
9. Run appropriate method. Fig. 1 legend outlines the event sequences for analytical and semi-prep columns. One important consideration is the extended period at beginning of the run where buffer B remains constant at 1%. This is done to help prevent the peptide from eluting prematurely. If that occurs, the peptide typically elutes during this initial wash period and should be collected. The presence of peptide can then be confirmed using MALDI (Section 2.4).

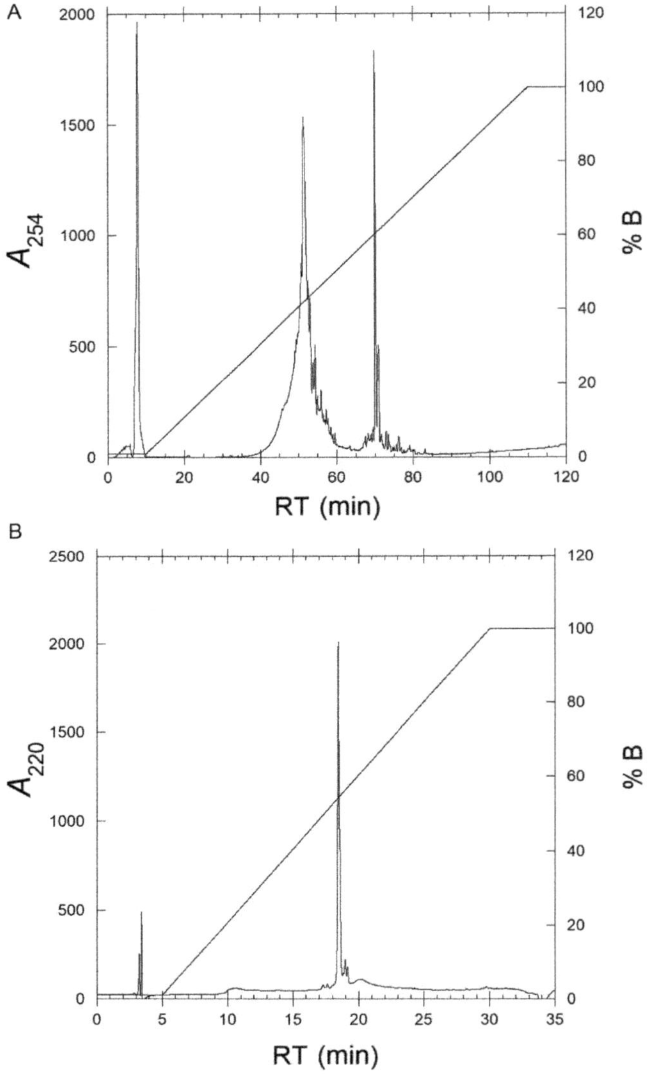

Fig. 1 (A) Crude semipreparative chromatogram of a-factor precursor peptide. The absorbance was monitored at 254 nm to observe aromatic groups and a flow rate of 3 mL/min. Gradient increased from 1% to 100% buffer B over 100 min with 10-min delay at the beginning of method. Fractions were collected between 40 and 60 min at 1-min intervals. (B) Analytical chromatogram of purified a-factor precursor. The absorbance was monitored at 220 nm to show all contaminants and a flow rate of 1 mL/min. Gradient increased from 1% to 100% buffer B over 25 min with 5-min delay at the beginning of method.

10. Unfarnesylated a-factor elutes between 60% and 80% buffer B, depending on the size of the column, with larger columns having lower retention times. Because of this, it is recommended to collect all fractions in that region and keep the ones exhibiting the correct m/z ratio (as measured by MALDI-MS). HPLC chromatograms of the crude peptide and purified material are shown in Fig. 1A and B, respectively.
11. Pool together the fractions displaying the correct m/z ratio based on desired purity as determined by MALDI or analytical HPLC integration (>90% required for NMR analysis, and >70% is required for farnesylation reaction). For pure fractions, proceed to step 12. For impure fractions, proceed to step 13.
12. Transfer pooled pure fractions to a preweighed glass vial and remove excess organic solvent using a stream of N_2. This step can take several hours. Once most of organic solvent is removed, the solution should be cloudy white in appearance. Flash-freeze in liquid N_2 and then place on a lyophilizer overnight (2 days if NMR experiments are to be run on sample to maximize water removal). Weigh the vial at the end to obtain final mass of the purified material. The peptide should appear as a white powder. 6–8 mg of peptide can be obtained at the end depending on purity of crude peptide. Final purified yields in the 60%–80% range can be expected.
13. If the volume of impure fractions is sufficiently small, it can be directly reinjected into the HPLC. Otherwise, remove the solvent using a gentle stream of N_2 and lyophilize as described in step 12, then dissolve the resulting powder in 1:1 buffer A/buffer B and reinject into the HPLC.

2.5.4 Notes

1. The purified peptide should have improved solubility (relative to the crude peptide) and can be dissolved in 1:1 buffer A/buffer B, DMSO, or 4:2:1 DMF/1-BuOH/0.1% aqueous TFA in a concentration of up to 5 mg/mL.
2. Once the peptide is solubilized, it is possible to accurately quantify its amount using the Ellman's assay for free thiols using previously published procedure (Riddles, Blakeley, & Zerner, 1983).
3. Multiple rounds of purification may be required to obtain highly pure peptide. This can be determined via MALDI analysis as well as analytical HPLC.
4. The peptide should be stored at 2–8°C for short periods of time (1–2 weeks), or at −20°C for extended periods.

5. For semi-prep HPLC, monitoring at 254 nm wavelength is recommended, as aromatic groups absorb at this wavelength. This simplifies the spectrum and prevent oversaturating the detector.
6. For analytical HPLC, monitoring at 220 nm wavelength is recommenced, as amide bonds absorbs in this region of the spectrum, hence possible contaminants will be observed.
7. Analytical chromatograms of HPLC fractions can be obtained by simply injecting 100 μL directly onto an analytical column and running the analytical method.

2.6 Farnesylation of a-Factor

This step is performed after the peptide synthesis and resin cleavage, since the highly acidic conditions for global peptide deprotection and cleavage cause isomerization and degradation of the isoprenoid group (Diaz-Rodriguez & Distefano, 2017). The reaction employed here uses zinc acetate dihydride to facilitate the nucleophilic reaction between the cysteinyl thiol moiety and the alkyl halide of *trans,trans*-farnesyl bromide (Scheme 4). To overcome solubility problems and their effects on alkylation rate, a mixture of several different solvents was used (Diaz-Rodriguez et al., 2015). Those conditions

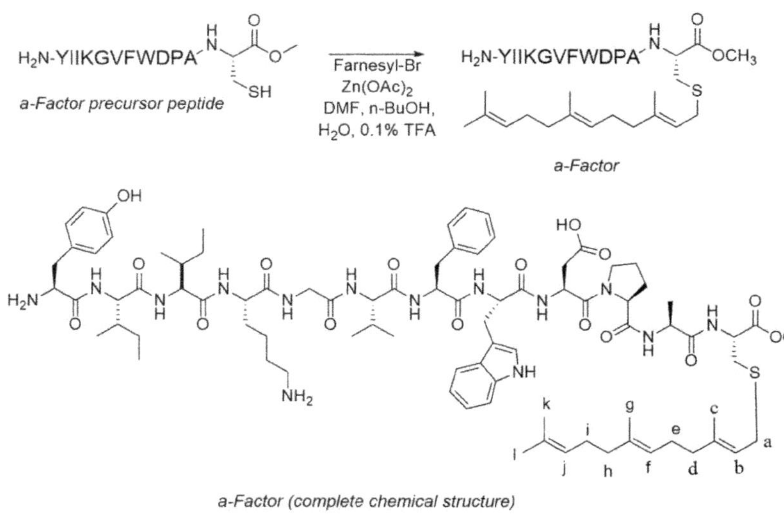

Scheme 4 Farnesylation of a-factor precursor peptide to yield a-factor.

along with the use of partially purified a-factor precursor allowed the reaction to proceed efficiently.

2.6.1 Equipment

- Standard synthetic chemistry labware

2.6.2 Reagents

- Purified or partially purified a-factor precursor
- *trans,trans*-farnesyl bromide
- $Zn(OAc)_2 \cdot H_2O$
- Degassed solvent blend: 4:2:1 DMF/1-BuOH/0.1% aqueous TFA

2.6.3 Procedure

1. Dissolve 10 mg (7.0 μmol) peptide in 3–10 mL of solvent blend. The amount required will vary depending on amount of impurities present. The smallest amount possible should be used to simplify the subsequent purification steps. The starting peptide needs to be at least partially purified (>70%), as crude peptide will likely have solubility issues.
2. Add 6.5 mg (5.0 equiv.) of $Zn(OAc)_2 \cdot H_2O$ in 0.5 mL H_2O containing 0.1% TFA to the solution.
3. Dilute 9.5 μL (5.0) equiv. of farnesyl bromide in 0.5 mL of DMF, then add it dropwise to the above solution while stirring. This step is critical to ensure that the farnesyl bromide remains soluble. Watch closely for any persisting undissolved materials. If observed, add more of the solvent blend as needed in 0.1 mL increments.
4. Stir under nitrogen for 2–4 h and then check for the reaction completion using MALDI-MS or analytical HPLC. The reaction time may be extended to obtain maximal conversion. Once the reaction is complete, it can be purified using HPLC.

2.6.4 Notes

1. Farnesyl bromide is air and moisture sensitive, so its exposure to those elements should be limited as much as possible. Unused material should be stored under nitrogen at 2–8°C.

2.7 HPLC Purification of Farnesylated a-Factor

In this step, care must be taken to dilute the solution containing the crude peptide appropriately to minimize the potential for premature elution. This issue is similar to that noted in step 2.5. To avoid this, dilute the crude reaction mixture, equilibrate the column with buffer A, and add an extended hold period in the HPLC method with buffer A to prevent premature elution.

2.7.1 Equipment

- HPLC instrument
- Semi-prep RP C-18 HPLC column (9.4 × 250 mm column used)
- Analytical RP C-18 HPLC column (4.6 × 250 mm column used)
- Sonicator
- Lyophilizer

2.7.2 Reagents

- HPLC grade H_2O with 0.1% TFA (buffer A)
- HPLC grade CH_3CN with 0.1% TFA (buffer B)
- HPLC grade CH_3OH

2.7.3 Procedure

1. Using a Falcon tube, dilute the crude reaction mixture at least twofold with a 50:50 buffer A/B mixture. This is done to prevent premature elution due to the large amount of DMF used in the farnesylation reaction.
2. Centrifuge at 4000 RPM for 5 min to remove any particulates in the reaction mixture.
3. Inject onto the HPLC column and run appropriate method. See Section 2.5.3 steps 8 and 9.
4. The farnesylated peptide typically elutes in the range of 75%–85% buffer B. Collect all peaks in this region and confirm the presence of the desired product by MALDI-MS. HPLC chromatograms of the crude peptide and purified material are shown in Fig. 2A and B, respectively.
5. After pooling the appropriate fractions (>90% purity as determined by MALDI or analytical HPLC integration), evaporate the organic solvent using a stream of N_2. The resulting solution should be cloudy when all the solvent is removed. This typically requires several hours. Once completely removed, flash-freeze using liquid N_2 and then lyophilize for at least 2 days to remove all water for subsequent NMR analysis.

2.7.4 Notes

1. More than one round of chromatography may be required to obtain highly pure a-factor since some impurities elute with similar retention times. If that is observed by MALDI-MS, partially pure fractions can be combined, lyophilized, and rechromatographed. Some a-factor fails to adsorb to the column and elutes in the 1% buffer B hold phase due

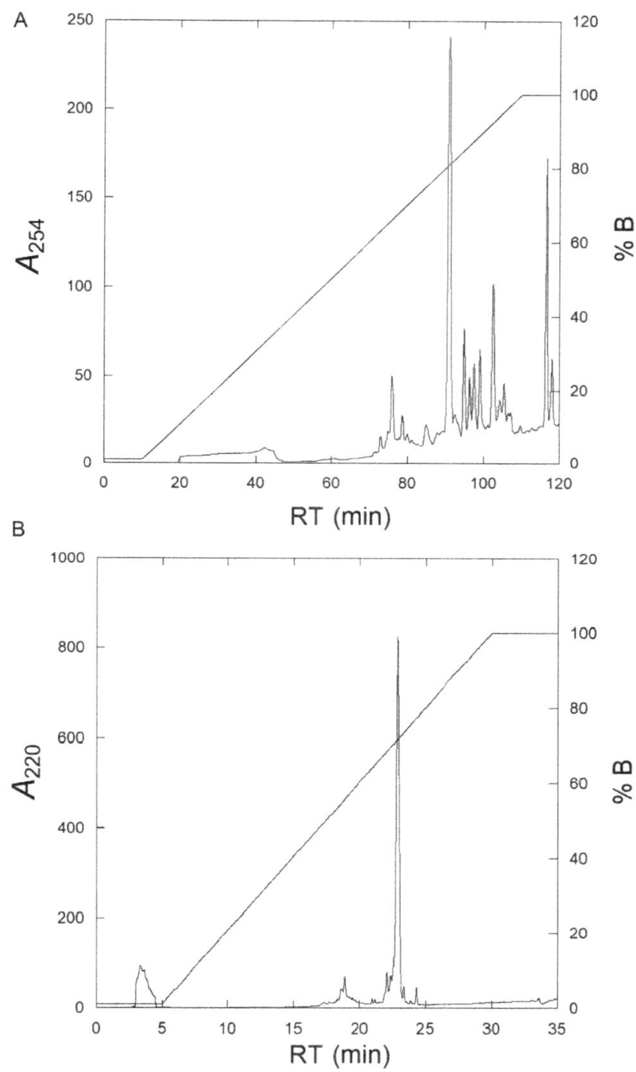

Fig. 2 (A) Crude semipreparative chromatogram of the farnesylation reaction. The absorbance was monitored at 254 nm to observe aromatic groups with a flow rate of 2 mL/min. Fractions were collected between 80 and 100 min at 1-min intervals. (B) Analytical chromatogram of purified a-factor precursor. The absorbance was monitored at 220 nm to show all contaminants with a flow rate of 1 mL/min.

to the DMF used for the farnesylation reaction. If a significant amount is observed (via MALDI-MS analysis), it can also be recombined with other impure fractions and then repurified.

3. ANALYSIS OF A-FACTOR

3.1 MALDI-MS Analysis

MALDI-MS allows the presence of the desired compound to be determined in a reliable manner. MALDI was chosen over ESI for two reasons; first, it generally affords monocharged species, which simplifies the analysis, especially when disulfide bonds were present (Nadler et al., 2017). Second, MALDI-MS tends to give more reproducible ionization of a-factor (compared with ESI).

3.1.1 Equipment

- MALDI Instrument (AB SCIEX MALDI-TOF 5800 was used)
- MALDI-MS plate
- Microcentrifuge
- Micropipettes

3.1.2 Reagents

- 50:50 buffer A/ buffer B mixture
- α-Cyano-4-hydroxycinnamic acid

3.1.3 Procedure

1. In a microcentrifuge tube, thoroughly mix a spatula-full of matrix in 1 mL of 50:50 buffer A/buffer B mixture, then centrifuge for 3 min at maximum speed. This will yield a saturated solution of matrix.
2. If DMF is present in the peptide sample, dilute it fivefold in 50:50 buffer A/buffer B mixture.
3. Spot 0.5 μL of the matrix solution on the MALDI-MS plate, followed by 0.5 μL of sample. Mix by pipetting up and down a few times. Add another 0.5 μL of matrix solution and pipette up and down repeatedly until small crystals are observed. Leave for a few minutes to dry. If DMF was present in solution a small drop of liquid (instead of crystals) may still be present after 10 min. If that is the case, add 1 μL of the 50:50 buffer A/buffer B mixture, mix by pipetting up and down and then leave to dry again. Repeat until spot appears completely dry. Alternatively, dilute 0.5 μL of initial sample with 2 μL of 50:50 buffer A/buffer B mixture and spot 0.5 μL of newly diluted sample.

4. Acquire MALDI-MS spectrum using positive reflector mode. Make sure that the mass range encompasses at least double the peptide's mass to observe any dimerized peptide (disulfide-linked dimer).
5. a-Factor typically appears as a mixture of $[M+H]^+$, $[M+Na]^+$, and $[M+K]^+$ ions. Other ions may be observed however including $[M+CH_3OH+H]^+$, $[M+CH_3CN+H]^+$, $[M+2Na-H]^+$, and $[M+CH_3CN+Na]^+$.
6. The correct m/z ratio for a-factor can be calculated using programs such as ChemDraw that compute the monoisotopic mass. Sample MALDI-MS spectra of purified unfarnesylated and farnesylated a-factor are shown in Fig. 3A and B, respectively.

3.1.4 Notes

1. A calibrant should always be used to ensure accurate measurements. The following calibration mixture was used: Leucine enkephalin ($[M+H]^+$ $m/z=556.2771$), Bradykinin 2–9 ($[M+H]^+$ $m/z=904.4680$), Angiotensin I ($[M+H]^+$ $m/z=1296.6850$), Glu-fibrinogen ($[M+H]^+$ $m/z=1570.6770$), and ACTH 18–39 ($[M+H]^+$ $m/z=2465.1990$).
2. The matrix solution can be stored on the benchtop and reused. However, a blank MALDI-MS spectrum of the matrix solution alone should be acquired each time to ensure that it has not been contaminated, and also to compare to the peptide's spectrum to avoid misidentifying the matrix peaks as contaminants. Special care should be taken to avoid labeling matrix clusters as contaminations (Smirnov et al., 2004).
3. In the case of DTT reduction, the crude peptide spectrum will show the $[M+Na]^+$ ion as the predominant species while the other two ($[M+H]^+$ and $[M+K]^+$) may not be observed or only be observed at low levels.
4. A good practice is to spot each sample twice (at different locations on the plate) and acquire spectra of both. This is to prevent false-negative results where some compounds fail to ionize due to inconstant cocrystallization with the matrix.

3.2 NMR Analysis

For the complete chemical characterization of the peptide, four homonuclear (^{1}H, ^{1}H) and heteronuclear (^{1}H, ^{13}C) 2D NMR experiments were necessary in addition to the basic ^{1}H 1D experiment. COSY and TOCSY were used to analyze the short and long range ^{1}H–^{1}H coupled spin systems, while HSQC and HMBC were used to analyze the short and long range ^{1}H–^{13}C coupled spin systems. The full assignment of farnesylated a-factor

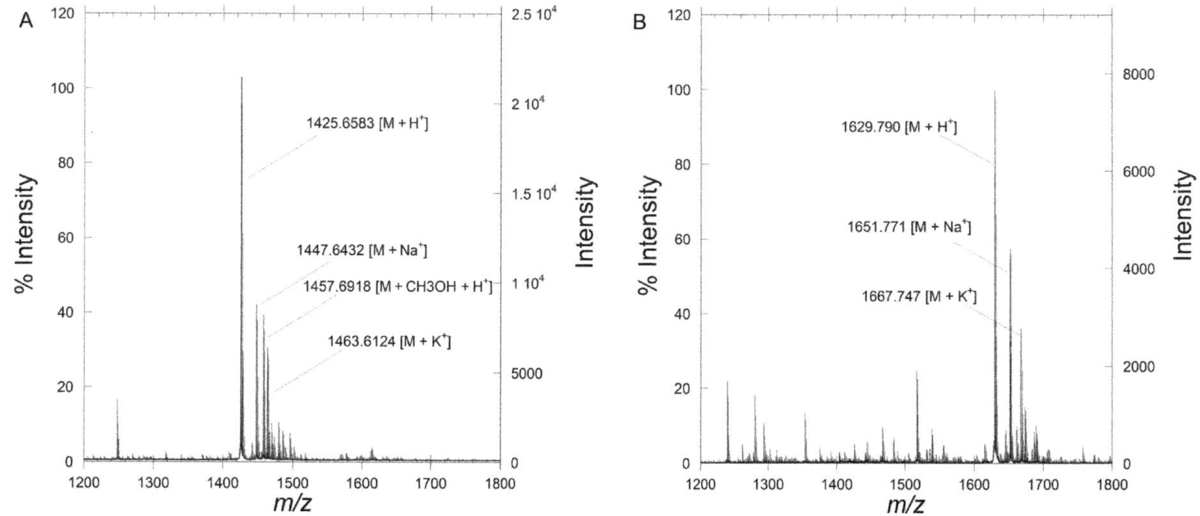

Fig. 3 (A) Example MALDI spectrum of purified a-factor precursor. (B) Example MALDI spectrum of purified farnesylated a-factor. Notice the ionic charged states.

was achieved, allowing unambiguous structural analysis of a-factor and a-factor precursors including specific stereochemical properties. For example, the TOCSY experiment allows for analysis and rough quantification of the extent of epimerization of the C-terminal cysteine residue (Diaz 2015), as the presence of L and D cysteine gives rise to significantly different chemical shifts.

3.2.1 Equipment

- DMSO-d_6-matched Shigemi NMR tube
- Hamilton syringe

3.2.2 Reagents

- DMSO-d_6

3.2.3 Sample Preparation

- Dissolve 1.5 mg peptide in 300 μL DMSO-d_6 for a final concentration of 3 m*M*. Concentrations as low as 0.7 m*M* have been shown to give sufficient NMR signal (Diaz-Rodriguez et al., 2015), but higher concentrations will yield better spectra with stronger signal.

3.2.4 NMR Experiments

All acquisition for the experiments reported here was done using a Bruker Avance 700-MHz spectrometer with 5 mm ^{1}H–^{13}C–^{15}N TXI cryoprobe. TopSpin was used to analyze all spectra.

3.2.4.1 One-Dimensional ^{1}H NMR (1D ^{1}H NMR)

Acquire 1D proton NMR data. This can be extremely useful as a final quick analytical tool, since it can be compared to known spectra to judge the quality of the sample before acquiring more time-consuming two-dimensional experiments. A sample ^{1}H 1D NMR spectrum is shown in Fig. 4. Assignment of the ^{1}H 1D spectrum of a-factor has been previously reported (Gounarides et al., 1991), as well as the ^{1}H 1D spectrum of the farnesyl moiety bonded to a factor (Anderegg, Betz, Carr, Crabb, & Duntze, 1988).

3.2.4.2 COSY

Acquire COSY spectrum to gain primary assignment of ^{1}H–^{1}H coupled spin systems. The COSY spectrum will only show proton–proton coupled

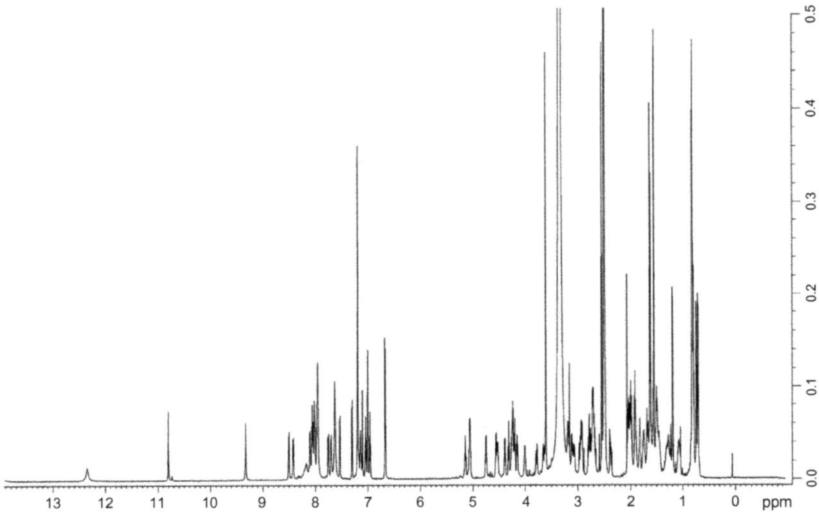

Fig. 4 1D ^{1}H NMR spectrum of farnesylated a-factor in DMSO-d_6. The water peak is observed at 3.33, and DMSO peak is observed at 2.5 and was used for all calibrations. Comparison to known 1D spectra offers a quick and reliable analytical method for determining the presence of correct peptide as well as purity. Spectra were acquired with 64 scans using 1 s recycle delay, 30-degree pulse, 3.1 s acquisition time with a sweep width of 15 ppm. Data were Fourier transformed with 0.3 Hz line broadening applied and manually phase-corrected. All acquisitions were done on a Bruker Avance 700-MHz spectrometer with a 5-mm TXI cryoprobe.

signals through less than three bonds. The assigned spectrum of a-factor is shown in Fig. 5. The parameters used in the experiment are provided in the figure legend. Assignments are based on the 1D ^{1}H spectrum.

3.2.4.3 TOCSY

Acquire TOCSY spectra to observe long range ^{1}H–^{1}H coupling (Bax & Davis, 1985). The TOCSY experiment shows long range interactions between coupled spins up to five continuous bonds. Assignments are based on the COSY spectrum and the 1D ^{1}H spectrum. The parameters used in the experiment are described in Fig. 6.

3.2.4.4 HSQC

Acquire phase sensitive HSQC using Echo\Antiecho-TPPI gradient selection to gain primary assignment of the backbone (Boyer, Johnson, & Krishnamurthy, 2003; Palmer, Cavanagh, Wright, & Rance, 1991).

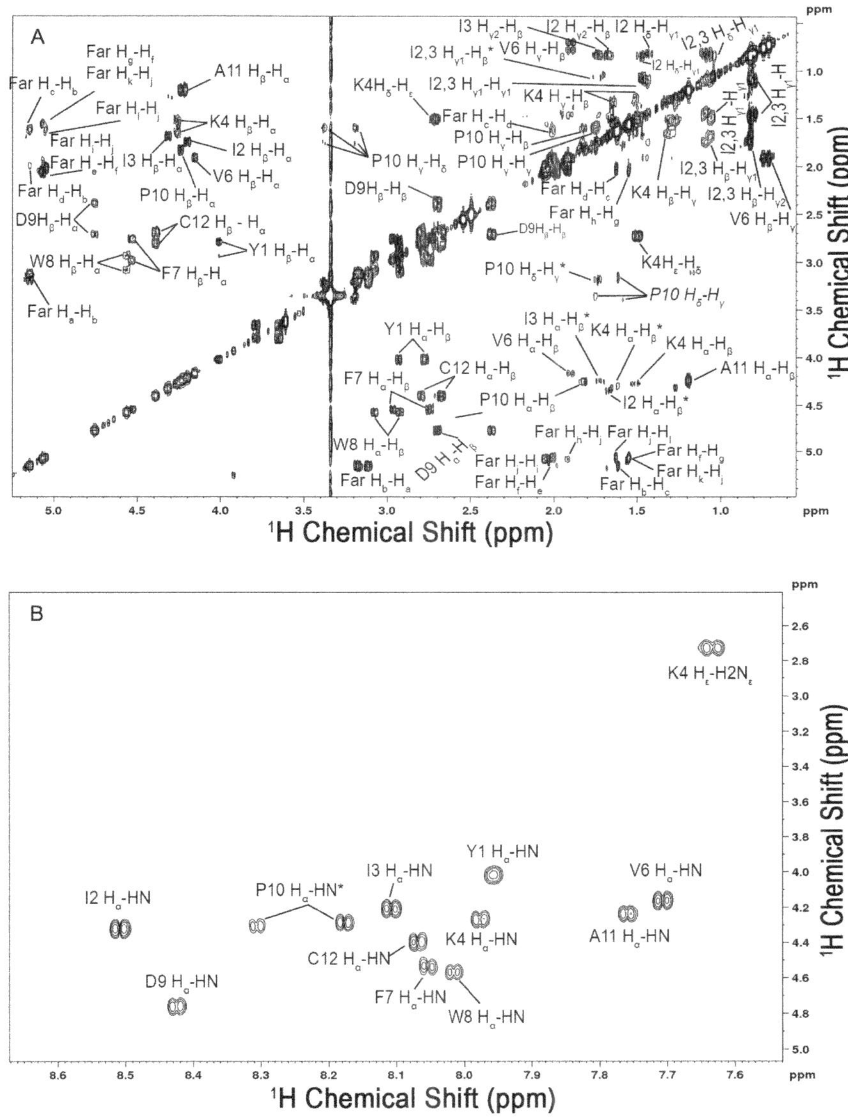

Fig. 5 2D COSY NMR spectrum of farnesylated a-factor in DMSO-d_6, showing ^{1}H–^{1}H short range coupled spins (i.e., three bonds or less). Assignments are shown for each peak in the spectrum. Peaks labeled with *asterisk* are shown at a twofold to fourfold higher contour level. (A) Assigned aliphatic region. (B) Assigned backbone region. Spectrum was acquired using 64 scans/increment, 16 dummy scans, 0.8 s recycle delay, and a spectral widths of 15 ppm. 6300 × 256 points were collected in the direct and indirect dimensions, respectively. Data were zero-filled with 8192 × 1024 points and Fourier-transformed with a shifted sine-bell squared function with 1.00 × 0.30 Hz line broadening applied. 2 × Linear prediction was applied in the indirect dimension. Sine bell shift was 0 × 0 and Gaussian max position 0 × 0.1.

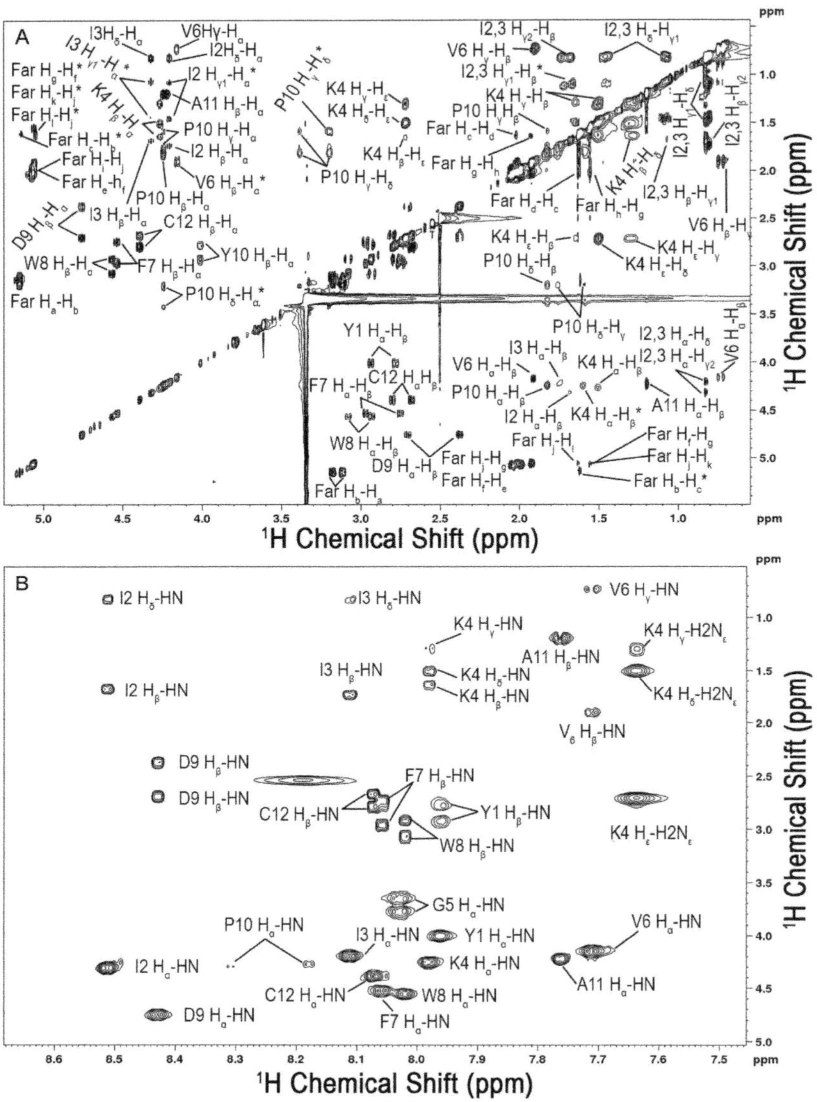

Fig. 6 2D TOCSY NMR spectrum of farnesylated a-factor in DMSO-d_6 showing ^1H–^1H long range coupled spins (up to five bonds). Assignments are shown adjacent to each peak. Peaks labeled with *asterisk* are shown at a twofold to fourfold higher contour level. (A) Assigned aliphatic region. (B) Assigned backbone region. ROESY peaks appear as negative resonances and are omitted from the spectrum for clarity. Spectral parameters were identical to the COSY spectrum except that the sine bell shift was 2×2 and the Gaussian max position was 1×1.

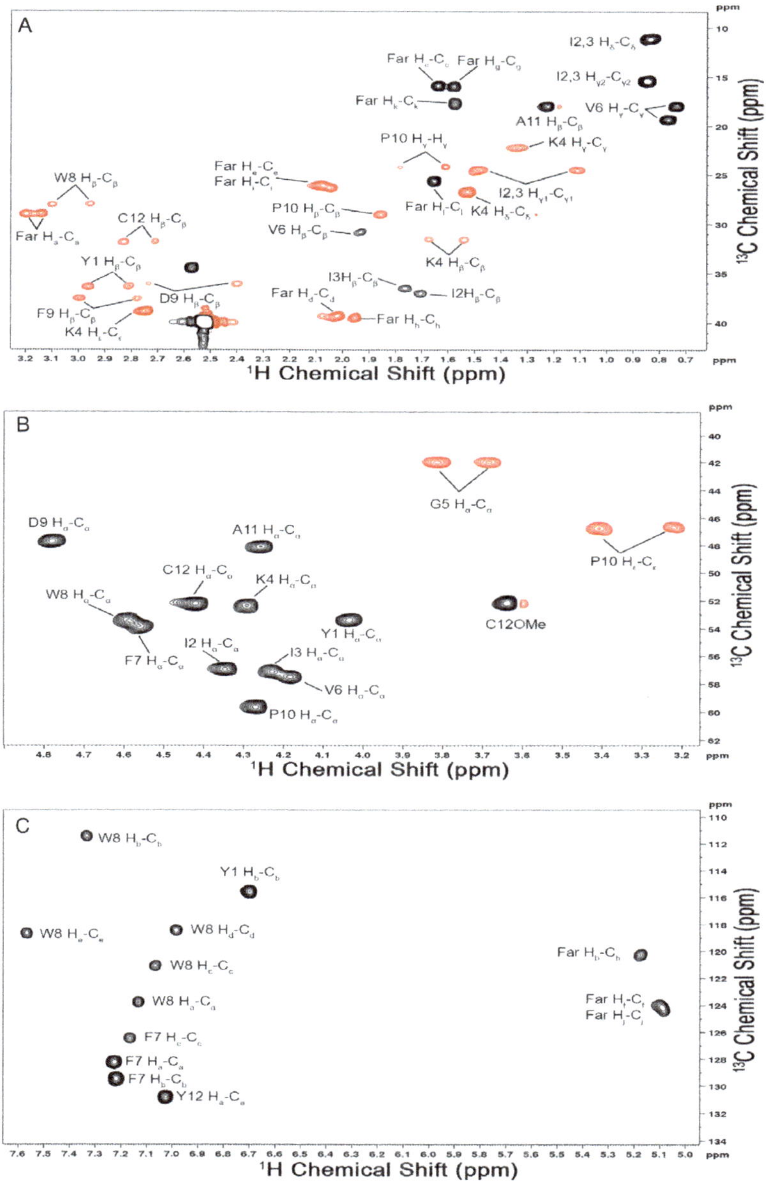

Fig. 7 See figure legend on next page.

This type of experiment only observes one bond carbon–hydrogen coupling (Schleucher et al., 1994). Peaks arising from CH_1 or CH_3 groups appear in the positive region, while peaks from CH_2 appear in the negative region. The spectrum can be manually phased. The assignments are based on the 1D ^1H spectrum and assisted by the ^{13}C chemical shift values predicted using ChemDraw. The assigned spectrum of a-factor is shown in Fig. 7. The parameters used for the data acquisition are described in the figure legend.

3.2.4.5 HMBC

Acquire phase sensitive HMBC using an echo/antiecho gradient selection with a threefold low pass J-filter to suppress single bond correlations and obtain long range interactions (Cicero, Barbato, & Bazzo, 2001). In this experiment, coupled spin systems up to three continuous carbon–carbon bonds were observed. The assignments are based on the 1D ^1H and HSQC spectra. An assigned spectrum of a-factor is shown in Fig. 8. The parameters used in the experiment are described in the figure legend.

4. SUMMARY

Using the side-chain anchoring method described here, it is possible to obtain a-factor in good yield and high purity. Most of the difficulties in the process arise from poor solubility, due to the presence of impurities and disulfide-linked dimers in the crude a-factor precursor. Sequential treatment with DTT and appropriate solvents as well as excess of TFA can be used to reduce the disulfide bond and keep the peptide solubilized for partial purification using HPLC. The excess TFA used, however, does interfere with adsorption of the peptide on the HPLC column. Sample dilution with aqueous solvent prior to injection, extended column equilibration, and an

Fig. 7 2D ^1H, ^{13}C-HSQC NMR spectrum of farnesylated a-factor in DMSO-d_6 showing ^{13}C–^1H short range coupled spin systems (via single bond). Assignments are shown adjacent to each peak. Peaks labeled with *asterisk* are shown at a twofold to fourfold higher contour level. Peaks with positive phasing (*black*) indicate a CH_3 or CH_1 moiety, while peaks with negative phasing (*red*) indicate a CH_2 moiety. (A) Assigned aliphatic region. (B) Assigned backbone region. (C) Assigned aromatic region. Spectrum was acquired using 32 scans/increment, 32 dummy scans, 1.5 s recycle delay, and a spectral widths of 15 ppm × 165 ppm. 2048 × 1024 points were collected in the directly and indirectly detected dimensions, respectively. Data were zero-filled with 4096 × 2048 points and Fourier-transformed with a shifted sine-bell squared function with 1.00 × 0.30 Hz line broadening applied. 3 × Linear prediction was applied in the indirect dimension. Sine bell shift was 2 × 2 and Gaussian max position 0 × 0.1.

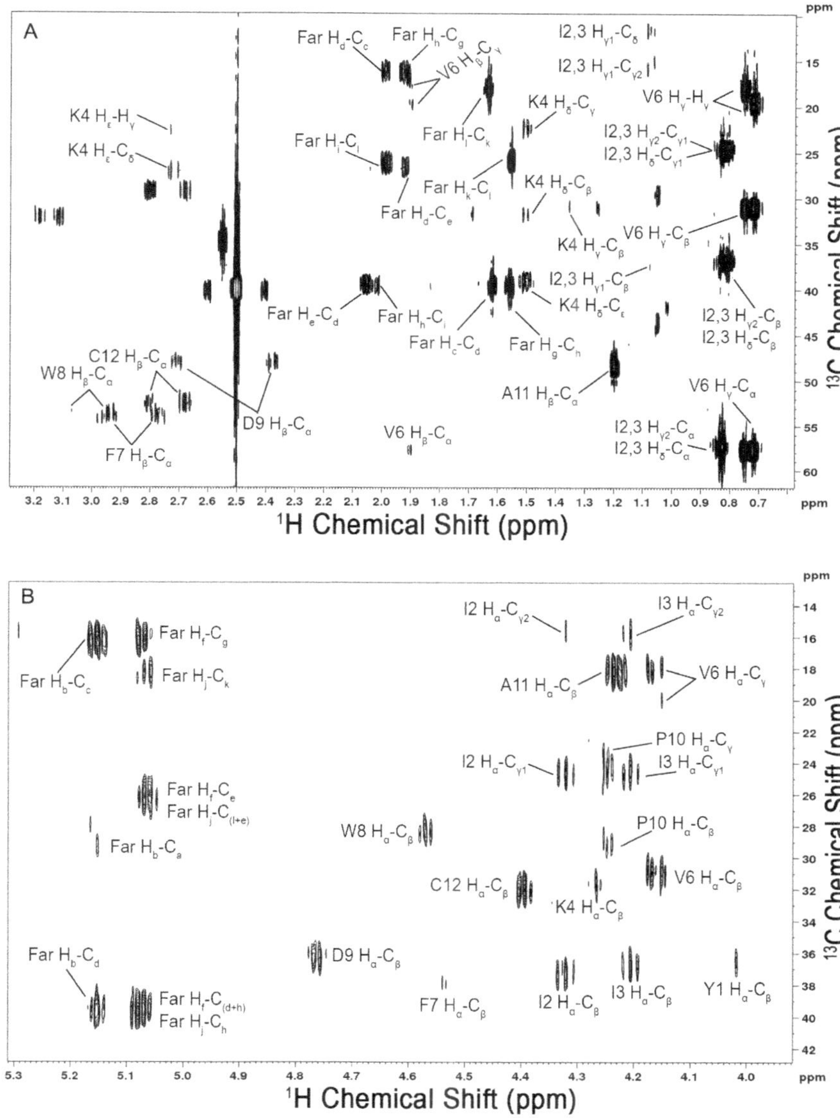

Fig. 8 2D ^{1}H, ^{13}C HMBC NMR spectrum of farnesylated a-factor in DMSO-d$_6$ showing ^{13}C–^{1}H long range coupled spin systems (up to five bonds). Assignments are shown adjacent to each peak. (A) Assigned aliphatic region. (B) Assigned backbone region. The spectrum was acquired using 32 scans/increment, 16 dummy scans, 0.8 s recycle delay, and a spectral widths of 15 ppm × 222 ppm. 5594 × 265 points were collected in the directly and indirectly detected dimensions, respectively. Data were zero-filled with 16,384 × 2048 points and Fourier-transformed with a shifted sine-bell squared function with 1.00 × 1.00 Hz line broadening applied. 3 × Linear prediction was applied in the indirect dimension. Sine bell shift was 4 × 4 and Gaussian max position was 0.5 × 0.5.

aqueous hold period all help ensure proper binding for purification. Partially purified a-factor precursor can then be farnesylated under mildly acidic conditions using farnesyl bromide and $Zn(OAc)_2$. Solubility here is again problematic, as farnesyl bromide is soluble only in organic solvents, while $Zn(OAc)_2$ requires aqueous conditions. Here a DMF/BuOH/aqueous TFA solvent blend is used to maintain a homogeneous mixture while pre-solubilized reactants are added sequentially. Finally, once a-factor has been farnesylated, it can be purified to homogeneity via multiple cycles of HPLC. Throughout the synthesis, MALDI-MS can be used to identify the presence of the correct products, as well as supplementing analytical HPLC for determining purity. MALDI-MS was chosen over ESI-MS methods as it simplified the analysis due to the production of singly charged species, as well as better ionization of a-factor and precursor. Once purified peptide is obtained, 2D NMR analysis provides an unambiguous analytical tool that can be used to inspect a-factor on the atomic level. COSY and TOCSY experiments provide the assignment of ^1H–^1H coupled spin systems, while HSQC and HMBC provide the assignments of ^1H–^{13}C coupled spin systems. While the methods described here are specifically for the preparation and characterization of a-factor, they can be used to prepare essentially any peptide that includes a C-terminal prenylated cysteine.

ACKNOWLEDGMENTS

Mass spectrometry analysis was performed at The University of Minnesota Department of Chemistry Mass Spectrometry Laboratory (MSL), supported by the Office of the Vice President of Research, College of Science and Engineering, and the Department of Chemistry at the University of Minnesota, as well as The National Science Foundation (NSF, Award CHE-1336940). The content of this chapter is the sole responsibility of the authors and does not represent endorsement by the MSL or NSF. This work was supported in part by the National Institutes of Health (RF1AG056976, GM084152, and GM106082) and by the National Science Foundation (CHE-1308655). NMR analysis was performed at Minnesota NMR center. Funding for NMR instrumentation was provided by the Office of the Vice President for Research, the Medical School, the College of Biological Science, NIH, NSF, and the Minnesota Medical Foundation.

REFERENCES

Anderegg, R. J., Betz, R., Carr, S. A., Crabb, J. W., & Duntze, W. (1988). Structure of Saccharomyces cerevisiae mating hormone a-factor. Identification of S-farnesyl cysteine as a structural component. *The Journal of Biological Chemistry*, *263*(34), 18236–18240.

Barany, G., Han, Y., Hargittai, B., Liu, R. Q., & Varkey, J. T. (2003). Side-chain anchoring strategy for solid-phase synthesis of peptid acids with C-terminal cysteine. *Biopolymers—Peptide Science Section*, *71*(6), 652–666.

Bax, A., & Davis, D. G. (1985). MLEV-17-based two-dimensional homonuclear magnetization transfer spectroscopy. *Journal of Magnetic Resonance*, *65*(2), 355–360.

Boyer, R. D., Johnson, R., & Krishnamurthy, K. (2003). Compensation of refocusing inefficiency with synchronized inversion sweep (CRISIS) in multiplicity-edited HSQC. *Journal of Magnetic Resonance*, *165*(2), 253–259.

Caldwell, G. a., Naider, F., & Becker, J. M. (1995). Fungal lipopeptide mating pheromones: A model system for the study of protein prenylation. *Microbiological Reviews*, *59*(3), 406–422.

Casey, P. J., & Seabra, M. C. (1996). Protein prenyltransferases. *The Journal of Biological Chemistry*, *271*(10), 5289–5292 [American Society for Biochemistry and Molecular Biology].

Chakrabarti, D., Silva, T. D., Barger, J., Paquette, S., Patel, H., Patterson, S., et al. (2002). Protein farnesyltransferase and protein prenylation in Plasmodium falciparum. *The Journal of Biological Chemistry*, *277*(44), 42066–42073.

Chan, W. C., & White, P. D. (2000). Fmoc solid phase peptide synthesis: A practical approach. In B. D. Hames (Ed.), *The practical approach*. Oxford: Oxford University Press.

Cicero, D. O., Barbato, G., & Bazzo, R. (2001). Sensitivity enhancement of a two-dimensional experiment for the measurement of heteronuclear long-range coupling constants, by a new scheme of coherence selection by gradients. *Journal of Magnetic Resonance*, *148*(1), 209–213.

Diaz-Rodriguez, V., & Distefano, M. D. (2017). A-Factor: A chemical biology tool for the study of protein prenylation. *Current Topics in Peptide and Protein Research*, *18*, 133–151.

Diaz-Rodriguez, V., Ganusova, E., Rappe, T. M., Becker, J. M., & Distefano, M. D. (2015). Synthesis of peptides containing C-terminal esters using Trityl side-chain anchoring: Applications to the synthesis of C-terminal ester analogs of the Saccharomyces cerevisiae mating pheromone a-factor. *Journal of Organic Chemistry*, *80*(22), 11266–11274.

Diaz-Rodriguez, V., Mullen, D. G., Ganusova, E., Becker, J. M., & Distefano, M. D. (2012). Synthesis of peptides containing C-terminal methyl esters using trityl side-chain anchoring: Application to the synthesis of a-factor and a-factor analogs. *Organic Letters*, *14*(22), 5648–5651.

Eckert, G. P., Hooff, G. P., Strandjord, D. M., Igbavboa, U., Volmer, D. A., Müller, W. E., et al. (2009). Regulation of the brain isoprenoids farnesyl- and geranylgeranylpyrophosphate is altered in male Alzheimer patients. *Neurobiology of Disease*, *35*(2), 251–257 [Elsevier Inc].

Eissler, S., Kley, M., Bächle, D., Loidl, G., Meier, T., & Samson, D. (2017). Substitution determination of Fmoc-substituted resins at different wavelengths. *Journal of Peptide Science*, *23*(10), 757–762.

García-Martín, F., Bayo, N., Cruz, L. J., Bohling, J., & Albericio, F. (2006). Chlorotrityl chloride (CTC) resin as a convenient reusable protecting group. *Understanding Biology Using Peptides*, *9*(3), 220–221.

Gounarides, J. S., Broido, M. S., Xue, C. B., Becker, J. M., & Naider, F. R. (1991). The conformation of a-factor is not influenced by the S-prenylation of Cys12. *Biochemical and Biophysical Research Communications*, *181*(3), 1125–1130.

Han, Y., Albericio, F., & Barany, G. (1997). Occurrence and minimization of cysteine racemization during stepwise solid-phase peptide synthesis. *The Journal of Organic Chemistry*, *62*(13), 4307–4312.

Hosokawa, A., Wollack, J. W., Zhang, Z., Chen, L., Barany, G., & Distefano, M. D. (2007). Evaluation of an alkyne-containing analogue of farnesyl diphosphate as a dual substrate for protein-prenyltransferases. *International Journal of Peptide Research and Therapeutics*, *13*(1–2), 345–354.

Kaiser, E., Colescott, R. L., Bossinger, C. D., & Cook, P. I. (1970). Color test for detection of free terminal amino groups in the solid-phase synthesis of peptides. *Analytical Biochemistry*, *34*(2), 595–598.

Kassai, H., & Fukada, Y. (2011). Farnesylation versus Geranylgeranylation in G-protein-mediated light signaling. *Enzyme*, *29*, 125–145.

Li, H., Kuwajima, T., Oakley, D., Nikulina, E., Hou, J., Yang, W. S., et al. (2016). Protein prenylation constitutes an endogenous brake on axonal growth. *Cell Reports*, *16*(2), 545–558 [ElsevierCompany].

Mullen, D. G., Kyro, K., Hauser, M., Gustavsson, M., Veglia, G., Becker, J. M., et al. (2011). Synthesis of a-factor peptide from Saccharomyces cerevisiae and photoactive analogues via Fmoc solid phase methodology. *Bioorganic and Medicinal Chemistry*, *19*(1), 490–497.

Nadler, W. M., Waidelich, D., Kerner, A., Hanke, S., Berg, R., Trumpp, A., et al. (2017). MALDI versus ESI: The impact of the ion source on peptide identification. *Journal of Proteome Research*, *16*(3), 1207–1215.

Ochocki, J. D., & Distefano, M. D. (2013). Prenyltransferase inhibitors: Treating human ailments from cancer to parasitic infections. *MedChemComm*, *4*(3), 476–492.

Palmer, A. G., Cavanagh, J., Wright, P. E., & Rance, M. (1991). Sensitivity improvement in proton-detected two-dimensional heteronuclear correlation NMR spectroscopy. *Journal of Magnetic Resonance*, *93*(1), 151–170.

Palsuledesai, C. C., Ochocki, J. D., Kuhns, M. M., Wang, Y. C., Warmka, J. K., Chernick, D. S., et al. (2016). Metabolic labeling with an alkyne-modified isoprenoid analog facilitates imaging and quantification of the prenylome in cells. *ACS Chemical Biology*, *11*(10), 2820–2828.

Riddles, P. W., Blakeley, R. L., & Zerner, B. (1983). Reassessment of Ellman's reagent. *Methods in Enzymology*, *91*, 49–60.

Rowell, C. A., Kowalczyk, J. J., Lewis, M. D., & Garcia, A. M. (1997). Direct demonstration of geranylgeranylation and farnesylation of Ki-Ras in vivo. *The Journal of Biological Chemistry*, *272*(22), 14093–14097.

Schleucher, J., Schwendinger, M., Sattler, M., Schmidt, P., Schedletzky, O., Glaser, S. J., et al. (1994). A general enhancement scheme in heteronuclear multidimensional NMR employing pulsed field gradients. *Journal of Biomolecular NMR*, *4*(2), 301–306.

Sinensky, M. (2000). Recent advances in the study of prenylated proteins. *Biochimica et Biophysica Acta—Molecular and Cell Biology of Lipids*, *1484*(2–3), 93–106.

Smirnov, I. P., Zhu, X., Taylor, T., Huang, Y., Ross, P., Papayanopoulos, I. A., et al. (2004). Suppression of α-cyano-4-hydroxycinnamic acid matrix clusters and reduction of chemical noise in MALDI-TOF mass spectrometry. *Analytical Chemistry*, *76*(10), 2958–2965.

Suazo, K. F., Schaber, C., Palsuledesai, C. C., Odom John, A. R., & Distefano, M. D. (2016). Global proteomic analysis of prenylated proteins in plasmodium falciparum using an alkyne-modified isoprenoid analogue. *Scientific Reports*, *6*, 1–11.

Vervacke, J. S., Funk, A. L., Wang, Y. C., Strom, M., Hrycyna, C. A., & Distefano, M. D. (2014). Diazirine-containing photoactivatable isoprenoid: Synthesis and application in studies with isoprenylcysteine carboxyl methyltransferase. *Journal of Organic Chemistry*, *79*(5), 1971–1978.

Zhang, S., Sperlich, B., Li, F., Al-ayoubi, S., Chen, H., Zhao, Y., et al. (2017). Phosphorylation weakens but does not inhibit membrane binding and clustering of K-Ras4B. *ACS Chemical Biology*, *12*(6), 1703–1710.

CHAPTER NINE

The Preparation and Solution NMR Spectroscopy of Human Glycoproteins Is Accessible and Rewarding

Adam W. Barb[1], Daniel J. Falconer, Ganesh P. Subedi

Roy J. Carver Department of Biochemistry, Biophysics and Molecular Biology, Iowa State University, Ames, IA, United States

[1]Corresponding author: e-mail address: abarb@iastate.edu

Contents

Abstract

The majority of proteins excreted by human cells and borne at the cell surface are modified with carbohydrates. Glycoproteins mediate a wide range of processes and adopt fundamental roles in many diseases. The carbohydrates covalently attached to proteins

Methods in Enzymology, Volume 614
ISSN 0076-6879
https://doi.org/10.1016/bs.mie.2018.08.021

during maturation in the cell directly impact protein structure and function as integral and indispensable components. However, the ability to study the structure of glycoproteins to high resolution was historically limited by technical barriers including a limited availability of appropriate recombinant protein expression platforms, limited methods to generate compositional homogeneity, and difficulties analyzing glycoprotein composition. Furthermore, glycoproteins and in particular the glycan moieties themselves often exhibit a high degree of conformational heterogeneity. Solution NMR spectroscopy is a powerful tool to study biological macromolecules that is capable of characterizing mobile elements of molecules with atomic-level resolution. Methods to express glycoproteins, incorporate stable isotope labels, and analyze glycoproteins have recently opened new avenues to prepare and investigate glycoproteins. These methods are accessible to many laboratories with experience expressing and purifying proteins from prokaryotic expression hosts.

1. INTRODUCTION

Glycoproteins represent a staggeringly small fraction of proteins analyzed by high-resolution techniques despite high concentrations at the cell surface and critical roles in many human diseases. Though one-half or more of all human proteins are predicted to contain a carbohydrate chain linked through an asparagine residue (N-linked glycan; Apweiler, Hermjakob, & Sharon, 1999), only ~2% of all human protein structures in the Protein Data Bank show at least the first three residues of what is likely a much larger N-glycan (14% of these are the crystallizable fragment of immunoglobulin G (IgG Fc), as of May 2018). Many reasons account for the dramatic underrepresentation in the structural database and in structure/function studies. These factors include: challenges associated with preparing glycoproteins, difficulties crystallizing glycoproteins or resolving disordered glycans, and the requirement of specialized techniques to characterize glycoprotein composition. Solution NMR spectroscopy is particularly well suited to explore the structure and function of glycoproteins. Furthermore, recent advances in protein expression and glycoprotein analyses surmount these barriers. The goal of this chapter is to describe practical considerations to prepare human glycoproteins with appropriate glycans for analysis by multinuclear solution NMR spectroscopy.

Mammals target many proteins to the secretory pathway with a ~20 amino acid signal peptide at the N-terminus that localizes the translating ribosome to an endoplasmic reticulum (ER) transmembrane pore complex (Moremen, Tiemeyer, & Nairn, 2012). The nascent polypeptide chain is imported into the ER lumen during which time the oligosaccharyltransferase activity associated with the pore complex recognizes a three amino acid sequence of [Asn/any nonproline residue/serine or threonine] and transfers a large 14-residue carbohydrate chain from a dolichol diphosphate donor to the Asn sidechain (Fig. 1). Modified polypeptides then fold in the lumen of the ER where the N-glycan remodeling continues. Most glycoproteins then proceed to the Golgi where those designated for secretion are exported to the surface of the cell. During this transport, other types of modification may occur, including O-linked *N*-acetylgalactosamine (O-GalNAc) modifications to Ser or Thr residues in flexible regions (Fig. 1). N- and O-glycans are further modified during transport and the types and extent of remodeling are due in large part to the enzymes expressed in the Golgi. As a result, most glycoproteins exhibit significant compositional heterogeneity.

Glycans are integral to protein function and harbor epitopes for specific receptors, provide stability, assist folding, impact oligomerization, and change the structure and function of proteins (reviewed extensively: Varki, 2017). Furthermore, blocking N-glycosylation is fatal to human cells in culture. Given the essential role glycans adopt in protein function, it is crucial to develop methods to simplify glycoprotein production and analysis. Furthermore, the appropriate investigation of disease-related factors often requires proteins in a form as similar to the native form as possible, including relevant modifications (Baum & Crocker, 2009). The development of multiple powerful expression and analytical tools provides new opportunities to probe glycoproteins using solution NMR spectroscopy. Here we describe the techniques utilized by our lab and the various considerations of targeted glycoprotein studies. Here we present strategies to prepare and analyze glycoproteins by solution NMR spectroscopy that utilize advances in glycoprotein expression using cultured human cells, strategies to produce homogenous glycoproteins with stable isotope labels, advances in mass spectrometry to analyze glycan composition, and specific NMR experiments to probe glycans. As such, the references cited note details of the individual techniques (with references to videos, where available).

The timing of this review should also mark a transition in glycoprotein studies. The recent development of simplified expression and analysis techniques imparts capabilities on the experimentalist that were previously accessible to only a few labs in the world. As a result, studies of glycoproteins are poised to reward the ambitious experimentalists who recognize a new opportunity to characterize previously inaccessible targets. These techniques

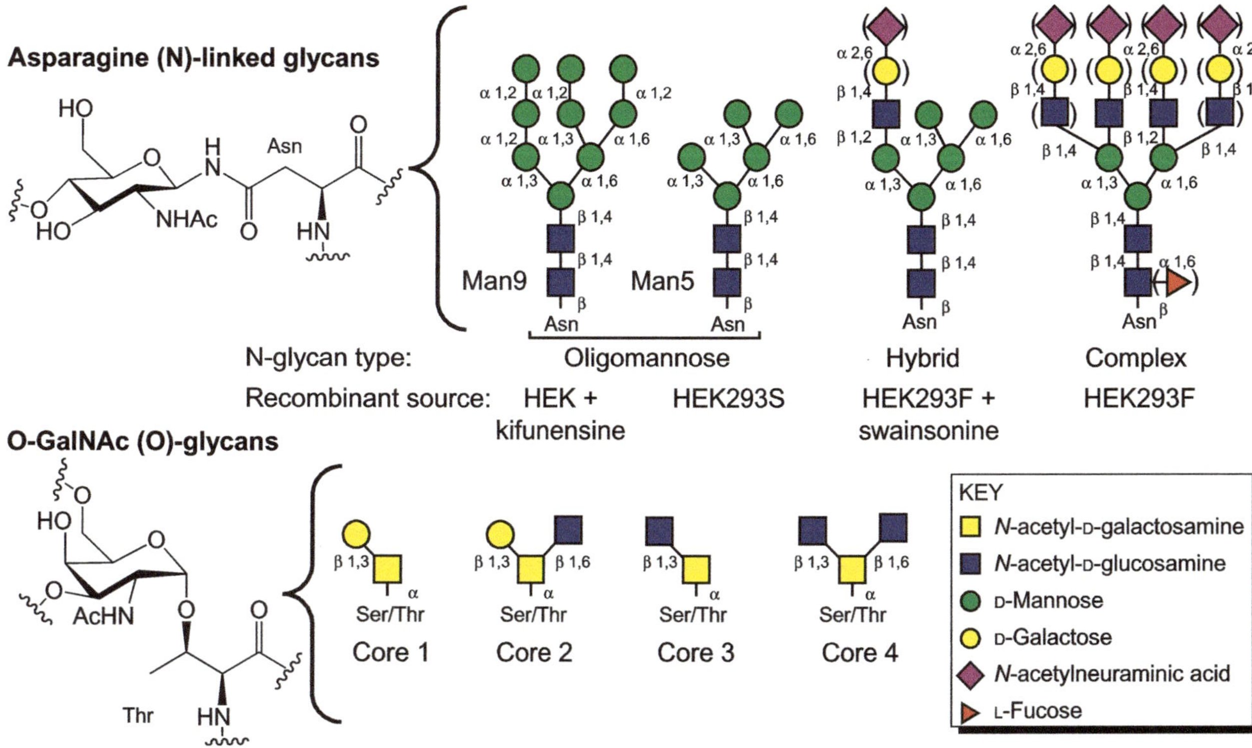

Fig. 1 N- and O-linked carbohydrates are two common modifications found on cell surface, secreted, Golgi, and ER proteins. Above, the structure of a glucosamine residue linked through an Asn residue to the protein and common N-glycan forms. Below, an O-GalNAc glycan linked through a Thr residue. Core structures, shown in cartoon forms, are often modified beyond the composition shown here.

are poised to promote a revolution in the structural biology of appropriately processed human proteins that will form the foundation of relevant models for disease processes.

2. GLYCOPROTEIN PRODUCTION

2.1 Suitable Expression Hosts

The vast majority of isotopically enriched proteins analyzed by solution NMR spectroscopy are expressed using *Escherichia coli*. This system, however, fails for glycoprotein expression in multiple key ways. *E. coli* does not contain the mammalian glycosylation machinery and thus does not synthesize glycoproteins. Mammalian chaperones assist protein folding but are missing in prokaryotic hosts. Lastly, the prokaryotic cytosol is reductive and incompatible with disulfide bond formation that occurs in the oxidative endoplasmic reticulum of eukaryotes. Currently, no holistic prokaryotic expression strategy to produce uniformly labeled mammalian glycoproteins exists, though multiple groups are engineering *E. coli* to surmount this limitation (Lee et al., 2017; Schein et al., 1992; Valderrama-Rincon et al., 2012; Wang & Amin, 2014). Eukaryotic microbes, namely, the yeasts, have an analogous oxidative secretory system with glycosylation machinery. Unfortunately, yeasts synthesize glycans that are very different from mammalian glycans though efforts are ongoing to correct this deficiency (Amano et al., 2008; Varki et al., 2017; Wang & Lomino, 2012). Thus, at present, mammalian protein expression to obtain appropriate glycoforms is most effectively performed using a mammalian host.

2.2 Human Embryonic Kidney (HEK)293 Cells

HEK293 cells provide appropriate glycosylation machinery and support high-yield expression. Though mammalian cells grow in considerably more complex media than *E. coli*, the development of rapid transient transfection methods for suspension cultures greatly reduces preparation times compared to stable transfections of adherent cells (Backliwal, Hildinger, Chenuet, et al., 2008; Wurm & Bernard, 1999). Cells for suspension cultures are commercially available, including HEK293F, and work with commercial vectors (ThermoFisher). Recent advances by Moremen and coworkers at the University of Georgia greatly expanded the system for secretory pathway proteins (Barb, Meng, et al., 2012; Moremen et al., 2018). A complete protocol for this expression system is available with an accompanying video

(Subedi, Johnson, Moniz, Moremen, & Barb, 2015). The equipment required to culture these cells includes a humidified, heated CO_2 shaker, a sterile BSL-2 biological safety cabinet, a centrifuge to pellet cultured cells, and a microscope or other device to count cells. HEK293F cells grow much slower than *E. coli* and divide only every 20–24 h. Despite the significant differences in handling *E. coli* and HEK293F cells, the latter are robust, forgiving, and can even be vortexed. It is important to note that handling HEK293 cells requires a higher standard for safety and sterility compared to handling *E. coli*. Not only must the cultures be protected from contamination, the *scientist* must be protected because human cells can harbor human diseases. Enhanced safety training is strongly advised, including training for handling human tissues and fluids plus other certifications required by the sponsoring institute.

2.3 Transient Transfection of HEK293

Transient transfection allows a faster route from plasmid preparation to protein expression because the time-consuming task of selecting stable transfectants is avoided. Plasmid DNA is combined with a transfection reagent like the cationic polymer polyethyleneimine (PEI) to precipitate DNA on the cell surface (Longo, Kavran, Kim, & Leahy, 2013). These plasmids contain an *E. coli* replication sequence plus appropriate selectable markers, thus the DNA for transfection is easily prepared using standard plasmid purification procedures. One unusual aspect of the transfection, in comparison to *E. coli* transformation, is the use of relatively high concentrations of plasmid DNA (~2.5 μg DNA/mL transfection) with a threefold mass excess of PEI. The use of actively dividing cells with high viability (>95%) increases protein yield. Following addition of the DNA, cells are incubated for 24 h, at which point the culture is diluted with an equal volume of medium containing 4.4 m*M* valproic acid (a histone deacetylase inhibitor that prevents loss of the transfected DNA) (Backliwal, Hildinger, Kuettel, et al., 2008). Protein expression usually follows the dilution step and cultures maintain a high degree of viability for 4–6 days. It is advisable to harvest the medium containing the secreted protein once cell viability drops below 50% (as judged by trypan blue staining). One limiting aspect of this expression system is proteolysis by enzymes released from apoptotic or lysed cells. Sensitive proteins may require culture harvest at an earlier time point. We have observed limited proteolysis of a few unstructured residues at the IgG1 Fc C-terminus and the degradation of N- and C-terminal poly-His tags (data not shown). The pGen2 vector used in our lab contains an N-terminal GFP tag that allows for easy protein expression monitoring (Subedi et al., 2015). With highly expressing proteins like GFP-hCD16a (~200 mg/L), the culture medium becomes visibly green on the day following culture dilution. Furthermore, some expressed proteins may not be released by the cells. This scenario is evident when the cell pellet, but not the medium, is green. In the latter example it is advisable to screen different protein constructs to identify one that will secrete soluble protein.

2.4 Stable Isotope Labeling in HEK293

HEK293F cells will grow in a chemically defined medium that contains multiple sources of carbon and nitrogen atoms, unlike many mammalian cell lines that require serum, a complex mixture of chemicals, growth factors, and proteins. The use of a chemically defined medium allows for substitution with isotope-labeled sugars or amino acids that are incorporated into the expressed glycoprotein (Dutta, Saxena, Schwalbe, & Klein-Seetharaman, 2012). Labeling can be achieved in one of two ways. It is possible to add a labeled amino acid or carbohydrate directly to the standard expression medium. This works well for [$^{13}C_U$]-glucose, though the percentage of label at any site will be directly proportional to the amount of label in the medium. For example, we routinely add 3 g/L [$^{13}C_U$]-glucose to a commercial medium that contains 4 g/L glucose for a final [^{13}C]-glycan content of roughly 40%. The primary advantage of this labeling strategy is the ease of medium preparation; no further adjustment to the commercial medium is required. One disadvantage of this strategy is that mass spectrometry-based analysis of glycan composition, as discussed in Section 4, will be complicated by a mixture of isotopologues. We have scant experience with adding labeled amino acids to a complete medium, and it is unclear how much or if label scrambling will occur.

Complete substitution of unlabeled amino acid(s) or glucose in the medium is achieved by purchasing or preparing a specially formulated medium lacking the specific labeling target. It is possible to purchase medium completely devoid of amino acids and carbohydrates to prepare any custom labeling scheme. In this case, we reconstitute the medium with 100 mg/L of each amino acid except Gln which is added at 1 g/L as well as 3 g/L glucose. Next, the medium is adjusted to an osmolarity of 300 mOsm/L with sodium chloride then sterilized before use. This strategy has proven effective with selective or uniform ^{13}C, ^{15}N labeling of amino acids plus glucose and often results in higher protein expression when compared to the commercial medium, likely due to a higher concentration of many amino acids. In our lab, yields depend on the protein expressed, ranging from 10 to 200 mg/L.

3. STRATEGIES TO OBTAIN GLYCOPROTEIN COMPOSITIONAL HOMOGENEITY

Glycoproteins are modified by glycosyltransferases and glycosylhydrolases in the ER and Golgi; however, the degree of remodeling for each glycan on each protein is not explicitly defined by a template (Moremen et al., 2012). This template-independent synthesis leads to potentially enormous compositional variability that is evident as a smear in an SDS-PAGE gel instead of a single, tight band. This section contains a description of multiple methods available to obtain homogeneity that manipulate N-glycans during protein expression, postpurification, or a combination of both. These methods have advanced glycoprotein studies, however, it is not always clear what types of glycans are found on glycoproteins in the human body, and proteins expressed by HEK293F cells do not necessarily contain the correct types of N-glycans, thus, caution is suggested (Patel, Roberts, Subedi, & Barb, 2018). Another aspect that is not addressed by any of these techniques is the accessibility of the N-glycosylation site to the oligosaccharyltransferase. Asparagine residues in the extreme N-terminus or within 50 residues of the C-terminus may not be efficiently modified (Kelleher & Gilmore, 2006).

3.1 The Gnt1$^-$ HEK293 Cell Line

One variation of the HEK293F cell line, HEK293S, contains the deletion of a key enzyme that catalyzes a branchpoint in N-glycan biosynthesis, Gnt1 (Reeves, Callewaert, Contreras, & Khorana, 2002; Stanley, Narasimhan, Siminovitch, & Schachter, 1975) (for a complete review of N-glycan biosynthesis, see Varki et al., 2017). The result is the expression of proteins with predominantly Man5 oligomannose N-glycans, though a small percentage of Man6, Man7, Man8, and Man9 forms should be expected (Fig. 1).

3.2 Inhibitors of Glycosylation Steps

Small molecules that inhibit specific N-glycan processing steps likewise restrict the manifold of N-glycan compositions. HEK293F cells, with a full complement of processing enzymes, can be treated with kifunensine to produce predominantly Man9 oligomannose N-glycans (Fig. 1). Swainsonine addition inhibits another mannosidase, Golgi mannosidase 2, and results in the production of glycans with predominantly hybrid type N-glycans (Fig. 1). Both of these inhibitors are relatively inexpensive, potent (5–10 μ*M*), and well tolerated. Likewise, fucosylation can be effectively reduced by 90%–95% by adding 100–250 μ*M* 2-deoxy-2-fluoro-fucose to the culture medium.

3.3 Glycan Remodeling Following Purification

Once expressed, glycans are often accessible to glycan-modifying enzymes in vitro. This approach provides two advantages: glycan homogeneity as well as a mechanism to introduce selective carbohydrate labels with labeled sugar nucleotides. Multiple methods to prepare labeled sugar nucleotides are published, including UDP-[^{13}C]-galactose, [^{13}C, ^{15}N]-CMP-*N*-acetylneuraminic acid, and [^{13}C, ^{15}N]-UDP-*N*-acetylglucosamine (Azurmendi et al., 2007; Barb, 2015; Barb & Prestegard, 2011; Macnaughtan et al., 2008; Yamaguchi et al., 1998). Many of these methods were developed to remodel glycoproteins with predominantly complex-type biantennary glycans with a varying level of galactose and *N*-acetylneuraminic acid residues following purification from human serum or HEK293F expression (Fig. 2). In principle, these techniques are applicable to many other N-glycoprotein N-glycans but have not been evaluated using complex-type N-glycans with more than two branches.

The first in vitro remodeling step is conducted with a commercially available neuraminidase to efficiently remove *N*-acetylneuraminic acids residues from the nonreducing termini. Neuraminidase-treated glycans generally remain inhomogeneous, but can be further treated in a galactosylation reaction catalyzed by commercially available bovine lactose synthase or trimmed further with a β-galactosidase (Fig. 2). The galactosylation reaction is rapid, easily completed, and will produce a homogenous N-glycan, depending on variability in the starting material. However, the β-galactosidase reaction has proven more challenging in our lab. Though β-galactosidase enzymes are commercially available, we screen multiple enzyme preparations to identify a single supply that will efficiently remove galactose residues from IgG1 Fc. As sources and suppliers change, we have repeated this exercise many times in the past decade.

Galactosylated N-glycans are suitable substrates for ST6Gal-I catalyzed sialylation (Macnaughtan et al., 2008). ST6Gal-I expressed to high yield in HEK293F cells with an N-terminal GFP fusion (Barb, Meng, et al., 2012).

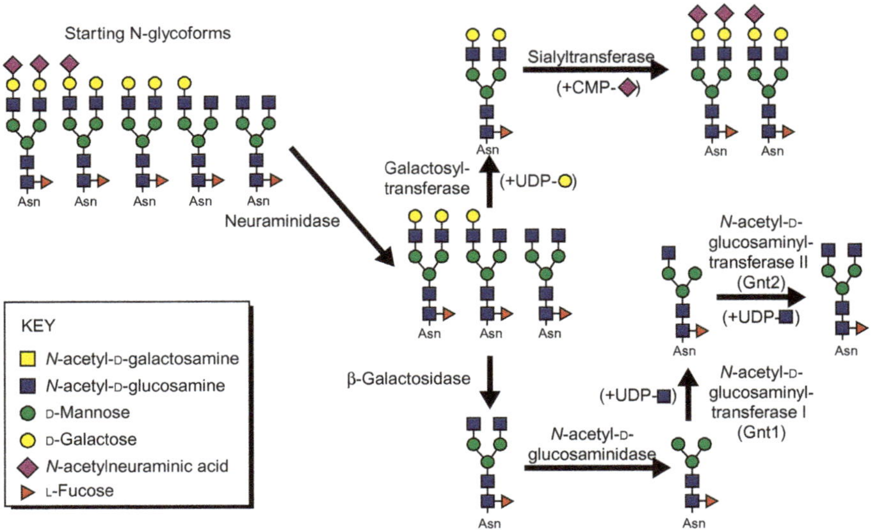

Fig. 2 N-glycan homogeneity can be achieved through postpurification remodeling. The predominant N-glycoforms found on hIgG1 Fc expressed from HEK293F cells can be remodeled with specific enzymes and substrates. The glycans are predominantly a complex-type biantennary N-glycan with varying levels of galactose and sialic acid incorporation (shown on the *left side* of the figure). NMR active isotopes can be incorporated using the glycosyltransferase enzymes noted above. Also note the paucimannose structure in the *bottom right* of the figure with two *N*-acetylglucosamine and three mannose residues (in this example, the paucimannose N-glycan contains a core fucose residue).

It is well known that ST6Gal-I sialylates the galactose residue on the Manα1–3Manβ branch of a complex-type biantennary N-glycan with 10-fold greater activity than the galactose residue on the Manα1–6Manβ branch (Barb, Meng, et al., 2012; Paulson, Rearick, & Hill, 1977). We were able to achieve high levels of IgG1 Fc with two sialic acids on each glycan only by using high amounts of ST6Gal-I (15 mol substrate:1 mol enzyme). The reaction is incubated for multiple days and washed with fresh CMP-*N*-acetylneuraminic acid each day to remove the CMP hydrolysis product that also inhibits the reaction. This is the most challenging glycosyltransferase reaction we have yet to encounter with IgG1 Fc as a substrate.

Shorter N-glycans can be formed by further digestion with a commercially available *N*-acetylhexosaminidase to generate paucimannose N-glycans (which may or may not contain a core fucose residue; defined in Fig. 2). These glycans can be extended by one *N*-acetylglucosamine residue attached to the Manα1–3Manβ residue using Gnt1, expressed from

HEK293F cells (Barb, 2015). Though the Man5 oligomannose N-glycan is the native substrate for this reaction, Gnt1 efficiently modifies the paucimannose N-glycan (containing two *N*-acetylglucosamine and three mannose residues). A second *N*-acetylglucosamine branch can be added to the Manα1–6Manβ residue with Gnt2.

The Man5 N-glycans from HEK293S expression (Fig. 1) are effective acceptor substrates for Gnt1 to create a Man5 N-glycan with an *N*-acetylglucosamine residue on the Manα1–3Manβ residue (Barb, 2015). We have not identified a suitable mannosidase that expresses in sufficient quantities to convert the Man5-GlcNAc N-glycan in vitro into the cognate Gnt2 substrate, a Man3-GlcNAc N-glycan.

4. COMPOSITIONAL ANALYSIS OF GLYCOPROTEINS

One critical aspect of glycan remodeling is assessing compositional heterogeneity arising from incomplete remodeling reactions in vivo and in vitro. Though many techniques are available, we use predominantly mass spectrometry-based methods due to rapidity and precision. In addition to confirming sample homogeneity, N-glycan processing during recombinant protein expression is a useful reporter of glycan accessibility. Analyses of N-glycan composition before enzymatic remodeling has the potential to directly support observations made by NMR spectroscopy (Subedi, Hanson, & Barb, 2014). For example, N-glycans that stabilize protein structure often form van Der Waals contacts with amino acids at the protein surface. These interactions restrict N-glycan processing and restrict enzymatic N-glycan remodeling in vivo and in vitro.

4.1 Electrospray Ionization Mass Spectrometry of Intact Glycoproteins

One advantage of electrospray ionization mass spectrometry (ESI-MS) is the ability to analyze intact glycoproteins rapidly with minimal sample preparation. Generally, a small amount of glycoprotein (1 μg) is loaded onto a C4 or C8 reversed phase column and eluted with an acetonitrile gradient into a mass spectrometer. Chromatography eliminates sample preparation steps including desalting, and the instrument time is often brief (15–20 min per sample). Intact mass spectrometry generates an envelope of peaks with different mass-to-charge ratios (m/z) for each protein as shown in Fig. 3A. Each different protein mass will generate a unique envelope, thus, the number of distinct glycoprotein masses that can be resolved in a single

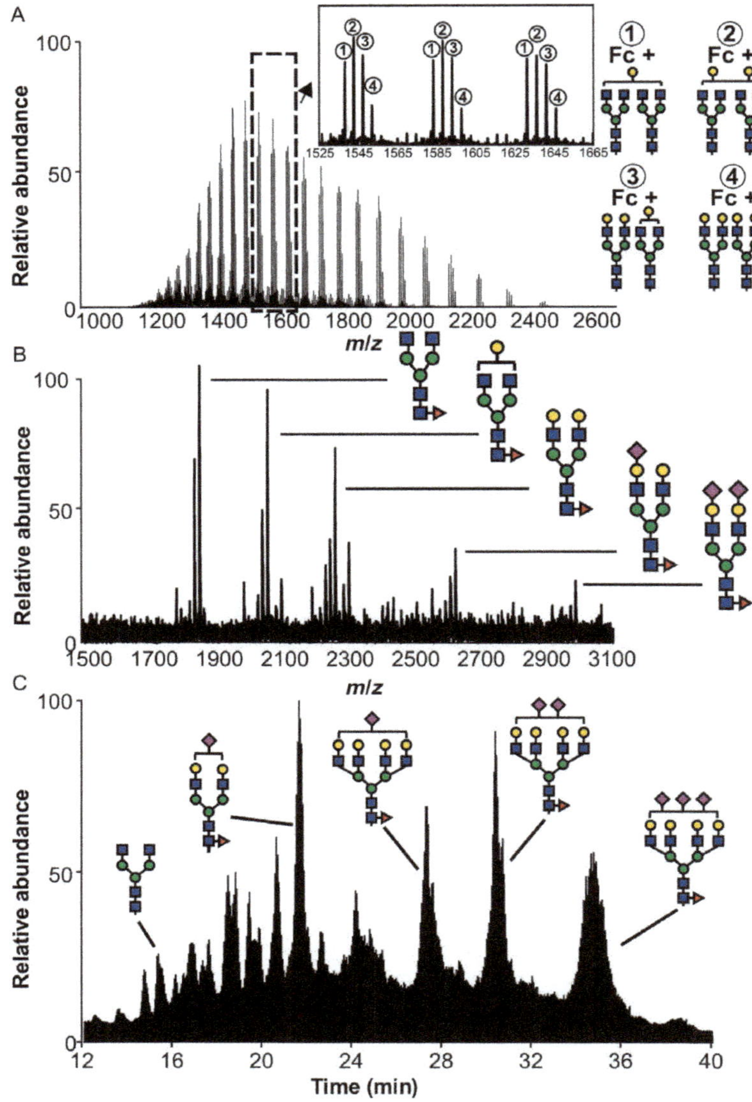

Fig. 3 N-glycan composition as probed with mass spectrometry. (A) Total ion current from an ESI-MS analysis of intact hIgG1 Fc, a homodimer containing two N-glycans. (B) Total ion current from MALDI-MS of permethylated hIgG1 Fc F243S N-glycans. (C) Total ion current from an ESI-MS analysis of procainamide-derivatized N-glycans from recombinant CD16a.

experiment is somewhat limited. As a result, this approach is the most effective when there are only a small number of predominant forms as observed for IgG Fc. A spectrum can be deconvoluted manually or with multiple software platforms to provide the intact mass of each species with potentially very high resolution (± 1 Da at ~ 50 kDa). One drawback to this technique is that all modifications to a single protein contribute to the intact mass; oxidation, alkylation, proteolysis, etc., introduce challenges to match the observed mass with a calculated composition. ESI-MS of intact glycoproteins works very well when assessing the efficacy of an in vitro glycan remodeling reaction, e.g., where ± 162 Da (a hexose) is easily confirmed. This approach is of limited applicability with isotope-labeled proteins that are more challenging to deconvolute, particularly in the case of label incorporation at lower percentages ($<95\%$).

4.2 Matrix-Assisted Laser Desorption Ionization Mass Spectrometry

N-glycans are directly observable using matrix-assisted laser desorption ionization mass spectrometry (MALDI-MS). First, 25–200 μg of glycoprotein is proteolyzed then digested with PNGaseF, a commercially available enzyme that specifically cleaves N-glycans from Asn side chains. Peptides are removed by passing the mixture over a C18 column to collect the flow-through which contains N-glycans. Next, released N-glycans are permethylated in a chemical reaction using anhydrous sodium hydroxide in dimethyl sulfoxide with iodomethane (Anumula & Taylor, 1992). A video protocol using related methods is available (Zaare, Aguilar, Hu, Ferdosi, & Borges, 2016). Following a two-phase extraction with methylene chloride and water, N-glycans are spotted on a MALDI plate with dihydroxybenzoic acid as the matrix. Permethylation replaces all labile hydrogens attached to nitrogen and oxygen residues with a methyl group, eliminating charges on the N-glycan. MALDI generates sodium or potassium adducts that are detectable as singly charged species with a TOF mass analyzer (Fig. 3B). These data are relatively simple to analyze as the predominant species only contain carbohydrate residues with a single metal ion and a single charge, though the methyl modifications must be accounted for. Permethylated N-glycans can be detected using multiple platforms, but we find MALDI-MS rapid and sufficiently sensitive. Another advantage of this approach is that even very basic instruments will produce reliable data. One disadvantage of the approach is the relatively lengthy preparation, which may take 2–3 days though many samples can be analyzed in parallel.

MALDI-MS of permethylated N-glycans is not as sensitive as the analysis of procainamide derivatized N-glycans by ESI-MS, as described below. It should also be noted that modification with sialic acids, including *N*-acetylneuraminic acid, is labile in weak acids including acetic acid. Thus, it is important to avoid low pH as much as possible to accurately assess sialylation.

4.3 Analysis of N-glycans by Hydrophilic Interaction Chromatography (HILIC)-ESI-MS/MS

HILIC-ESI-MS/MS is the standard technique used by our laboratory to characterize the compositional heterogeneity of recombinant proteins due to high sensitivity and precision (Fig. 3C). This technique requires only a small amount of glycoprotein (1–5 μg). In lieu of permethylation, N-glycans are derivatized at the reducing end by reacting with procainamide to form a Schiff base that is reduced with sodium cyanoborohydride (Patel et al., 2018). The reaction mixture is next directly injected onto a HILIC column (Waters) in 75% acetonitrile, 25% H_2O, 0.1% formic acid, and 0.01% trifluoroacetic acid, and eluted with an increasing linear water gradient. HILIC separates glycans by size, with larger glycans eluting later in the profile. Raw data are analyzed with BYONIC (Protein Metrics) and validated manually by matching the monoisotopic peak with the calculated mass. Retention time provides an additional validation and correlates linearly with N-glycan mass (Patel et al., 2018). Analysis of N-glycans using a Q-Exactive Orbitrap mass spectrometer (Thermo Scientific) provided mass accuracy greater than 0.01 Da. Fragmentation in the second MS dimension can confirm the composition and provide insight regarding the configuration of individual N-glycan species (core vs branch fucosylation, etc.). These methods are straightforward and easily validated.

Studies of proteins with multiple N-glycans per polypeptide chain may require added sophistication to analyze each N-glycan separately. The analysis of glycopeptides provides more information than the analysis of released N-glycans, including what type of N-glycans are found at specific sites in the protein. Glycoproteomics approaches are widely applied to analyze glycopeptides (Kolarich, Jensen, Altmann, & Packer, 2012; Plomp, Bondt, de Haan, Rombouts, & Wuhrer, 2016). Though linking glycans and peptides introduces complexity into the analysis as the critical data are found in MS^2 spectra, multiple programs are available to guide the analysis including BYONIC and pGlyco2.0 (Liu et al., 2017). Another potentially challenging aspect of glycopeptide analysis involves identifying appropriate proteases and reaction conditions to separate the individual N-glycans. Once digested, the mixture of peptides and glycopeptides can be analyzed using the HILIC separation described above. Peptides separate based on N-glycan composition with aglycosylated peptides eluting very early from the column. C18 reversed-phase columns may also be used, but this approach often requires glycopeptide purification prior to chromatography. A simple clean-up procedure reduces the ion-suppressing effects of nonglycosylated peptides which are predominant and often ionize more efficiently than glycopeptides.

5. SOLUTION NMR OF LABELED GLYCOPROTEINS

NMR spectra of isotope-labeled glycoproteins are collected with the same pulse sequences applied to most [^{13}C, ^{15}N]-labeled proteins. Here we describe considerations for NMR analyses of glycoproteins with labeled backbone residues and labeled carbohydrates.

5.1 Analysis of Glycoproteins With Backbone Labels

N-glycans promote folding and stability in secreted glycoproteins, therefore many glycoproteins are stable at high temperatures. IgG1 Fc, for example, is stable for weeks to months at 50°C. N-glycans, however, can restrict protein tumbling in solution, leading to broader linewidths and reduced signal in 3d experiments requiring multiple coherence transfer steps. We often observe a benefit to using TROSY-based experiments (Pervushin, Riek, Wider, & Wuthrich, 1997) even in moderately sized proteins without benefit of perdeuteration. Unfortunately, deuterium labeling to reduce spoiling of the TROSY-effect due to dipolar interactions with remote ^{1}H has not yet been demonstrated using mammalian cells. It is assumed mammalian cells are unlikely to survive in high percentages of $^{2}H_2O$. Supplementing expression medium with [^{2}H, ^{13}C, ^{15}N]-labeled amino acids is very expensive, tolerated by insect cells and potentially applicable to protein expression in HEK293 (Franke et al., 2018). We have observed improvements in spectral quality by removing N-glycans that do not contribute or contribute minimally to protein structure and retaining those that are essential for protein structure and function. Nonessential N-glycans are removed by mutating the Asn residue to Gln. In our experience, as discussed later, removing nonessential N-glycans does reduce protein stability to a moderate degree for CD16a.

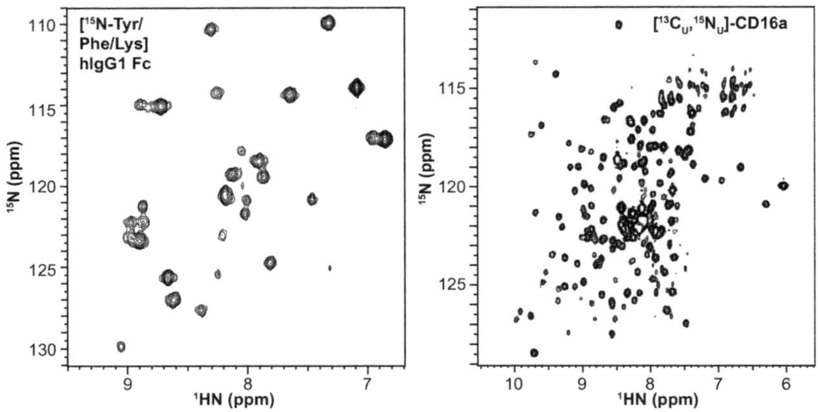

Fig. 4 Glycoproteins expressed at high yield with high percentages of ^{15}N and ^{13}C incorporated into the polypeptide backbone. (*left*) Selective labeling to incorporate ^{15}N into the amides of Tyr, Phe, and Lys residues of glycosylated IgG1 Fc, a ~50 kDa homodimer with one N-glycan per chain, observed in a 2d HSQC-TROSY collected at 50°C. (*right*) Complete ^{15}N and ^{13}C labeling of a CD16a variant with two N-glycans at N45 and N162 observed in a 2d projection of a 3d HNCA-TROSY experiment collected at 30°C. The CD16a variant is a ~23 kDa monomer. These spectra were collected 16.4T.

Example spectra from isotope-labeled glycoproteins expressed in HEK293F cells are shown in Fig. 4. IgG1 Fc is a ~50 kDa homodimer composed of two identical polypeptide chains, each with an N-glycan attached at residue 297. The IgG1 Fc N-glycan does not contribute substantially to the protein rotational correlation time (Subedi & Barb, 2015). [$^{15}N_{amide}$-Tyr/Phe/Lys]-IgG1 Fc produced from HEK293F cells provides a clear fingerprint of the protein in an ^{15}N-HSQC-TROSY spectrum, with minimal evidence for metabolic scrambling of the labeled amino acids. The 37 Tyr, Phe, or Lys residues in this construct provide 29 clear peaks in a ^{15}N-HSQC spectrum. Kato and coworkers assigned the protein backbone using traditional triple resonance experiments (Yagi et al., 2015).

The extracellular antibody-binding domain of CD16a/Fc γ receptor 3a is a smaller monomeric protein of ~23 kDa with as many as five N-glycans. Mutating three asparagine residues to prevent N-glycosylation at N38, N74, and N169 does not reduce antibody-binding affinity but does reduce the stability of the protein. The result is a soluble CD16a variant with two N-glycans at N45 and N162 that provides well-dispersed NMR spectra (Fig. 4B) (Ferrara, Stuart, Sondermann, Brunker, & Umana, 2006). An HNCA spectrum of [^{13}C, ^{15}N]-CD16a also shows clear evidence for efficient resonance transfer and high levels of label incorporation using the uniform labeling method described in Section 2.4 (Fig. 4B). Thus, these strategies to label amino acids are suitable for targeted studies with selective labeling or global studies of structure and motion using standard solution NMR pulse sequences.

Glycan truncation is another approach to limit the negative impact of large N-glycans on protein tumbling without mutating asparagine residues. Glycan truncation is achievable by expressing protein with the HEK293S cell line that produces only oligomannose N-glycans that can be trimmed by endoglycosidase F to a single asparagine-linked *N*-acetylglucosamine. The preservation of stability upon truncation is likely due to the predominant stabilizing effect of the first *N*-acetylglucosamine residue; Hanson et al. demonstrated that the first monosaccharide residue contributed roughly 2/3 of the energy toward stabilizing CD2 when compared to the stabilization provided by the full-length N-glycan (Hanson et al., 2009). Thus, in some cases, truncating the N-glycan may allow sufficient sensitivity for successful backbone assignment through traditional triple-resonance experiments.

5.2 Analysis of Glycoproteins With Uniformly Labeled Carbohydrates

Expressing glycoproteins in standard culture medium (containing [^{12}C]-glucose) supplemented with [^{13}C]-glucose provides an effective and relatively simple strategy to rapidly assess N-glycans using NMR (Subedi & Barb, 2015; Subedi, Falconer, & Barb, 2017). This strategy predominantly incorporates [^{13}C] into the glycans with little evidence for scrambling into amino acids (Fig. 5A). It is likely the high concentration of amino acids in the medium prevents metabolic scrambling. The advantage of this approach arises from the unique spectral signature of each N-glycan: the anomeric carbon (C1 for aldohexoses like glucose, galactose, and mannose) resonates at a greater frequency ($\sim$100 ppm) compared to other carbohydrate carbons at (50–80 ppm). However, the anomeric carbon of the *N*-acetylglucosamine residue, attached through a nitrogen atom to the Asn sidechain, resonates at roughly 80 ppm with a ^{1}H1 chemical shift around 5.0 ppm. Though there are many anomeric protons and carbons in a glycan, there is only one N-linked anomeric carbon per N-glycan. Furthermore, the location proximal to the polypeptide provides a single site for each N-glycan that is positioned to report differences arising from interactions with the amino acid residues and variations in amino acid type. These Asn-linked correlations can be easily distinguished in a 2d ^{1}H–^{13}C HSQC spectrum of CD16a (Fig. 5A). Furthermore, the signals of interest appear in a region of the

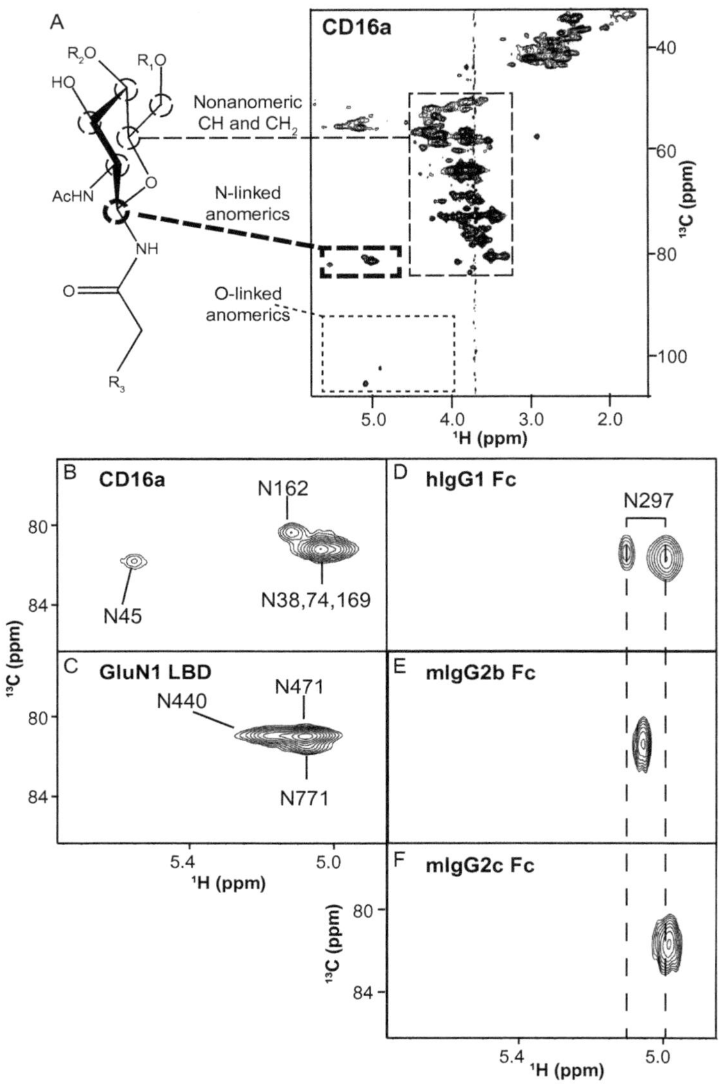

Fig. 5 Proteins with ^{13}C-labeled N-glycans provide a single, unique signal for each N-glycan. (A) A ^{13}C-HSQC of [^{13}C-glycan] CD16a with 5 N-glycans shows predominantly signal corresponding to carbohydrate and potentially limited scrambling to amino acids. Of the three distinct regions noted in the spectrum, cross-peaks from the anomeric asparagine-linked H1–C1 atoms are separated from the other signals, with one coherence for each N-glycan residue (expanded in B). (C–F) Similar analyses of four glycoproteins yield comparable results. These spectra were collected 16.4T.

spectrum with few signals from amino acids. This glycoprotein labeling and pulse sequence strategy provide an N-glycan fingerprint for each glycoprotein, similar to a 2d ^{1}H–^{15}N HSQC for backbone amide correlations in proteins.

Significant differences in peak positions arising from Asn-linked $^{1}H1$–$^{13}C1$ correlations have proven valuable to identify specific N-glycans in proteins that relate to protein function. CD16a contains as many as five N-glycans that show three distinct N-glycan peaks (Fig. 5B). Mutating Asn residues to Gln revealed which peaks belonged to which N-glycans (Subedi et al., 2017). As mentioned earlier, the two N-glycans that are important for antibody-binding affinity, N45 and N162, show distinct peaks that are separate from peaks arising from the N38, N74, and N169 glycans. The unique environment for each peak was destroyed by proteolysis causing the peaks to collapse upon the large intense peaks formed by N38, N74, and N169. Thus, the chemical environment of the Asn-linked $^{1}H1$ and $^{13}C1$ atoms of the N45- and N162-glycans is likely influenced by surrounded polypeptide residues. A similar analysis of the GluN1 ligand binding domain from the *N*-methyl-D-aspartate receptor (NMDAR) with three N-glycans revealed similar differences that could be destroyed through proteolysis.

Analysis of IgG Fc provided further insight into the chemical environment surrounding the glycan at the point of attachment. Human IgG1 Fc binds tightly to CD16a and shows two clear signals of differing intensities and linewidths originating from the Asn-linked *N*-acetylglucosamine residue (Fig. 5D) (Subedi & Barb, 2015). The appearance of two peaks for a single H1–C1 correlation likely results from slow conformational exchange. Interestingly, two IgG1 Fc forms that cannot bind CD16a, D265A, and proteolyzed IgG1 Fc, showed only a single peak that matched the position of the minor peak observed for intact IgG1 Fc. This result indicates the conformation contributing to the more intense peak on the right hand side of the intact IgG1 Fc spectrum is consistent with a conformation that binds receptor. The same analytical approach was applied to mouse IgG2b Fc and IgG2c Fc. Unlike human Fc, the mouse proteins showed only one peak for the Asn-linked *N*-acetylglucosamine residue; however, the peak from IgG2c Fc was found further to the right and mIgG2c bound mouse Fc γ receptor IV with ~fivefold greater affinity that mIgG2b Fc, consistent with the stabilized conformation observed in hIgG1 Fc (Falconer & Barb, 2018).

These results demonstrate that peaks corresponding to the Asn-linked *N*-acetylglucosamine residue provide a wealth of information regarding individual N-glycan behavior in many different glycoproteins.

5.3 Analysis of Glycoproteins With Specifically Labeled Carbohydrates

Labeling glycoprotein glycans with [^{13}C, ^{15}N]-labeled carbohydrate residues following purification greatly simplifies spectral complexity and assists peak assignment. Our laboratory has used this approach extensively to study the IgG1 Fc N-glycans (Barb, 2015; Barb, Ho, Flanagan-Steet, & Prestegard, 2012; Barb, Meng, et al., 2012; Barb & Prestegard, 2011; Subedi et al., 2014). A wealth of labeling strategies and nuclei are accessible though labeled sugar nucleotides are time-consuming or expensive to obtain. Furthermore, some glycosyltransferase reactions are challenging to optimize to achieve complete protein labeling. One example of postpurification labeling is the addition of [^{13}C2]-galactose to the IgG1 Fc N-glycans. Spin relaxation and relaxation-dispersion measurements demonstrated that the galactose residues were mobile, in contrast to previous interpretations of structures determined by X-ray crystallography (Barb & Prestegard, 2011). Furthermore, adding *N*-acetylglucosamine, galactose, and *N*-acetylneuraminic acid residues to the Fc N-glycan branches reduced carbohydrate motion after each addition (Barb, 2015; Barb, Meng, et al., 2012). Glycan labeling after purification represents an orthogonal approach to study the structure, function, and interactions of glycoprotein glycans. This strategy often benefits from the greater motion, and thus narrower lines, experienced by residues distal to the reducing end point of attachment in large proteins. This property is in contrast to backbone atoms in structured regions or select carbohydrate residues located closer to the glycan reducing end.

6. SUMMARY

Glycoproteins should no longer represent a forbidden territory to structural biologists and NMR spectroscopists due to the extraordinary advances in glycoprotein expression platforms, MS-based analytic methods, and NMR strategies. The goal to accurately recapitulate native disease processes will push laboratories to manufacture proteins that mimic glycan composition as well as behavior as closely as possible. Future steps toward understanding these complex molecules in human and animal health will require no less.

ACKNOWLEDGEMENT

This material is based upon work supported by the National Institutes of Health under Award No. R01 GM115489 (NIGMS) and the Roy J. Carver Department of Biochemistry,

Biophysics & Molecular Biology at Iowa State University. Any opinions, findings, and conclusions or recommendations expressed in this material are those of the authors and do not necessarily reflect the views of the National Institutes of Health.

REFERENCES

Amano, K., Chiba, Y., Kasahara, Y., Kato, Y., Kaneko, M. K., Kuno, A., et al. (2008). Engineering of mucin-type human glycoproteins in yeast cells. *Proceedings of the National Academy of Sciences of the United States of America, 105*(9), 3232–3237. https://doi.org/10.1073/pnas.0710412105.

Anumula, K. R., & Taylor, P. B. (1992). A comprehensive procedure for preparation of partially methylated alditol acetates from glycoprotein carbohydrates. *Analytical Biochemistry, 203*(1), 101–108.

Apweiler, R., Hermjakob, H., & Sharon, N. (1999). On the frequency of protein glycosylation, as deduced from analysis of the SWISS-PROT database. *Biochimica et Biophysica Acta, 1473*(1), 4–8.

Azurmendi, H. F., Vionnet, J., Wrightson, L., Trinh, L. B., Shiloach, J., & Freedberg, D. I. (2007). Extracellular structure of polysialic acid explored by on cell solution NMR. *Proceedings of the National Academy of Sciences of the United States of America, 104*(28), 11557–11561. https://doi.org/10.1073/pnas.0704404104.

Backliwal, G., Hildinger, M., Chenuet, S., Wulhfard, S., De Jesus, M., & Wurm, F. M. (2008). Rational vector design and multi-pathway modulation of HEK 293E cells yield recombinant antibody titers exceeding 1 g/l by transient transfection under serum-free conditions. *Nucleic Acids Research, 36*(15), e96. https://doi.org/10.1093/nar/gkn423.

Backliwal, G., Hildinger, M., Kuettel, I., Delegrange, F., Hacker, D. L., & Wurm, F. M. (2008). Valproic acid: A viable alternative to sodium butyrate for enhancing protein expression in mammalian cell cultures. *Biotechnology and Bioengineering, 101*(1), 182–189. https://doi.org/10.1002/bit.21882.

Barb, A. W. (2015). Intramolecular N-glycan/polypeptide interactions observed at multiple N-glycan remodeling steps through [(13)C,(15)N]-N-acetylglucosamine labeling of immunoglobulin G1. *Biochemistry, 54*(2), 313–322. https://doi.org/10.1021/bi501380t.

Barb, A. W., Ho, T. G., Flanagan-Steet, H., & Prestegard, J. H. (2012). Lanthanide binding and IgG affinity construct: Potential applications in solution NMR, MRI, and luminescence microscopy. *Protein Science, 21*(10), 1456–1466. https://doi.org/10.1002/pro.2133.

Barb, A. W., Meng, L., Gao, Z., Johnson, R. W., Moremen, K. W., & Prestegard, J. H. (2012). NMR characterization of immunoglobulin G Fc glycan motion on enzymatic sialylation. *Biochemistry, 51*(22), 4618–4626. https://doi.org/10.1021/bi300319q.

Barb, A. W., & Prestegard, J. H. (2011). NMR analysis demonstrates immunoglobulin G N-glycans are accessible and dynamic. *Nature Chemical Biology*, 7(3), 147–153. https://doi.org/10.1038/nchembio.511.

Baum, L. G., & Crocker, P. R. (2009). Glycoimmunology: Ignore at your peril! *Immunological Reviews, 230*(1), 5–8. https://doi.org/10.1111/j.1600-065X.2009.00800.x.

Dutta, A., Saxena, K., Schwalbe, H., & Klein-Seetharaman, J. (2012). Isotope labeling in mammalian cells. *Methods in Molecular Biology, 831*, 55–69. https://doi.org/10.1007/978-1-61779-480-3_4.

Falconer, D. J., & Barb, A. W. (2018). Mouse IgG2c Fc loop residues promote greater receptor-binding affinity than mouse IgG2b or human IgG1. *PLoS One, 13*(2), e0192123. https://doi.org/10.1371/journal.pone.0192123.

Ferrara, C., Stuart, F., Sondermann, P., Brunker, P., & Umana, P. (2006). The carbohydrate at FcgammaRIIIa Asn-162. An element required for high affinity binding to non-fucosylated IgG glycoforms. *The Journal of Biological Chemistry*, *281*(8), 5032–5036. https://doi.org/10.1074/jbc.M510171200.

Franke, B., Opitz, C., Isogai, S., Grahl, A., Delgado, L., Gossert, A. D., et al. (2018). Production of isotope-labeled proteins in insect cells for NMR. *Journal of Biomolecular NMR*, *71*, 173–184. https://doi.org/10.1007/s10858-018-0172-7.

Hanson, S. R., Culyba, E. K., Hsu, T. L., Wong, C. H., Kelly, J. W., & Powers, E. T. (2009). The core trisaccharide of an N-linked glycoprotein intrinsically accelerates folding and enhances stability. *Proceedings of the National Academy of Sciences of the United States of America*, *106*(9), 3131–3136. https://doi.org/10.1073/pnas.0810318105.

Kelleher, D. J., & Gilmore, R. (2006). An evolving view of the eukaryotic oligosaccharyltransferase. *Glycobiology*, *16*(4), 47R–62R. https://doi.org/10.1093/glycob/cwj066.

Kolarich, D., Jensen, P. H., Altmann, F., & Packer, N. H. (2012). Determination of site-specific glycan heterogeneity on glycoproteins. *Nature Protocols*, 7(7), 1285–1298. https://doi.org/10.1038/nprot.2012.062.

Lee, K. B., Nam, D. H., Nuhn, J. A. M., Wang, J., Schneider, I. C., & Ge, X. (2017). Direct expression of active human tissue inhibitors of metalloproteinases by periplasmic secretion in *Escherichia coli*. *Microbial Cell Factories*, *16*(1), 73. https://doi.org/10.1186/s12934-017-0686-9.

Liu, M. Q., Zeng, W. F., Fang, P., Cao, W. Q., Liu, C., Yan, G. Q., et al. (2017). pGlyco 2.0 enables precision N-glycoproteomics with comprehensive quality control and one-step mass spectrometry for intact glycopeptide identification. *Nature Communications*, *8*(1), 438. https://doi.org/10.1038/s41467-017-00535-2.

Longo, P. A., Kavran, J. M., Kim, M. S., & Leahy, D. J. (2013). Transient mammalian cell transfection with polyethylenimine (PEI). *Methods in Enzymology*, *529*, 227–240. https://doi.org/10.1016/B978-0-12-418687-3.00018-5.

Macnaughtan, M. A., Tian, F., Liu, S., Meng, L., Park, S., Azadi, P., et al. (2008). ^{13}C-sialic acid labeling of glycans on glycoproteins using ST6Gal-I. *Journal of the American Chemical Society*, *130*(36), 11864–11865. https://doi.org/10.1021/ja804614w.

Moremen, K. W., Ramiah, A., Stuart, M., Steel, J., Meng, L., Forouhar, F., et al. (2018). Expression system for structural and functional studies of human glycosylation enzymes. *Nature Chemical Biology*, *14*(2), 156–162. https://doi.org/10.1038/nchembio.2539.

Moremen, K. W., Tiemeyer, M., & Nairn, A. V. (2012). Vertebrate protein glycosylation: Diversity, synthesis and function. *Nature Reviews. Molecular Cell Biology*, *13*(7), 448–462. https://doi.org/10.1038/nrm3383.

Patel, K. R., Roberts, J. T., Subedi, G. P., & Barb, A. W. (2018). Restricted processing of CD16a/Fc gamma receptor IIIa N-glycans from primary human NK cells impacts structure and function. *The Journal of Biological Chemistry*, *293*(10), 3477–3489. https://doi.org/10.1074/jbc.RA117.001207.

Paulson, J. C., Rearick, J. I., & Hill, R. L. (1977). Enzymatic properties of beta-D-galactoside alpha2 leads to 6 sialytransferase from bovine colostrum. *The Journal of Biological Chemistry*, *252*(7), 2363–2371.

Pervushin, K., Riek, R., Wider, G., & Wuthrich, K. (1997). Attenuated T2 relaxation by mutual cancellation of dipole-dipole coupling and chemical shift anisotropy indicates an avenue to NMR structures of very large biological macromolecules in solution. *Proceedings of the National Academy of Sciences of the United States of America*, *94*(23), 12366–12371.

Plomp, R., Bondt, A., de Haan, N., Rombouts, Y., & Wuhrer, M. (2016). Recent advances in clinical Glycoproteomics of immunoglobulins (Igs). *Molecular & Cellular Proteomics*, *15*(7), 2217–2228. https://doi.org/10.1074/mcp.O116.058503.

Reeves, P. J., Callewaert, N., Contreras, R., & Khorana, H. G. (2002). Structure and function in rhodopsin: High-level expression of rhodopsin with restricted and homogeneous N-glycosylation by a tetracycline-inducible N-acetylglucosaminyltransferase I-negative

HEK293S stable mammalian cell line. *Proceedings of the National Academy of Sciences of the United States of America, 99*(21), 13419–13424. https://doi.org/10.1073/pnas.212519299.

Schein, C. H., Boix, E., Haugg, M., Holliger, K. P., Hemmi, S., Frank, G., et al. (1992). Secretion of mammalian ribonucleases from *Escherichia coli* using the signal sequence of murine spleen ribonuclease. *The Biochemical Journal, 283*(Pt. 1), 137–144.

Stanley, P., Narasimhan, S., Siminovitch, L., & Schachter, H. (1975). Chinese hamster ovary cells selected for resistance to the cytotoxicity of phytohemagglutinin are deficient in a UDP-N-acetylglucosamine—Glycoprotein N-acetylglucosaminyltransferase activity. *Proceedings of the National Academy of Sciences of the United States of America, 72*(9), 3323–3327.

Subedi, G. P., & Barb, A. W. (2015). The structural role of antibody N-glycosylation in receptor interactions. *Structure, 23*(9), 1573–1583. https://doi.org/10.1016/j.str.2015.06.015.

Subedi, G. P., Falconer, D. J., & Barb, A. W. (2017). Carbohydrate-polypeptide contacts in the antibody receptor CD16A identified through solution NMR spectroscopy. *Biochemistry, 56*(25), 3174–3177. https://doi.org/10.1021/acs.biochem.7b00392.

Subedi, G. P., Hanson, Q. M., & Barb, A. W. (2014). Restricted motion of the conserved immunoglobulin G1 N-glycan is essential for efficient FcgammaRIIIa binding. *Structure, 22*(10), 1478–1488. https://doi.org/10.1016/j.str.2014.08.002.

Subedi, G. P., Johnson, R. W., Moniz, H. A., Moremen, K. W., & Barb, A. (2015). High yield expression of recombinant human proteins with the transient transfection of HEK293 cells in suspension. *Journal of Visualized Experiments, 106*, e53568. https://doi.org/10.3791/53568.

Valderrama-Rincon, J. D., Fisher, A. C., Merritt, J. H., Fan, Y. Y., Reading, C. A., Chhiba, K., et al. (2012). An engineered eukaryotic protein glycosylation pathway in *Escherichia coli. Nature Chemical Biology, 8*(5), 434–436. https://doi.org/10.1038/nchembio. 921.

Varki, A. (2017). Biological roles of glycans. *Glycobiology, 27*(1), 3–49. https://doi.org/10.1093/glycob/cww086.

Varki, A., Cummings, R. D., Esko, J. D., Stanley, P., Hart, G. W., Aebi, M., et al. (2017). *Essentials of glycobiology* (3rd ed.). Cold Spring Harbor, New York: Cold Spring Harbor Laboratory Press.

Wang, L. X., & Amin, M. N. (2014). Chemical and chemoenzymatic synthesis of glycoproteins for deciphering functions. *Chemistry & Biology, 21*(1), 51–66. https://doi.org/10.1016/j.chembiol.2014.01.001.

Wang, L. X., & Lomino, J. V. (2012). Emerging technologies for making glycan-defined glycoproteins. *ACS Chemical Biology*, 7(1), 110–122. https://doi.org/10.1021/cb200429n.

Wurm, F., & Bernard, A. (1999). Large-scale transient expression in mammalian cells for recombinant protein production. *Current Opinion in Biotechnology, 10*(2), 156–159.

Yagi, H., Zhang, Y., Yagi-Utsumi, M., Yamaguchi, T., Iida, S., Yamaguchi, Y., et al. (2015). Backbone (1)H, (13)C, and (15)N resonance assignments of the Fc fragment of human immunoglobulin G glycoprotein. *Biomolecular NMR Assignments, 9*(2), 257–260. https://doi.org/10.1007/s12104-014-9586-7.

Yamaguchi, Y., Kato, K., Shindo, M., Aoki, S., Furusho, K., Koga, K., et al. (1998). Dynamics of the carbohydrate chains attached to the Fc portion of immunoglobulin G as studied by NMR spectroscopy assisted by selective ^{13}C labeling of the glycans. *Journal of Biomolecular NMR, 12*(3), 385–394.

Zaare, S., Aguilar, J. S., Hu, Y., Ferdosi, S., & Borges, C. R. (2016). Glycan node analysis: A bottom-up approach to Glycomics. *Journal of Visualized Experiments* (111). https://doi.org/10.3791/53961.

CHAPTER TEN

Nonuniform Sampling for NMR Spectroscopy

Scott Robson*, Haribabu Arthanari*,†, Sven G. Hyberts*, Gerhard Wagner*,1

*Department of Biological Chemistry and Molecular Pharmacology, Harvard Medical School, Boston, MA, United States
†Dana Farber Cancer Institute, Harvard Medical School, Boston, MA, United States
[1]Corresponding author: e-mail address: gerhard_wagner@hms.harvard.edu

Contents

Methods in Enzymology, Volume 614
ISSN 0076-6879
https://doi.org/10.1016/bs.mie.2018.09.009

Abstract

Nonuniform sampling was first proposed more than 40 years ago as an alternate method for sampling two-dimensional NMR data was initially pursued by only a small number of scientists. However, it has been gradually adopted after it was shown that major gains in measuring time and spectrum resolution can be obtained. Furthermore, migration of NMR software to the Unix environment facilitated development of new processing tools, and there is now a selection of programs available that yield high-quality reconstructions of NUS data. Moreover, it became obvious that recording high-resolution 3D and 4D protein NMR spectra at the resolution provided by modern high-field instruments was not possible with uniform sampling. It has become apparent that sparse, low dynamic-range NMR spectra, in particular, the triple resonance experiments are all best recorded with NUS. Optimal sampling schedules can yield benefits with respect to detecting weak peaks in high dynamic-range spectra and therefore a careful use of 2D HSQC-like spectra of mixed concentrations of small molecules is feasible. It is not yet clear whether crowded high dynamic-range spectra, such as NOESYs with many cross-peaks, benefit from NUS. On the other hand, NUS appears to be the best option for recording such high dynamic-range spectra if crowding can be reduced by shorten NOESY mixing times and/or by reducing overlap through isotopic labeling in high-resolution 3D and 4D experiments. Here we discuss principals and applications of the method, trying to provide a wide overview, but are biased by our own approaches. Thus, we will obviously miss important developments elsewhere and do not claim to be comprehensive.

ABBREVIATIONS

ASR artifact to signal ratio
DFT discrete Fourier transform
FFT fast Fourier transform
hmsIST Harvard Medical School iterative soft thresholding
IHT iterative hard thresholding
IST iterative soft thresholding
L_0 "norm" number of nonzero entries in a vector
L_1 norm sum of absolute values in a vector
L_2 norm square root of the sum of absolute values in a vector squared
MDD multidimensional decomposition
NMR nuclear magnetic resonance
NOE nuclear Overhauser effect
NOESY NOE spectroscopy
nuDFT nonuniform discrete Fourier transform
NUS nonuniform sampling
PGS Poisson-gap sampling
rmsd root mean square difference
SMILE sparse multidimensional iterative lineshape enhanced
SNR signal-to-noise ratio

1. INTRODUCTION

Nuclear magnetic resonance (NMR) spectroscopy of biological macromolecules detects signals from NMR-sensitive nuclei and uses them to examine properties of the studied systems. Thus, NMR spectra provide a plethora of sensors that can be used to reveal properties and functional aspects of molecular systems.

Conventional sampling in the indirect dimensions to the full transverse relaxation time T_2 is typically no problem for 2D NMR spectra. However, uniform sampling the indirect dimensions of 3D and 4D spectra up to T_2 or even just 0.5 T_2 would take unreasonable amounts of time, years for 4D, but can be achieved in a week with nonuniform sampling (NUS). However, NUS requires optimized sampling schedules, and reconstruction software that is fast and minimizes the risk of creating artifacts. Although NUS is now widely accepted, creating optimal sampling schedules is still a topic of current research. In addition, there are multiple reconstruction software packages available based on different strategies and assumptions about expected line shapes.

Recent developments in NUS cover different approaches of sampling strategies and data reconstruction methods. An excellent recent overview of the theoretical background and all the intricacies of different approaches can be found in (Mobli & Hoch, 2014). Here we will focus on practical implications and guidance to users who want to apply this approach. While we try to cover work from other groups we focus on techniques developed in our laboratory, which are also widely used by many other groups. However, we do not provide a comprehensive overview of the field.

2. DEVELOPMENT OF NUS METHODS

The first attempt of NUS was to acquire only a subset of points but sample more points at the beginning where S/N is high and fewer toward the end according to an exponential weighting scheme. This exponential "nonlinear sampling" was proposed as an alternative for acquiring two-dimensional NMR spectra (Barna, Laue, Mayger, Skilling, & Worrall, 1987). Reconstruction of the interrupted free induction decays was performed by a maximum entropy method (MEM), which was originally developed for analyzing incomplete and noisy radio and X-ray astronomy data (Gull & Daniell, 1989). MEM claimed to increase both sensitivity and resolution of noisy NMR spectra (Sibisi, Skilling, Brereton, Laue, & Staunton, 1984). MEM was then adapted to reconstruct incomplete time-domain data at high resolution. This initial publication touched the essential benefits and problems of NUS that are still valid today, which are to develop alternate sampling methods, enable high resolution, optimize sensitivity, and develop high-fidelity processing software that diminishes sampling artifacts.

In subsequent years, NUS, previously called nonlinear sampling, was primarily pursued by Hoch and coworkers (Donoho, Johnstone, Stern, & Hoch, 1990; Hoch, 1989; Hoch & Stern, 1996) who provided a comprehensive treatise of this field at that time and included an outline of iterative soft thresholding (IST) and many other fundamental aspects of NUS and NMR data processing in general. The lab pursued several early applications of NUS, such as reconstructing 2D double-quantum filtered NMR spectra of proteins (Schmieder, Stern, Wagner, & Hoch, 1993, 1994). They developed a dedicated software package, the Rowland NMR Toolkit, RNMRTK, which was for a long time the only package that could reconstruct NUS data (Li, Stern, & Hoch, 1998). It was written during the Digital Enterprise Corporation VAX/VMS era but did not fit well into the emerging and now more commonly used Unix environment. Nevertheless, it was shown that complete sets of six triple resonance experiments, which were needed for full backbone assignment of a 11 kDa cytoplasmic domain of a membrane protein, could be recorded within 32 h using NUS and reconstructed with RNMRTK at high fidelity; in contrast, conventional uniform sampling would have taken 146 h (Rovnyak, Frueh, et al., 2004). The demonstration of the tremendous time saving was impressive and helped NUS to become more generally adopted.

An important step to facilitate reconstructing of NUS data was the development of NMR processing software in the Unix environment, such as NMR Pipe (Delaglio et al., 1995), which could readily be used to implement NUS processing tools. This stimulated development of processing scripts based on this software package and included the forward maximum entropy routine (Hyberts, Frueh, Arthanari, & Wagner, 2009) and a number of other reconstruction programs.

3. RECONSTRUCTION PROGRAMS

Programs for reconstruction of NUS data are key for utilizing this technology. Here we present a necessarily incomplete overview of software available for this task.

3.1 Nonuniform Discrete Fourier Transform

It is entirely possible to transform NUS data by using a regular discrete Fourier transform (DFT) (a nuDFT). Skipped samples are set to zero and the padded data set subjected to regular DFT. Padding of the data underrepresents the set of basis functions in the Fourier sum. Because the set is not orthogonal, those in the set interfere with each other and result in artifacts that complicated spectral interpretation. However, the strongest signal present in the time domain will still be the most prominent in the frequency domain, so a quick DFT of a padded NUS data set can permit some interpretation of the spectrum and a check on the performance of a running experiment. nuDFT has been used in 4D C,C nuclear Overhauser effect (NOE) data (Stanek, Podbevsek, Kozminski, Plavec, & Cevec, 2013). Many of the methods discussed later rely on the prominence of frequency domain signals above the artifacts of the nuDFT. An example of processing a spectrum with nuDFT is shown in Fig. 1A (random sampling) and Fig. 1B (Poisson-gap sampling (PGS) (see below)). Clearly choice of schedule impacts this first step greatly and impacts the quality of reconstruction (see Section 4). An example of this is shown in Fig. 1C and D where the same random and PGS schedules are used with Harvard Medical School iterative soft thresholding (hmsIST) reconstruction. The success of reconstruction in this case appears to depend highly on the schedule and the nuDFT.

3.2 Clean/Scrub

Historically, the CLEAN algorithm of data reconstruction was the first method to attempt to reconstruct a spectrum from NUS data (Högbom, 1974). Originally designed for image reconstruction in the astronomical field, this method uses the nuDFT and subsequent identification of peaks which are then iteratively subtracted from the data. Additional improvements to the algorithm were suggested by Zhou and coworkers (Coggins, Werner-Allen, Yan, & Zhou, 2012; Coggins & Zhou, 2008).

3.3 Maximum Entropy

The principal of a MEM was first used with NMR data to improve S/N of regularly acquired data (Sibisi et al., 1984) and later used for the reconstruction of NUS data in two dimensions (Barna et al., 1987). Here the entropy is $S = -\int f(\nu)\log f(\nu)d\nu$, where f is normalized to $\int f(\nu)d\nu = 1$, and ν are the values in the frequency spectrum. The method reconstruction as an inverse problem where trial frequency domain spectra are proposed and compared in the time domain to sampled data. The difference is minimized under a

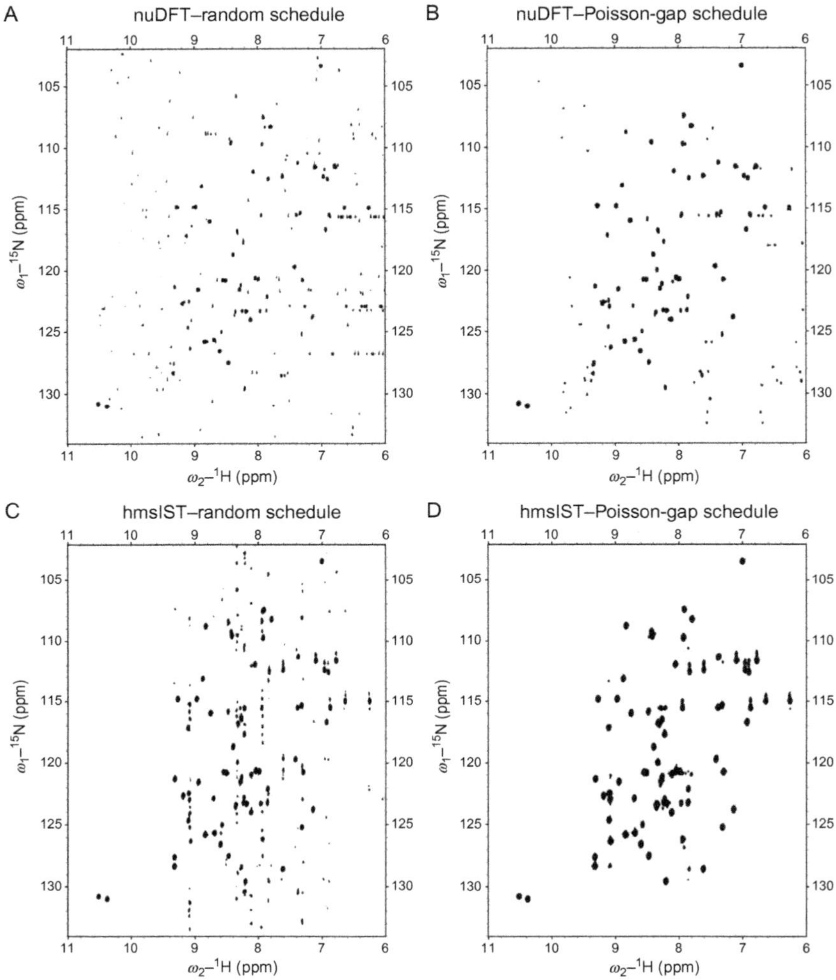

Fig. 1 Interplay of schedule and the nuDFT. A comparison of a nuDFT on a sampled ^{1}H–^{15}H HSQC spectrum of GB1 with (A) a random schedule vs (B) a Poisson-gap schedule. The impact of the noisier nuDFT vs the cleaner one is seen on the hmsIST reconstruction with the random schedule (C) vs the Poisson-gap schedule (D).

maximum entropy constraint in information content. A variation of the method was popularized as "MaxEnt" and incorporated into the Rowland NMR Toolkit (Hoch & Stern, 1996). The method requires setting three parameters Co, def, and lambda, which for most users were hard to manually determine. Recently, automatic determination of these parameters became available (Mobli, Maciejewski, Gryk, & Hoch, 2007a, 2007b). MaxEnt particularly suffers nonlinearity with weaker signals being underrepresented in.

This is especially problematic for NOE spectroscopy (NOESY) data analysis and other quantitative NMR spectra-like relaxation or J coupling measurements, although calibration of peak heights can be done by injection of calibration peaks of known intensity (Schmieder, Stern, Wagner, & Hoch, 1997). Note that here and in the following, the normalization of the integral over the spectrum to 1 has been abandoned.

3.4 Multidimensional Decomposition

Multidimensional decomposition (MDD) (Orekhov, Ibraghimov, & Billeter, 2001; Tugarinov, Kay, Ibraghimov, & Orekhov, 2005) assumes a full frequency domain spectrum can be modeled as a sum of components written as a sum of components written as a direct product of 1D line shapes. A target function is defined as the L_2 norm of the difference between the model in the time domain and the sampled data, only at sampled points. This target function is minimized. No assumptions are made about line shapes. MDD then approximates the spectrum as a sum of components of not yet known line shapes. It defines a target function written as the square of the difference between the spectrum S and a sum of components but only evaluated at the sampled points. The target function is then minimized using principle component analysis which needs line shapes and amplitudes of the multidimensional peaks. This approach has been further developed since.

3.5 Forward Maximum Entropy

The forward maximum (FM) entropy routine minimizes a target function in the form of the negative entropy. The initial target function $T(f) = \sum_{k=1}^{N} |f_k| \log |f_k|$ is calculated for the Fourier transformed sparsely sampled free induction decay with zeros in the missing time points. Next the missing time points are changed by $\pm\delta$ followed for a new calculation of the target function, and a multidimensional gradient of the target function with respect to the missing data points is calculated. The target function is then minimized with the Polak–Ribiere conjugant gradient method. During all minimization steps the originally measured time domain data points are kept unchanged. This approach was successful in reconstructing high-resolution 2D NMR spectra of collections of metabolites (Hyberts et al., 2007), or 3D triple resonance spectra of proteins (Hyberts et al., 2009). However, it was slow for reconstructing high-resolution NUS 3D and not fast enough to process high-resolution 4D protein NMR spectra. As NUS appears most useful for acquiring high-resolution multidimensional

NMR spectra the slow reconstruction speed was a major hurdle and essentially a show stopper for this approach. Note that in FM and in MaxEnt, the normalization of the integral over the spectrum to 1 as used in MEM has been abandoned.

3.6 Sparse Multidimensional Iterative Lineshape Enhanced

Recently a new algorithm, sparse multidimensional iterative lineshape-enhanced (SMILE) reconstruction was released (Ying, Delaglio, Torchia, & Bax, 2017). SMILE operates by making reasonable underlying assumptions about NMR data. Namely, that all signals detected arise from exponentially decaying sinusoids of a predetermined phase (Lorentzian line shapes). Like other methods, SMILE starts by executing a nuDFT on zero-padded data. The spectrum is peak picked and peaks filtered through a collection of empirically derived criteria. Peak height is estimated from the nuDFT spectrum while linewidth is simulated as a series of R_2 values between $1/(4T)$ and $3/T$, where T is the entire acquisition time in each dimension. The optimal R_2 match to the data is selected. The simulated peak is reverse Fourier transformed and subtracted from the initial data set, which then undergoes another round of peak discovery. SMILE is entirely integrated into the nmrPipe pipeline although it processes data as entire planes (2D spectra), cuboid (3D spectra), and hypercuboids (4D spectra), instead of serially like nmrPipe. This allows SMILE to make efficient use of computer memory and less disk activity. When RAM memory is limited, as in dealing with large 4D hypercuboids, SMILE operates on an adjacent series of 3D cuboids, in effect simulating a region of the 4D hypercuboid. The benefits of working with the entire spectra in memory, rather than as indirect lines (2D spectra), planes (3D spectra), or cuboids (4D spectra) seem to be improvements in peak shapes, an idea that could be more fully explored by other algorithms.

3.7 Compressed Sensing Methods

Compressed sensing attempts to recover a spectrum from NUS data by posing the problem explicitly as an underdetermined linear system. The presence of only sparse signals ensures that a solution can be found. The condition of sparsity is enforced by minimizing the number of nonzero frequency components. The L_0 norm of the frequency space simply counts these nonzero components but minimization of L_0 is computationally difficult. The method of iterative reweighted least squares (Kazimierczuk &

Orekhov, 2011) approximates the L_0 norm. However, minimization of the L_1 norm, the sum of all the frequency component intensities, should in theory provide an exact and sparse solution (Candes, Romberg, & Tao, 2006) and can be achieved with efficient algorithms. Stern et al. have shown that the IST algorithm performs an L_1 norm minimization efficiently in $n \times \log(n)$ time (Stern, Donoho, & Hoch, 2007). The CS methods of IST, iterative hard thresholding (IHT) (Blumensath & Davies, 2008) and the YALL1 algorithm (Yang & Zhang, 2011) have been compared directly using 40% sampled NOE spectra, with the Nietlispach group claiming IHT being perhaps a slightly better algorithm under these test conditions (Bostock, Holland, & Nietlispach, 2012). The methods of compressed sensing have also been extensively explored by the Byrd group. In this work, an L_1 norm (L1), a reweighted L_1 norm (IRL1), and a Gaussian-SL0 regularization were applied to nonuniform data and directly compared. They conclude that L1 minimization is the most efficient, but in terms of signal linearity the Gaussian-SL0 method is superior (Sun, Gill, Li, Huang, & Byrd, 2015). This measure of accuracy, however, comes at a computational price. Their software package, NESTA-NMR, can utilize all three methods and can be incorporated into a traditional NMRPipe script.

3.8 IST and hmsIST

When realizing that FM reconstruction was too slow to reconstruct NUS 4D spectra, we turned our attention to the Iterative thresholding strategy (Drori, 2007). Indeed, our initial implementation, hmsIST, can reconstruct 3D and 4D spectra in reasonable amounts of time, with typical 3D data sets reconstructed on modest hardware like laptop computers in minutes to tens of minutes (Hyberts, Milbradt, Wagner, Arthanari, & Wagner, 2012). hmsIST uses the nmrPipe processing chain, so all the functionality of apodization, zero filling, phase correct, and others can be applied in the familiar domain of nmrPipe. hmsIST also only requires two parameters to be set, a threshold level and number of iterations, both of which have intuitive meanings. Additionally, extension of data beyond the last sampled point is easy and gives good results (Hyberts, Robson, & Wagner, 2017). Importantly, hmsIST does not require correct phases to be applied to indirect dimensions before reconstruction. When incorrect phases are used with other methods, reconstruction must be started again with the correct phases, while hmsIST only applies phase correction during the relatively fast Fourier transform (FFT) step after reconstruction.

hmsIST implements the IST algorithm in a straightforward way. First, the direct dimension is Fourier transformed and phased for all sampled points. For each point along the frequency domain of the direct dimension, the sparsely sampled indirect matrix is assembled and padded with zeros where no sample exists. For a 3D experiment this constitutes a 2D plane. For a 4D experiment this matrix is a cube. A nuDFT of these data is executed, yielding a spectrum with signals mixed with artifacts. A threshold is calculated as a fraction of the highest peak. Anything above this threshold is added to a spectral accumulation space. Everything below is reverse Fourier transformed back into the time domain. Points in this new time domain that correspond to unsampled points are again set to zero and the cycle repeats. This process is iterated with signal accumulating in the spectral accumulation space. With each iteration, the artifacts from the strongest signals are slowly reduced as intensity from those signals are transferred to the spectral accumulation space. The process is continued until a preset number of iterations takes place or the L_2 norm of the residual no longer changes.

3.9 Practical Aspects to hmsIST Reconstruction

A common question has been, how many iterations are necessary and what level of thresholding is appropriate? In our experience, aggressive thresholding (<0.90) may lead to signal intensity loss while too few iterations will not reconstruct signals toward the noise level, resulting in an artificially smooth baseline. In our experience, a threshold level of 0.98 and iterations on the order of 300–400 are satisfactory. Although if time can be spent in computation we will select many iterations (say 10,000) and a gentler threshold of 0.99 or 0.995 and have hmsIST automatically terminate once the L_2 norm of the residual no longer changes. Typical data usually converges before 500–1000 iterations under these circumstances.

Using the earlier parameters, we find a typical 2D NUS HSQC spectrum can be reconstructed in a few seconds. The reconstruction program is inserted directly into a typical nmrPipe transformation script, just after transposition after the direct dimension is transformed. When done this way, hmsIST can only run on one CPU core (one thread), but since reconstruction is so fast this problem is incidental. A typical 3D triple resonance spectrum (e.g., 32×64 complex points in the indirect dimensions) takes approximately 10 min to reconstruct and Fourier transform when run on a quad core Intel i7 CPU with hyperthreading (eight threads at once). Memory usage is low as only eight hypercomplex indirect planes are held in memory at once. Our protocol has historically used the GNU Parallel perl script to manage the number of processes run simultaneously. We have more recently introduced a python script which manages processes using its internal multiprocessing library. This may be more convenient as most systems come preinstalled with a python interpreter but not the GNU parallel program. Reconstruction of 4D data is computationally more intensive as cubes, rather than planes need to be held in memory and iteratively transformed. Depending on the resolution of the 3D indirect cubes it may be possible to max-out memory if eight of them are held in memory (one for each thread). In these cases, the number of threads will need to be limited. Cube transformation takes longer than plane transformation so systems with large numbers of cores and large memory facilitate parallelization of processing. With 32 hyperthreads and 512 GB of RAM we can reconstruct typical 4D spectra in a few hours. A 4D ^{13}C NOESY with indirect cubes of $128 \times 333 \times 128$ complex points and 333 direct points took approximately 2 days to reconstruct. Arguably a major bottleneck in reconstruction is read/write times to disk. These can be partially overcome by use of local scratch SSD disks instead of network file systems.

One of the major obstacles to new users is the assortment of shell and nmrPipe scripts required to process spectra. Recently, we have coded a command line python program, masterHI, that simplifies hmsIST processing into four steps: (1) conversion of data to nmrPipe format, (2) phasing and Fourier transform of the direct dimension, (3) reconstruction of the spectrum, and finally (4) phasing and Fourier transform of the indirect dimensions. Each of these steps is executed by masterHI writing the appropriate processing script and executing it. masterHI will autodetect many of the options required for reconstruction, but these can usually be overwritten on the command line. masterHI parallelizes reconstruction using the abovementioned python framework. The scripts generated by masterHI remain in the processing directory so they can be "hand-tweaked" by an experienced user and rerun. Details for masterHI can be found at this web address: http://github.com/nomadiq/masterHI.

4. NUS SCHEDULES

Key to NUS reconstruction is the selection of sampling schedules that allow a faithful reconstruction and/or detection of signals that would be obtained with uniform sampling, but at much shorter time. The latter point suggests that NUS has the biggest impact on recording 3D and 4D spectra and less on 2D experiments. Nevertheless, sampling schedules are typically first evaluated for one indirect dimension, which has implications to higher dimensions as well.

There are an immense number of possible sampling schedules. For example, randomly selecting 256 out of 1024 points can be done in C_{256}^{1024} (1024 choose 256) ways or more than 10^{248} combinations, which obviously cannot be exhaustively explored. Thus, a schedule is usually selected according to a random or weighted random algorithm. To achieve this, random number generation with standard C library functions like *drand48*, which only promises a randomization to 2^{48} ways, or just 10^{14} combinations, is used to produce seed values. If a specific seed number yields a sampling schedule with good results, the random schedule can be reproduced using the same seed number. Thus, it should be possible to create libraries of good random sampling schedules that yield good results (Mobli & Hoch, 2014). However, the possibilities of schedule generation are limited by seed value ranges and time constraints to fully explore the massive sampling space.

4.1 Criteria for Evaluating Sampling Schedules

Key points to consider are (i) to minimize sampling artifacts relative to the experimental noise, (ii) how sparse can the sampling schedule be, and (iii) how far should the decaying coherences be sampled relative to the relaxation time of the decaying coherences. Furthermore, there is the question whether different types of spectra require different designs of sampling schedules? There are broadly two types of spectra to consider. Low dynamic-range spectra including HSQC and most triple resonance spectra where peaks have similar intensities and the goal is to obtain precise peak positions but faithful reproduction of peak heights is a lesser issue and the effects of aggressive sparse sampling may be tolerated. Then there are high dynamic-range experiments, NOESYs or relaxation-type experiments, and good reconstruction of peak heights is more important. Also, NOE experiments are typically less sparse with signals and therefore require more samples to accurately detect. Thus, less aggressively sparse sampling schedules may be appropriate.

4.2 Fidelity of Sampling Schedules With the L_2 Norm and the Artifact to Signal Ratio (ASR)

To predict the performance of sampling schedules our analysis consists of simulating a single exponentially decaying signal, applying a NUS schedule and reconstructing the signal with hmsIST. We then use two methods to evaluate the fidelity of sampling schedules. The L_2 norm is essentially the root mean square difference (rmsd) between the reconstructed NUS spectrum and the uniformly sampled spectrum transformed with FFT. This yields an overall evaluation of sampling schedules. For experiments with high-dynamic range we are also concerned with small artifacts due to the interrupted sampling schedules and measure the ratio of the largest artifact to signal ratio (ASR) or $\Delta L_2/L_2$. In Fig. 2, we select 256 of 1024 sampling points and evaluate a subset of typically used sampling schedules. For both panels the lowest values represent the best schedules. While all examined sampling strategies can yield very good schedules, the chance to find the seed number that yields the best schedules depends strongly on the sampling strategy.

4.2.1 Uniformly Random Sampling

In this case, the sampling points are selected uniformly random. Using the drand48 C function to generate random numbers, a point is selected between $t=0$ and $t=t_{max}$. After the random picking, one typically excludes it before picking the next point randomly until the desired sparse number of time points is selected. We created random sampling schedules with 100 different seed numbers for the drand function and measured the L_2 norm (rmsd) between reconstruction and the FFT of the full signal. We find that the rmsd varies over a large range of values depending on the seed number (Fig. 2A, *red crosses*). Similarly, we find that the height of the maximal ASR also varies over a large range (Fig. 2B, *red crosses*), and the probability that a selected seed number yields an optimal schedule is rather low.

4.2.2 Exponentially Weighted Random Sampling

Here the random probability function of sampling is altered by exponential weighting that biases shorter evolution times over longer evolution times. This is motivated by the fact that the signal is stronger at the early part of the decay and resembles the matched filter of apodization functions. This exponentially random-weighted sampling schedule was first used by Barna et al. (1987).This schedule performs significantly better than uniformly random sampling (Fig. 2, *green triangles*).

4.2.3 Poisson-Gap Sampling

When working with different sampling schedules we realized that it is desirable to control the sizes of gaps between sampling points. Empirically, large

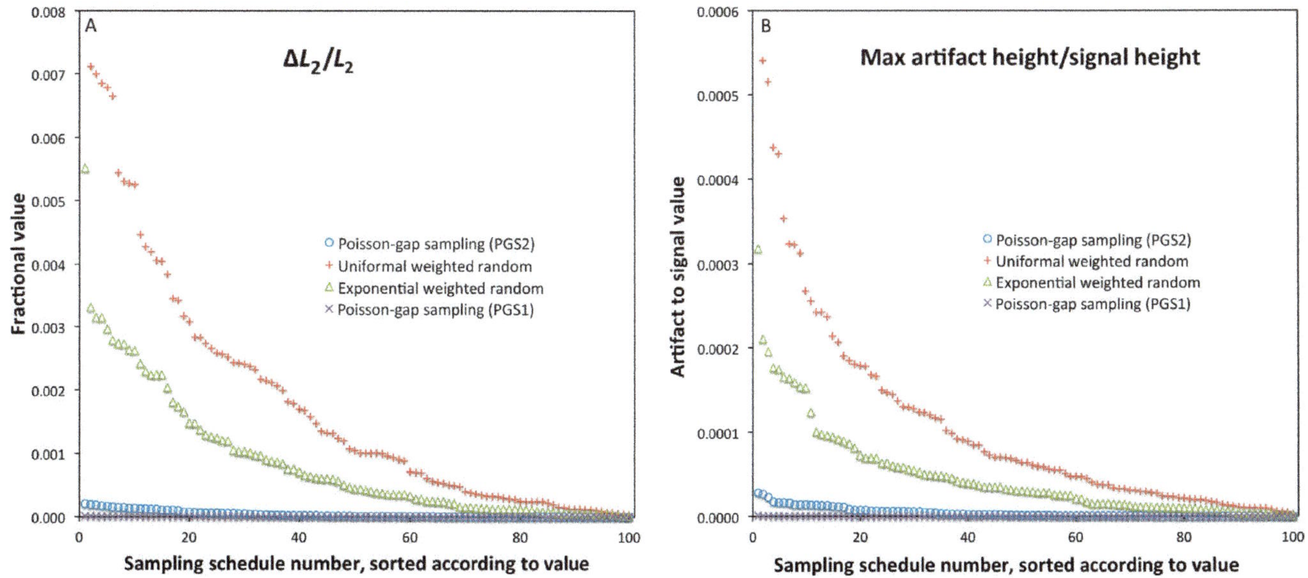

Fig. 2 (A) Variation of L_2 norms after hmsIST reconstruction for a set of 100 seed numbers/NUS schedules for selecting 256 out of 1024 time points for uniformly random, exponentially weighted random, PGS1 and PGS2 algorithms. Here $L_2 = \sqrt{\sum_i \left(x_i^{\text{US}}\right)^2}$ and $\Delta L_2 = \sqrt{\sum_i \left(x_i^{\text{US}} - x_i^{\text{NUS}}\right)^2}$. (B) The same comparison for max artifact to signal ratio, ASR.

gaps were more detrimental to the fidelity of the reconstruction, and especially gaps at the beginning and the end of the schedule had the greatest impact. We concluded that schedules would benefit from controlling the gap sizes according to a discrete distribution with values always greater than zero, as gaps must always be a positive integer. The Poisson distribution fulfills this requirement perfectly (Hyberts, Takeuchi, & Wagner, 2010). Here samples are selected by producing a gap from one (the first) point to the next with a gap size determined by the Poisson distribution. An average value is selected so the chosen number of points match the selected sparsity. We also described how the expectation value of the Poisson distribution could be varied by a sinusoidal function by weighting the selection of gap lengths by the first quarter or the first half of a sinus function, termed PGS2 and PGS1, respectively. PGS1 minimizes the gap lengths at the beginning and the end of the evolution times while PGS2 minimizes the gap sizes primarily at the beginning. Indeed, we find that using this approach yields L_2 norms that are less dependent on the choice of the seed number (Fig. 2, *blue circles and purple crosses*).

Sinusoidal-weighted Poisson-gap sampling has been shown to yield lower amounts of artifacts than uniformly random and exponentially random-weighted sampling schedules, at least when used in conjunction with hmsIST (Hyberts, Arthanari, Robson, & Wagner, 2014). Fig. 3 shows a difficult example of reconstruction. This 3D ^{15}N-dispersed NOESY was recorded with a long mixing time of 150 ms and is an example of a high-dynamic range problem where weak peaks are close to strong NOESY peaks, in particularly those close to auto (diagonal) peaks. Using a modest 50% sampling level (64 acquired of 128 data points in the indirect proton dimension was acquired) we find our Poisson-gap schedules perform much better with greater artifact suppression.

4.3 Sampling Schedules in Two and Three Indirect Dimensions

For multiple indirect dimensions, weighted sampling uses weighting functions like the exponential to define the probability of sampling as a dot product of the weightings in each orthogonal dimension. Because our method is based on gaps and not weightings, per se, our approach has instead been to interweave 1D Poisson-gap lines along each orthogonal direction (Hyberts, Arthanari, & Wagner, 2012). In the literature, this method has received some criticism as there is no guarantee that this will produce the gap distributions over two or more dimensions that it does in one dimension. In Fig. 4

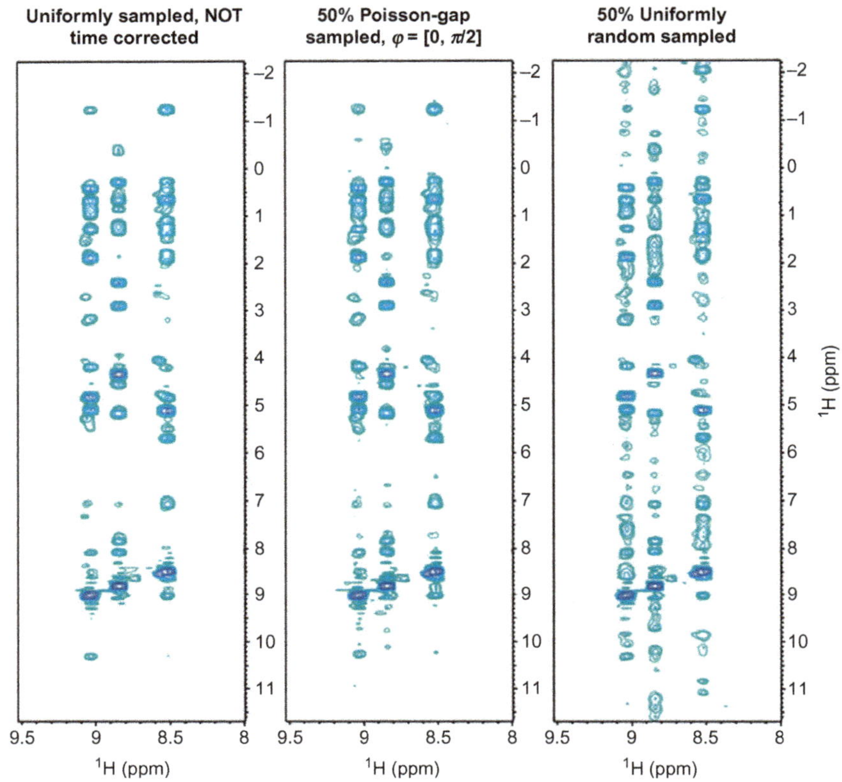

Fig. 3 Comparison of a plane of a 3D ^{15}N-dispersed NOESY of GB1 containing three NOE strips. *Left*: High-resolution uniformly sampled spectrum. *Middle*: 50% NUS spectrum extracted from the US data set (*left*) using PGS2 Poisson-gap sampling. Note this corresponds to 50% of the measurement time and the *left* and *middle* spectra are not time equivalent. *Right*: 50% NUS spectrum extracted from the US data set (*left*) using uniformly random sampling. Note that this spectrum exhibits numerous spurious peaks.

(*left*), we show a histogram of the "Manhattan" or L_1 distance (gap) sizes for 10 randomly generated unweighted Poisson-gap schedules over a 64×64 plane with 200 points selected for each schedule (~4.8% sampling). Note, the Manhattan distance is the fewest number of integer steps to reach a point's nearest neighbor where steps are only permitted parallel to the directions of the dimensions. This strictly yields integer step sizes. Distances were then offset by one ("1" was subtracted from each distance) so adjacent points would have a distance of zero. The distribution of these gaps resembles a Poisson distribution with an average value of 2.3, although not perfectly (Fig. 4, *right*). Since the Poisson distribution was originally only used to

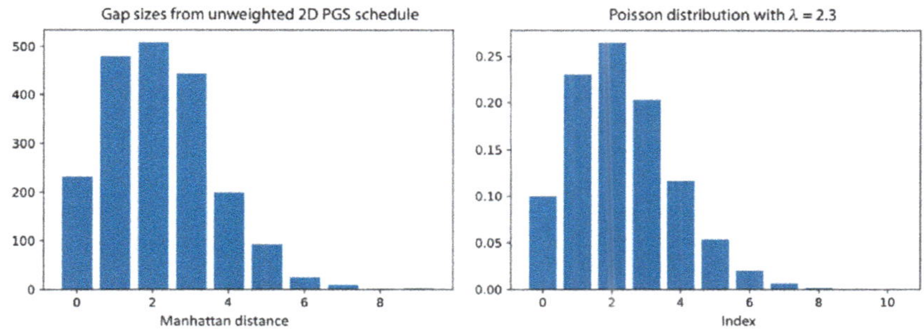

Fig. 4 Comparison of empirically measured Manhattan distances in a 2D weaved unweighted PGS schedule vs the Poisson distribution for an average value of 2.3. The Manhattan distances are derived from 10 repeats of 64×64 schedules with 200 samples selected for each schedule.

constrain gap sizes to positive integers that remain close to a mean value and not for some other specific property of the Poisson distribution, it is in our opinion that the Manhattan gap sizes seen in these unweighted 2D schedules fulfill our original wish to constrain gap sizes and that our weaving process achieves this goal.

5. BENEFITS OF NUS

5.1 NUS Permits Recording of 3D and 4D NMR Spectra at High Resolution

To sample 3D and 4D spectra to evolution times around the transverse relaxation times of the decaying coherences is only possible with NUS. An early example of the gain in resolution is shown in Fig. 5, where time-equivalent 3D HNCO spectra for a 50 kDa protein domain were recorded with uniform and nonuniform sampling. With uniform sampling, 50 and 25 consecutive complex points were sampled in the nitrogen and carbon dimensions, respectively—a total of 1250 points. With nonuniform (nonlinear) sampling, the same number of points were randomly spread over 400 and 100 indirect time points in the nitrogen and carbon dimensions, respectively, covering a time grid of 40,000 points with only 3% of points measured experimentally. In this way, the decaying coherences are sampled to far longer decay times. The gain in resolution is dramatic. While the US HN–C′ strip plots are cluttered from peaks of neighboring planes, the NUS strips are clean, and the H^N–N planes also demonstrate the gain in resolution. This example in Fig. 5 was processed with the FM-reconstruction method, which we found was not fast enough to reconstruct high-resolution

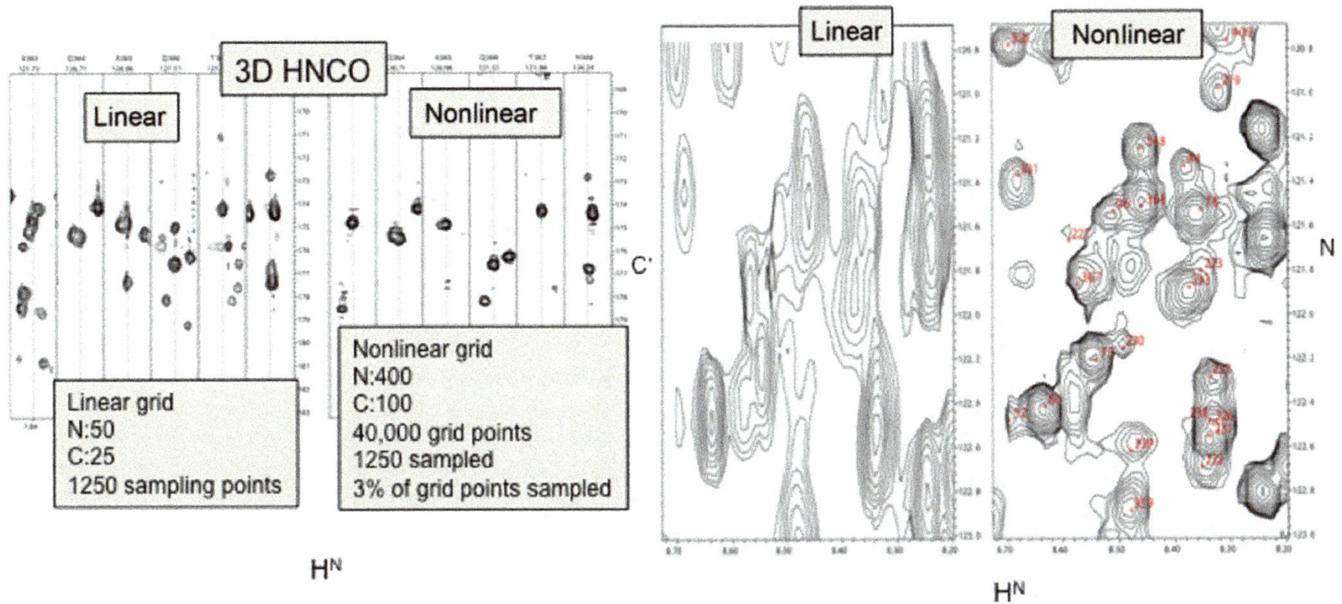

Fig. 5 NUS yields high resolution in indirect dimensions and eliminates peak break through between planes of 3D spectra. Comparison of time-equivalent US and NUS HNCO spectra of the EntF C domain. *Left*: comparison of H^N–C' strips for equal numbers of 1250 sampling points but distributed uniformly over 1250, or nonuniformly over 40,000 grid points. *Right*: same comparison of a section of an H^N–N plane. *Adapted with permission from Hyberts, S. G., Arthanari, H., Robson, S. A., & Wagner, G. (2014). Perspectives in magnetic resonance: NMR in the post-FFT era.* Journal of Magnetic Resonance, 241, *60–73. https://doi.org/10.1016/j.jmr.2013.11.014.*

4D spectra (Hyberts et al., 2014, 2009). 4D reconstruction became possible with the more efficient hmsIST reconstruction software (Hyberts, Milbradt, et al., 2012).

5.2 How Far to Sample and Extrapolate With NUS

The dependence of the signal-to-noise ratio (SNR) on the length of the acquired FID has been discussed in detail before and is sketched in Fig. 6. The heights of Lorentzian signals are proportional to the integral over the envelope of the free induction decay calculated from zero to the maximum evolution time, τ. The noise is proportional to $\tau^{1/2}$. This yields a maximum at $1.256 \times T_2$ as shown in Fig. 6A. Measuring data beyond this point increases resolution but lowers the SNR (Rovnyak, Hoch, Stern, & Wagner, 2004). Thus, it was suggested that NUS should optimally also be recorded to about $1.2 \times T_2$ in the indirect dimensions. This was further developed with theoretical methods by the Rovnyak group (Palmer et al., 2015). On the other hand, the amount and behavior of noise in reconstruction of NUS data is complex and likely depends on the reconstruction method.

To understand the effect of noise and sampling length experimentally, we recorded a high-resolution uniformly sampled HSQC spectrum with 512 complex indirect points in the carbon dimension of GB1 as a small model protein. We then extracted subsets of 64 increments distributed over different ranges of evolution times using Poisson-gap sampling and hmsIST for reconstruction (Hyberts, Milbradt, et al., 2012; Hyberts et al., 2010). All extracted NUS time domain data sets were then extrapolated to $2 \times T_2$, and the resulting time domain data were reconstructed with hmsIST. The selection of time domain data is illustrated in Fig. 6B. The lowest row represents the case of uniformly sampling (100%) the first 64 points up to $0.25 \times T_2$ and extrapolating with hmsIST to $2 \times T_2$ (additional 448 points). The average peak height in the HSQC (Signal) is divided by the maximum pixel (noise) ichosen from the first row of the 2D plane that has no peaks. The ratio of the signal to the tallest noise peak is used to define sensitivity. This sensitivity is plotted as the first point in Fig. 6C for three different noise levels. This is then repeated 28 times by selecting 64 measured increments nonuniformly over increasingly longer evolution times and plotting the sensitivity for these additional NUS data sets in Fig. 6C. The figure indicates that for these time-equivalent spectra and the processing method used here the maximum sensitivity is obtained when spreading experimentally measured increments up

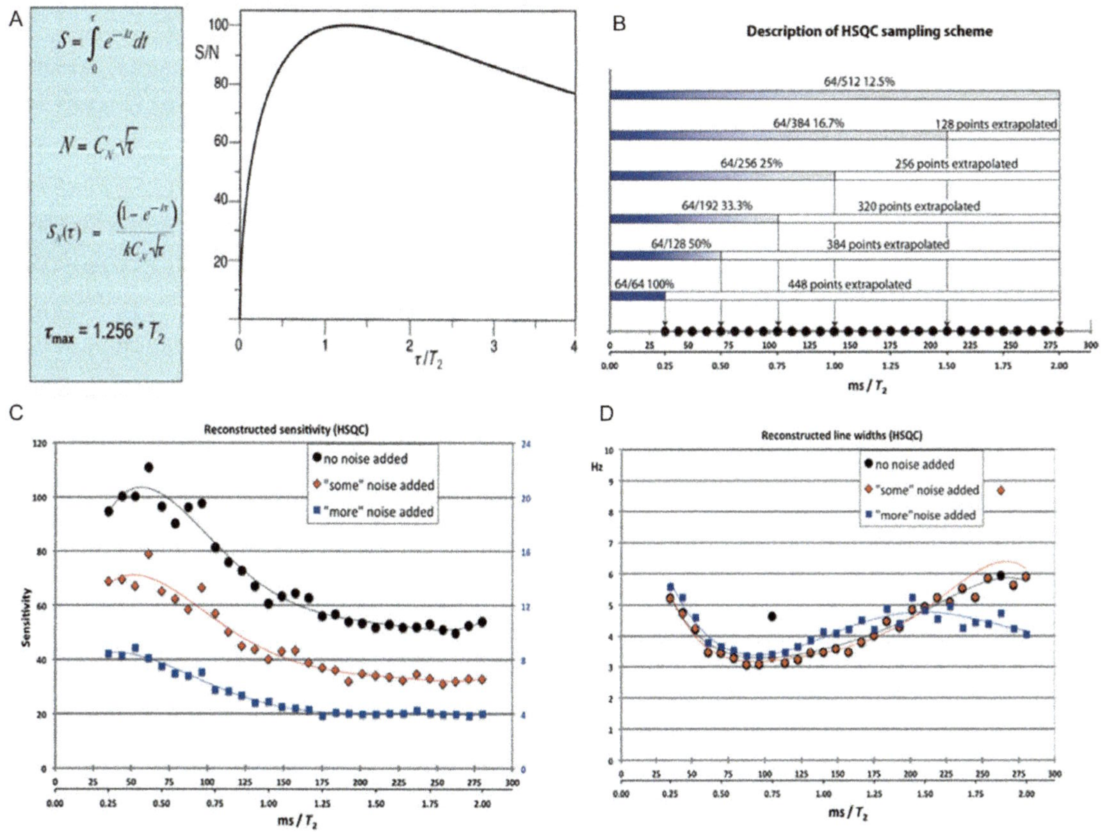

Fig. 6 See figure legend on next page.

to approximately $0.5 \times T_2$ (Fig. 6B). Similarly we find the reconstructed line width reaches a broad minimum around $0.6 \times T_2$ (Fig. 6D). Furthermore, the highest accuracy of peak position is also reached here (Hyberts et al., 2017). Therefore, we obtain best sensitivity, resolution and peak accuracy when sampling nonuniformly up to $0.5 \times T_2$ and extrapolate with hmsIST to $2 \times T_2$.

5.3 Can NUS Increase Sensitivity?

While it has become obvious that NUS permits recording of 3D and 4D NMR spectra to very high resolution not achievable with US, it has not been clear whether it also can enhance sensitivity. To explore this, we designed an empirical procedure for estimating the detection sensitivity of NMR spectra. The method is applicable to US as well as NUS data (Hyberts, Robson, & Wagner, 2013) and visualized in Fig. 7. Here, synthetic signals of relative strengths from 0.01 to 10 were created, and 100 different sets of white Gaussian noise were added. An automatic peak picker was used to test whether the tallest peak picked is at the correct position and at least $\sqrt{2}$ taller than the next tallest pixel. If this is the case it is counted as a

Fig. 6 Estimate of the signal-to-noise ratio (SNR) depending on length of acquisition (τ/T_2). (A) Estimate for uniform sampling and uniform noise. (B) Strategy for estimating the sensitivity and line widths in time-equivalent NUS spectra where 64 measured data points were spread over increasingly large ranges of evolution times and finally extrapolated to $2 \times T_2$. Each *black point* represents one NUS scenario with the longest NUS experimental evolution time of an experiment placed on the horizontal axis. (C) Simulation of sensitivity in NUS spectra without and with two levels of artificial Gaussian random noise added to the experimental spectra. The time-domain data were extrapolated to $2 \times T_2$ and reconstructed with hmsIST, and the SNR was measured as an average over all HSQC cross-peaks of GB1. Note that the noise is not *white* anymore. The maximum sensitivity is obtained at about $0.4 \times T_2$, as the latest experimentally measured increment, and tends to move to lower evolution times with more noise. (D) Effect on line widths the range of experimentally measured increments. *Sharpest lines* are found with experimentally measured increments up to around $0.6 \times T_2$. To compromise between sensitivity and resolution, we conclude that it is best to sample experimentally to around $0.5 \times T_2$. It seems wise to use the T_2 of the *broadest lines* one wants to detect as these are the most difficult to detect. *Adapted from Hyberts, S. G., Robson, S. A., & Wagner, G. (2017). Interpolating and extrapolating with hmsIST: Seeking a tmax for optimal sensitivity, resolution and frequency accuracy.* Journal of Biomolecular NMR, *68(2), 139–154. doi: 10.1007/s10858-017-0103-z; Rovnyak, D., Hoch, J. C., Stern, A. S., & Wagner, G. (2004). Resolution and sensitivity of high field nuclear magnetic resonance spectroscopy.* Journal of Biomolecular NMR, *30(1), 1–10. https://doi.org/10.1023/B:JNMR.0000042946.04002.19.*

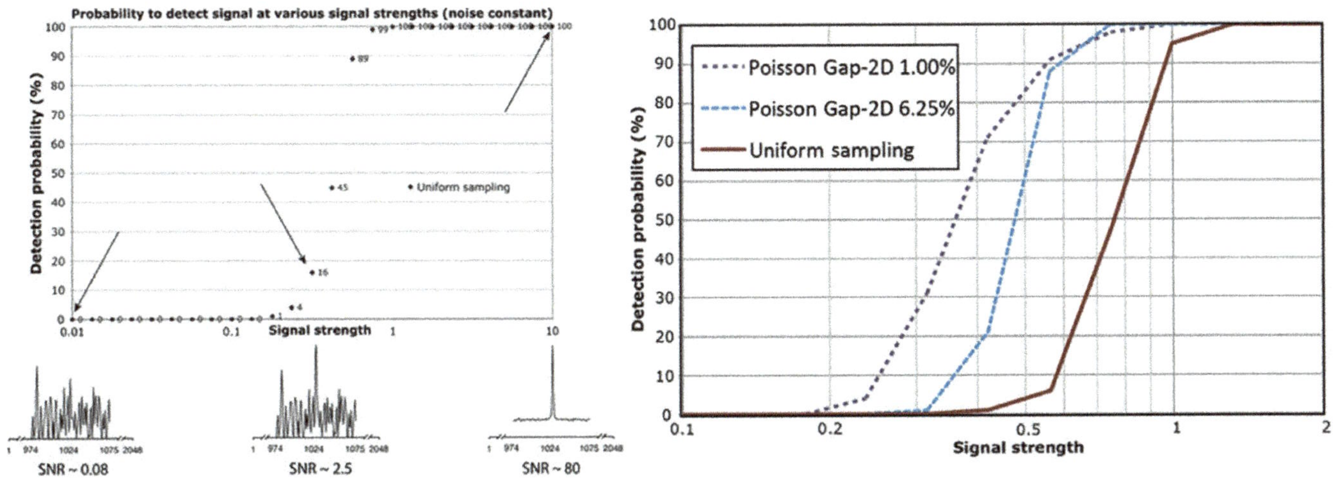

Fig. 7 Procedure for estimating the detection sensitivity of NMR spectra. *Left*: Logarithmic scale plot of signals strengths of simulated *Lorentzian lines*. In 100 cases, different Gaussian noise is added and typical resulting spectra are shown underneath with SNR of ~0.08, ~2.5, and ~80. The number of correctly picking the correct peak out of 100 noise immersions yields the detection probability and exhibits a *sigmoidal curve* over a logarithmic abscissa. *Right*: Application of the method for two indirect dimensions (3D spectrum). Clearly spreading the measured time points uniformly over the two indirect dimensions has lowest sensitivity while 6.25% and 1.00% sampling increases the probability of detecting weak peaks dramatically. Note that these results depend on the sampling strategy and were obtained here with Poisson-gap sampling (PGS2) and reconstruction with hmsIST (Hyberts et al., 2013).

successful detection. If the peak can be successfully identified *n* times out of the 100 spectra with different noise sets, it is assigned a detection probability of *n*%. This is illustrated with the left panel of Fig. 7 for uniform sampling. The three arrows indicate that a signal of strength 0.01 will never be detected (0% detection probability), the peak with strength 10 will always be detected (100%), whereas the signal in the middle has been correctly identified in 16 out of the 100 cases with different noise and is assigned a 16% detection probability.

This approach was applied to a comparison of time-equivalent uniform and nonuniform sampling with one or two indirect dimensions (Hyberts et al., 2013). In Fig. 7, *right*, we show results for a 3D experiment with two indirect dimensions and PGS (Hyberts et al., 2010). Two sampling sparsities are compared with uniform sampling. Here 6.25% sampling means that eight times more scans can be acquired for each increment at a subset of 6.25% of the indirect points, and for 1% sampling. For 1% sparsity, 100 times more scans per increment are acquired at 1% of the grid points. The figure shows, for example, that a signal of 0.5 relative strength has about a 5% chance to be detected with uniform sampling but has a 60% and 85% probability to be detected with 6.25% and 1.0% sparsity, respectively, using time-equivalent spectra. However, this sensitivity gain depends on the sampling strategy. Here PGS2 Poisson-gap sampling was used, which favors sampling at sorter evolution times where signals are stronger. The benefit of detecting weak signals shown in Fig. 7, *right* is more modest with only a single indirect dimension (2D spectra) but is expected to be even larger for 4D spectra (Hyberts et al., 2013).

5.4 Experiments That Benefit Most From NUS-Enhanced Resolution

Triple resonance experiments have benefited the most from NUS collection for two reasons: (i) They usually have fewer and consequently sparsely distributed peaks and therefore respond well to sparse sampling. (ii) The long INEPT transfer steps mean these experiments are generally weak, but under perdeuterated conditions, nuclei relax slowly during their evolution periods so large resolution gains can be realized with sparse sampling while weak signals can be "dug out" of the noise by using signal averaging in the same time a linear spectrum would take to acquire, or even faster. For example, we recently used NUS to record a 3D HNCA extending the acquisition time to ~100 ms, collecting 750 complex points in the indirect ^{13}C dimension. Here 9% of the indirect dimension was sampled, where 5216 complex points were selected from a Nyquist grid of 75×750 ($^{15}N \times ^{13}C$), or a total of 56,250, complex points. This HNCA experiment combined with the mixed-pyruvate labeling strategy allowed the assignment of 85% of 42 kDa protein (MBP) where the total acquisition time was ~100 h (compared to ~1100 h for a linear experiment which would have been prohibitive) (Robson et al., 2018).

5.5 Benefits for Sensitivity-Limited Samples, Low-Gamma Nuclei, and IDPs

Experiments that detect low gamma nuclei like $^{13}C/^{15}N$ detected experiments suffer from low sensitivity, as do experiments that detect natural abundance $^{13}C/^{15}N$ nuclei. We found NUS very useful in conjunction with low gamma detection experiments. Heavy nuclei relax more slowly than ^{1}H so direct detection automatically acquires this dimension to high resolution; however, this is countered by lower sensitivity. Sparse sampling combined with increases in signal averaging enhances the signal level and has permitted assignment of proteins through proline-rich regions in IDPs (Chhabra et al., 2018). NUS has been successfully used to record ^{1}H–^{13}C HMQC of an 150 kDa antibody using natural abundance (Arbogast, Brinson, & Marino, 2015). In the case of sensitivity-limited samples, the time savings obtained by using NUS should be directed to collecting more scans in fewer increments for sensitivity.

6. IMPLEMENTATION OF NUS EXPERIMENTS

NUS can readily be implemented in both Varian/Agilent and Bruker instruments. A detail description of the setup procedure is outlined in a previous review (Hyberts et al., 2014). There are two aspects to consider while executing NUS experiments on the spectrometers (1) generation of a sampling schedule. This could be done on the fly at the spectrometer or generated by a standalone program and transferred to the spectrometer computer. The latter gives the user a choice in the flavor of schedule that can be used. (2) Processing of data at the spectrometer. Bruker software (>Topspin 3.0) gives the user the ability to process NUS data on the spectrometer while recording (a license might be required). The user has the option of using either MDD (Orekhov & Jaravine, 2011) or CS (Kazimierczuk & Orekhov, 2011), both these software were developed by the Orekhov group. For the CS algorithm, options IST and IRLS, with or without virtual echo are available. The schedule generator used in Topspin is based on the program "nussampler" developed by the Orekhov group (Orekhov, Ibraghimov, & Billeter, 2003). It allows the choice of various sampling schemes and weighing of the schedule based on T_2 or J coupling.

It is possible to use PGS directly on a Bruker spectrometer with a schedule generator from the Wagner group (http://gwagner.med.harvard.edu/intranet/hmsIST/gensched_new.html) or upon request. This macro creates and integrates a PGS scheduled on the fly for any 2D, 3D, and 4D experiments at the spectrometer computer. In addition, hmIST reconstruction can also be executed on the fly at the Bruker spectrometer to process NUS data and display with the Bruker software. The associated software and au programs can be obtained freely with an email request to the Wagner lab. It should be noted that these programs are compiled for the UNIX-operating system and currently efforts are in progress to compile these programs for the Windows version of Topspin.

In Agilent/Varian systems there is a similar schedule generator that takes inputs of T_2 and coupling constraints and there is also the option to incorporate a user created schedule. In both the Bruker and Agilent/Varian systems the setup of NUS experiments is simple and requires only clicking a few options. The user will setup a uniform experiment first, then decide how far to sample, what percentage of the indirect Nyquist grid to sample and what flavor of the sampling schedule to use. Once this is decided, the setup is straightforward. Attention should be paid to constant-time limitations, relaxation, and homonuclear-coupling constrains.

7. CONCLUSION

The power of NUS has now widely been recognized. The field is still developing and there are numerous open questions. Although we have tried to provide an overview of other techniques we mainly focused on methods developed in our laboratory. We have mainly used the hmsIST software to address how to maximize general fidelity of reconstruction, suppress artifacts, maximize sensitivity. Thus, this chapter represents our view of the field.

Generally, it has clearly been demonstrated that 3D and 4D experiments will absolutely need NUS if the resolution power of modern high-field instruments should be fully exploited. Thus, the largest impact of NUS is in the recording of these higher dimensional experiments, and it appears that they are least prone to artifacts. Furthermore, techniques that require very high resolution for interpretation of data, such as those using specific alternate labeling with pyruvate precursors, can be recorded with NUS at high resolution within a reasonable amount of time.

Generally, spectra of low dynamic range, including most common triple resonance experiments, where peaks have similar intensities and spectra contain fewer peaks, are very suitable for NUS and allow low density sampling. Experiments with high dynamic range, including NOESYs, and experiments where peak heights are to be measured, should be recorded less sparse. Reconstruction quality of NUS spectra depends on the number of signals to be constructed, and the sparsity should be correspondingly adjusted. Thus, NOESY spectra with shorter mixing times will be easier to reconstruct.

As a rule of thumb, we have good experience when using 20% sparsity per indirect dimension, which relates to 4% and 0.8% for 3D and 4D spectra, respectively, and has been shown in several published studies. Ideally, we record experimentally increments with NUS to the maximum $0.5 \times T_2$, and extrapolate with hmsIST to about $2 \times T_2$.

All sampling schedules should include much randomness. We observed that sampling schedules that try to avoid large gaps at the beginning of the evolution periods yield less artifacts than uniformly random sampling. A preliminary principle component analysis suggested that the first and the second point should always be sampled in each indirect dimension to minor folding artifacts.

It is desirable to predict the performance of a sampling schedule prior to do the experiment. To achieve this, we simulate a single exponentially decaying line, sample it with our strategy and randomness defined by a particular seed number. Evaluating the quality of the reconstruction for one or several simulated signals should make it possible to create libraries of optimal schedules that can be plugged in when setting up experiments.

The development and application of NUS procedures together with new labeling techniques and pulse sequence development has extended the capability of NMR spectroscopy significantly for more efficient studies of larger proteins and measuring dynamic parameters. We anticipate that this approach will be more widely adopted and implemented for convenient use.

ACKNOWLEDGMENTS

This work was supported by the NIH grant GM47467. We thank Dr. Paul Coote for fruitful discussions on aspects of pyruvate labeling and ^{13}Cb decoupling.

REFERENCES

Arbogast, L. W., Brinson, R. G., & Marino, J. P. (2015). Mapping monoclonal antibody structure by 2D 13C NMR at natural abundance. *Analytical Chemistry, 87*(7), 3556–3561. https://doi.org/10.1021/ac504804m.

Barna, J. C. J., Laue, E. D., Mayger, M. R., Skilling, J., & Worrall, S. J. P. (1987). Exponential sampling, an alternative method for sampling in two-dimensional NMR experiments. *Journal of Magnetic Resonance, 73*, 69–77.

Blumensath, T., & Davies, M. E. (2008). Iterative thresholding for sparse approximations. *Journal of Fourier Analysis and Applications, 14*(5), 629–654. https://doi.org/10.1007/s00041-008-9035-z.

Bostock, M. J., Holland, D. J., & Nietlispach, D. (2012). Compressed sensing reconstruction of undersampled 3D NOESY spectra: Application to large membrane proteins. *Journal of Biomolecular NMR, 54*(1), 15–32. https://doi.org/10.1007/s10858-012-9643-4.

Candes, E. J., Romberg, J., & Tao, T. (2006). Robust uncertainty principles: Exact signal reconstruction from highly incomplete frequency information. *IEEE Transactions on Information Theory, 52*(2), 489–509.

Chhabra, S., Fischer, P., Takeuchi, K., Dubey, A., Ziarek, J. J., Boeszoermenyi, A., et al. (2018). (15)N detection harnesses the slow relaxation property of nitrogen: Delivering enhanced resolution for intrinsically disordered proteins. *Proceedings of the National Academy of Sciences of the United States of America, 115*(8), E1710–E1719. https://doi.org/10.1073/pnas.1717560115.

Coggins, B. E., Werner-Allen, J. W., Yan, A., & Zhou, P. (2012). Rapid protein global fold determination using ultrasparse sampling, high-dynamic range artifact suppression, and time-shared NOESY. *Journal of the American Chemical Society, 134*(45), 18619–18630. https://doi.org/10.1021/ja307445y.

Coggins, B. E., & Zhou, P. (2008). High resolution 4-D spectroscopy with sparse concentric shell sampling and FFT-CLEAN. *Journal of Biomolecular NMR, 42*(4), 225–239. https://doi.org/10.1007/s10858-008-9275-x.

Delaglio, F., Grzesiek, S., Vuister, G. W., Zhu, G., Pfeifer, J., & Bax, A. (1995). NMRPipe: A multidimensional spectral processing system based on UNIX pipes. *Journal of Biomolecular NMR, 6*(3), 277–293.

Donoho, D. L., Johnstone, I. M., Stern, A. S., & Hoch, J. C. (1990). Does the maximum entropy method improve sensitivity? *Proceedings of the National Academy of Sciences of the United States of America, 87*(13), 5066–5068.

Drori, I. (2007). Fast minimization by iterative thresholding for multidimensional NMR spectroscopy. *EURASIP Journal on Advances in Signal Processing, 2007*(1), 020248. https://doi.org/10.1155/2007/20248.

Gull, S. F., & Daniell, G. J. (1989). Image reconstruction from incomplete and noisy data. *Nature, 272*, 686–690.

Hoch, J. C. (1989). Modern spectrum analysis in nuclear magnetic resonance: Alternatives to the Fourier transform. *Methods in Enzymology, 176*, 216–241.

Hoch, J. C., & Stern, A. S. (1996). *NMR data processing*. New York: John Wiley & Sons.

Högbom, J. A. (1974). Aperture synthesis with a non-regular distribution of interferometer baselines. *Astronomy and Astrophysics Supplement, 15*, 417–426.

Hyberts, S. G., Arthanari, H., Robson, S. A., & Wagner, G. (2014). Perspectives in magnetic resonance: NMR in the post-FFT era. *Journal of Magnetic Resonance, 241*, 60–73. https://doi.org/10.1016/j.jmr.2013.11.014.

Hyberts, S. G., Arthanari, H., & Wagner, G. (2012). Applications of non-uniform sampling and processing. In M. Billeter & V. Orekhov (Eds.), *Novel sampling approaches in higher dimensional NMR* (pp. 125–148). Berlin, Heidelberg: Springer Berlin Heidelberg.

Hyberts, S. G., Frueh, D. P., Arthanari, H., & Wagner, G. (2009). FM reconstruction of non-uniformly sampled protein NMR data at higher dimensions and optimization by distillation. *Journal of Biomolecular NMR, 45*(3), 283–294. https://doi.org/10.1007/s10858-009-9368-1.

Hyberts, S. G., Heffron, G. J., Tarragona, N. G., Solanky, K., Edmonds, K. A., Luithardt, H., et al. (2007). Ultrahigh-resolution (1)H-(13)C HSQC spectra of metabolite mixtures using nonlinear sampling and forward maximum entropy reconstruction. *Journal of the American Chemical Society, 129*(16), 5108–5116.

Hyberts, S. G., Milbradt, A. G., Wagner, A. B., Arthanari, H., & Wagner, G. (2012). Application of iterative soft thresholding for fast reconstruction of NMR data non-uniformly sampled with multidimensional Poisson gap scheduling. *Journal of Biomolecular NMR, 52*(4), 315–327. https://doi.org/10.1007/s10858-012-9611-z.

Hyberts, S. G., Robson, S. A., & Wagner, G. (2013). Exploring signal-to-noise ratio and sensitivity in non-uniformly sampled multi-dimensional NMR spectra. *Journal of Biomolecular NMR, 55*(2), 167–178. https://doi.org/10.1007/s10858-012-9698-2.

Hyberts, S. G., Robson, S. A., & Wagner, G. (2017). Interpolating and extrapolating with hmsIST: Seeking a tmax for optimal sensitivity, resolution and frequency accuracy. *Journal of Biomolecular NMR, 68*(2), 139–154. https://doi.org/10.1007/s10858-017-0103-z.

Hyberts, S. G., Takeuchi, K., & Wagner, G. (2010). Poisson-gap sampling and forward maximum entropy reconstruction for enhancing the resolution and sensitivity of protein NMR data. *Journal of the American Chemical Society, 132*(7), 2145–2147. https://doi.org/10.1021/ja908004w.

Kazimierczuk, K., & Orekhov, V. Y. (2011). Accelerated NMR spectroscopy by using compressed sensing. *Angewandte Chemie (International Ed. in English), 50*(24), 5556–5559. https://doi.org/10.1002/anie.201100370.

Li, K. B., Stern, A. S., & Hoch, J. C. (1998). Distributed parallel processing for multidimensional maximum entropy reconstruction. *Journal of Magnetic Resonance, 134*(1), 161–163.

Mobli, M., & Hoch, J. C. (2014). Nonuniform sampling and non-Fourier signal processing methods in multidimensional NMR. *Progress in Nuclear Magnetic Resonance Spectroscopy, 83*, 21–41. https://doi.org/10.1016/j.pnmrs.2014.09.002.

Mobli, M., Maciejewski, M. W., Gryk, M. R., & Hoch, J. C. (2007a). An automated tool for maximum entropy reconstruction of biomolecular NMR spectra. *Nature Methods, 4*(6), 467–468. https://doi.org/10.1038/nmeth0607-467.

Mobli, M., Maciejewski, M. W., Gryk, M. R., & Hoch, J. C. (2007b). Automatic maximum entropy spectral reconstruction in NMR. *Journal of Biomolecular NMR, 39*(2), 133–139. https://doi.org/10.1007/s10858-007-9180-8.

Orekhov, V. Y., Ibraghimov, I. V., & Billeter, M. (2001). MUNIN: A new approach to multi-dimensional NMR spectra interpretation. *Journal of Biomolecular NMR, 20*(1), 49–60.

Orekhov, V. Y., Ibraghimov, I., & Billeter, M. (2003). Optimizing resolution in multidimensional NMR by three-way decomposition. *Journal of Biomolecular NMR, 27*(2), 165–173.

Orekhov, V. Y., & Jaravine, V. A. (2011). Analysis of non-uniformly sampled spectra with multi-dimensional decomposition. *Progress in Nuclear Magnetic Resonance Spectroscopy, 59*(3), 271–292. https://doi.org/10.1016/j.pnmrs.2011.02.002.

Palmer, M. R., Suiter, C. L., Henry, G. E., Rovnyak, J., Hoch, J. C., Polenova, T., et al. (2015). Sensitivity of nonuniform sampling NMR. *The Journal of Physical Chemistry B, 119*(22), 6502–6515. https://doi.org/10.1021/jp5126415.

Robson, S. A., Takeuchi, K., Boeszoermenyi, A., Coote, P. W., Dubey, A., Hyberts, S., et al. (2018). Mixed pyruvate labeling enables backbone resonance assignment of large proteins using a single experiment. *Nature Communications, 9*(1), 356. https://doi.org/10.1038/s41467-017-02767-8.

Rovnyak, D., Frueh, D. P., Sastry, M., Sun, Z. Y., Stern, A. S., Hoch, J. C., et al. (2004). Accelerated acquisition of high resolution triple-resonance spectra using non-uniform sampling and maximum entropy reconstruction. *Journal of Magnetic Resonance, 170*(1), 15–21. https://doi.org/10.1016/j.jmr.2004.05.016.

Rovnyak, D., Hoch, J. C., Stern, A. S., & Wagner, G. (2004). Resolution and sensitivity of high field nuclear magnetic resonance spectroscopy. *Journal of Biomolecular NMR, 30*(1), 1–10. https://doi.org/10.1023/B:JNMR.0000042946.04002.19.

Schmieder, P., Stern, A. S., Wagner, G., & Hoch, J. C. (1993). Application of nonlinear sampling schemes to COSY-type spectra. *Journal of Biomolecular NMR, 3*(5), 569–576.

Schmieder, P., Stern, A. S., Wagner, G., & Hoch, J. C. (1994). Improved resolution in triple-resonance spectra by nonlinear sampling in the constant-time domain. *Journal of Biomolecular NMR, 4*(4), 483–490.

Schmieder, P., Stern, A. S., Wagner, G., & Hoch, J. C. (1997). Quantification of maximum-entropy spectrum reconstructions. *Journal of Magnetic Resonance, 125*(2), 332–339. https://doi.org/10.1006/jmre.1997.1117.

Sibisi, S., Skilling, J., Brereton, R. G., Laue, E. D., & Staunton, J. (1984). Maximum entropy signal processing in practical NMR spectroscopy. *Nature, 311*, 446. https://doi.org/10.1038/311446a0.

Stanek, J., Podbevsek, P., Kozminski, W., Plavec, J., & Cevec, M. (2013). 4D non-uniformly sampled C,C-NOESY experiment for sequential assignment of 13C, 15N-labeled RNAs. *Journal of Biomolecular NMR, 57*(1), 1–9. https://doi.org/10.1007/s10858-013-9771-5.

Stern, A. S., Donoho, D. L., & Hoch, J. C. (2007). NMR data processing using iterative thresholding and minimum l(1)-norm reconstruction. *Journal of Magnetic Resonance, 188*(2), 295–300. https://doi.org/10.1016/j.jmr.2007.07.008.

Sun, S., Gill, M., Li, Y., Huang, M., & Byrd, R. A. (2015). Efficient and generalized processing of multidimensional NUS NMR data: The NESTA algorithm and comparison of regularization terms. *Journal of Biomolecular NMR, 62*(1), 105–117. https://doi.org/10.1007/s10858-015-9923-x.

Tugarinov, V., Kay, L. E., Ibraghimov, I., & Orekhov, V. Y. (2005). High-resolution four-dimensional 1H-13C NOE spectroscopy using methyl-TROSY, sparse data acquisition, and multidimensional decomposition. *Journal of the American Chemical Society, 127*(8), 2767–2775. https://doi.org/10.1021/ja044032o.

Yang, J., & Zhang, Y. (2011). Alternating direction algorithms for ℓ_1-problems in compressive sensing. *SIAM Journal on Scientific Computing, 33*(1), 250–278. https://doi.org/10.1137/090777761.

Ying, J., Delaglio, F., Torchia, D. A., & Bax, A. (2017). Sparse multidimensional iterative lineshape-enhanced (SMILE) reconstruction of both non-uniformly sampled and conventional NMR data. *Journal of Biomolecular NMR, 68*(2), 101–118. https://doi.org/10.1007/s10858-016-0072-7.

CHAPTER ELEVEN

Exploring Protein Conformational Landscapes Using High-Pressure NMR

Julien Roche*, Catherine A. Royer†,[1], Christian Roumestand‡

*Roy J. Carver Department of Biochemistry, Biophysics and Molecular Biology, Iowa State University, Ames, IA, United States

†Department of Biological Sciences, Rensselaer Polytechnic Institute, Troy, NY, United States

‡Centre de Biochimie Structural CNRS Université de Montpellier UMR, Montpellier, France

[1]Corresponding author: e-mail address: royerc@rpi.edu

Contents

Abstract

Protein conformational landscapes define their functional properties as well as their proteostasis. Hence, detailed mapping of these landscapes is necessary to understand and modulate protein conformation. The combination of high pressure and NMR provides a particularly powerful approach to characterizing protein conformational transitions. First, pressure, because its effects on protein structure arise from elimination of solvent excluded void volume, represents a more subtle perturbation than chemical denaturants, favoring the population of intermediates. Second, the residue-specific and multifaceted nature of NMR observables informs on many local structural properties of proteins, aiding in the characterization of intermediate and excited states.

Methods in Enzymology, Volume 614
ISSN 0076-6879
https://doi.org/10.1016/bs.mie.2018.07.006

1. INTRODUCTION

The sequences of extant biological macromolecules (proteins and nucleic acids) have evolved to allow their functional properties to confer a requisite competitive advantage on the organisms in which they are found, for the environmental conditions in which these organisms live. While homologues of the same proteins across all organisms in which they are produced adopt the same or quite similar structures, their global stabilities, and perhaps even more importantly, the degree to which they populate intermediate or excited state structures can vary considerably. These excited states or intermediate structures can play central roles in biomolecular function. For example, in the case of proteins, local unfolding or complex dissociation may be required for or coupled to interaction with ligand or substrate. Posttranslational modification probability likewise may differ significantly for different conformational states of proteins. Such modifications can impact protein localization or turnover, and hence modulate their functional and/or proteostasis properties. It is increasingly recognized that selective pressure on entire conformational landscapes has modulated protein and nucleic acid sequences. However, while the understanding of how sequence impacts global biomolecular stability has progressed a great deal in recent years, much less is known about how sequence modulates these conformational landscapes. In this chapter, we discuss high-pressure NMR as a particularly useful methodology for exploring biomolecular conformational landscapes, and the effects of changes in sequence thereon. We will focus on protein folding landscapes, with some discussion of protein association. However, this methodology can be applied to nucleic acid conformational landscapes as well.

Much of the information obtained to date concerning protein folding mechanisms comes from folding/unfolding experiments performed in vitro. Several perturbation methods can be used to modulate protein stability: adding chemical denaturants (urea, guanidinium chloride, GuHCl) or stabilizers (osmolytes such as sucrose or glycerol), changing the pH, or modifying the thermodynamic parameters of the sample (pressure, temperature). While most protein folding studies have used chemical denaturants, pH, or temperature, pressure presents a number of advantages and can provide complementary information to that obtained using the more common perturbation approaches. The temperature dependence of protein stability is complex, with a strong temperature dependence of both the enthalpy and entropy change upon folding due to the large difference in heat capacity between the folded and unfolded states (Privalov & Gill, 1988). Chemical denaturation (or stabilization) is directly related to the preferential interaction (or exclusion) of the additive, and thus depends upon the absolute difference in exposed surface area between the folded and unfolded states. This in turn is simply correlated with the size of the protein (Myers, Pace, & Scholtz, 1995), as is the change in heat capacity. Changing pH can have interesting effects, as the change in stability depends on the local pK_a values of ionizable residues. Pressure, while relatively little employed, is actually a very straightforward perturbation. Pressure-induced unfolding occurs because the molar volume of the folded states of proteins is larger than that of their unfolded states. We demonstrated recently (Roche, Caro, et al., 2012; Rouget et al., 2011) that this effect was primarily due to the existence within folded protein structures, of solvent excluded void volume that is eliminated in the unfolded form. The difference in volume, ΔV_f, is present at atmospheric pressure. All that pressure does, per Le Chatelier's principle, is to shift the $F \leftrightarrow U$ equilibrium toward the lower volume, unfolded state. Given this mechanism for pressure effects on protein stability, its effects, like pH, can be "local." That is, for comparable stabilities, protein regions or domains that are less well packed will be more pressure sensitive than more tightly packed regions. Another consequence of these volumetric effects is that single amino acid substitutions can drastically alter susceptibility to pressure (Ando et al., 2008; Roche, Caro, et al., 2012). In contrast to pH, denaturants, and high temperature, which are highly disruptive, pressure will only act to eliminate internal void volume, such that pressure unfolded states may be less disrupted than those observed using these other methods of denaturation (Rouget et al., 2011; Zhang et al., 2018).

Thus, in spite of some difficulties in implementation, most of which have been adequately addressed in available commercial instrumentation, pressure represents a method of choice to study protein folding. Generally, unfolding by pressure is reversible, most prominently because high pressure disfavors intermolecular protein interactions, such that irreversible aggregation is most often avoided. Moreover, pressure is a continuous variable, requiring a single sample for the entire unfolding (and refolding) profile. This reversibility gives access to a large panel of thermodynamic parameters specific to the folding/unfolding reaction (Akasaka, Kitahara, & Kamatari, 2013; Kamatari, Kitahara, Yamada, Yokoyama, & Akasaka, 2004). Pressure perturbation is generally used in conjunction with spectroscopic methods, such as circular dichroism, fluorescence, or IR spectroscopy (Dellarole & Royer, 2014). These observables yield global information on the evolution of the tertiary or secondary structure during protein unfolding. Due to its extreme sensitivity to structural perturbations, nuclear magnetic resonance (NMR) constitutes an attractive alternative observable, particularly in conjunction with a continuous variable such as pressure. Although 1D NMR can provide global information on protein folding thermodynamics, the multiple probes available through multidimensional NMR give access to local, residue-specific information. Thus, the combination of high pressure and NMR constitutes a powerful tool that can lead to new knowledge concerning the role of residue packing in protein stability, of conformational fluctuations in water penetration, or that can be used to describe folding intermediates or other details of the conformational landscape, which cannot be accessed by global spectroscopic observables.

2. MONITORING "GLOBAL" FOLDING/UNFOLDING REACTIONS WITH 1D HIGH-PRESSURE NMR SPECTROSCOPY

The NMR resonances (or chemical shifts) of magnetically active nuclei (^{1}H, ^{15}N, ^{13}C) in a protein are extremely sensitive to the through-space effects of neighboring groups: this leads to the good dispersion usually observed in the NMR spectrum of a folded protein. When the protein unfolds, these through-space effects are largely eliminated: the NMR resonances become almost uniquely sensitive to through-bond effects between covalently bound atoms. As a consequence, the NMR spectrum collapses and becomes poorly resolved (Fig. 1). For instance, since they are borne by the same chemical groups, all the ^{1}H resonances of the methyl groups, or of the amide groups of the amino acids in a protein will give rise to almost identical resonances, with only slight variations due to differences in the chemical nature of the 20 amino acid side chains. Thus, it is possible to monitor the high-pressure denaturation of a protein by following the evolution of its 1D NMR spectrum, as depicted in Fig. 1 for Titin I27 domain, a protein for which folding has been extensively studied. As generally observed, folded and unfolded species are in slow exchange with regard to the NMR timescale: we do not observe chemical shift variations, indicative of fast exchange, as a function of pressure, but rather the disappearance of resonances belonging to the native form, with the concomitant appearance of new peaks that correspond to the spectrum of the unfolded species with a similar chemical environment for all residues. Note than coupling high-pressure and NMR raises complicated technological issues. The

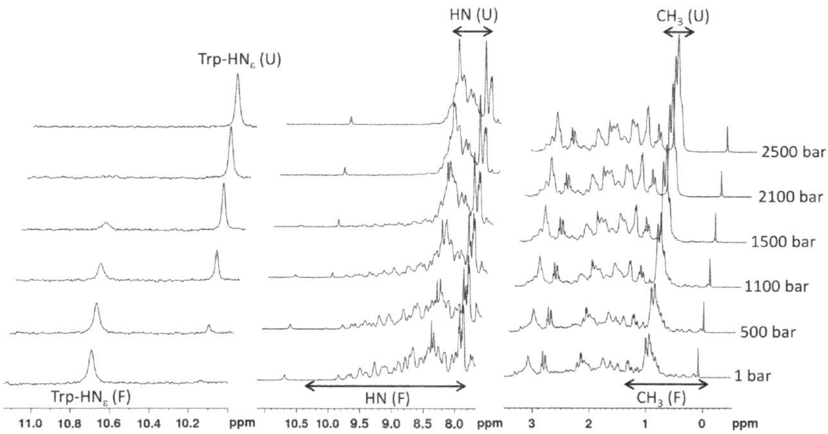

Fig. 1 Monitoring the pressure unfolding of Titin I27 domain with 1D NMR spectroscopy. Stacked plot of 1D spectra (amide and aliphatic regions) recorded at 600 MHz on a sample of Titin I27 domain (1 m*M* in Tris buffer pH 7.0, 298 K, 1.7 *M* GuHCl) as a function of pressure. The regions corresponding to HN amide and CH_3 methyl group resonances in the folded (*F*) and unfolded (*U*) states are indicated with *arrows*. The decrease of the resonances corresponding to the HN of Trp-34 in the folded state with the concomitant increase of the same resonances in the unfolded state is also shown (*left*).

experimental device should of course support high pressure but must be nonmagnetic and permeable to radiofrequency. A satisfying commercial solution has emerged, based on the use of ceramic tubes (zirconium oxide) developed by the Wand group (Peterson & Wand, 2005) and now available commercially (Daedalus Innovations™). These tubes can support up to 300 MPa (3 kbar), usually enough to unfold protein. If proteins are too stable to be unfolded by pressure over this range, one can adjust the protein stability, generally by shifting the temperature or adding low concentration of chaotropic reagents. For instance, in the case of Titin I27 domain (Fig. 1), 1.7 *M* of GuHCl has been added to the protein solution in order to allow complete unfolding in the range 1–2500 bar. The ceramic high-pressure NMR tubes can be used in combination with any commercial NMR probe and maintain a sensitivity of ~50% relative to that obtained with classical 3 mm Borosilicate or Pyrex glass tubes.

Often, the decrease of the composite peaks corresponding to the folded (*F*) conformation, or the increase of peaks corresponding to the unfolded states (*U*), with pressure follows a sigmoidal curve characteristic of a two-state equilibrium (Fig. 2), in the form of:

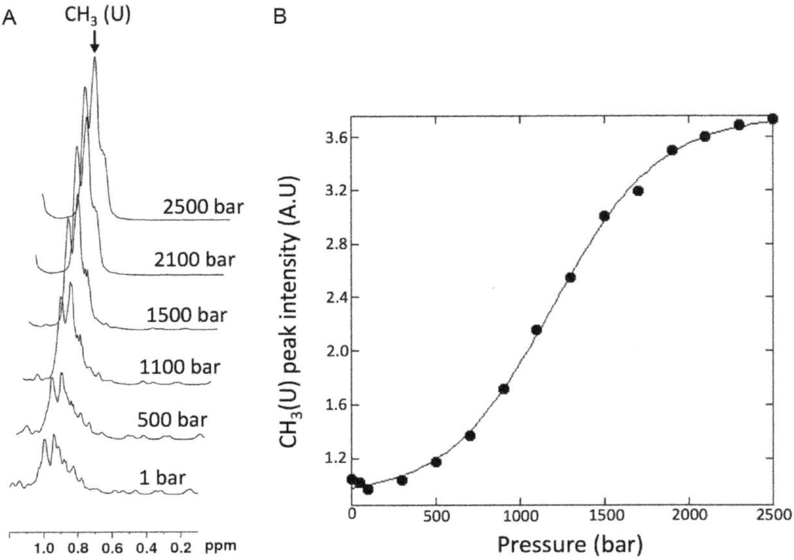

Fig. 2 Measuring steady-state thermodynamic parameters form 1D NMR spectroscopy. (A) Stacked plot of the methyl resonance region of 1D spectra recorded on a sample of Titin I27 domain as a function of pressure (same conditions as stated in Fig. 1). The *arrow* indicates the line corresponding to the methyl resonances in the unfolded species (*U*). (B) Fit of the sigmoidal increase of the methyl resonances of the unfolded state with pressure using Eq. (6).

$$F \underset{k_f}{\overset{k_u}{\rightleftarrows}} U$$

with a characteristic equilibrium constant of:

$$K_{\text{eq}} = \frac{k_f}{k_u} = \frac{[U]}{[F]} \tag{1}$$

where k_f and k_u stand for the folding and the unfolding rate constants. K_{eq} can be also expressed from the Boltzmann equation as:

$$K_{\text{eq}} = \exp\left(\frac{-\Delta G_{\text{eq}}}{RT}\right) \tag{2}$$

where the free energy change can be expressed as a Taylor expansion, truncated at the second order term:

$$\Delta G_{\text{eq}} = G_U - G_F = \Delta G^0 + \Delta V^0(p - p_0) - \frac{1}{2}\Delta\beta V^0(p - p_0)^2 + \dots \tag{3}$$

Here ΔG and ΔG^0 are the Gibbs free energy changes from F to U at pressure p and p_0 ($p_0 = 0.1$ MPa), respectively; ΔV^0 is the partial molar volume change; $\Delta\beta$ ($= -(1/V^0) * \delta V/\delta p$) is the change in compressibility coefficient, R is the gas constant, and T is the absolute temperature. For proteins the differences in compressibility between native and denatured states are negligible (Ravindra & Winter, 2003). Hence, the expression of ΔG simplifies to:

$$\Delta G_{\text{eq}} = \Delta G^0 + \Delta V^0 (p - p_0) \tag{4}$$

The observable will be I, the intensity (or integral) of a peak corresponding to either the folded species or of the unfolded species. In the case of Titin I27 domain, we chose to follow the increase of a peak corresponding to methyl groups corresponding to the unfolded species (Fig. 2). Thus, the equilibrium constant can be written as:

$$K_{\text{eq}} = \frac{[U]}{[F]} = \frac{I_F - I}{I - I_U} \tag{5}$$

where, for a given residue, I_F stands for the intensity of the corresponding NMR line in the folded spectrum at 1 bar ($I_F = I_{\text{min}}$), while I_U corresponds to the intensity of the same line at high pressure, when the protein is fully unfolded ($I_U = I_{\text{max}}$) for the unfolded state methyl peaks. Combining this equation with Eqs. (2) and (4) gives the characteristic equation for a two-state equilibrium:

$$I = \frac{I_F + I_U e^{-[\Delta G^0 + (p - p_0)\Delta V^0]/RT}}{1 + e^{-[\Delta G^0 + (p - p_0)\Delta V^0]/RT}} \tag{6}$$

Fitting the data plotted in Fig. 2 for Titin I27, we can extract "global" values for ΔV^0 of unfolding (−75 mL/mol) and ΔG^0 (2.5 kcal), under the conditions of the study (pH 7, 25°C, 1.7 *M* GuHCl). Generally, uncertainties on the recovered parameters depend upon, not only the quality of the data, but also on how well-defined are the low- and high-pressure plateau regions. Typically, uncertainty on the free energy and volume changes of unfolding are between 5% and 10%.

The parameters ΔV^0 and ΔG^0 are characteristics of the system at equilibrium, at atmospheric pressure. A complete understanding of the protein folding/unfolding phenomenon requires a temporal description of the reaction to obtain information concerning the folding barrier. Such a description relies on the measurement of kinetic parameters, after perturbation of the thermodynamic equilibrium between the folded and unfolded conformers of the protein. In addition, these studies yield the rates of folding and unfolding under native conditions and the activation volume measured between the unfolded state or the folded state and the transition state ensemble (TSE), $\Delta V_f^{\ddagger}$ and $\Delta V_u^{\ddagger}$, respectively.

When applied to the complete unfolding/folding transition, high-pressure can slow down the rate of folding (Vidugiris, Markley, & Royer, 1995), and also possibly unfolding (Rouget et al., 2011), by several orders of magnitude due to the very large volumes of activation involved in these reactions. Thus, a reaction completed in few seconds at atmospheric pressure can take several minutes to even a few hours to relax at higher pressure. For example, Δ + PHS SNase at pressure above 1 kbar exhibits residue-specific relaxation times greater than 10 h (Roche, Caro, et al., 2012). It becomes thus feasible to monitor the unfolding kinetics using 1D real-time NMR. In the case of Titin I27, we used pseudo-2D experiments to measure the relaxation times after different p-jump (Fig. 3A and B). They consist in a series of 1D NMR spectrum recorded at a constant repeating rate. After Fourier transform along the acquisition dimension, one can extract from the matrix obtained a column at a chemical shift corresponding, for instance, to the methyl groups of the unfolded states. In that case, we observe an exponential rise of this resonance with time, from which the relaxation time ($\tau = 1/(k_u + k_f)$) can be extracted. It is possible to extract the values of k_u and k_f, as well as those of the activation volume of unfolding $\Delta V_u^{\ddagger}$ (or folding, $\Delta V_f^{\ddagger}$) by measuring this relaxation time $\tau_{(p)}$, after different p-jumps in the pressure range where unfolding appears. The p-jump amplitude should be sufficient to observe a measurable intensity change for the NMR resonance but should remain moderate to avoid any imbalance between the folding and the unfolding reactions: an excessive positive p-jump, for instance, will favor the unfolding reaction at the expense of the folding reaction. For Titin I27, we have used pressure jumps of 200 bar, which is about 10% of the pressure range needed to fully unfold the protein (2000 bar).

With $\tau_{(p)} = 1/(k_u + k_f)$ and $k_f(p) = k_{f0}e^{-p\Delta V_f^{\ddagger}/RT}$ and $k_u(p) = k_{u0}e^{-p\Delta V_u^{\ddagger}/RT}$, $\tau_{(p)}$ can be rewritten as:

$$\tau_{(p)} = \left[k_{u0} e^{\left(\frac{-p\Delta V_{u0}^{\mp}}{RT}\right)} + k_{f0} e^{\left(\frac{-p\Delta V_{f0}^{\mp}}{RT}\right)} \right]^{-1} \tag{7}$$

Knowing the value of ΔV_{eq}, the volume difference between the folded and unfolded states measured at equilibrium ($\Delta V_{\text{eq}} = \Delta V_f - \Delta V_u$), and

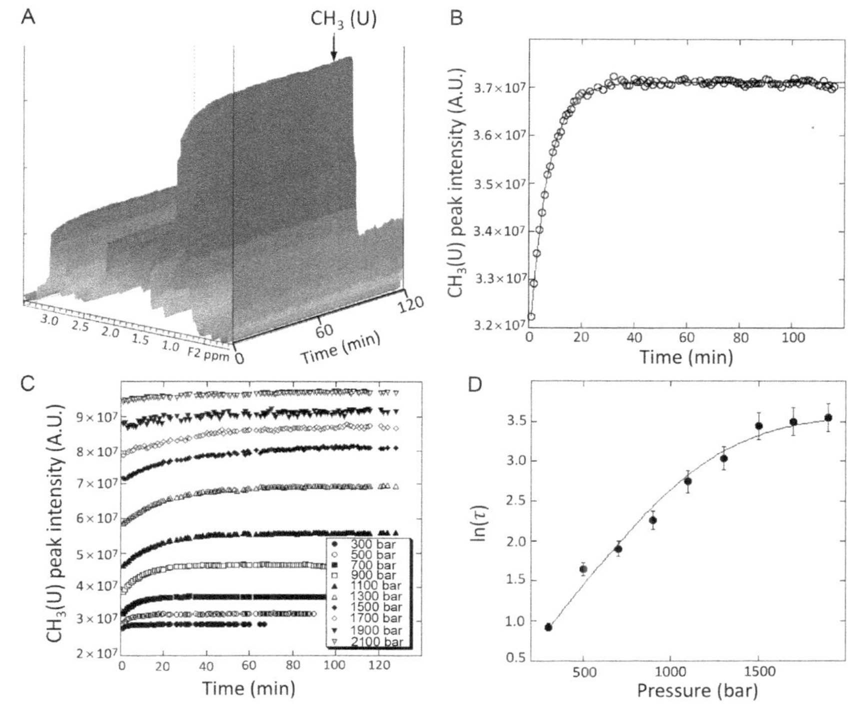

Fig. 3 Measuring the unfolding relaxation time after a *p*-Jump 1D NMR spectroscopy. (A) Series of 1D spectra of Titin I27 domain (same conditions as stated in Fig. 1) recorded with time after a 200 bar *p*-Jump (500 → 700 bar). Sixty spectra of 2 min each were recorded over a period of 2 h. The position of the resonance corresponding to methyl groups in the unfolded species is indicated with an *arrow*. (B) Fit of the exponential rise of this methyl resonance, allowing the measurement of the relaxation time, τ, at 700 bar. (C) Exponential rises of the resonance corresponding to the methyl groups in the unfolded states after successive 200 bar *p*-Jumps between 300 and 2100 bar, the pressure range where Titin I27 unfolds. Relaxation times $\tau_{(p)}$ can be measured from these experiments for the different pressures. (D) "Chevron plot" of the natural logarithm of τ at different pressures: the fit with Eq. (8) allows to extract the kinetic constants k_u and k_f, as well as the activation volume of folding $\Delta V^{\ddagger}_{f0}$ or of unfolding $\Delta V^{\ddagger}_{u0}$ at atmospheric pressure.

$K_{eq} = k_f/k_u$, one can constrain the fit and decrease the number of parameters. Eq. (7) becomes:

$$\tau_{(p)} = \left[k_{u0} e^{\left(\frac{-p\Delta V^{\mp}_{u0}}{RT}\right)} + k_{u0} K_{eq} e^{\left(\frac{-p\left(\Delta V_{eq} + \Delta V^{\mp}_{u0}\right)}{RT}\right)} \right]^{-1} \tag{8}$$

Only two variables need to be fitted: k_{u0} (the unfolding rate at atmospheric pressure) (or folding, k_{f0}) and $\Delta V^{\ddagger}_{u0}$ (the activation volume for unfolding at atmospheric pressure) or ΔV_{0f} (Fig. 3C). Generally, it is advantageous to fit for the parameters (folding or unfolding) which are most affected by pressure. The fit is usually performed on a plot of $\ln(\tau)$ as a function of pressure that presents the characteristic pattern of a "chevron plot" (Fig. 3D). In the case of Titin I27, we measured an activation volume $\Delta V^{\ddagger}_{f}$ close to the equilibrium ΔV^{0} value, indicating that the molar volume of the transition state is quite close to that of the folded state, suggesting a dehydrated TSE where most of the native cavities are still present. The intrinsic limitation of this approach resides in the fact that while NMR has high spatial resolution, its time resolution is limited: the recording time of a 1D NMR spectrum ranges from few seconds to 1–2 min, depending on the sample concentration. Thus, such experiments can be used only in the case of proteins having slow relaxation times, above 5–10 min. In the case of protein having fast relaxation times (<1 s), other NMR approaches are available, mainly based on 2D exchange spectroscopy techniques. They will be discussed in details in a later section.

3. MONITORING "LOCAL" PROTEIN FOLDING/UNFOLDING REACTION WITH HIGH-PRESSURE 2D NMR SPECTROSCOPY: EXPLORING THE FOLDING LANDSCAPE

In many of the studies reported in the literature, and whatever the spectroscopy used (Fluorescence, IR, CD, or NMR), the folding/unfolding reaction of a protein can be approximate by a two-state model, excluding the possibility of any folding intermediates in the folding energy landscape of a protein. This is of course a rough approximation: most globular proteins are known to deviate from a true two-state folding mechanism and an ideal global cooperativity by populating en route folding intermediate states, underlying the concept of "foldons" (Englander, Mayne, & Krishna, 2007; Lindberg & Oliveberg, 2007). This apparent discrepancy is essentially due to the low stability of the intermediate states which are often only transiently populated, with no detectable equilibrium population. Moreover, often the spectroscopic observable gives access only to global "apparent" values for thermodynamic parameters (ΔG^{0} and ΔV^{0} or m-value). Deviations from two-state behavior manifest as smaller values for the thermodynamic parameters, yet without prior knowledge of the true global stability and cooperativity, such deviations from two-state behavior are often not recognized. For example, in the case of Titin I27, ΔG^0 and ΔV^0 values measured from the pressure dependence of the resonance corresponding to the unfolded state methyl groups or to the indole NH of the tryptophan side chain in the denatured state are significantly different (72 ± 8 vs 83 ± 5 mL/mol), indicating deviation from true two-state behavior. Multiple probes allow a more complete description of the folding/unfolding process.

Multidimensional NMR spectroscopy is able to provide residue-specific information, through correlation spectroscopy involving nuclei located on the peptide backbone (^{1}H and ^{15}N of amide groups, Hα and Cα), allowing a residue-specific analysis of the folding process. Thus, the atomic resolution offered by NMR experiments provides an intrinsic multiprobe approach to assess the degree of protein folding cooperativity, which is otherwise difficult to characterize using techniques such as circular dichroism or fluorescence. Amide protons offer ideal probes to monitor the unfolding reaction: each amino acid (with the exception of proline) bears an NH group, and the corresponding proton resonances represent the most resolved region of the proton spectrum of a protein. This resolution can be considerably enhanced through homonuclear (COSY, TOCSY, NOESY) or heteronuclear ([^{1}H–^{15}N]-HSQC) experiments. In addition, their acidic character makes these amide protons well suited for proton/deuteron exchange measurements, a property that has been extensively used to evaluate the local stability of a protein, with a possible association with high pressure, as will be discussed further. Series of [^{1}H–^{15}N] HSQC can also be used to monitor the dissociation of protein complexes induced by pressure. This was applied for the HIV-1 protease for which the monomer–dimer exchange occurs on a slow time scale (Ishima, Torchia, & Louis, 2007). At low enough concentration, both the dimer and monomer cross-peaks can be simultaneously monitored as a function of pressure and the intensity profiles of interfacial residues can be fitted to calculate the relative population of dimer, monomer, and unfolded states at each pressure (Louis & Roche, 2016). These experiments allow the determination of the binding affinity of the dimer at atmospheric pressure and the volume change associated with the pressure-induced dissociation of the dimer (Louis & Roche, 2016).

3.1 Tracking Intermediate States in the Protein Folding Landscape

The pressure-induced unfolding reaction, which occurs on a slow timescale with respect to NMR, is often analyzed by monitoring the decrease in volume or intensity of native state cross-peaks in a series of [^{1}H–^{15}N]-HSQC

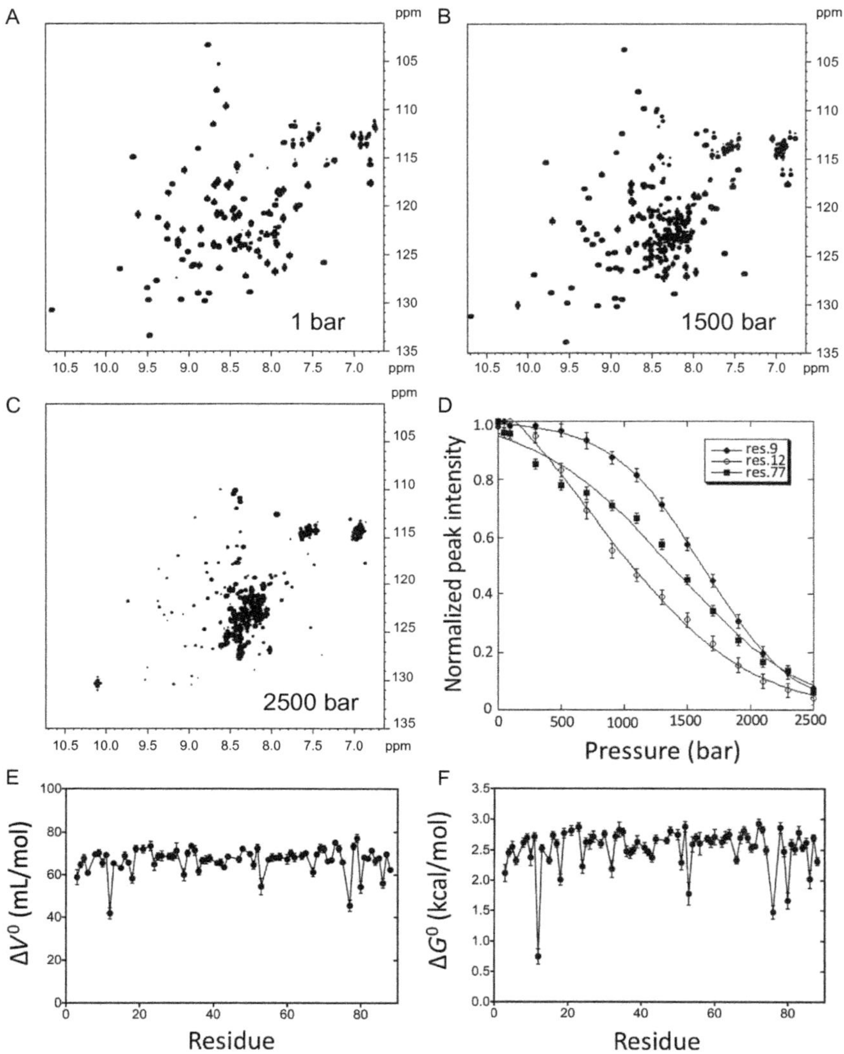

Fig. 4 2D NMR detected high pressure unfolding of Titin I27 domain. (A–C) Examples of [^{1}H–^{15}N] HSQC NMR spectra recorded on a ^{15}N-labeled sample of Titin I27 at different pressures as indicated. (D) Examples of three residues exhibiting distinct unfolding profiles. (E) Residue-specific ΔV^0 and (F) ΔG^0 measured at equilibrium from pressure unfolding of Titin I27 domain.

experiments recorded at increasing pressure (Fig. 4A–C). Of course, to extract the most useful information from these HP 2D NMR experiments, the cross-peaks must be assigned to specific residues in the protein sequence. Care must be taken that these measurements be recorded when the equilibrium between native/unfolded populations has been reached after a pressure-jump: the pseudo-2D experiments described earlier can be readily used to evaluate the value of the relaxation time τ needed to reach equilibrium. Moreover, since the intensity (or integral) of each amide NH cross-peak is taken to represent the population of a conformational ensemble in which that amide resonance is in the chemical environment of the folded state, it is important to minimize loss of intensity due to experimental artifacts. In particular, it is important to choose a delay time between scans that is ~fivefold longer than the longitudinal relaxation time, T_1. Thus, T_1 should be measured (e.g., by inversion recovery) at atmospheric pressure for each sample at the temperature and under the solution conditions of the HP experiment to be performed. While T_1 can vary strongly with temperature, we have found that for proteins and nucleic acids proton T_1 values are pressure independent. Thus, the T_1 measurement need not be made as a function of pressure.

The residue-specific pressure-dependent peak volume or intensity values can be analyzed by individually fitting using Eq. (6), yielding residue-specific apparent ΔV_u^0 and ΔG_u^0 values (Roche, Caro, et al., 2012; Roche, Dellarole, Royer, & Roumestand, 2015; Roche, Louis, Bax, & Best, 2015) (Fig. 4D). Unlike the composite 1D methyl proton resonances of the unfolded state or fluorescence or CD, these 2D (usually ^{15}N–^{1}H HSQC) residue-specific profiles generally correspond to apparent two-state transitions, at least locally. This is because the resonance is either at the frequency observed for the folded state or it is not, in which case the intensity of the "folded state" peak decreases. Note that these "folded state" peaks shift somewhat with pressure due to compression and small conformational changes in the "folded state" ensemble. These shifts have been investigated for many proteins (for example, see Akasaka, 2006), and it has been shown recently that to a large extent these shifts are due to solvent-induced polarization of the backbone (Frach et al., 2016).

In contrast, here we are concerned with the disappearance of the "folded state" peaks as the protein populates conformational states in which the chemical environment of the residues is more strongly modified, leading to a loss of "folded state" peak intensity. Note that the high-pressure plateau of the folded state (whether it is attained or not in the experiment) should be equal to zero since at some pressure all conformers in which a residue is in its "folded state" environment should be eliminated. Thus, even if that high-pressure plateau is not attained due to experimental limitations of the apparatus, for example, in many cases fixing the high-pressure plateau value (I_U in Eq. 6) to a value of zero in the fits allows for a reasonable estimation of the apparent thermodynamic parameters. Note that usually one fits the decrease of the cross-peaks for residues in the "folded state," instead of the sigmoidal increase of cross-peaks in the "unfolded state." This is because (i) the NMR spectra of the folded protein are usually assigned, which is rarely the case for the unfolded species and (ii) the NMR spectra of folded protein show usually a better dispersion, facilitating the analysis. In some cases, the unfolded state peaks can be accessed. We have done so for the unfolded state indole NH peak of for tryptophan in single tryptophan containing proteins (Zhang et al., 2018), and as noted above in the 1D spectrum of Titin I27. This information is quite useful as this peak is generally in slow exchange with that corresponding to the folded state. Moreover, both of these profiles (Trp indole "*F*" and "*U*") can be compared to that obtained using fluorescence from the same residue. Such comparisons can bring to light the population of intermediate species. In addition, the unfolded state peaks for glycine residues are often also easy to quantify. Even in the absence of assignments for these peaks, and depending on the protein's sequence, they can yield useful information when compared to the loss of "folded state" glycine peaks (Zhang et al., 2018).

Large variations in the ΔV^0 and ΔG^0 values measured for different residues of a given protein typically reflect departure from a two-state unfolding transition and inform on the potential presence of intermediate states as observed for Δ+PHS staphylococcal nuclease (Snase), Titin I27, and a variant of the pp32 leucine-rich repeat protein in which the N-terminal capping motif has been deleted (pp32-ΔNcap) (Louis & Roche, 2016; Roche, Caro, et al., 2012; Zhang et al., 2018). In the case of Titin I27, for example, while most of the residues in the Titin I27 protein exhibit a similar ΔV^0 for unfolding of ≈ -70 mL/mol, in some areas of the protein the measured residue-specific ΔV^0 values fall below this average value (<55 mL/mol) (Fig. 4E), meaning that some regions of the protein unfold before others, suggesting the presence of folding intermediates, i.e., partially folded conformers having some degree of stability, in the protein energy landscape (Fig. 4F).

Structural and energetic information about the folding intermediates can also be extracted from the pressure-dependent multidimensional NMR data (de Oliveira & Silva, 2015; Roche, Caro, et al., 2012; Roche, Dellarole, et al., 2012) using the following procedure inspired by the analysis method used by Munoz and coworkers to characterize the thermal unfolding of the downhill protein BBL (Naganathan & Munoz, 2008; Sadqi, Fushman, & Muñoz, 2006). After normalizing the residue-specific denaturation curves obtained from the amide cross-peak intensity decays measured on the HSQC experiments recorded at variable pressure (see Fig. 5D), the value of 1 for a given cross-peak ($I=I_F=1$) can be associated with a probability of 1 (100%) for the corresponding residue to be in the native state, with all the native contacts present. Similarly, a residue for which, at the same pressure, the corresponding cross-peak has disappeared ($I=I_U=0$) from the HSQC spectrum has a probability equal to 0 to be in a native state: it belongs to an unfolded state where all the native contacts are lost. Now, we consider two residues i and j in an intermediate situation where the probability to be in a folded state are $P(i)$ and $P(j)$, respectively, at a given pressure. If these two residues are in contact in the native state, we assume the probability $P(i,j)$ to be in contact at this pressure to be given by the geometric mean (Fossat et al., 2016) of the individual probabilities $P(i,j)=[P(i)\times P(j)]^{0.5}$. Pressure-dependent contact maps can be constructed based on the contact map of the folded protein. These folded state contact maps are constructed from the 3D crystal or NMR structure of the protein by measuring all contacts between different atoms: usually, only the distances between Cα atoms of the different residues are used. These contacts are then plotted in a diagonal diagram with the protein sequence numbering on the X and Y axis). It is now possible to use a color code (for example) in order to report

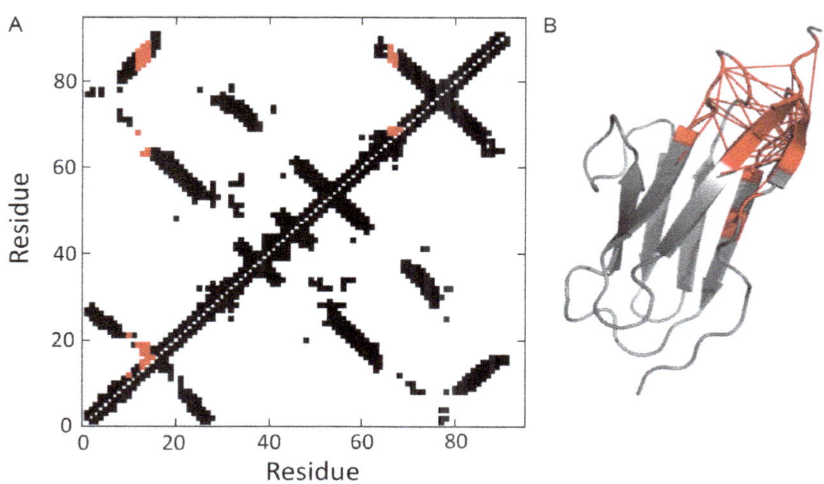

Fig. 5 Pressure denaturation of Titin I27 domain. (A) Contact maps built from the NMR structure of Titin I27 domain. The contacts above the diagonal have been colored in *red* when contact probabilities (P_{ij}) lower than 0.5 are observed at a pressure of 800 bar. (B) *Ribbon representations* of the structure of I27 where the *red ticks* represent contacts that are weakened ($P_{ij}\leq 0.5$) at 800 bar.

in this contact map the pressure-dependent probability of contact, as defined earlier (Fig. 5).

The strong deviation from two-state behavior in the HSQC-detected high-pressure unfolding of pp32 ΔN-cap (Zhang et al., 2018) provides another example of deducing some of the structural properties of folding intermediates using fractional contacts (Fig. 6). The N-terminal capping motif (yellow in Fig. 6A) provides stability to the WT protein (Dao, Majumdar, & Barrick, 2014), but CD-detected urea melts of the ΔN-cap variant showed no evidence for the strong deviation from two-state

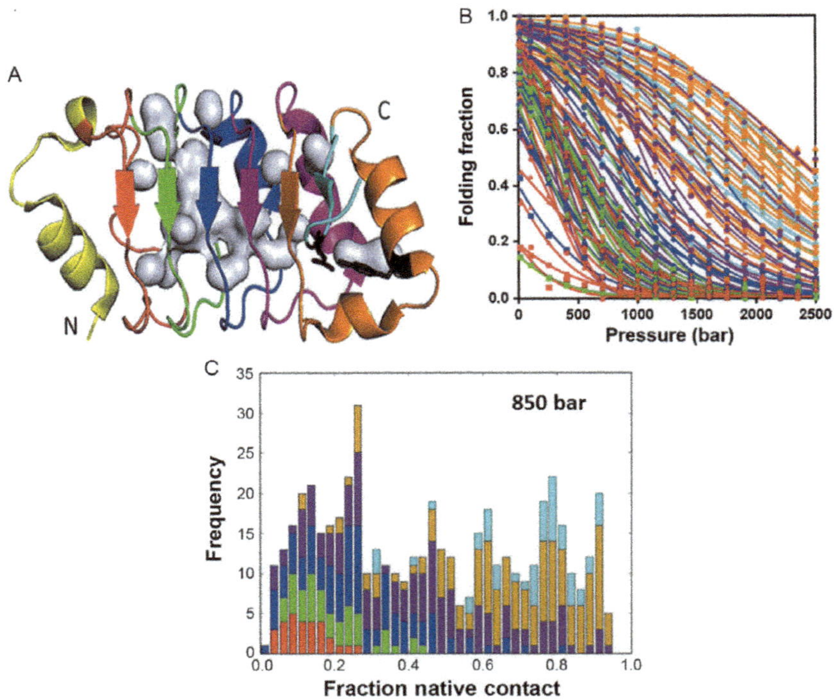

Fig. 6 HSQC-detected pressure denaturation of pp32 ΔN-cap. (A) Schematic of the protein in which the repeats and capping motifs are color coded. The N-cap (*yellow*) is deleted in the pp32 ΔN-cap variant. (B) Pressure-dependent loss of amide HSQC peak intensities for pp32 ΔN-cap. *Curves* are colored following the color scheme in (A). (C) Fractional contact histogram for pp32 ΔN-cap at 850 bar. *Bars* are color coded for the repeat according to the scheme in (A). Interfacial contacts are reported twice. *Results are taken from Zhang Y., Berghaus M., Klein S., Jenkins K., Zhang S., McCallum S.A., et al. (2018) High-pressure NMR and SAXS reveals how capping modulates folding cooperativity of the pp32 leucine-rich repeat protein.* Journal of Molecular Biology 430, *1336–1349.*

behavior exhibited in the pressure-dependent HSQC profiles of the "folded state" amide NH peaks (Fig. 6B). The fractional contact values at 850 bar represented as histograms colored per repeat (Fig. 6C) reveal that the N- to C-terminal stability gradient already apparent in WT pp32 (Fossat et al., 2016) is reinforced by deletion of the N-cap, leading to the population of multiple intermediate species. While the intensity of the "folded state" peaks for residues in the first repeat (red) has disappeared by 1 kbar, residues in the C-terminal repeat and capping motif (brown and cyan) retain fractional "folded state" peak intensity even at 2500 bar. The composite "unfolded state" methyl peak intensity in the 1D proton spectrum did not begin to increase until above 1 kbar. Hence the intermediates populated present molten globule-like properties. This illustrates the distinction between true thermodynamic ensembles, and the locally "folded" and "unfolded" states detected by HP HSQC. The residue-specific transitions observed for "folded state" HSQC peaks represent the transition from a true thermodynamic ensemble in which that residue's amide resonates at its native state frequency, and one in which it does not. In the case of pp32 ΔN-cap, for residues in the N-terminal repeat 1, this is a transition from the true folded state to one in which repeat 1 is disordered, but the others are not. The transition observed for residues in the C-terminal capping motif is from an intermediate ensemble in which repeats 1–3 are disordered, while those in the C-terminus are not to the high-pressure unfolded state.

Fractional contact maps determined from pressure unfolding experiments can also be combined with Go-model simulations to obtain a structural representation of the conformal ensemble populated at high pressure. Native contact lists can be obtained from a contact map and used for molecular modeling, with the possibility to "weight" the probabilities of forming a native contact calculated at a given pressure. A large number of contact lists will be then generated, instead of one. For example, a contact between two residues associated to a probability of 0.8 will be randomly added to 80 lists out of 100. For example, if the contact probability is only 0.4, the corresponding contact will be present in 40 lists over 100. One independent MD simulation is then run for each contact list and the conformations generated by all the simulations are collectively analyzed. In the case of the protein Δ + PHS SNase, such analysis allowed us to identify a population of conformers corresponding to a folding intermediate where the C-terminal α-helix is disrupted whereas the N-terminal β-barrel maintains its native structure (Fig. 7).

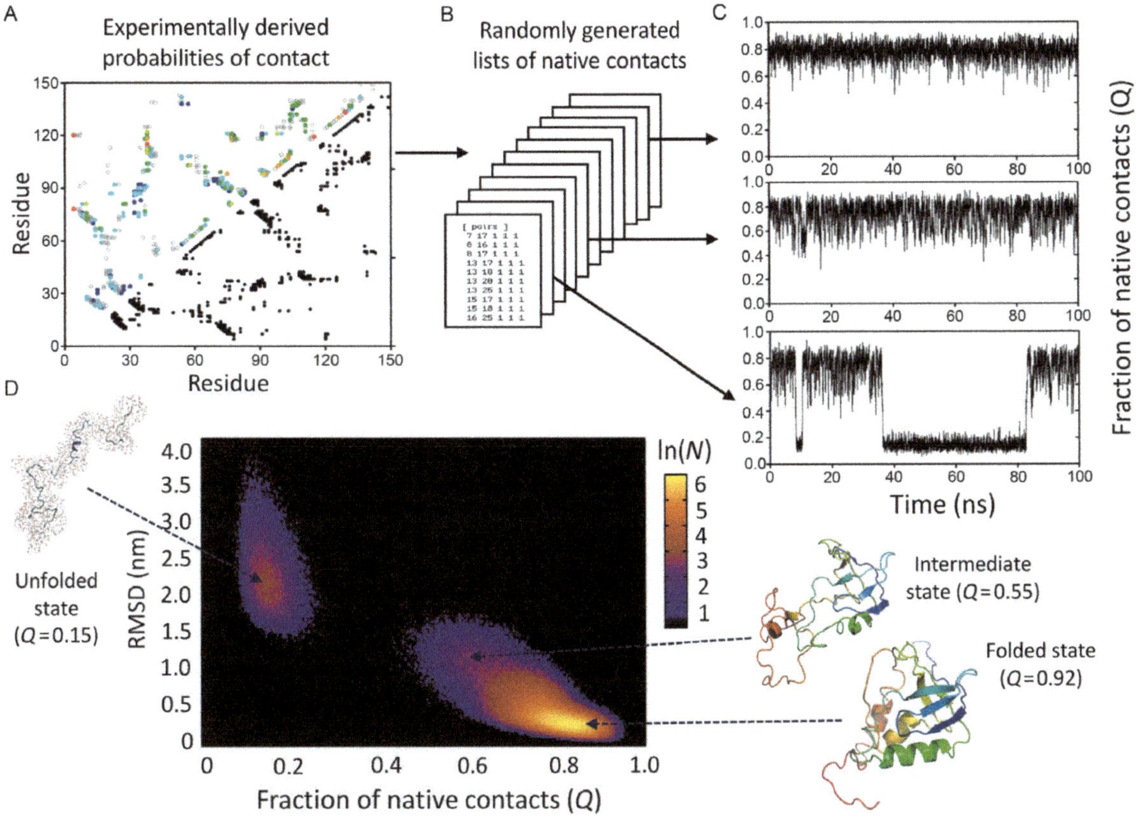

Fig. 7 Reconstruction of pseudo free-energy map from experimentally derived contact maps. (A and B) The contact map calculated at 800 bar from the residue-specific denaturation curves of Δ + PHS SNase in 1.8 *M* GuHCl is used to generate a list of native contacts. (C) These lists of native contacts are then used as input for a series of independent Go-model simulation. (D) The conformations generated by these simulations are collectively analyzed to reconstitute the free-energy landscape of the protein at a given pressure. In the example shown here, we observed the presence of a distinct intermediate state with a partially unfolded C-terminal subdomain (Roche, Caro, et al., 2012).

3.2 High-Pressure H/D Exchange

The early steps of unfolding can also be detected using H/D exchange measurements, a well-established NMR technique developed by Englander and coworkers (Roder, Elove, & Englander, 1988) and designed to identify distinct local stabilities in globular proteins. In such experiments, protonated samples are typically lyophilized and then quickly dissolved in D_2O buffer just prior to measurements. Series of HSQC spectra are then recorded and the decrease of individual cross-peak intensity is monitored over time, reflecting the exchange of amide protons with the deuterated buffer (Fig. 8). The hydrogen/deuterium exchange reaction is classically described as a two-step reaction characterizing a transient structural opening reaction of the protein that exposes an amine proton to the solvent (Englander & Kallenbach, 1983):

$$\text{Close}\,(H) \underset{k_{\text{cl}}}{\overset{k_{\text{op}}}{\longleftrightarrow}} \text{Open}\,(H) \overset{k_{\text{rc}}}{\longleftrightarrow} \text{Open}\,(D)$$

The steady-state exchange rate is given by:

$$k_{\text{ex}} = \frac{k_{\text{op}} k_{\text{rc}}}{k_{\text{op}} + k_{\text{cl}} + k_{\text{rc}}}$$

where the rate of exchange, k_{ex}, is calculated from the rate of the opening and closing reactions, k_{op} and k_{cl}, respectively, and the rate of exchange for a fully exposed amide proton, k_{rc}. Most commonly, exchange in proteins below pH 9 occurs through the so-called EX2-limit where $k_{\text{cl}} \gg k_{\text{rc}}$ and the H/D exchange rate is defined as:

$$k_{\text{ex}} = \frac{k_{\text{op}}}{k_{\text{cl}}} k_{\text{rc}} = k_{\text{op}} k_{\text{rc}}$$

In the EX2 condition, the H/D exchange rate provides the free energy for the underlying structural opening reaction:

$$\Delta G_x = -RT \ln K_{\text{op}} \text{ and PF} = \frac{1}{K_{\text{op}}} = \frac{k_{\text{rc}}}{k_{\text{ex}}}$$

The rate constant k_{ex} can be readily measured for each residue from the decrease in intensity of their corresponding cross-peak in the $[^1\text{H}-^{15}\text{N}]$-HSQC after dissolving the sample in D_2O buffer (Fig. 9). The effective equilibrium opening constant and corresponding protection factor (PF) is then obtained by normalizing k_{ex} by the intrinsic rate of exchange)

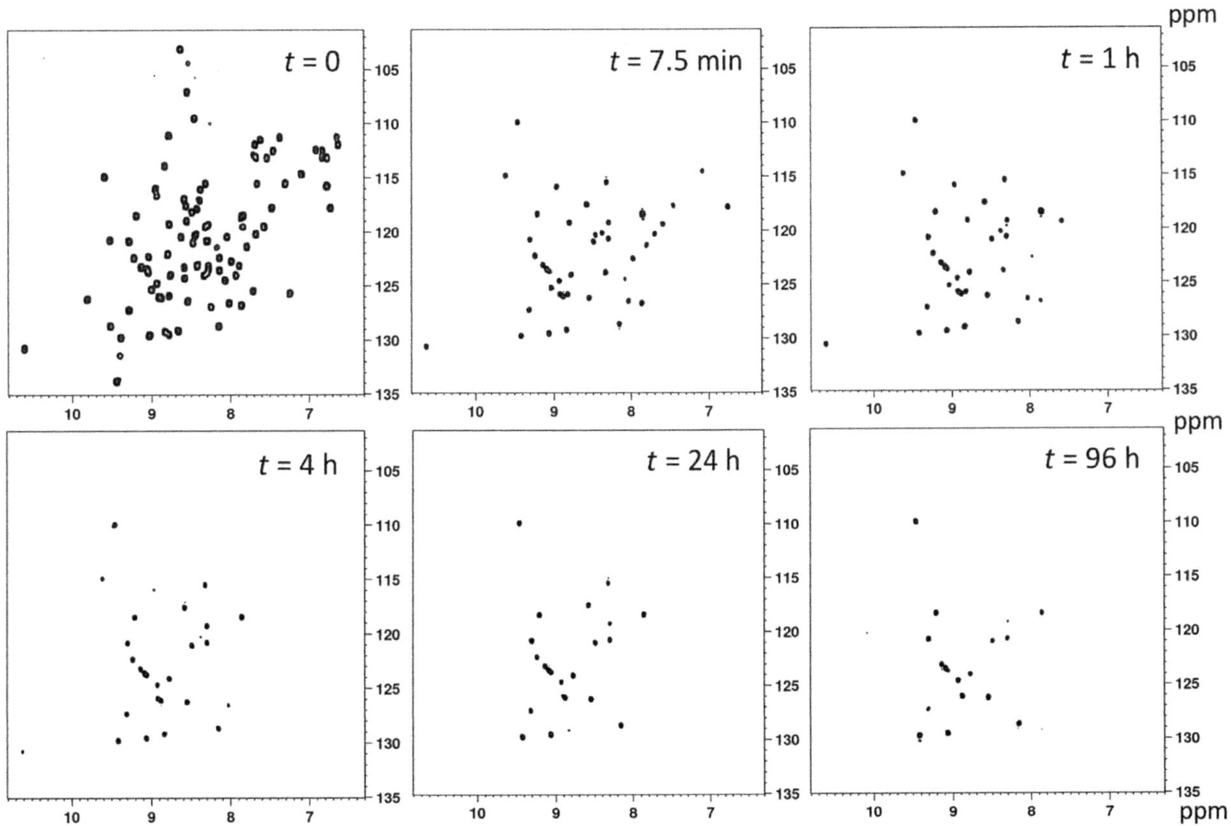

Fig. 8 Proton/deuteron amide exchange measurement on Titin I27 domain at atmospheric pressure. [^{1}H–^{15}N]-HSQC (7.5 min recording time) were recorded at 298 K over a period of 4 days after dissolving a freshly lyophilized sample of Titin I27 (same buffer conditions as stated in Fig. 1) in D_2O. The spectrum at time $t = 0$ represents the reference spectrum recorded in H_2O before freeze-drying.

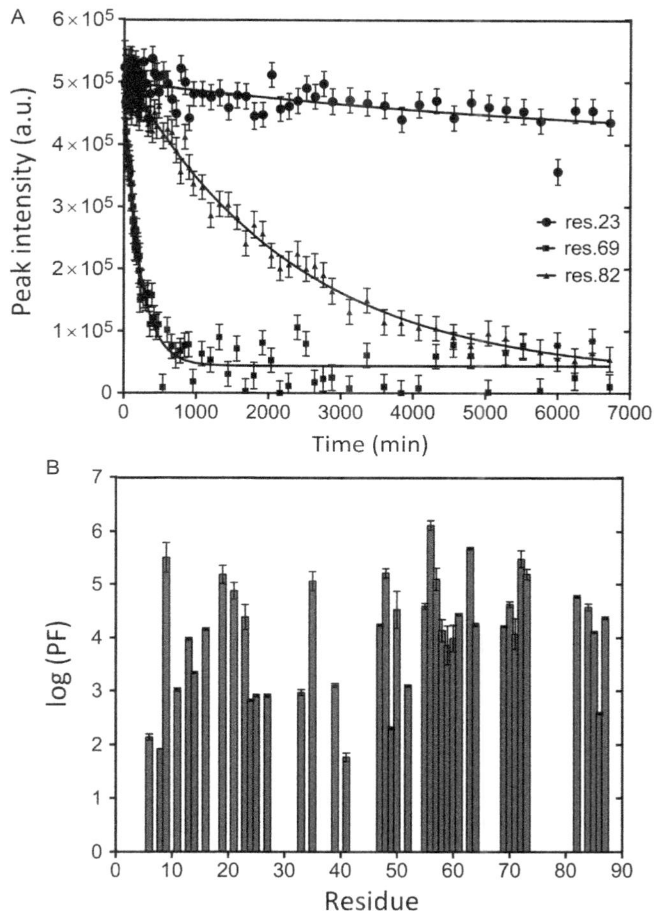

Fig. 9 Measurement of amide protecting factors (PF) from proton/deuteron exchange on Titin I27 domain at atmospheric pressure. (A) Measurement of the exchange rate constant k_{ex} on [^{1}H–^{15}N]-HSQC for three representative residues. (B) Plot of the protecting factors (PF) vs the protein sequence.

(Bai, Sosnick, Mayne, & Englander, 1995). The rates of exchange of fully solvent-exposed amide protons (k_{rc}) are determined for each amino acid on the basis of model peptide studies and depend on the neighboring residues types, the pH of the buffer and temperature at which the experiment is recorded (Bai, Milne, Mayne, & Englander, 1993). Fuentes and Wand combined for the first time the H/D exchange experiments with pressure perturbation to examine the energetics of the apocytochrome b562, introducing for this purpose an additional correction to account the effect of pressure on the k_{rc} (Fuentes & Wand, 1998). With increasing pressure, they observed a systematic increase in the rate of exchange or equivalently, a decrease in the calculated protection factors. Apparent volume changes for exchange (ΔV_x) were estimated from the linear dependence of the free energy of exchange with pressure ($\Delta G_x(p) = \Delta G_x^0 + p\Delta V_x$). Using this method, they were able to identify three regions with distinct subglobal cooperative stabilities and pressure sensitivities (Fuentes & Wand, 1998). High-pressure H/D exchange experiments were also used in the case of the cavity variants of SNase showing considerable variations in the measured ΔV_x for the mutants, compared to the flat ΔV_x profile obtained for the pseudo wild type Δ+PHS (Roche, Dellarole, et al., 2012). In good agreement with equilibrium unfolding data, these H/D experiments revealed the major remodeling of the folding free-energy landscape in response to the introduction of cavity-creating mutations.

4. CHARACTERIZATION OF THE TSE IN THE FOLDING REACTION

As for steady-state measurement, pressure unfolding kinetics can benefit from multidimensional NMR, giving a residue-per-residue description of the TSE. In theory, such a description could be reached using the method proposed by Roche, Caro, et al. (2012) described earlier, based on the analysis of the probability of contacts between residues as determined from the denaturation curves measured at a residue level. Indeed, Fossat et al. (2016) successfully used this approach to fully map the entire folding landscape of the leucine-rich repeat protein, pp32 (Anp32), combining pressure-dependent site-specific ^{1}H–^{15}N HSQC data with coarse-grained molecular dynamics simulations. The results obtained using this equilibrium approach demonstrate that the main barrier to folding of pp32 is quite broad and lies near the unfolded state, with structure apparent only in the C-terminal region. Nevertheless, this approach, based on steady-state experiments, failed to provide information on the rate constants k_u and k_f associated to the folding process.

As with steady-state studies, a local description of the kinetic parameters and of the TSE should be, in principle, attainable through multidimensional NMR spectroscopy. Nevertheless, while multidimensional NMR experiments provide residue-specific structural information, their time resolution is limited (10–40 min for HSQC spectra). In cases for which folding is intrinsically sufficiently slow and folding activation volumes sufficiently large (Royer, 2008; Vidugiris et al., 1995), pressure-jump can be combined with classical 2D NMR experiments, such as ^{1}H–^{15}N HSQC, to study folding kinetics, yielding local folding and unfolding rates for nearly every residue in a protein. This was done, for instance, on SNase and a series of variants with extremely slow relaxation times (up to 12h) (Roche et al., 2013).

For proteins with faster relaxation times (<1 h), conventional ^{1}H–^{15}N HSQC can be replaced by 2D SOFAST-HMQC experiments (Schanda & Brutscher, 2005, Schanda, Forge, & Brutscher, 2007), that can be acquired in few tens of seconds. Such experiments were used to measure the local residue-specific relaxation times for several cavity-containing mutants of SNase. These experiments allow a residue-specific description of the TSE and of its molar volume, and thus internal cavity volume, relative to the folded and unfolded states (Fig. 10). In the case of the I92A SNase cavity-mutant, we also showed that the TSE molar volume was sensitive to the concentration of guanidinium chloride (Roche et al., 2013): increasing the concentration of this chaotropic reagent promoted the selection of higher volume conformers in the TSE, confirming the so-called Hammond effect (Matouschek & Fersht, 1993; Pappenberger, Saudan, Becker, Merbach, & Kiefhaber, 2000). In addition, the local description of the TSE revealed that cavity hydration concerns different regions of the protein, depending on the guanidinium concentration. The obvious limitation of the method presented above to obtain kinetics parameters of the folding reaction is the low time resolution of 2D NMR spectroscopy, even though this drawback has been at least partially circumvented by methodological developments (Frydman, Scherf, & Lupulescu, 2002; Gal, Schanda, Brutscher, & Frydman, 2007): HSQC-type experiments can be now recorded in less than a second, extending significantly the application of real-time spectroscopy.

The above approaches remain unsatisfactory to study subsecond folding kinetics, which is the case for most small globular proteins. In such cases, high-pressure relaxation dispersion experiments can be used to uncover the rate of exchange and thermodynamic properties of a two-state or three-state folding process on a millisecond timescale (Bezsonova, Korzhnev, Prosser, Forman-Kay, & Kay, 2006; Korzhnev et al., 2006; Tugarinov, Libich, Meyer, Roche, & Clore, 2015). Zhang et al. introduced the use of high-pressure ZZ-exchange experiments to investigate the transition states of protein folding, obtaining residue-specific folding rates for the two autonomous N-terminal and C-terminal domains of the ribosomal protein L9 (Zhang et al., 2016). These ZZ-exchange experiments also provide a

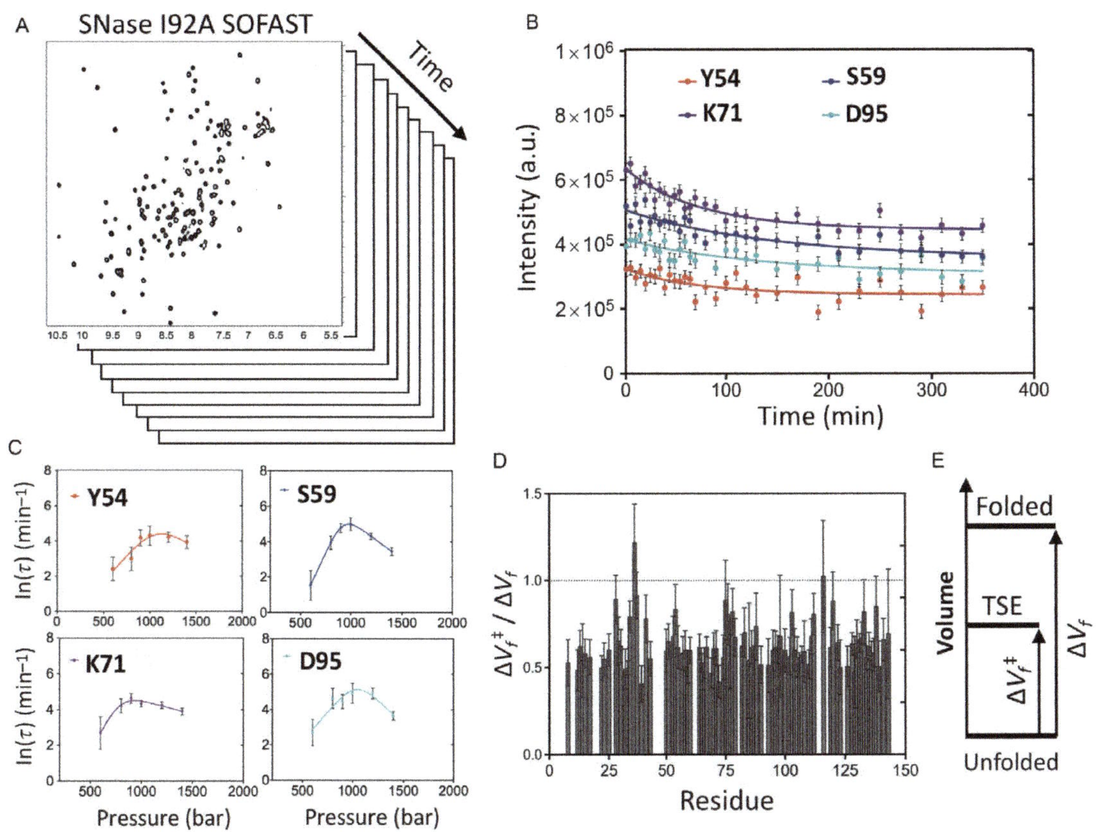

Fig. 10 Residue-specific hydration state of the transition state ensemble (TSE) of SNase I92A variant obtained through *p*-jump kinetic measurements. (A) Time series of SOFAST HMQC were recorded after 200 bar pressure jumps. (B) The intensity profiles of individual amide cross-peaks are then analyzed and fitted to obtain the value of $\tau_{(p)} = 1/(k_u + k_f)$ at a given pressure. (C) The plots of $\ln(\tau)$ vs pressure are finally fitted for each amide bond to extract the value of activation volume for folding ($\Delta V_f^{\ddagger}$). (D) The ratio of ΔV^0 at equilibrium to $\Delta V_f^{\ddagger}$ measured for SNase I92A are plotted as a function of the amino sequence, showing an average ratio of ca. 0.6 (E), indicating that the creation of a large cavity in the

relatively easy method to assign the $^{1}H/^{15}N$ chemical shifts of the pressure-induced denatured states (Zhang et al., 2016). Recently, Charlier et al. introduced a novel method for measuring the rate of exchange and chemical shifts of the folded, intermediate, and unfolded states induced by switching pressure on a millisecond timescale (Charlier et al., 2018). The new pressure-jump NMR apparatus is based on a spectrometer-controlled valve governing a reservoir of hydraulic fluid that can be rapidly and repeatedly opened/closed during the course of a 2D or 3D NMR experiment (Alderson, Charlier, Torchia, Anfinrud, & Bax, 2017; Charlier et al., 2018). This system is reminiscent of the pressure-jump apparatus originally designed by Kremer et al. for introducing pulsed pressure perturbation in 1D and 2D NMR experiments (Kremer et al., 2011).

5. CONCLUSIONS

This chapter describes several NMR experiments that can be used in conjunction with high hydrostatic pressure to populate and characterize partially folded, excited states on protein folding landscapes. The experiments described range from simple one-dimensional proton spectra to more complex multidimensional kinetic schemes. NMR offers a myriad of further possibilities. We have not discussed characterization of pressure effects on protein dynamics by ^{2}H methyl relaxation and ^{15}N amide relaxation (Fu et al., 2012) or ^{13}C–^{1}H HSQC experiments to monitor local disruption of protein cores by the observation of folded state methyl group resonances. Residual dipolar couplings and NOEs can be measured at high pressure (Fu & Wand, 2013; Kitahara, Yokoyama, & Akasaka, 2005; Roche, Dellarole, et al., 2015; Roche, Louis, et al., 2015; Sibille, Dellarole, Royer, & Roumestand, 2014), providing complementary structural constraints on the intermediates populated. Nor have we described here the use of NMR-based DOSY diffusion measurements to characterize the intermolecular association of GMP at high pressure (Gao et al., 2017). Because pressure acts locally on biomolecules, leading to increased population of intermediates and excited states relative to other perturbation approaches, and because the NMR toolbox coupled with pressure offers such a wide range of approaches to probe multiple and complementary physical properties of biomolecules, we expect this approach to yield increasingly deeper insight into biomolecular sequence, structure–function relationships.

REFERENCES

Akasaka, K. (2006). Probing conformational fluctuations of proteins by pressure perturbation. *Chemical Reviews*, *106*, 1814–1835.

Akasaka, K., Kitahara, R., & Kamatari, Y. O. (2013). Exploring the folding energy landscape with pressure. *Archives of Biochemistry and Biophysics*, *531*, 110–115.

Alderson, T. R., Charlier, C., Torchia, D. A., Anfinrud, P., & Bax, A. (2017). Monitoring hydrogen exchange during protein folding by fast pressure jump NMR. *Journal of the American Chemical Society*, *139*, 11306–11309.

Ando, N., Barstow, B., Baase, W. A., Fields, A., Matthews, B. W., & Gruner, S. M. (2008). Structural and thermodynamic characterization of T4 lysozyme mutants and the contribution of internal cavities to pressure denaturation. *Biochemistry*, *47*, 11097–11109.

Bai, Y., Milne, J. S., Mayne, L., & Englander, S. W. (1993). Primary structure effects on peptide group hydrogen exchange. *Proteins*, *17*, 75–86.

Bai, Y., Sosnick, T. R., Mayne, L., & Englander, S. W. (1995). Protein folding intermediates—Native state hydrogen exchange. *Science*, *269*, 192–197.

Bezsonova, I., Korzhnev, D. M., Prosser, R. S., Forman-Kay, J. D., & Kay, L. E. (2006). Hydration and packing along the folding pathway of SH3 domains by pressure-dependent NMR. *Biochemistry*, *45*, 4711–4719.

Charlier, C., Alderson, T. R., Courtney, J. M., Ying, J., Anfinrud, P., & Bax, A. (2018). Study of protein folding under native conditions by rapidly switching the hydrostatic pressure inside an NMR cell. *Proceedings of the National Academy of Sciences of the United States of America*, *115*, E4169–E4178.

Dao, T. P., Majumdar, A., & Barrick, D. (2014). Capping motifs stabilize the leucine-rich repeat protein PP32 and rigidify adjacent repeats. *Protein Science*, *23*, 801–811.

de Oliveira, G. A., & Silva, J. L. (2015). A hypothesis to reconcile the physical and chemical unfolding of proteins. *Proceedings of the National Academy of Sciences of the United States of America*, *112*, E2775–E2784.

Dellarole, M., & Royer, C. A. (2014). High-pressure fluorescence applications. *Methods in Molecular Biology*, *1076*, 53–74.

Englander, S. W., & Kallenbach, N. R. (1983). Hydrogen exchange and structural dynamics of proteins and nucleic acids. *Quarterly Reviews of Biophysics*, *4*, 521–655.

Englander, S. W., Mayne, L., & Krishna, M. M. G. (2007). Protein folding and misfolding: Mechanism and principles. *Quarterly Reviews of Biophysics*, *40*, 287–326.

Fossat, M. J., Dao, T. P., Jenkins, K., Dellarole, M., Yang, Y. S., McCallum, S. A., et al. (2016). High-resolution mapping of a repeat protein folding free energy landscape. *Biophysical Journal*, *111*, 2368–2376.

Frach, R., Kibies, P., Bottcher, S., Pongratz, T., Strohfeldt, S., Kurmann, S., et al. (2016). The chemical shift baseline for high pressure NMR spectra of proteins. *Angewandte Chemie International Edition*, *55*, 8757–8760.

Frydman, L., Scherf, T., & Lupulescu, A. (2002). The acquisition of multidimensional NMR spectra within a single scan. *Proceedings of the National Academy of Sciences of the United States of America*, *99*, 15858–15862.

Fu, Y., Kasinath, V., Moorman, V. R., Nucci, N. V., Hilser, V. J., & Wand, A. J. (2012). Coupled motion in proteins revealed by pressure perturbation. *Journal of the American Chemical Society*, *134*, 8543–8550.

Fu, Y., & Wand, A. J. (2013). Partial alignment and measurement of residual dipolar couplings of proteins under high hydrostatic pressure. *Journal of Biomolecular NMR*, *56*, 353–357.

Fuentes, E. J., & Wand, A. W. (1998). Local stability and dynamics of apocytochrome b562 examined by the dependence of hydrogen exchange on hydrostatic pressure. *Biochemistry*, *37*, 9877–9883.

Gal, M., Schanda, P., Brutscher, B., & Frydman, L. (2007). UltraSOFAST HMQC NMR and repetitive acquisition of 2D protein spectra at Hz rates. *Journal of the American Chemical Society*, *129*, 1372–1377.

Gao, M., Harish, B., Berghaus, M., Seymen, R., Arns, L., McCallum, S. A., et al. (2017). Temperature and pressure limits of guanosine monophosphate self-assemblies. *Scientific Reports*, *7*, 9864.

Ishima, R., Torchia, D. A., & Louis, J. M. (2007). Mutational and structural studies aimed at characterizing the monomer of HIV-1 protease and its precursor. *The Journal of Biological Chemistry*, *282*, 17190–17199.

Kamatari, Y. O., Kitahara, R., Yamada, H., Yokoyama, S., & Akasaka, K. (2004). High-pressure NMR spectroscopy for characterizing folding intermediates and denatured states of proteins. *Methods*, *34*, 133–143.

Kitahara, R., Yokoyama, S., & Akasaka, K. (2005). NMR snapshots of a fluctuating protein structure: Ubiquitin at 30 bar-3 kbar. *Journal of Molecular Biology*, *347*, 277–285.

Korzhnev, D. M., Bezsonova, I., Evanics, F., Taulier, N., Zhou, Z., Bai, Y., et al. (2006). Probing the transition state ensemble of a protein folding reaction by pressure-dependent NMR relaxation dispersion. *Journal of the American Chemical Society*, *128*, 5262–5269.

Kremer, W., Arnold, M., Munte, C. E., Hartl, R., Erlach, M. B., Koehler, J., et al. (2011). Pulsed pressure perturbations, an extra dimension in NMR spectroscopy of proteins. *Journal of the American Chemical Society*, *133*, 13646–13651.

Lindberg, M. O., & Oliveberg, M. (2007). Malleability of protein folding pathways: A simple reason for complex behaviour. *Current Opinion in Structural Biology*, *17*, 21–29.

Louis, J. M., & Roche, J. (2016). Evolution under drug pressure remodels the folding free-energy landscape of mature HIV-1 protease. *Journal of Molecular Biology*, *428*, 2780–2792.

Matouschek, A., & Fersht, A. R. (1993). Application of physical organic chemistry to engineered mutants of proteins: Hammond postulate behavior in the transition state of protein folding. *Proceedings of the National Academy of Sciences of the United States of America*, *90*, 7814–7818.

Myers, J. K., Pace, C. N., & Scholtz, J. M. (1995). Denaturant m values and heat capacity changes: Relation to changes in accessible surface areas of protein unfolding. *Protein Science*, *4*, 2138–2148.

Naganathan, N., & Munoz, V. (2008). Determining denaturation midpoints in multiprobe equilibrium protein folding experiments. *Biochemistry*, *47*, 6752–6761.

Pappenberger, G., Saudan, C., Becker, M., Merbach, A. E., & Kiefhaber, T. (2000). Denaturant-induced movement of the transition state of protein folding revealed by high-pressure stopped-flow measurements. *Proceedings of the National Academy of Sciences of the United States of America*, *97*, 17–22.

Peterson, R. W., & Wand, J. A. (2005). Self-contained high-pressure cell, apparatus, and procedure for the preparation of encapsulated proteins dissolved in low viscosity fluids for nuclear magnetic resonance spectroscopy. *The Review of Scientific Instruments*, *76*, 094101.

Privalov, P. L., & Gill, S. J. (1988). Stability of protein structure and hydrophobic interaction. *Advances in Protein Chemistry*, *39*, 191–234.

Ravindra, R., & Winter, R. (2003). On the temperature-pressure free-energy landscape of proteins. *Chemphyschem*, *4*, 359–365.

Roche, J., Caro, J. A., Norberto, D. R., Barthe, P., Roumestand, C., Schlessman, J. L., et al. (2012). Cavities determine the pressure unfolding of proteins. *Proceedings of the National Academy of Sciences of the United States of America*, *109*, 6945–6950.

Roche, J., Dellarole, M., Caro, J. A., Guca, E., Norberto, D. R., Yang, Y. S., et al. (2012). Remodeling of the folding free-energy landscape of staphylococcal nuclease by cavity-creating mutations. *Biochemistry*, *51*, 9535–9546.

Roche, J., Dellarole, M., Caro, J. A., Norberto, D. R., Garcia, A. E., Garcia-Moreno, B., et al. (2013). Effect of internal cavities on folding rates and routes revealed by real-time pressure-jump NMR spectroscopy. *Journal of the American Chemical Society*, *135*, 14610–14618.

Roche, J., Dellarole, M., Royer, C. A., & Roumestand, C. (2015). Exploring the protein folding pathway with high-pressure NMR: Steady-state and kinetics studies. *Sub-Cellular Biochemistry*, *72*, 261–278.

Roche, J., Louis, J. M., Bax, A., & Best, R. B. (2015). Pressure-induced structural transition of mature HIV-1 protease from a combined NMR/MD simulation approach. *Proteins: Structure, Function, and Bioinformatics*, *83*, 2117–2123.

Roder, H., Elove, G. A., & Englander, S. W. (1988). Structural characterization of folding intermediates in cytochrome-c by H-exchange labeling and proton NMR. *Nature*, *335*, 700–704.

Rouget, J. B., Aksel, T., Roche, J., Saldana, J. L., Garcia, A. E., Barrick, D., et al. (2011). Size and sequence and the volume change of protein folding. *Journal of the American Chemical Society*, *133*, 6020–6027.

Royer, C. A. (2008). The nature of the transition state ensemble and the mechanisms of protein folding: A review. *Archives of Biochemistry and Biophysics*, *469*, 34–45.

Sadqi, M., Fushman, D., & Muñoz, V. (2006). Atom-by-atom analysis of global downhill protein folding. *Nature*, *442*, 317–321.

Schanda, P., & Brutscher, B. (2005). Very fast two-dimensional NMR spectroscopy for real-time investigation of dynamic events in proteins on the time scale of seconds. *Journal of the American Chemical Society*, *127*, 8014–8015.

Schanda, P., Forge, V., & Brutscher, B. (2007). Protein folding and unfolding studied at atomic resolution by fast two-dimensional NMR spectroscopy. *Proceedings of the National Academy of Sciences of the United States of America*, *104*, 11257–11262.

Sibille, N., Dellarole, M., Royer, C., & Roumestand, C. (2014). Measuring residual dipolar couplings at high hydrostatic pressure: Robustness of alignment media to high pressure. *Journal of Biomolecular NMR*, *58*, 9–16.

Tugarinov, V., Libich, D. S., Meyer, V., Roche, J., & Clore, G. M. (2015). The energetics of a three-state protein folding system probed by high-pressure relaxation dispersion NMR spectroscopy. *Angewandte Chemie, International Edition*, *54*, 11157–11161.

Vidugiris, G. J. A., Markley, J. L., & Royer, C. A. (1995). Evidence for a molten globule-like transition state in protein folding from determination of activation volumes. *Biochemistry*, *34*, 4909–4912.

Zhang, Y., Berghaus, M., Klein, S., Jenkins, K., Zhang, S., McCallum, S. A., et al. (2018). High-pressure NMR and SAXS reveals how capping modulates folding cooperativity of the pp32 leucine-rich repeat protein. *Journal of Molecular Biology*, *430*, 1336–1349.

Zhang, Z., Kitazawa, S., Peran, I., Stenzoski, N., McCallum, S. A., Raleigh, D. P., et al. (2016). High pressure ZZ-exchange NMR reveals key features of protein folding transition states. *Journal of the American Chemical Society*, *138*, 15260–15266.

CHAPTER TWELVE

CS-ROSETTA

Santrupti Nerli*,†, Nikolaos G. Sgourakis*,1

*Department of Chemistry and Biochemistry, University of California Santa Cruz, Santa Cruz, CA, United States
†Department of Computer Science, University of California Santa Cruz, Santa Cruz, CA, United States
[1]Corresponding author: e-mail address: nsgourak@ucsc.edu

Contents

Abstract

Chemical Shift-Rosetta (CS-Rosetta) is an automated method that employs NMR chemical shifts to model protein structures de novo. In this chapter, we introduce the terminology and central concepts of CS-Rosetta. We describe the architecture

Methods in Enzymology, Volume 614
ISSN 0076-6879
https://doi.org/10.1016/bs.mie.2018.07.005

and functionality of automatic NOESY assignment (AutoNOE) and structure determination protocols (Abrelax and RASREC) within the CS-Rosetta framework. We further demonstrate how CS-Rosetta can discriminate near-native structures against a large conformational search space using restraints obtained from NMR data, and/or sequence and structure homology. We highlight how CS-Rosetta can be combined with alternative automated approaches to (i) model oligomeric systems and (ii) create NMR-based structure determination pipeline. To show its practical applicability, we emphasize on the computational requirements and performance of CS-Rosetta for protein targets of varying molecular weight and complexity. Finally, we discuss the current Python interface, which enables easy execution of protocols for rapid and accurate high-resolution structure determination.

ABBREVIATIONS

AA amino acids
BMRB biological magnetic resonance bank
CCD cyclic coordinate descent
CEN centroid
CS chemical shift
EC evolutionary coupling
HM homology modeling
IO input/output
JD job distributor
KDA kilo dalton
MPI message passing interface
NEF NMR exchange format
NESG Northeast Structural Genomics
NMR nuclear magnetic resonance
NOE nuclear overhauser enhancement
PCS pseudocontact shift
PDB protein data bank
PRE paramagnetic relaxation enhancement
RASREC resolution-adapted structural recombination
RCI random coil index
RDC residual dipolar coupling
SPARTA shifts predicted from analogy in residue type and torsion angle
TALOS torsion angle likelihood obtained from shift and sequence similarity

1. INTRODUCTION

Determination of high-resolution protein structures from amino acid sequence alone remains as one of the most challenging problems in computational biology (Bender et al., 2016). This is attributed to the time and computational power required for exploration of large and continuous 3D conformational search space. Toward addressing this bottleneck, software suites such as Rosetta (Leaver-Fay et al., 2011) employ efficient heuristic algorithms alongside physically realistic energy functions to minimize the search space. While these energy functions are sufficient to drive the structure calculations toward a native fold for proteins of smaller size ($\leq$15 kDa), restraints provided by experimental data are required for modeling larger and more complex molecules. A method that stems from this idea is Chemical Shift-Rosetta (or CS-Rosetta) (Shen et al., 2008). CS-Rosetta employs NMR (nuclear magnetic resonance) chemical shifts, which are highly informative probes that report on secondary and tertiary structures, and dynamics of proteins (Berjanskii & Wishart, 2017), to aid the search for near-native structures. Alongside chemical shifts, CS-Rosetta makes use of structural restraints derived from alternative NMR experiments and/or evolutionarily related proteins to model challenging targets with high accuracy.

CS-Rosetta contains two algorithms, Abrelax (ab initio relax) (Shen et al., 2008) and RASREC (resolution-adapted structural recombination) (Lange & Baker, 2012), that each utilize torsion angle, and short- and long-range distance restraints derived from NMR chemical shifts, residual dipolar couplings (RDCs) (Raman et al., 2012) and nuclear Overhauser enhancements (NOEs) to model protein structures up to 40 kDa (Lange et al., 2012). Compared to Abrelax, RASREC makes use of additional optimization strategies together with the support for advanced parallelization scheme toward efficient and accurate modeling of larger protein structures. CS-Rosetta also enables rapid and automated assignment of NOESY data through the AutoNOE (Automatic NOESY assignment) (Lange, 2014) module. A brief overview of CS-Rosetta has been previously described (see Guerry & Herrmann, 2011; Nerli, McShan, & Sgourakis, 2018). Here, we provide a detailed description of the algorithms and architecture on the current implementation of CS-Rosetta. Throughout the text, the `typewriter` font is used to represent keywords, which denote external method parameters, software executables, C++ programming language classes, and function names within CS-Rosetta.

2. TERMINOLOGIES

In this section, we highlight the terminologies and concepts that are fundamental to the Rosetta framework with a focus on CS-Rosetta. Whereas the Rosetta suite supports a wide range of biomolecular systems, the methodology, algorithms, and energy functions of CS-Rosetta are currently only applicable to proteins.

2.1 Energy

The main driving force of Rosetta is its energy function (Alford et al., 2017; Leaver-Fay et al., 2013, 2011; O'Meara et al., 2015). The energy function quantifies the static conformational state of a protein, which can be utilized to distinguish native-like structures from alternative folded and unfolded proteins (Alford et al., 2017). This function is a linear combination of terms that capture physically realistic structural features. Specifically, these terms describe a set of potentials that report on inter/intramolecular interactions, solvation, hydrogen bonding, electrostatics, side chain rotamer orientation, and dihedral preferences. During the structure modeling process, Rosetta can modify the contribution of each energy term using predefined (score3.wts, talaris2013.wts, ref2015.wts, and others) or customizable weights. This procedure may help steer the calculations toward a global minimum in the energy landscape for proteins of interest. We use the terms, energy function, energy, score, and score function synonymously in this chapter.

2.2 Pose

In Rosetta, a biomolecular system is represented using a C++ container class known as a Pose (Leaver-Fay et al., 2011). Pose holds information on the chemical composition (protein amino acid sequence), degrees of freedom (variable factors of a conformation), kinematics, and energy of the molecular system. For protocols such as ab initio (where structures are modeled solely from amino acid sequence information) (Bradley, Misura, & Baker, 2005), degrees of freedom of a Pose can be derived from an extended polypeptide chain. In alternative protocols, degrees of freedom are derived from a reference structure input manually by the user. Moves (see Section 2.5) are applied to a Pose during the process of sampling diverse conformations while searching for the native (or global optimum with respect to the score function). If the conformation of a Pose is modified using any mover, then its score function is updated accordingly.

2.3 Fold Tree

Each amino acid within a protein chain in Rosetta is characterized by two degrees of freedom known as internal coordinates (given by the torsion angles, ϕ and ψ) (Bradley & Baker, 2006; Leaver-Fay et al., 2011).

A small perturbation in the local conformation of a protein chain, induced by a mover (see Section 2.5), leads to the propagation of changes globally, called the lever arm effect. To eliminate this effect, a protein chain is denoted as a directed, acyclic graph called the fold tree, which is attached to the Pose. In a fold tree, the nodes indicate amino acids of a protein and the edges represent peptide bonds, which imply that any residue i is connected to its neighbors $i - 1$ and $i + 1$ by edges. Alongside peptide bonds, through-space bonds (or long-range connections) can be added as edges in this tree without the violation of acyclicity property of the graph. To preserve this property, an insignificant peptide edge is deleted, and this introduction of gap in the fold tree is called a chain-break or a jump. These edges govern the propagation of conformational changes during a move. The fold tree with discontinuous peptide edges is referred to as broken-chain fold tree.

2.4 Centroid and Full-Atom Representation

Protocols that require extensive sampling such as ab initio (Bradley et al., 2005) switch to coarse-grained representation, where each residue in a protein chain contains five backbone atoms (N, H, C^{α}, C, O) and one side chain centroid atom (CEN) characterized by its amino acid type. This simplified representation is also referred to as low-resolution mode. The use of centroid mode during early stages of structure determination boosts the performance of sampling methods by reducing the degrees of freedom introduced through side chain dihedral angles (Shen et al., 2008).

On the other hand, a full-atom representation (also referred to as high-resolution mode) includes backbone atoms in addition to all side chain atoms. The full-atom representation mode is typically used during the refinement stages of structure determination protocols. In this chapter, we interchangeably use terms for low-resolution with centroid and high-resolution with full-atom.

2.5 Mover

A mover is an entity that can modify or analyze any given Pose (Leaver-Fay et al., 2011). CS-Rosetta's structure modeling protocols are built using several movers, which are referenced throughout this manuscript.

2.6 Topology Broker

Topology broker is a framework that enables construction of complicated simulation systems (Bender et al., 2016; Porter, Weitzner, & Lange, 2015).

Broker guides the resampling protocols used for structure determination by providing them information about (i) movers to apply, (ii) chain-breaks to add, (iii) representation modes to switch to, and (iv) construction of refined fold tree. Any knowledge about the protein target including structural restraints (from NMR data or any other source) or rigid regions in homologous structures from databases such as PDB (or protein data bank) (Berman et al., 2000) is retained by the broker in the form of claimers. Specifically, claimers define the accepted and unaccepted degrees of freedom for a given Pose. For instance, if rigid regions are defined for a Pose, then the broker will not accept any changes in those regions. Therefore, broker allows rapid protocol development, since the methods do not have to keep track of constraints that must to be satisfied by the system explicitly when perturbing the conformation.

3. PRELIMINARY REQUIREMENTS

CS-Rosetta makes use of Rosetta's basic de novo structure prediction algorithm termed fragment assembly (Simons, Kooperberg, Huang, & Baker, 1997). In a general fragment assembly approach, a set of small structural segments are selected from a library constructed using high-resolution X-ray structures and inserted into the protein backbone being modeled. Therefore, fragments of high quality are essential to obtain accurate structures (Nerli et al., 2018). In CS-Rosetta, NMR chemical shifts are used in the fragment selection process, allowing for higher accuracy in sampling near-native segments (Nerli et al., 2018). To select such fragments, Robetta server (Kim, Chivian, & Baker, 2004) or CS-Rosetta's fragment picker (Gront, Kulp, Vernon, Strauss, & Baker, 2011) (see Section 9) can be used. Before using CS-Rosetta's fragment picker, the following steps must be carried out.

3.1 Determination of Structurally Rigid Regions

Due to the observation that chemical shifts capture local structural features (Mielke & Krishnan, 2009), they are used to define rigid and well-conformed regions (typically composed of secondary structural elements such as α-helices and β-sheets) that aid the fragment selection process (Shen et al., 2008). In contrast, disordered or flexible regions of a protein (for example, at the N- and C-termini) exhibit higher levels of structural diversity and therefore decrease the efficiency of sampling protocols. To overcome this bottleneck, CS-Rosetta makes use of TALOS-N (torsion angle likelihood obtained from shift and sequence similarity) (Shen & Bax, 2013), which takes as input the amino acid sequence and NMR chemical shifts to characterize the regions of rigidity and flexibility using random coil index (RCI). Using the predicted RCI values, the protein sequence is trimmed along the termini to eliminate any flexible end regions.

3.2 Fragment Picking

CS-Rosetta uses a fragment picker module built within Rosetta to construct high-quality sequence-specific fragment library using chemical shifts and a protein structure database called "vall" (shipped with both Rosetta software suite and CS-Rosetta toolbox) (Gront et al., 2011). The vall database consists of multiple sets (referred as chunks) of protein chains from X-ray structures without any chain-breaks. Each chunk is split into sets of fragments of required size with overlapping amino acid sequences in a "sliding window" fashion to cover the entire protein chain. The selection of polypeptide segments of the target protein from vall into the fragment library is governed heavily by the secondary structural elements predicted using TALOS-N (Shen & Bax, 2013). The plausible candidate fragments selected from vall are scored based on target sequence profile, chemical shift compatibility, secondary structure similarity, the presence of residues in the allowed regions of Ramachandran map, and RMSD of ϕ and ψ dihedral angles. In addition, the fragment picker allows flexibility to choose structures from which the segments are derived. For instance, homologous structures can be excluded during fragment library construction, which is typically desired while running benchmark calculations for method development. For structure calculation using CS-Rosetta, 200 fragments of 3 and 9 amino acids in length are picked for each residue position in the target sequence. The sampled fragments in these libraries define the degrees of freedom (or backbone torsion angles) a polypeptide chain can adopt during the structure calculation process. Rosetta applies these degrees of freedom to a target chain being modeled using the fragment insertion mover. The larger fragments (nine AA) are used during early exploration phases of structure calculation, whereas small fragments (three AA) are utilized predominantly in the later refinement stages (see Section 4).

4. METHODS

In this section, we describe Abrelax, RASREC, and AutoNOE methods. In addition, we present a list of files used by each method that contain all the required parameters, which can be tuned, whereas a detailed description of the parameters themselves is available at the CS-Rosetta website (see Section 9).

4.1 Abrelax

Ab initio relax (or Abrelax) in CS-Rosetta is a standard protocol that uses as input the amino acid sequence and NMR chemical-shift-guided segments and assembles them using a fragment sampler to model near-native protein structures (Fig. 1, violet) (Shen et al., 2008). This protocol is invoked through a mover called the "`AbrelaxMover`," which consists of an initialization step and an apply step. The initialization step creates instances of protocols used for idealization, fragment sampling, loop closure, and structure refinement. The apply step is utilized to invoke each of the protocols using the instances created during the initialization step. The `AbrelaxMover` is initiated via BrokerMain, a file that contains the main function used to start the protocol.

Fragment sampling is carried out in the low-resolution mode in four distinct stages (I, II, IIIa, IIIb, and IV) (Fig. 1, violet). During each stage, fragment insertion moves are performed on the Pose. This process involves selecting a candidate segment from the fragment library at a given position and adjusting the torsion angles of the target protein chain to match the angles from the segment. The resulting Pose is evaluated using stage-specific score terms and weights (Fig. 1). The low-scoring candidates at each stage are retained for further sampling. During stage I, 2000 fragment insertion moves with segments of AA length 9 are performed using Monte Carlo method with a score term to penalize any steric clashes introduced during the moves. In stage II, the same number of moves are performed with score terms, including the hydrophobic core, pairwise intramolecular interactions, and secondary structure packing to capture the communication between residue-pairs. In stage III, 10 iterations of 2000 fragment insertions are used to promote sampling of both local (IIIa) and nonlocal (IIIb) β-strand pairings with the help of bias used for scoring. In the last stage of this protocol, 4000 cycles of small fragment insertion moves are carried out, and the resulting models are scored using terms contributed heavily by radius of gyration to form closely packed structures. These low-resolution structures further undergo postfragment sampling stages that include (i) loop closure, (ii) conversion to full-atom representation, (iii) energy refinement using relax (full-atom refinement of structures in the Rosetta force field) (Tyka et al., 2011). For the assessment of the resulting high-resolution models, experimental chemical shifts are compared with predicted chemical shifts from the models using SPARTA (shifts predicted from analogy in residue type and torsion angle) (Shen & Bax, 2007). The deviation between two chemical shifts is used to adjust the energy function (Shen et al., 2008).

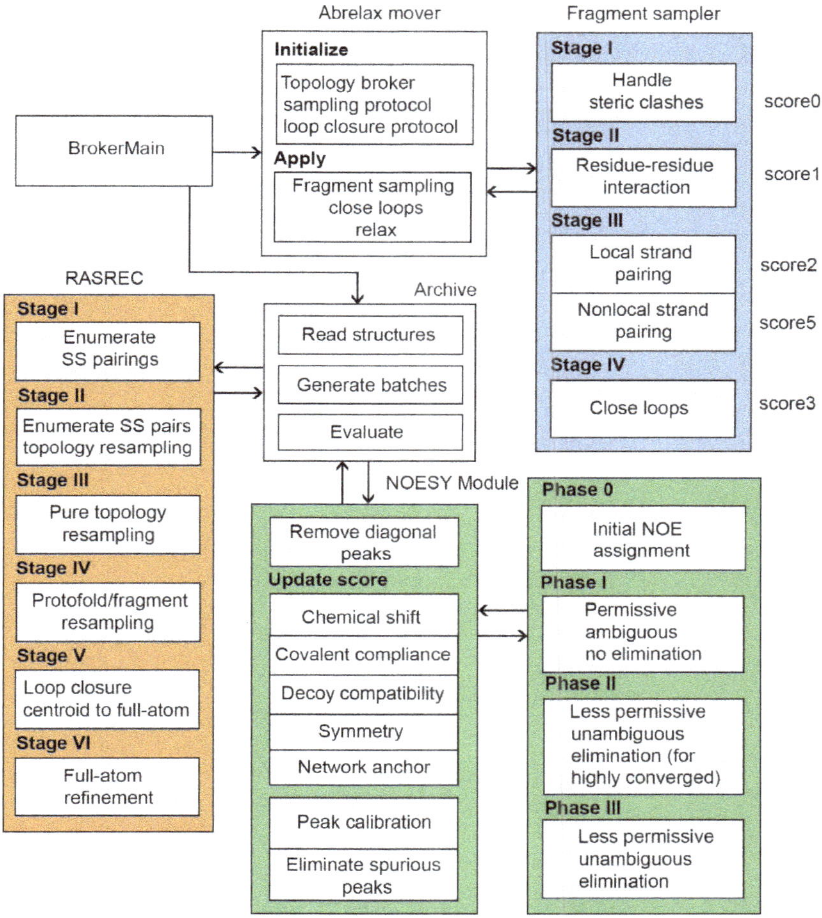

Fig. 1 Overview of CS-Rosetta. BrokerMain marks the start of any CS-Rosetta protocol. Abrelax is triggered from the AbrelaxMover via BrokerMain and contains two representative steps, (i) a setup step that is used to initialize topology broker, fragment sampler, and loop closure protocols, and (ii) an apply step, which calls the ab initio fragment sampler (*violet*) to obtain low-resolution models that undergo loop closure and high-resolution structure refinement using relax. A fragment sampler (and assembly) runs through four stages (*violet*): (i) stage I: large fragment insertion moves are performed and any steric clashes are prevented using score weights provided by score0, (ii) stage II: fragment insertion moves are combined with score weights (score1) to account for residue-pair interactions, (iii) stage III: fragment insertion moves are carried out to promote local (stage IIIa) and nonlocal (stage IIIb) β-strand pairings with the help of weights defined in score2 and score5, and (iv) stage IV: small fragment insertion moves are performed primarily to close loops. In addition, score weight for radius of gyration is increased (in score3). RASREC is an iterative method that performs structure calculation across six resampling stages (*tan*). In stage I, β-strand pairs are exhaustively

(Continued)

The Abrelax protocol provides tuning parameters to increment the iteration cycles, modify the weights of score terms that are applied during each stage, and skip any phase if deemed necessary. Setting up structure calculations using Abrelax is made easy with the help of automated setup scripts (see Section 9). These scripts are used to initialize method parameters to standard values (based on a set of benchmark calculations carried out previously; Raman et al., 2012; Shen et al., 2008). These parameters are defined in files known as flag files. Here, we provide a brief description of flag files used during Abrelax calculations.

flags_abrelax typically consists of (i) option to include H^{α} atoms during centroid mode (`-chemical::patch_selectors CENTROID_HA`). This option is incorporated if input NOE or RDC restraints contain H^{α} atoms, (ii) path to the topology broker setup file (`-broker::setup`), (iii) path to chemical shift file for the purpose of evaluating output structures (`-evaluation::chemical_shifts`) if specified by the user, and (iv) a parameter to describe the number of structures to generate (`-out::nstruct`).

flags_denovo is used to define (i) weight correction terms for scores contributed by radius of gyration, and residues present in α-helices and loops, (ii) loop closure options such as `-overwrite_filter_scorefxn` to force

Fig. 1—Cont'd enumerated. During stage II, β-strand pairs are enumerated again and β-topologies are resampled. Stage III is used to carry out only β-topology resampling. During stage IV, folds are intensified (also known as protofold resampling) and fragments are resampled from low-resolution models. Loops are closed and the resulting structures are converted from centroid to full atom in stage V. In the last stage (stage VI), the final structural models are refined in the Rosetta force field. Archive in Rosetta is used for the purpose of communication between parallelizable unit (MPI-based job distributor) and RASREC algorithm. Based on the status of the job distributor, archive can read decoys, setup batches of runs or check if the calculation is complete. AutoNOE iteratively assigns NOEs and performs structure calculation using RASREC (*green*). It interacts with RASREC across all the stages of resampling. AutoNOE updates NOE assignment parameters at three distinct phases (*green, right*). Phase I is designed to be more permissive meaning; more ambiguous NOE restraints are assigned, and no restraints are eliminated. Phase II is less permissive, where unambiguous NOE restraints are favored and elimination of violated restraints is allowed in the highly converged regions. Phase III is the most rigid, where unambiguous NOE restraints are generated and violated restraints are eliminated. The assigned NOE list generated at the start of each resampling phase is filtered by eliminating all the diagonal cross-peaks. The NOEs are further scored and refined using chemical shift compatibility, covalent compliance, agreement with decoys (obtained from RASREC), symmetric assignments, and network anchoring. The intensities of cross-peaks of assigned NOEs are calibrated to obtain upper distance bounds for restraints followed by the removal of spurious peaks.

specified weights rather than predefined weights that are applied to filter models when closing loops, (iii) fold tree chain-break options (for example, `-jumps::ramp_chainbreaks`, which is used to tune the bias applied to chain-break energy terms across multiple stages of simulation), (iv) logging information in the event of errors, and (v) fully qualified paths to input fragment and amino acid sequence files.

flags_nmr_patches file includes information about patch files, which describe weights for specific energy terms to be used during different stages of ab initio (`-abinitio::stageXX_patch`, where XX represents stages of the Abrelax protocol), full-atom relax (`-score::patch`), and loop closure (for instance, `-patch_filter_scorefxn` used to filter models when closing loops).

nmr_relax_patch defines `atom_pair_constraint` and `rdc` weight terms used to bias restraint-based scores (see Section 6) provided typically by NOE and RDC measurements that are applied during structure modeling and refinement.

nmr_patch is similar to and contains same set of parameters as the *nmr_relax_patch* file. The primary difference is that they are applied during different stages of structure calculation. For ab initio stages and loop closure, *nmr_patch* is used. For full-atom relax, *nmr_relax_patch* is used. Their usage can be found within *flags_nmr_patches*.

setup_init.tpb is a topology broker file, which contains definition of claimers, such as `ConstraintClaimer`, that is applied to the Pose during centroid and full-atom stages. The `ConstraintClaimer` defines degrees of freedom through structural restraints. The claimer definitions specify fully qualified paths to the constraint files and weights that are applied during scoring.

After setting up the protocol for Abrelax calculations, a directory consisting of input, run, and test folders is created. The input folder contains chemical shift, constraint, and topology broker files. The test and run folders carry all the flag files (listed in *flag_list.txt*) that will be used during calculation. While the test folder is similar to the run folder, it is used mainly to check if the calculations proceed without any interruption or errors. Abrelax can be run using the executable, `minirosetta.mpi.<ext>` (where `<ext>` is platform and compiler dependent) in parallel on a multiprocessor system using the MPI (message passing interface) framework. Here, each CPU performs independent exploration and refinement steps. The MPI framework (i) enables seamless communication between processors independent of platforms used, (ii) manages real-time distribution of workload across computer clusters, and (iii) ensures safe resource management. During the execution of the protocol, calculated structural models are appended to "*decoys.out*," a compact representation used to optimize storage space and is also referred to as silent file in Rosetta. As the protocol progresses, it writes information messages, warnings, and errors in log files, which can be used to troubleshoot in the event of failure or premature termination.

4.2 RASREC

RASREC is an iterative Abrelax protocol, which exploits the features of intermediate structures to build self-consistent core of a protein that resembles the native fold (Lange & Baker, 2012). To accomplish this task, RASREC employs several optimization strategies and embeds them into distinct phases consisting of exploration, feature reuse (or resampling), and structure refinement (Blum, Jordan, & Baker, 2010; Bradley & Baker, 2006; Brunette & Brock, 2005, 2008; Qian et al., 2007). A subset of the optimization strategies is dedicated toward sampling nonlocal β-sheets due to the complexity involved in modeling them (Bradley & Baker, 2006).

RASREC is invoked through the Archive framework (see Section 5) via BrokerMain and runs across six stages (Fig. 1, tan). During early stages of the protocol, secondary structures are used to determine β-strand pairings, which allow sampling of diverse β-sheet topologies (Bradley & Baker, 2006). The strand pairings are represented using a fold tree, where the pair edges are added into the tree through fragment insertion moves. For this purpose, segments of β-strands are obtained from a database of β-sheets from high-resolution X-ray structures. Later, the fragments (both β-strand fragments and those from sequence-specific fragment library) are assembled using the four stages of ab initio. The low-scoring models are recycled to produce new libraries of structural fragments for use in the subsequent stages of assembly (Blum et al., 2010). The folds of the low-resolution models in the later stages are guided by the trajectories of successful candidates from early stages (Brunette & Brock, 2005). Following these steps, loops are closed in the low-resolution models, which are converted to full-atom representation mode and refined in the Rosetta force field to produce physically realistic structures.

RASREC is a highly parallelized protocol that makes use of the MPI framework. Specifically, it utilizes a workload manager to distribute batches of structure calculation jobs across computer clusters during multiple stages of resampling and refinement. The resampling stages are divided into four centroid (I–IV) and two full-atom (V and VI) modes (Fig. 1, tan). In the first two stages, secondary structural elements obtained from NMR chemical shift-guided fragments of length three (see Section 3) are used to enumerate all possible β-strand pairings. All the consecutive residues assumed to form β-sheets referred to as β-residues are considered in pairs to account for all possible peptide plane orientations and pleating. The use of nonlocal β-residues in these stages introduces chain-breaks in the fold tree (Bradley & Baker, 2006), which allows sampling of diverse and nonlocal β-sheet topologies. From these stages, a set of batches and two centroid pool directories are created. Each batch directory consists of intermediate structural models computed using ab initio (stages I–IV) together with enumerated strand–strand pairing data and custom parameter flag files used for calculation. The stage-specific centroid pool directories contain low-scoring decoys from respective phases and a STATUS file presenting structure acceptance history into the pool. Stages II and III resample β-topologies from a set of decoys found in the centroid pool of previous stages. The β-strand pairings are scored using features such as contact order (residue–residue interactions that are close in space), sequence separation (between residues located within strands), and frequency of occurrence of the strand pairs in decoys. Recurring β-strand pairings with high contact order are favored. The score contributions of selected strand pairs are summed to give the total score for the β-topology. In stage IV, the protein fold is guided by a restricted conformational search. Specifically, the annealing trajectories of successful candidates are used to restart the structure calculation from stage II (Brunette & Brock, 2005). Besides sampling complex nonlocal β-folds, other structural features are resampled during stages III to VI from the segments of low-resolution conformations (from early stages) determined using fragment assembly. The centroid pool directories (from stages III and IV) are populated with structural models for the later stages. In stages V and VI, incomplete low-resolution models from previous stages (due to the use of broken-chain fold tree to sample nonlocal β-strands) are closed. The chain-break (or loop) closure is carried out using two strategies, (i) idealization and (ii) loop rebuilding.

Loop closure by idealization attempts to bridge the broken chain by selecting physically realistic values of bond lengths and angles for each atom pair in that region. However, the introduction of such values may result in radical changes in the fold tree. To avoid these changes, high weight restraints are added to govern the lever arm effects and to ensure that the perturbations are < 0.1 Å. Nonetheless, it is not always guaranteed that this procedure will not disturb the identified fold, in which case RASREC opts for a second strategy to close loops. This strategy involves rebuilding the residues of the loop by performing fragment assembly followed by loop closure using cyclic coordinate descent (CCD) (Canutescu & Dunbrack, 2003; Wang, Bradley, & Baker, 2007). The score terms applied in the course of this stage rely on the number of residues to be rebuilt. Whereas chain-break energy terms are used while building smaller subunits (<10 residues), additional terms contributed by residue-pair interaction are applied for larger loops.

The RASREC method setup is similar to that of the Abrelax protocol, where a directory consisting of flag files is created. In addition to the parameters used by Abrelax, RASREC makes use of flag files listed later.

flags_iterative is used to define control flags that are applied during different stages of resampling. For instance, `-iterative:fullatom` enables the method to run the full-atom structure modeling and refinement stages (V, VI), `-iterative:accept_ratio` describes the cutoff value to be used for switching to next stage. Here, if the number of structures chosen as plausible candidates falls below the cutoff, then the stage is assumed to be complete.

super_quick_relax_patch.txt is similar to *nmr_patch* or *nmr_relax_patch* files that contain weight definitions for scoring restraints derived typically from NOE or RDC measurements. The primary use of this file is to provide constraint score weights for quick side chain repacking within individual batch files.

flags_fullatom defines the parameters that can be used during the process of (i) closing loops such as `-loops:idealize_before_loop_close` (idealizes the structure before attempting to close loops) and (ii) structure refinement using relax.

flags_cs_rescore contains path to chemical shift file used for model evaluation together with weights for chemical shift score term applied during both centroid and full-atom modes. Bias terms (such as `-iterative:centroid_before_quickrelax_weight` and `-iterative:fullatom_after_quickrelax_weight`) are also included for the purpose of filtering models before adding them to the structure pool for subsequent stages.

nmr_pool_patch has weight terms applied to restraint-based scores and chain-breaks. These scores are used to filter models before adding them into the structure pool.

flags_rasrec is the flag file used in RASREC instead of *flags_abrelax* (used by Abrelax). This file includes reference to *flags_cs_rescore* and topology broker (*setup_init.tpb*) files.

Unlike Abrelax, RASREC creates a set of target-dependent batch directories alongside stage-specific directories at the end of each of the six phases. The structures from the centroid pool stages are added into *centroid_pool_stageXX* directories, where XX = {1, 2, 3, 4}. Similarly, the full-atom decoys from stages V and VI are added to *fullatom_pool_stageYY* directories, where YY = {7, 8} (even though they are stages V and VI, they are called 7 and 8). The final structural models are available in the form of a silent file (*decoys.out*) in *fullatom_pool* folder within the run directory.

To highlight the difference between Abrelax and RASREC, we calculated the structure of 20 kDa Northeast Structural Genomics (NESG) consortium target, SgR145 (Wunderlich et al., 2004). Both the structure calculations were setup using NMR chemical shift-guided fragments and previously measured NOE restraints (Lange et al., 2012) (BMRB ID 16806). We sampled a total of 1000 structures using Abrelax, whereas RASREC runs produced nine batches consisting of 100 decoys each before convergence. We found that the lowest energy models generated using Abrelax and RASREC exhibit C^{α} RMSDs of 2.4 and 1.6 Å from the native structure (PDB ID 2N47; Tang et al., 2015), respectively (Fig. 2). Moreover, the accuracy of side chain dihedral placement (χ^1) of buried residues calculated using 10 $Å^2$ solvent accessible surface area cutoff with respect to the reference (PDB ID 2N47) is approximately 65% and 69% in the lowest energy Abrelax and RASREC models, respectively (Fig. 2A and B). The convergence of 10 lowest energy structures generated using RASREC is much higher than the Abrelax models (Abrelax: 43% and RASREC: 76%) (Fig. 2C). Most importantly, the structures sampled using RASREC are far superior compared to Abrelax models as indicated by the energy funnel, suggesting that the optimization strategies implemented in RASREC exhibit great promise in finding near-native structures rapidly (Fig. 2D).

4.3 AutoNOE

Alongside de novo structure prediction, CS-Rosetta offers an algorithm, AutoNOE, for automated assignment of 2D, 3D, or 4D NOESY cross-peaks (Lange, 2014). The NOESY cross-peaks result from coupling of two proton nuclei with different chemical shift values during magnetization transfer (Keeler, 2010). The process of labeling these cross-peaks in the NOESY spectra is known as NOE assignment. Whereas information about atom labels can be obtained from a known structure or an assigned resonance list, AutoNOE makes use of assigned chemical shift list to carry out the NOE

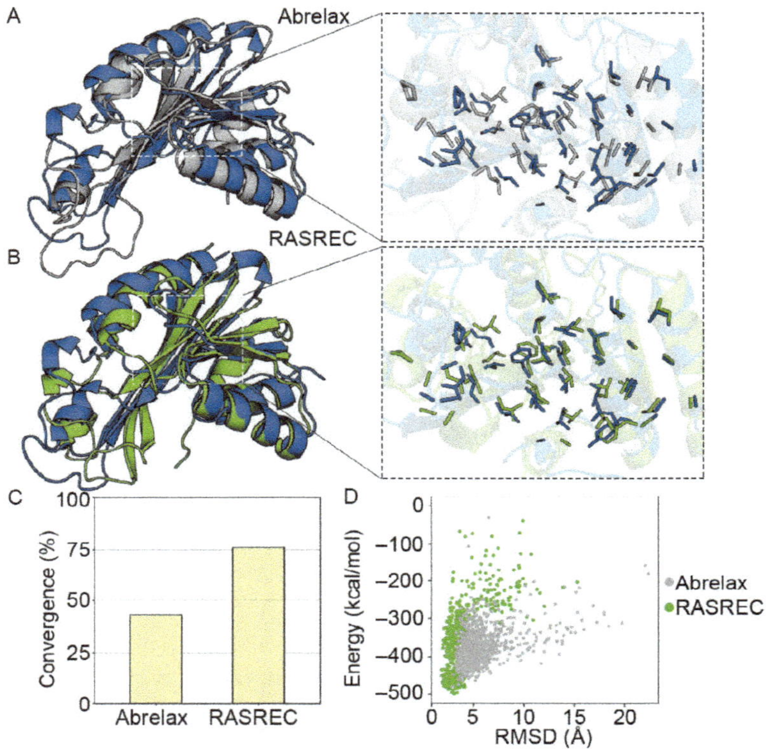

Fig. 2 Comparison of structural models generated by Abrelax and RASREC. Structures of 182 AA NESG protein target, SgR145, computed using Abrelax and RASREC methods. Lowest energy model of SgR145 determined using (A) Abrelax (*gray*, *left*) and (B) RASREC (*green*, *left*), superimposed with a reference structure (*blue*, PDB ID 2N47). Side chain placements of the buried residues in the lowest energy model are shown on the *right*. (C) Convergence of 10 lowest energy structures computed using Abrelax and RASREC for residues in structurally rigid regions (43% and 76%, respectively). (D) Plot showing Rosetta energy computed using ref2015.wts (Alford et al., 2017) (in kcal/mol) and RMSD to native (PDB ID 2N47) (in Å) of the structures generated over different stages of Abrelax (*gray*) and RASREC (*green*) protocols. Structures of SgR145 were computed from fragments derived using previously measured chemical shifts and NOE restraints (Lange et al., 2012) (BMRB ID 16806). Buried residues were selected using 10 Å^2 solvent accessibility surface area in the native (PDB ID 2N47). Structurally rigid regions were identified as regions consisting of α-helices and β-sheets.

assignment task. These assigned NOESY cross-peaks give rise to interproton distance restraints that are employed by structure determination methods to guide the conformational search. In CS-Rosetta, the NOE assignment (or NOESY) module is inspired largely by alternative automated tools for NOE assignment including ARIA (Linge, Habeck, Reiping, & Nilges, 2003; Nilges, 1993, 1997), AutoStruct (Huang et al., 2003), and CANDID (Herrmann, Güntert, & Wüthrich, 2002). AutoNOE takes as input the assigned NMR resonances together with unassigned NOESY cross-peak lists and generates a set of NOESY assignments from which it derives NOE distance restraints for structure calculation within RASREC.

The objective of the NOESY module is to assign proton atom labels from the input chemical shift list to the NOE cross-peaks such that the values of frequencies in both these lists match within the user specified tolerance levels (Fig. 1, green). To achieve this, the method starts by mapping input chemical shift frequencies to the unassigned NOESY cross-peaks. The initial mapping results in a set of ambiguously assigned NOEs due to potential overlap in the spectra (Nilges, 1993, 1997). These ambiguous NOEs are filtered based on a series of criteria. At first, the diagonal peaks (these are the peaks, which have same frequency for the proton and carbon dimensions) are eliminated. Second, a set of scores are computed to determine the confidence in the assignments. Third, NOE cross-peak assignments are filtered using the thresholds determined based on the confidence scores. Finally, NOE cross-peak intensities are calibrated to generate distance restraints.

4.3.1 NOE Assignment Confidence Scores

For any NOE assignment, a set of descriptors are scored. These scores strengthen the confidence in the atom label that AutoNOE assigns to a cross-peak (Fig. 1, green). A list of all the scores computed toward maximizing the confidence of an NOE assignment includes the following.

(i) *Chemical shift* score reports on the deviation between chemical shifts of dimensions in the NOESY cross-peak list with chemical shifts of atoms from the input list.

(ii) *Covalent compliance* score favors intraresidue and interresidue proton pairs that are within 5 Å from one another. This occurs when structures are available during sampling stages.

(iii) *Decoy compatibility* checks the compatibility of assigned NOEs and decoys generated during the intermediate stages of RASREC calculations using assigned NOEs. To perform such a validation, distances of assigned proton pairs are computed and further normalized using the cumulative distance of ambiguous assignments of the same cross-peak.

(iv) *Symmetry* score exploits the symmetry information of the NOE cross-peaks to determine whether they are reciprocated. In particular, each coupling gives rise to two cross-peaks that exhibit symmetry across the diagonal in the spectrum, and this information is utilized to increase the confidence of assigned NOEs.

(v) *Network anchoring* helps AutoNOE in reducing the number of assignment possibilities, which can arise when the peaks overlap (Herrmann et al., 2002). In order to accomplish this, assigned NOEs are weighed based on their participation in any network. A network of NOEs is found by looking for alternative residue *k* through which any two residues *i* and *j* are connected in space.

(vi) *Volume contribution* score of a cross-peak assignment is given by the combined contributions of other scores such as chemical shift, decoy compatibility, and either of symmetry, covalent compliance, weight of input structures, or network anchor scores. There is a cyclic dependency between volume contribution and other score terms due to which these terms are refined iteratively.

4.3.2 Formation of Distance Restraints

The labeled NOESY cross-peaks are used to create distance restraints that are employed by RASREC during structure determination. The restraints are scored with a flat bottom potential, meaning that, the distance between two protons is bounded and structures violating it are treated with a quadratic penalty term, which is added to the score function. The lower bound for a restraint is typically 1.5 Å (approximate van der Waals contact distance) (Lange, 2014). An upper bound on the distance for a cross-peak that has at least one assignment with volume contribution greater than the given threshold is derived using distance between protons and the deviation of the volume contribution from the cutoff.

Calibration of peak intensities: To generate upper distance bounds from the cross-peaks, their intensities are calibrated iteratively. For the purpose of calibration, protons are classified into backbone (H^{α} and H^{N}), β (H^{β}), methyl, and side chain groups, where each group has a different constant associated with it. These constants are derived based on the availability of intermediate decoys. If the decoys are unavailable, then the constants are calculated with a fixed upper distance bound of 3.8 Å. Whereas, if the decoys are available, then the calibration constants are tuned using 10%

of the upper distance bounds violated in the structures by a given cutoff (Herrmann et al., 2002; Lange, 2014).

Elimination of spurious peaks: Even in the presence of all the filters, there are cross-peaks that are unfit to produce structural restraints. Such cross-peaks are eliminated from further consideration. The criteria for elimination include (i) volume contribution (if assignments consistently have lower volume contribution score than expected), (ii) network anchor score (if the network anchor score is low) (Herrmann et al., 2002), and (iii) distance restraint violation (if many modeled structures violate the generated distance restraints).

The AutoNOE protocol parameters can be customized and are available in the form of flag files similar to the ones used in Abrelax and RASREC. Together with the flags used in RASREC, AutoNOE has a set of new parameters available in the following files.

noe_super_quick_relax.txt is a patch file that consists of weight term applied during side chain repacking.

flags_autoNOE is a method specific flag file similar to *flags_rasrec* or *flags_abrelax*. It consists of fully qualified paths to unassigned peak, chemical shift, and *flags_cs_rescore* files.

flags_phaseII and *flags_phaseIII* define the network anchor score cutoff parameters, which are applied to eliminate spurious cross-peaks. In addition, they define number of cycles (or iterations) to be used to assign NOEs (this value is typically set to 7, which is based on the number of cycles used in CANDID; Herrmann et al., 2002), and upper bound for the distance restraint, which is used for assignments that result in unreasonably high values during calibration.

flags_noe_assign includes parameters (i) to indicate that the centroid mode structures must include H^{α} atom (this flag is useful when assigning NOEs that contain H^{α} atoms, since they contribute significant number of restraints during structure calculation), (ii) that specify paths to *noe_super_quick_relax.txt*, which is used to repack side chains, (iii) to bias calibration of peak intensities, and (iv) to weigh distance restraints generated by AutoNOE.

flags_initnoe_filter can be used to specify parameter to skip network analysis via `-noesy:no_network` option. Currently, this option is turned off.

AutoNOE calculations can be set up with multiple score weights for distance restraints generated using the NOESY module. A typical setup creates cst_XX directories, where XX = {5, 10, 25, 50} (these are the default weights chosen based on the benchmark calculations, but any custom weights can be used; Lange, 2014; Zhang, Porter, Tripsianes, & Lange, 2014). Each cst_XX directory contains input, test, and run folders. At the start of AutoNOE calculations, an initial assignment step must be carried out by executing `initialize_assignments_phaseI.sh` available within the run directory. This initialization step generates a set of ambiguously assigned NOE restraints that will be used in the early stage of structure calculation runs in RASREC. AutoNOE interacts with RASREC during all the resampling stages (I–VI) and repeats stages III and IV (therefore, there are a total of eight stages). At the end of each resampling stage, 30 lowest energy conformers from the structure pool are used to filter restraints based on their compatibility. AutoNOE updates NOE assignment parameters at three distinct phases (I, II, and III) that are interleaved with structure calculation runs (Fig. 1, green). During Phase I, all the NOEs assigned by AutoNOE are allowed despite being highly ambiguous. In Phase II, all the ambiguous restraints in highly structured regions of the best scoring decoys are eliminated. All the ambiguous and unsatisfied restraints in top scoring structural models are removed during Phase III of this protocol, and the restricted lists of NOEs are used within RASREC. Once the runs have converged, *fullatom_pool* directory is populated with final decoys. The list of NOEs is subsequently obtained by executing `final_assignment.sh`, which generates restraints based on 10 lowest energy structures from *decoys.out*. Upon execution of `final_assignment.sh`, NOEs (*noe_auto_assign.cst*) together with a detailed description of confidence classes (high, medium, and low, where $0.7 \leq \text{high} \leq 1$, $0.45 \leq \text{medium} < 0.7$, and $0 \leq \text{low} < 0.45$) and corresponding scores are available (in *NOE_final.out*) for further analysis if required.

To highlight how the quality of structural models improves from one stage to another using AutoNOE and RASREC, we computed structures of 20 kDa target, α-lytic protease protein using NMR chemical shifts and two unassigned 4D NOESY peak lists (HCCH and HCNH) obtained from (Evangelidis et al., 2018) (BMRB ID 30322). The 10 lowest energy structures from each stage of the combined method show (i) that during the iterative AutoNOE assignment and RASREC modeling process, structures are refined to higher-resolution and convergence as the calculation progresses (Fig. 3A and B), and (ii) the synergy between structural convergence, NOE restraints satisfied by the decoys and RMSD to X-ray (PDB ID 1P01) (in Å) (Fig. 3A and B). Furthermore, it can be noted that the number of unique NOE restraints assigned by AutoNOE are self-consistent across all the stages of the protocol (Fig. 3D).

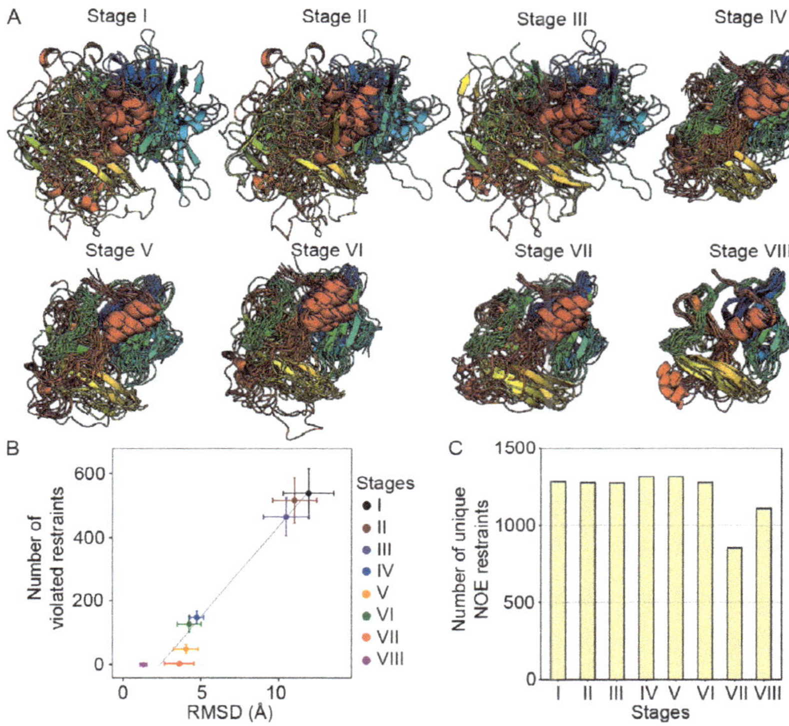

Fig. 3 NOESY cross-peak assignment by AutoNOE. Structural models of 20 kDa α-lytic protease (aLP) protein calculated using the combined AutoNOE-RASREC protocol. (A) Ten lowest energy structures of aLP generated at the end of each resampling stage of RASREC using the NOEs assigned by AutoNOE. (B) Plot showing the synergy between RASREC and AutoNOE as a function of the average number of violated constraints and RMSD to X-ray (PDB ID 1P01) (in Å) through RASREC stages I–VIII for the 10 lowest energy structures. (C) Total number of unique NOE restraints assigned by AutoNOE at the end of each stage of the RASREC protocol. The structures of aLP were calculated using previously measured NMR chemical shifts and two unassigned 4D HCCH and HCNH NOESY peak lists (Evangelidis et al., 2018) (BMRB ID 30322). All the structure diagrams were generated using PyMOL (https://pymol.org/2/).

5. ARCHITECTURE

In this section, we describe the core architecture of Abrelax, RASREC, and AutoNOE methods. These methods are modularized and follow object oriented design principles from ROSETTA3 in their implementation (Leaver-Fay et al., 2011). Rosetta is primarily divided into three main sets of libraries, one of which is the `protocols` library that encompasses all the algorithms used to interact with and/or modify the structures. The algorithms are separated into namespaces based on their functionality. CS-Rosetta's structure modeling methods (Abrelax and RASREC) live within the `abinitio` namespace, whereas NOE assignment module resides in the `noesy_assign` namespace and both are nested under the `protocols` library (Lange, 2014; Lange & Baker, 2012; Shen et al., 2008). All the C++ classes that are part of these protocols carry out several tasks. Below, we provide a brief overview.

5.1 Structure Modeling Methods

Abrelax can be accessed by the `broker_main()` procedure present in the `BrokerMain` source file (Fig. 4). This procedure creates an instance of the `AbrelaxMover` class and presents it to the MPI-based job distributor (or JD) that handles distribution of jobs in a parallelized manner and resides in the `jd2` namespace. The JD is responsible for assigning new jobs to the worker nodes (or processors) and run independent instances of `AbrelaxMover` on the Pose. JD works well when writing the output structures one at a time into a single file (*decoys.out*) and examining it before processing new jobs. However, it cannot handle unexpected file read/write (race condition) behaviors from processors that may occur during parallelization. To avoid this situation, MPI framework is used. In particular, `MPIFileBufJobDistributor` (a derived class from `JobDistributor`) decouples IO (input/output) and job distribution (Lange & Baker, 2012). A detailed working of the MPI framework in the context of CS-Rosetta is given in the supplementary information of Lange and Baker (2012). In this discussion, we have abstracted MPI-based job distributors with their parent class, `JobDistributor`.

The main sampling strategy of CS-Rosetta is in the `FragmentSampler` class, which provides implementations of methods that carry out stages I–IV of Abrelax. `ConstraintFragmentSampler` inherits from `FragmentSampler` and is used primarily to associate input constraints with the Pose and further overrides functions from its parent such as `do_stage1_cycles()` and `prepare_stageYY()` where $YY = \{1, 2\}$ to carry out stage-specific operations. After the completion of sampling, the Pose is passed to the topology broker, which ensures that it (Pose) respects all the claims defined in the *setup_init.tpb* file. The Pose is returned to `AbrelaxMover`, which (i) closes chain-breaks introduced during sampling, (ii) switches from centroid to full-atom representation, and (iii) performs energy refinement to produce realistic structure that is written into *decoys.out* file by the process managing IO.

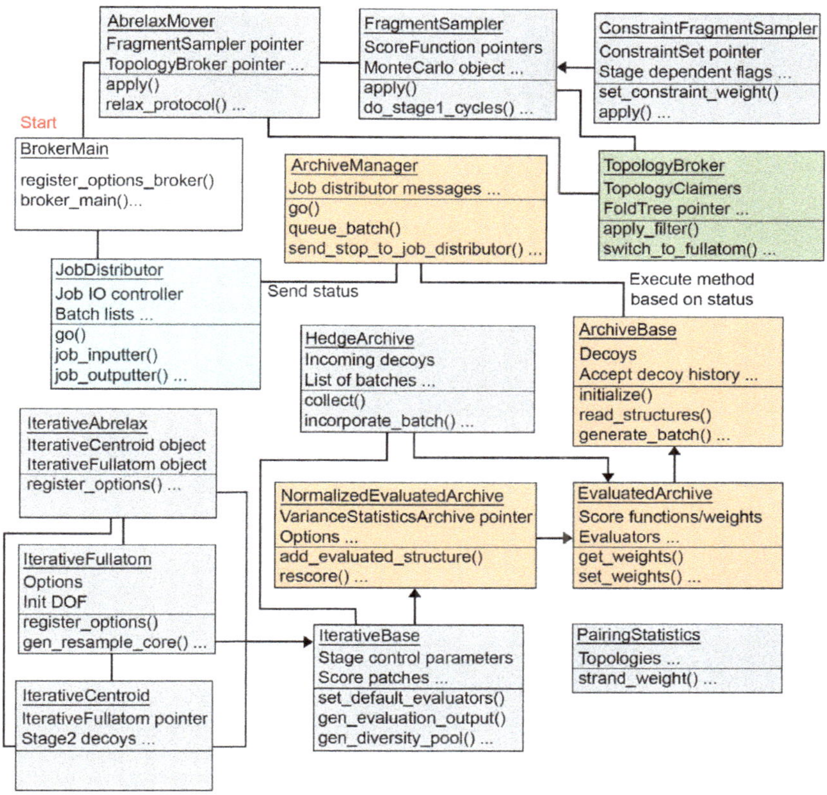

Fig. 4 Architecture diagram of Abrelax and RASREC. `BrokerMain` is the point of entry for both Abrelax and RASREC methods. The `broker_main()` method creates an instance of `AbrelaxMover` and presents it to the `JobDistributor`. During Abrelax calculations, `FragmentSampler` and its derived class `ConstraintFragmentSampler` perform four stages of resampling. The resulting model is passed to the `TopologyBroker` which ensures that the desired degrees of freedom are consistent with one another. In RASREC calculations, `JobDistributor` invokes `ArchiveManager`, a controller class responsible for setting up iterative protocols (here, we just abstract the MPI-based job distributors with a parent class `JobDistributor` to simplify the diagram). `ArchiveManager` receives status messages from `JobDistributor` about the current state of structure calculation runs, based on which, it decides the next course of actions and calls corresponding interfaces from `ArchiveBase`. `EvaluatedArchive` and `NormalizedEvaluatedArchive`, descendants of `ArchiveBase`, are used to evaluate incoming decoys and filter them based on their weighted sum of scores. `HedgeArchive` is used mainly to collect structures from batches for later use (in stage V). `IterativeBase`, derived from `NormalizedEvaluatedArchive`, is the main class that encompasses six stages of the RASREC protocol. `IterativeBase` is accessed in a stage-specific manner; stages I–IV are realized through `IterativeCentroid` and stages V and VI through `IterativeFullatom`. `IterativeAbrelax` is a wrapper

(Continued)

The RASREC protocol like Abrelax can be triggered from BrokerMain (Fig. 4). RASREC interacts extensively with the Archive framework, which lives in the archive namespace under jd2. Archive was developed primarily to facilitate interfaces from the iterative protocols (through their parent, ArchiveBase class) to the controller class, ArchiveManager, that communicates with the job distributor. In particular, ArchiveManager receives messages such as QUEUE_EMPTY and JOB_COMPLETION from the job distributor, based on which, it invokes necessary interfaces from ArchiveBase. When the protocol begins execution, the job distributor will examine the queue for any pending jobs (which is empty). Therefore, it sends QUEUE_EMPTY message to the ArchiveManager, which subsequently calls the generate_batch() method from the IterativeBase class, a descendent of ArchiveBase (through polymorphism). During the process of batch generation, *batch_XX* directory containing all the necessary parameter and protocol files is created, where XX is the batch number. ArchiveManager writes the status of the batch in a *BATCH_INFO* file within each batch directory, which is used to retrieve the status of a calculation if the protocol stops prematurely. Whenever the current batch is saturated, the job distributor sends JOB_COMPLETION message to the ArchiverManager, which forwards the newly generated decoys to the read_structures() method in IterativeBase. The read_structures() method filters and adds decoys to the structure pool. The main functionality of RASREC is implemented in IterativeBase, which calls procedures such as gen_enumerate_pairings() (stages I and II), gen_resample_topologies() (stages II and III), gen_resample_stage2() (stage IV), and gen_resample_fragments() (stages IV–VI) to perform stage-specific operations. Here, gen_resample_topologies() procedure invokes PairingStatistics to compute scores for β-strand pairs during

Fig. 4—Cont'd class for IterativeCentroid and IterativeFullatom. PairingStatistics is used to compute β-strand and β-topology-related scores during topology resampling. The *shaded boxes* represent classes. The *colors* indicate namespaces under the protocols library within which the classes reside (*gray*: abinitio, *tan*: archive under jd2, *blue*: jd2, and *green*: topology_broker). The *underlined text* within each box is the name of a class (except BrokerMain, which is not a class). A subset of data members is represented in an abstract manner by text above the line within the box. A small subset of nonoverriding methods is listed in each class below the line. *Solid arrows* represent inheritance relationships between classes. *Straight lines* connecting between classes show simple association, which means that, an object of one class is being used in the other. Only a representative subset of classes is shown in this diagram.

β-sheet topology resampling. `IterativeBase` is accessed using a wrapper class, `IterativeAbrelax`, which references `IterativeCentroid` (stages I–IV) and `IterativeFullatom` (stages V and VI) classes (both derive from `IterativeBase`) based on the representation mode of the protocol. The number of batch directories to be created is decided in real time, therefore, the job distributor, specifically `MPIArchiveJobDistributor` (a subclass of `MPIFileBufJobDistributor`) enables addition of batches to the job queue spontaneously. The decoys from batches are randomly selected and stored by `HedgeArchive` in *centroid_pool_hedge* directory. These archived structures are used during stage V of RASREC to promote extensive sampling of the full-atom energy function.

Evaluation of the decoys is carried out based on the energy terms weighted using custom or predefined bias by `EvaluatedArchive`, a derived class from `ArchiveBase` and `NormalizedEvaluatedArchive`, which inherits from `EvaluatedArchive`. The aggregation of weighted energy terms is represented by `score_final` tag for each structure in the silent file (*decoys.out*). In addition, the weighted sum of scores in the intermediate stages undergo on-demand updates and are made rapidly accessible under `_archive_select_score` tag in the silent file carrying structures to subsequent stages (or the structure pool).

5.2 NOE Assignment Module

The NOE assignment method within the `noesy_module` namespace can be invoked from a publicly available application, `r_noe_assign.<ext>` (where <ext> is platform and compiler dependent) or from RASREC through the `generate_batch()` method of the `IterativeBase` class (Lange, 2014). The public app is called by the CS-Rosetta toolbox (see Sections 4 and 9) during initialization and finalization of NOE assignments, whereas the `generate_batch()` method calls the NOESY module at the start of each different stage of the RASREC protocol (Fig. 1, green). The `assign()` interface used by both these points of entry marks the start of the NOE assignment algorithm and is available in `NoesyModule.impl.hh` (Fig. 5). Here, methods within the `CrossPeakList` class are called specifically to carry out score calculations during the assignment of NOESY cross-peaks. AutoNOE is designed to accept multiple unassigned cross-peak lists. Each input list is encapsulated within the `CrossPeakList` class that provides functionality to read/write from iostream/files, update relevant scores, calibrate peak lists, and generate centroid and full-atom restraints for use in RASREC.

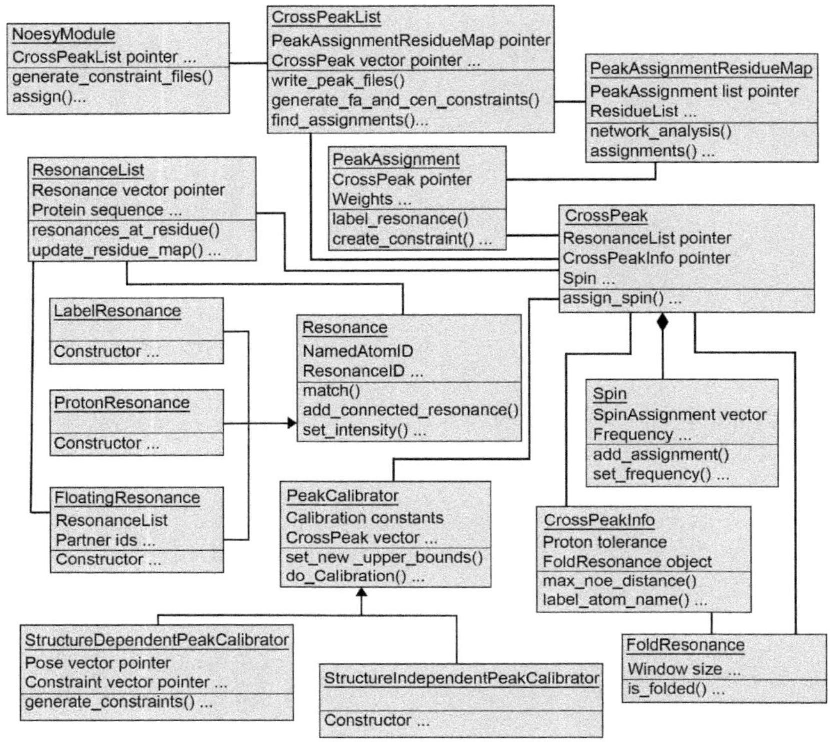

Fig. 5 Architecture diagram of AutoNOE. `NoesyModule` is the means to access NOE assignment protocol and contains a reference to the `CrossPeakList` class. The `CrossPeakList` class holds information of the entire list of NOE peaks using a vector of `CrossPeak` objects. In addition, it carries instances of `PeakAssignmentResidueMap` that maps an assigned peak (held by the `PeakAssignment` class) to its corresponding residue. The `CrossPeak` class contains a `Spin` class, which stores properties of a spin such as its identifier and chemical shift frequency. The `CrossPeak` class contains references to `ResonanceList` and a helper class, `CrossPeakInfo`, that stores cross-peak information. `ResonanceList` is used to store a list of `Resonance` objects read from the input chemical shift file. `Resonance` is inherited by `LabelResonance`, `ProtonResonance`, and `FloatingResonance`, which are instantiated based on the type of atom specified in the chemical shift file. `StructureDependentPeakCalibrator` and `StructureIndependentPeakCalibrator` inherit from the `PeakCalibrator` class, which associate with instances of `CrossPeak` and are used during calibration of peak intensities to produce upper distance bounds. The *shaded boxes* represent classes. The *colors* indicate namespaces under `protocols` library within which the classes reside (*gray*: `noesy_assign`). The *underlined text* within each box is the name of a class. A subset of data members is represented in an abstract manner by text above the line within the box. A small subset of nonoverriding methods is listed in each class below the line. *Solid arrows* represent inheritance relationships between classes. A *line with diamond* at its end-point represents composition (a class owns another class). *Straight lines* connecting between classes show simple association, which means that, an object of one class is being used in the other.

Each cross-peak in the list is associated with a set of properties, which are held by the `CrossPeak` class. These include proton spins mapped with chemical shift frequencies, peak intensity, cross-peak information (for each proton dimension such as, atom name and tolerance, which are stored as instances of the `CrossPeakInfo` class), and input resonance list. The `CrossPeak` class implements methods to find and assign atom labels to spins based on chemical shift frequencies and their tolerances. The `Spin` class contained within `CrossPeak` is used to encapsulate resonance frequency and spin assignment identifier which can be used to index into the resonance table. For the purposes of faster access to the assignments and residues (instead of referring from `CrossPeak` every time), `PeakAssignment` and `PeakAssignmentResidueMap` classes are used. `PeakAssignment` class stores score terms, spin identifiers, and resonance labels for each assignment of the cross-peak. `PeakAssignmentResidueMap` stores a mapping of the `PeakAssignment` list to the target residue list.

Calibration of peak intensities is handled separately by the `PeakCalibrator` class that exhibits simple association with `CrossPeak`. `PeakCalibrator` is inherited by `StructureDependentCalibrator` and `StructureIndependentCalibrator` which are invoked depending on the availability of decoys (this relies on the stage of the protocol) during calibration (see Section 4).

To store input chemical shifts and their atom labels, `ResonanceList` and `Resonance` classes are used. `ResonanceList` is the container class that keeps a list of resonances (each defined by `Resonance` class), identifiers for each resonance, and the corresponding residue number in the protein sequence. In particular, `ResonanceList` can be used to obtain tuples of chemical shift frequencies and respective atoms. `Resonance` encapsulates chemical shift id (or label; this is just an integer), frequency, tolerance, atom, residue, and atom-specific calibration flag (see Section 4.3). It is inherited by `ProtonResonance`, `LabelResonance`, and `FloatingResonance`. Depending on the type of atom encountered while reading chemical shift file, `ProtonResonance` (for protons) or `LabelResonance` (for other atoms associated with protons) is instantiated. In addition, if the given chemical shift frequencies lie in the folded region of the NMR spectra, then `FoldResonance` is utilized so that necessary corrections are applied while mapping the frequencies to the NOE cross-peaks.

6. INTEGRATION WITH STRUCTURAL RESTRAINTS

To model protein structures with novel or challenging folds, restraints derived from NMR experiments or homologous proteins that guide the conformation search toward a native become crucial. CS-Rosetta allows easy integration of such restraints to help the search that employs Monte Carlo methods. Restraints in Rosetta are used within the score functions, specifically, to favor or penalize the models based on their compatibility. All the protocols within CS-Rosetta accept restraint files and corresponding weight terms as input. These restraints are associated with the Pose being perturbed at all the time thus making them an integral part during the application of any mover. In this section, we highlight a representative subset of the types of structural restraints supported by CS-Rosetta.

6.1 NMR

NMR spectroscopy provides a means to obtain local and global structural information of proteins, which is later converted to distance or torsion angle restraints used to drive the structure modeling calculations. Local restraints are typically provided by NMR chemical shifts and NOEs (Anglister, Srivastava, & Naider, 2016; Güntert, 2009), whereas global information is obtained from RDC (Blackledge, 2005), PCS (pseudocontact shift) and PRE (paramagnetic relaxation enhancement) (Clore & Iwahara, 2009) restraints.

NMR chemical shifts are highly informative parameters that report on protein backbone and side chain conformations and are utilized heavily by NMR-based structure determination programs to model near-native structures (Berjanskii & Wishart, 2017). CS-Rosetta exploits NMR chemical shifts in two ways: first, to derive torsion angle restraints using TALOS-N (Shen & Bax, 2013), that are applied during the fragment selection process; second, to measure their consistency with SPARTA-predicted (Shen & Bax, 2007) chemical shift values obtained from simulated (intermediate or final) structural models. This fitness measure (deviation of experimental and predicted chemical shifts) given by chi-squared term is used to adjust the score function. During structure modeling in CS-Rosetta, the contribution of chemical shifts to the score function can be observed under `chem_shifts` column in the output file containing decoys. The corresponding weights for this term during centroid or full-atom modes can be customized using `-cenpool_chemicalshift_weight` and `-fapool_chemicalshift_weight` tags present in the flag files.

Chemical shifts are generally complemented with long-range distance restraints provided by NOEs (distance between atom-pairs that are close in space) (Güntert, 2003). While unassigned NOE peak lists can be used by AutoNOE algorithm to perform NOE assignment and structure calculation iteratively (see Section 4.3), it requires higher levels of resonance assignment completeness (>70% of resonances must be assigned) (Evangelidis et al., 2018). Therefore, when it is not feasible to obtain near-complete resonance assignments, structure determination algorithms (Abrelax and RASREC) can use all available chemical shifts together with NOE restraints assigned manually to model structures. NOE restraints are employed primarily to penalize the models that disagree with their distance bounds during structure calculation runs (Table 1). The amount of penalty applied is given by the `atom_pair_constraint` score term in the output file containing decoys. The weights used to compute `atom_pair_constraint` (represented by the same name in the patch files) can be tuned up or down as per custom requirements during different stages of calculations in centroid or full-atom modes. The NOE distance restraints typically start with the keywords, `AmbiguousNMRDistance` or `AtomPair` (or `NamedAtomPair`), and further contain definitions of atom-pairs, and penalty function type and its parameters (Table 1).

Whereas chemical shifts and NOEs are sufficient to model lower-molecular-weight complexes (<25 kDa), additional data providing information about orientation of local and global regions is generally necessary to determine structures of large proteins or oligomeric systems. Such information is available from RDC measurements, which are useful in describing the orientation of the amino acids in proteins with respect to the global alignment frame (or the external magnetic field) (Donald & Martin, 2009). In CS-Rosetta, backbone RDC measurements are used in the score functions to evaluate their fitness to the structural models generated during centroid and full-atom stages (Raman et al., 2012). Corresponding scores in the decoys file are represented using `rdc` and `rdc_filt` (`rdc_filt` = `rdc` score term * `rdc` weight term) terms. Whereas a bias on RDC scores is indicated by the `rdc` term in the patch files.

The presence of paramagnetic metal ions in proteins induce significant changes in chemical shift values resulting in long-range structural restraints known as PCS (Schmitz, Vernon, Otting, Baker, & Huber, 2012). PCS measurements can be combined with Abrelax to model protein structures with increased accuracy (Schmitz et al., 2012; Yagi et al., 2013). To perform PCS-based structure calculations for a target with one or multiple paramagnetic tags, preliminary models are determined using only the chemical shift-derived fragments. These models are used to tune target-specific PCS score weights for each paramagnetic tag, which are applied during low-resolution

Table 1 Function Types Typically Used in Constraints to Penalize Structural Models in CS-Rosetta

Bounded
Format:

```
AmbiguousNMRDistance atom1_label atom1_residue_no atom2_label
  atom2_residue_no BOUNDED lb ub sd annotation
```

Example:

```
AmbiguousNMRDistance H 185 QD2 175 BOUNDED 1.5 4.20 1.000 automatic
  NOE Peak
AmbiguousNMRDistance H18H23 BOUNDED 1.5 3.40 1.000 automatic NOE Peak
```

Sigmoid
Format:

```
AtomPair atom1_label atom1_residue_no atom2_label atom2_residue_no
  SIGMOID center slope annotation
```

Example:

```
AtomPair CB38 CB56 SIGMOID 8.00 1.00 #ContactMap: 1.35
AtomPair CB 118 CB 165 SIGMOID 8.00 1.00 #ContactMap: 1.22
```

SOGFUNC
Format:

```
AtomPair atom1_label atom1_residue_no atom2_label atom2_residue_no
  SOGFUNC no_of_functions mean1 sd1 wt1 ...
```

Example:

```
AtomPair CA 4 CA 23 SOGFUNC 2 8.980 2.897 0.265 18.300 7.300 0.735
AtomPair CA 8 CA 47 SOGFUNC 1 7.842 2.297 0.312
```

center, center of the SIGMOID function; *lb*, lower bound; *sd*, standard deviation; *slope*, slope of the SIGMOID function; *ub*, upper bound; *wt*, weight.
Rosetta provides a large set of function types, here we show only some examples of a representative subset used by CS-Rosetta (https://www.rosettacommons.org/docs/latest/rosetta_basics/file_types/constraint-file).

mode of the structure calculation runs using `-abinitio::stageXX_patch` option where XX = {1, 2, 3a, 3b, 4}. Since the PCS measurements are recorded primarily for backbone atoms, they are applied only during low-resolution modeling phase. Following low-resolution phase, models are refined using standard refinement protocols within Rosetta. These models are filtered using a combined score produced by PCS data (during low-resolution stage) and full-atom refinement.

PRE measurements can be used to obtain distances between C^{α} atoms of PRE-affected residues and C^{β} atoms of the residues to which a paramagnetic probe is attached. While NOEs generally provide distance restraints between protons that are <6 Å apart in space, PREs can be used similarly as bounded restraints (Table 1) providing information about atoms that are spaced between ~20–25 Å (Tamaki et al., 2016). Moreover, the weights of these distance restraints can be tuned up or down using standard patch file containing `atom_pair_constraint` term. Alternatively, CS-Rosetta can be combined with solvent accessible data acquired through PRE-measurements (sPRE-CS-Rosetta) (Hartlmüller, Göbl, & Madl, 2016). Such data as demonstrated by Hartlmüller et al. (2016) is used within scoring functions to validate modeled structures during low-resolution and high-resolution modes. The experimental data file containing sPRE values is input in a format similar to the chemical shift file format in TALOS. Score weight represented by `spre` term in a patch file is added to the various ab initio stages given by `-abinitio::stageXX_patch` option in the flag files, where XX = {1, 2, 3a, 3b, 4}. Simultaneously, more parameters used in scoring, such as grid resolution (used during prediction of sPRE values from models), nonstereospecificity (using sPRE values obtained by averaging over stereospecific atoms) in full-atom mode, distances between (i) paramagnetic agent and the nuclei and (ii) the outer boundary of paramagnetic grid and the protein, can be adjusted for custom targets.

6.2 Homologous Proteins

Restraints derived from homologous or evolutionarily related proteins provide complementary information alongside sparse experimental data. CS-Rosetta offers a mechanism to generate such restraints and further supports data obtained from external tools or webservers to model accurate structures (Nerli et al., 2018).

CS-HM-Rosetta (or Chemical Shift-Homology Modeling-Rosetta) (Thompson et al., 2012) is a protocol whose primary purpose is to generate restraints from homologous proteins that can be combined with NMR chemical shifts within main structure modeling methods (Abrelax and RASREC). Here, the protocol first performs multiple sequence alignment of a query protein to all the proteins in the PDB using HHpred webserver (Söding, Biegert, & Lupas, 2005) or HHsearch (Söding, 2005) stand-alone software. The resulting alignment profile is used by CS-HM-Rosetta to measure distances between C^{α} atoms of residues that are consistent in multiple template proteins and are validated using criteria including, local and global sequence similarities and buried surface (Thompson & Baker, 2011). This validation information is combined to produce interatomic distance restraints as a mixture of Gaussian probability densities. The restraints are given higher weights if they are observed between same pair of atoms across multiple templates. The combined Gaussian probability densities are represented using a SOGFUNC function in restraint definition (Table 1).

More recently, sequence covariance information has been used to derive structural restraints (termed evolutionary coupling (EC) restraints) to guide the search for native conformation (Balakrishnan, Kamisetty, Carbonell, Lee, & Langmead, 2011; de Juan, Pazos, & Valencia, 2013; Jones, Buchan, Cozzetto, & Pontil, 2012; Kamisetty, Ovchinnikov, & Baker, 2013; Marks, Hopf, & Sander, 2012; Ovchinnikov et al., 2017; Sułkowska, Morcos, Weigt, Hwa, & Onuchic, 2012). These restraints are generated by utilizing genomic data and applying statistical methods such as, pseudolikelihood maximization (Ekeberg, Lövkvist, Lan, Weigt, & Aurell, 2013) and/or direct information (Marks et al., 2011) to measure evolutionarily coupled residues that are closer in space. Currently, there are many webservers and stand-alone software tools available to predict such restraints (Jones et al., 2012; Kamisetty et al., 2013; Marks et al., 2012). These distance restraints can be used alone or combined with sparse NMR data and CS-Rosetta algorithms to model high-resolution structures. Toward this end, CS-Rosetta expects data to be in a distance restraint file format (Table 1). For instance, if the restraints are generated using EVFold (Marks et al., 2011, 2012), then Rosetta provides scripts (see Section 9) to convert them into Rosetta compatible constraint files (with Sigmoid function; Table 1) that can be used directly within structure calculation runs as demonstrated previously (Braun, Leman, & Lange, 2015).

7. COMBINED APPROACHES

CS-Rosetta can be combined with alternative automated approaches to model complex protein systems. One such approach is RosettaOligomers (Sgourakis et al., 2011). This approach makes use of NMR chemical shifts and NOE restraints to model monomeric subunits, which are subsequently combined using docking protocols within Rosetta. RosettaOligomers has been employed previously to model symmetric protein complexes with both simple and interleaved interfaces (Sgourakis et al., 2011). To model these systems, structural fragments for individual monomers are derived using chemical shifts. These fragments are used within Abrelax or RASREC to obtain models of monomeric subunits. Depending on the convergence of resulting monomers, docking is carried out using different protocols. Specifically, if the convergence is <3 Å, symmetric docking is performed (using SymDock.<ext>, where <ext> is platform and compiler specific) (André, Bradley, Wang, & Baker, 2007), otherwise, fold and dock is applied (using minirosetta.<ext> with a topology broker file, where <ext> is platform and compiler specific) (Das et al., 2009). Both the docking algorithms require symmetry information of the oligomeric system, which is provided via the symmetry definition file. This file contains information about how to score models and allowed degrees of freedom. To create such a file, Rosetta provides make_symmdef_file_denovo.py (if no reference structure is available) and make_symmdef_file.py (if reference structure is used to define symmetry) tools. The full description of parameters used to benchmark this method is given in supplementary information of Sgourakis et al. (2011).

CS-Rosetta relies heavily on increased levels of completeness and correctness of assigned NMR chemical shifts to label NOE cross-peaks or model high-resolution structures. However, assignment of chemical shifts is a labor-intensive process, which involves recording many (typically 6–10) NOESY spectra and labeling thousands of peaks within them with atom names to obtain desired levels of accuracy for small proteins (<15 kDa) (Evangelidis et al., 2018; Würz, Kazemi, Schmidt, Bagaria, & Güntert, 2017). To address this problem, many automated methods have emerged that can assign resonances of protein targets with higher molecular weights efficiently (Würz et al., 2017). Recently, an automated method, 4D-CHAINS, was developed (Evangelidis et al., 2018). This method makes use of only two complementary (4D-NOESY and TOCSY) spectra, yet assigns backbone and side chain chemical shifts with high accuracy (>95% for the benchmark targets used in that study). As demonstrated previously, this method can be combined with CS-Rosetta to assign NOEs and/or model structures with subatomic resolution (Evangelidis et al., 2018). Moreover, the assigned chemical shift files generated by 4D-CHAINS can be used directly to pick fragments and drive any CS-Rosetta calculations.

8. COMPUTATIONAL REQUIREMENTS

All the methods in CS-Rosetta have been tested extensively on Linux and MacOS (or UNIX-based) operating systems with 16–2048 parallel processors. Previous reports suggest that a minimum of 16 processors can be used to calculate structures of up to 200 AA in a reasonable amount of time (Nerli et al., 2018). Here, we demonstrate the performance of all the methods represented as a function of time used for structure calculation and their molecular weights (Fig. 6). To measure the running times of these methods, we calculated structures of four targets of weights ranging from 15.5 to 27.3 kDa (same benchmark set as used in Evangelidis et al., 2018). Each independent run was performed on 100 CPUs with fragments picked using manually assigned chemical shifts and NOE restraints obtained from 4D HCCH and HCNH NOESY spectra (Evangelidis et al., 2018). Moreover, the largest target was supplied with RDC restraints (Evangelidis et al., 2018). The structure calculation runs were carried out

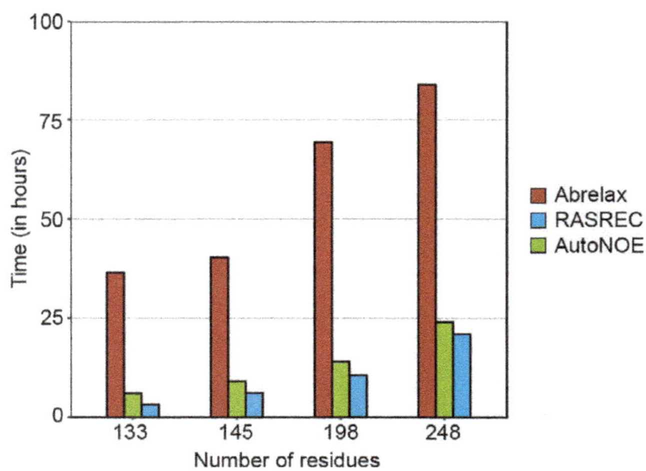

Fig. 6 Performance of Abrelax, RASREC, and AutoNOE algorithms. Performance of Abrelax (*red*), RASREC (*blue*), and AutoNOE (*green*) methods shown as a function of the number of residues and time (in hours) used to calculate structures of four benchmark targets (protein targets used in Evangelidis et al., 2018). Each bar in the plot represents independent structure calculation carried out in parallel using 100 CPUs. The structure calculation of all the targets was performed using same type of input constraints (Chemical shifts and NOEs alongside RDC measurements for the largest target obtained from Evangelidis et al., 2018; BMRB IDs 30322, 30325, 30326, and 30327) and executed until they reached convergence.

until all the methods reached convergence (>90% for 10 lowest energy structures). From all the calculations, we observed that RASREC is 5- to 10-folds faster than Abrelax. On the other hand, AutoNOE utilized approximately 3–4 h more than the RASREC calculations for all the targets in the benchmark set. These results are demonstrated on a specific computer cluster (AMD Opteron(tm), 2.4 GHz processor 6378) with a fixed number of CPUs. Using higher number of processors can reduce the running times of these protocols significantly (for instance, the authors of RASREC were able to calculate four structures per second with 2048 processors; Lange & Baker, 2012).

9. TOOLBOX

CS-Rosetta has many tuning parameters that can drive the calculations toward a native fold depending on the target of interest. While customization of all the parameters is allowed, CS-Rosetta provides automated setup scripts (or toolbox) where these parameters have been adjusted and predefined based on a set of benchmark calculations performed previously (Lange, 2014; Lange & Baker, 2012; Raman et al., 2012; Shen et al., 2008; Zhang et al., 2014). A series of function-specific flag files that define these parameters must be created before executing the methods (see Section 4). The CS-Rosetta toolbox, which has a Python interface, renders it easy to use and further allows flexibility in setting up these parameters for structure calculations without having to learn the complex details of Rosetta specifics. Besides setup scripts, the toolbox contains many helper programs that can be used to analyze outputs of structure calculation runs (Table 2). In addition, the toolbox consists of programs that allow rapid conversion of deposited NMR data in BMRB (Biological Magnetic Resonance Bank) (Ulrich et al., 2008) or NEF (NMR exchange format) (Gutmanas et al., 2015) file formats to Rosetta compatible file formats.

Software programs used to calculate distance restraints from homologous proteins using CS-HM-Rosetta (Thompson et al., 2012) are shipped with the toolbox. Structural restraints derived from ECs through separate web servers such as EVFold (Marks et al., 2011, 2012) can be converted into Rosetta compatible file formats using helper scripts available at rosetta/demos/protocol_capture/rasrec_evolutionary_restraints/ (in the Rosetta suite) (Braun et al., 2015). In addition, alternative web servers such as

Table 2 Contents of the CS-Rosetta Toolbox

Purpose	Scripts	Description
Install and setup	install, setup_run, setup_list, setup_autoNOE, setup_target, setup_display, list_setup	Used for the installation of toolbox and to set up structure calculation runs
Clean up files	clean_peak_file, clean_rdc_file	Clean input RDC and peak files
Renumbering	renumber_pdb, renumber_prot, renumber_rdc, renumber_talos, renumber_upl	Used to renumber input files after trimming flexible regions from chemical shift file (in TALOS format)
Chemical shift and RDC analysis scripts	get_rdc_tensor, process_prot_for_autoNOE, prot_completeness	Analyze input RDC values based on histogram method. Check resonance assignment completeness
Correction or addition of data	correctCSoffset, addFasta2Prot	Tune chemical shift offset. Add residue information to the chemical shift file (.prot)
Convert or extract data	prot2talos, talos2fasta, talos2prot, pred2rigid, pdb2fasta, convert_atomnames, fasta2seq, fasta2weight, rosetta2upl, silent_data, upl2rosetta, upl2fasta	Scripts used to extract information such as amino acid sequence from structures in PDB format, TALOS files or convert from one file format to another (e.g., prot2talos)
Map full-atom constraints to coarse grained constraints	cst_map_toCB, cst_map_toCEN	Convert constraints in full-atom representation to centroid or coarse grained (where side chains are characterized by C^{β} atoms) representations
Extract from BMRB	bmrb2fasta, bmrb2prot, bmrb2rdc, bmrb2talos	Extract data from BMRB files to use them within structure calculation runs
Extract from or convert to NEF	nef, nef2fasta, nef2peaks, nef2prot, nef2rdc, nef2restraint, nef2talos	Extract data from NEF files to use them for structure calculation. Convert input data to NEF file format for deposition into databases
Fragment picking	pick_fragments	Pick fragments using chemical shift information

Table 2 Contents of the CS-Rosetta Toolbox—cont'd

Purpose	Scripts	Description
Extracting structures	extract_decoys, pack_pdb, unpack_pdb	Extracting structures post structure calculation runs
Postcalculation analysis	annotate_target, autoNOE_select_final_run, extract_scores, median, noeout2txt	Scripts used to determine best constrained runs, top scoring decoys, and their statistics
CS-HM-Rosetta	convert_aln, get_pdbs, predict_distances	Predict distance restraints based on HHsearch or HHpred alignment file

A detailed description and usage of these scripts is available in the Application Reference section at https://csrosetta.chemistry.ucsc.edu.

GREMLIN (Kamisetty et al., 2013) generate EC restraints in required file format and can be used directly within CS-Rosetta calculations.

The toolbox is available for download at https://csrosetta.chemistry.ucsc.edu. Additional documentation about setting up the structure calculation runs and commands can be found at the same site.

10. CONCLUSIONS

CS-Rosetta offers a wide variety of capabilities in the field of de novo protein structure modeling with sparse NMR data. While the classical algorithm, Abrelax, works very well for smaller sized proteins, it has been superseded by RASREC that employs a variety of optimization strategies to model larger complexes. Alongside structure prediction, CS-Rosetta readily enables NOESY cross-peak assignment through the AutoNOE module. Together with unprecedented parallelization, CS-Rosetta's robust algorithms push the boundaries of molecular weight limits of proteins. The combined use of experimental and simulated restraints derived from ECs works very well with RASREC and holds promise toward designing proteins for therapeutics. Due to its adaptability, CS-Rosetta allows it to be integrated with alternative approaches to model oligomeric proteins or build an automated pipeline for NMR-based structure determination. Simultaneously, the modularized architecture of CS-Rosetta makes it easy to embed new methods to address more difficult systems or model other biomolecules.

ACKNOWLEDGMENTS

The authors would like to thank Oliver Lange, Robert Vernon, Ad Bax, and David Baker for helpful discussions over the years, and Andrew C. McShan for providing valuable feedback on this manuscript. This manuscript was supported in part by funds from the Intramural research program of the NIAID, NIH, a K-22 Career Development, and an R35 Outstanding Investigator Award to N.G.S. through NIAID(AI2573-01) and NIGMS (1R35GM125034-01), respectively.

REFERENCES

Alford, R. F., Leaver-Fay, A., Jeliazkov, J. R., O'Meara, M., DiMaio, F. P., Park, H., et al. (2017). The Rosetta all-atom energy function for macromolecular modeling and design. *Journal of Chemical Theory and Computation*, *13*(6), 3031–3048. https://doi.org/10.1021/acs.jctc.7b00125.

André, I., Bradley, P., Wang, C., & Baker, D. (2007). Prediction of the structure of symmetrical protein assemblies. *Proceedings of the National Academy of Sciences of the United States of America*, *104*(45), 17656–17661. https://doi.org/10.1073/pnas.0702626104.

Anglister, J., Srivastava, G., & Naider, F. (2016). Detection of intermolecular NOE interactions in large protein complexes. *Progress in Nuclear Magnetic Resonance Spectroscopy*, *97*(Suppl. C), 40–56. https://doi.org/10.1016/j.pnmrs.2016.08.002.

Balakrishnan, S., Kamisetty, H., Carbonell, J. G., Lee, S.-I., & Langmead, C. J. (2011). Learning generative models for protein fold families. *Proteins*, *79*(4), 1061–1078. https://doi.org/10.1002/prot.22934.

Bender, B. J., Cisneros, A., Duran, A. M., Finn, J. A., Fu, D., Lokits, A. D., et al. (2016). Protocols for molecular modeling with Rosetta3 and RosettaScripts. *Biochemistry*, *55*(34), 4748–4763. https://doi.org/10.1021/acs.biochem.6b00444.

Berjanskii, M. V., & Wishart, D. S. (2017). Unraveling the meaning of chemical shifts in protein NMR. *Biochimica Et Biophysica Acta*, *1865*(11 Pt. B), 1564–1576. https://doi.org/10.1016/j.bbapap.2017.07.005.

Berman, H. M., Westbrook, J., Feng, Z., Gilliland, G., Bhat, T. N., Weissig, H., et al. (2000). The protein data bank. *Nucleic Acids Research*, *28*, 235–242. https://doi.org/10.1093/nar/28.1.235.

Blackledge, M. (2005). Recent progress in the study of biomolecular structure and dynamics in solution from residual dipolar couplings. *Progress in Nuclear Magnetic Resonance Spectroscopy*, *46*(1), 23–61. https://doi.org/10.1016/j.pnmrs.2004.11.002.

Blum, B., Jordan, M. I., & Baker, D. (2010). Feature space resampling for protein conformational search. *Proteins: Structure, Function, and Bioinformatics*, *78*(6), 1583–1593. https://doi.org/10.1002/prot.22677.

Bradley, P., & Baker, D. (2006). Improved beta-protein structure prediction by multilevel optimization of nonlocal strand pairings and local backbone conformation. *Proteins: Structure, Function, and Bioinformatics*, *65*(4), 922–929. https://doi.org/10.1002/prot.21133.

Bradley, P., Misura, K. M. S., & Baker, D. (2005). Toward high-resolution de novo structure prediction for small proteins. *Science (New York, N.Y.)*, *309*(5742), 1868–1871. https://doi.org/10.1126/science.1113801.

Braun, T., Leman, J. K., & Lange, O. F. (2015). Combining evolutionary information and an iterative sampling strategy for accurate protein structure prediction. *PLoS Computational Biology*, *11*(12), e1004661. https://doi.org/10.1371/journal.pcbi.1004661.

Brunette, T. J., & Brock, O. (2005). Improving protein structure prediction with model-based search. *Bioinformatics (Oxford, England)*, *21*(Suppl. 1), i66–i74. https://doi.org/10.1093/bioinformatics/bti1029.

Brunette, T., & Brock, O. (2008). Guiding conformation space search with an all-atom energy potential. *Proteins: Structure, Function, and Bioinformatics*, *73*(4), 958–972. https://doi.org/10.1002/prot.22123.

Canutescu, A. A., & Dunbrack, R. L. (2003). Cyclic coordinate descent: A robotics algorithm for protein loop closure. *Protein Science: A Publication of the Protein Society*, *12*(5), 963–972.

Clore, G. M., & Iwahara, J. (2009). Theory, practice, and applications of paramagnetic relaxation enhancement for the characterization of transient low-population states of biological macromolecules and their complexes. *Chemical Reviews*, *109*(9), 4108–4139. https://doi.org/10.1021/cr900033p.

Das, R., André, I., Shen, Y., Wu, Y., Lemak, A., Bansal, S., et al. (2009). Simultaneous prediction of protein folding and docking at high resolution. *Proceedings of the National Academy of Sciences of the United States of America*, *106*(45), 18978–18983. https://doi.org/10.1073/pnas.0904407106.

de Juan, D., Pazos, F., & Valencia, A. (2013). Emerging methods in protein co-evolution. *Nature Reviews Genetics*, *14*(4), 249. https://doi.org/10.1038/nrg3414.

Donald, B. R., & Martin, J. (2009). Automated NMR assignment and protein structure determination using sparse dipolar coupling constraints. *Progress in Nuclear Magnetic Resonance Spectroscopy*, *55*(2), 101–127. https://doi.org/10.1016/j.pnmrs.2008.12.001.

Ekeberg, M., Lövkvist, C., Lan, Y., Weigt, M., & Aurell, E. (2013). Improved contact prediction in proteins: Using pseudolikelihoods to infer Potts models. *Physical Review E*, *87*(1), 012707. https://doi.org/10.1103/PhysRevE.87.012707.

Evangelidis, T., Nerli, S., Nováček, J., Brereton, A. E., Karplus, P. A., Dotas, R. R., et al. (2018). Automated NMR resonance assignments and structure determination using a minimal set of 4D spectra. *Nature Communications*, *9*(1), 384. https://doi.org/10.1038/s41467-017-02592-z.

Gront, D., Kulp, D. W., Vernon, R. M., Strauss, C. E. M., & Baker, D. (2011). Generalized fragment picking in Rosetta: Design, protocols and applications. *PLoS One*, *6*(8), e23294. https://doi.org/10.1371/journal.pone.0023294.

Guerry, P., & Herrmann, T. (2011). Advances in automated NMR protein structure determination. *Quarterly Reviews of Biophysics*, *44*(3), 257–309.

Güntert, P. (2003). Automated NMR protein structure calculation. *Progress in Nuclear Magnetic Resonance Spectroscopy*, *43*(3), 105–125. https://doi.org/10.1016/S0079-6565(03)00021-9.

Güntert, P. (2009). Automated structure determination from NMR spectra. *European Biophysics Journal*, *38*(2), 129–143. https://doi.org/10.1007/s00249-008-0367-z.

Gutmanas, A., Adams, P. D., Bardiaux, B., Berman, H. M., Case, D. A., Fogh, R., et al. (2015). NMR exchange format: A unified and open standard for representation of NMR restraint data. *Nature Structural & Molecular Biology*, *22*(6), 433–434. https://doi.org/10.1038/nsmb.3041.

Hartlmüller, C., Göbl, C., & Madl, T. (2016). Prediction of protein structure using surface accessibility data. *Angewandte Chemie International Edition*, *55*(39), 11970–11974. https://doi.org/10.1002/anie.201604788.

Herrmann, T., Güntert, P., & Wüthrich, K. (2002). Protein NMR structure determination with automated NOE assignment using the new software CANDID and the torsion angle dynamics algorithm DYANA. *Journal of Molecular Biology*, *319*(1), 209–227. https://doi.org/10.1016/S0022-2836(02)00241-3.

Huang, Y. J., Swapna, G. V. T., Rajan, P. K., Ke, H., Xia, B., Shukla, K., et al. (2003). Solution NMR structure of ribosome-binding factor a (RbfA), a cold-shock adaptation protein from Escherichia coli. *Journal of Molecular Biology*, *327*(2), 521–536. https://doi.org/10.1016/S0022-2836(03)00061-5.

Jones, D. T., Buchan, D. W. A., Cozzetto, D., & Pontil, M. (2012). PSICOV: Precise structural contact prediction using sparse inverse covariance estimation on large multiple sequence alignments. *Bioinformatics*, *28*(2), 184–190. https://doi.org/10.1093/bioinformatics/btr638.

Kamisetty, H., Ovchinnikov, S., & Baker, D. (2013). Assessing the utility of coevolution-based residue–residue contact predictions in a sequence- and structure-rich era. *Proceedings of the National Academy of Sciences of the United States of America*, *110*(39), 15674–15679. https://doi.org/10.1073/pnas.1314045110.

Keeler, J. (2010). *Understanding NMR spectroscopy* (2nd ed.). Wiley. Retrieved from https://www.wiley.com/en-us/Understanding+NMR+Spectroscopy%2C+2nd+Edition-p-9780470746080.

Kim, D. E., Chivian, D., & Baker, D. (2004). Protein structure prediction and analysis using the Robetta server. *Nucleic Acids Research*, *32*, W526–W531. [Web Server issue]. https://doi.org/10.1093/nar/gkh468.

Lange, O. F. (2014). Automatic NOESY assignment in CS-RASREC-Rosetta. *Journal of Biomolecular NMR*, *59*(3), 147–159. https://doi.org/10.1007/s10858-014-9833-3.

Lange, O., & Baker, D. (2012). Resolution-adapted recombination of structural features significantly improves sampling in restraint-guided structure calculation. *Proteins*, *80*(3), 884–895. https://doi.org/10.1002/prot.23245.

Lange, O. F., Rossi, P., Sgourakis, N. G., Song, Y., Lee, H.-W., Aramini, J. M., et al. (2012). Determination of solution structures of proteins up to 40 kDa using CS-Rosetta with sparse NMR data from deuterated samples. *Proceedings of the National Academy of Sciences of the United States of America*, *109*(27), 10873–10878. https://doi.org/10.1073/pnas.1203013109.

Leaver-Fay, A., O'Meara, M. J., Tyka, M., Jacak, R., Song, Y., Kellogg, E. H., et al. (2013). Scientific benchmarks for guiding macromolecular energy function improvement. *Methods in Enzymology*, *523*, 109–143. https://doi.org/10.1016/B978-0-12-394292-0.00006-0.

Leaver-Fay, A., Tyka, M., Lewis, S. M., Lange, O. F., Thompson, J., Jacak, R., et al. (2011). ROSETTA3: An object-oriented software suite for the simulation and design of macromolecules. *Methods in Enzymology*, *487*, 545–574. https://doi.org/10.1016/B978-0-12-381270-4.00019-6.

Linge, J. P., Habeck, M., Reiping, W., & Nilges, M. (2003). ARIA: Automated NOE assignment and NMR structure calculation. *Bioinformatics*, *19*(2), 315–316. https://doi.org/10.1093/bioinformatics/19.2.315.

Marks, D. S., Colwell, L. J., Sheridan, R., Hopf, T. A., Pagnani, A., Zecchina, R., et al. (2011). Protein 3D structure computed from evolutionary sequence variation. *PLoS One*, *6*(12), e28766. https://doi.org/10.1371/journal.pone.0028766.

Marks, D. S., Hopf, T. A., & Sander, C. (2012). Protein structure prediction from sequence variation. *Nature Biotechnology*, *30*, 1072–1080. https://doi.org/10.1038/nbt.2419.

Mielke, S. P., & Krishnan, V. V. (2009). Characterization of protein secondary structure from NMR chemical shifts. *Progress in Nuclear Magnetic Resonance Spectroscopy*, *54*(3–4), 141–165. https://doi.org/10.1016/j.pnmrs.2008.06.002.

Nerli, S., McShan, A. C., & Sgourakis, N. G. (2018). Chemical shift-based methods in NMR structure determination. *Progress in Nuclear Magnetic Resonance Spectroscopy*, *106–107*, 1–25. https://doi.org/10.1016/j.pnmrs.2018.03.002.

Nilges, M. (1993). A calculation strategy for the structure determination of symmetric dimers by 1H NMR. *Proteins: Structure, Function, and Bioinformatics*, *17*(3), 297–309. https://doi.org/10.1002/prot.340170307.

Nilges, M. (1997). Ambiguous distance data in the calculation of NMR structures. *Folding and Design*, *2*, S53–S57. https://doi.org/10.1016/S1359-0278(97)00064-3.

O'Meara, M. J., Leaver-Fay, A., Tyka, M. D., Stein, A., Houlihan, K., DiMaio, F., et al. (2015). Combined covalent-electrostatic model of hydrogen bonding improves structure prediction with Rosetta. *Journal of Chemical Theory and Computation*, *11*(2), 609–622. https://doi.org/10.1021/ct500864r.

Ovchinnikov, S., Park, H., Varghese, N., Huang, P.-S., Pavlopoulos, G. A., Kim, D. E., et al. (2017). Protein structure determination using metagenome sequence data. *Science*, *355*(6322), 294–298. https://doi.org/10.1126/science.aah4043.

Porter, J. R., Weitzner, B. D., & Lange, O. F. (2015). A framework to simplify combined sampling strategies in Rosetta. *PLoS One*, *10*(9), e0138220. https://doi.org/10.1371/journal.pone.0138220.

Qian, B., Raman, S., Das, R., Bradley, P., McCoy, A. J., Read, R. J., et al. (2007). High-resolution structure prediction and the crystallographic phase problem. *Nature*, *450*(7167), 259. https://doi.org/10.1038/nature06249.

Raman, S., Lange, O. F., Rossi, P., Tyka, M., Wang, X., Aramini, J., et al. (2012). NMR structure determination for larger proteins using backbone-only data. *Science*, *327*(5968), 1014–1018. https://doi.org/10.1126/science.1183649.

Schmitz, C., Vernon, R., Otting, G., Baker, D., & Huber, T. (2012). Protein structure determination from pseudocontact shifts using ROSETTA. *Journal of Molecular Biology*, *416*(5), 668–677. https://doi.org/10.1016/j.jmb.2011.12.056.

Sgourakis, N. G., Lange, O. F., DiMaio, F., André, I., Fitzkee, N. C., Rossi, P., et al. (2011). Determination of the structures of symmetric protein oligomers from NMR chemical shifts and residual dipolar couplings. *Journal of the American Chemical Society*, *133*(16), 6288–6298. https://doi.org/10.1021/ja111318m.

Shen, Y., & Bax, A. (2007). Protein backbone chemical shifts predicted from searching a database for torsion angle and sequence homology. *Journal of Biomolecular NMR*, *38*(4), 289–302. https://doi.org/10.1007/s10858-007-9166-6.

Shen, Y., & Bax, A. (2013). Protein backbone and sidechain torsion angles predicted from NMR chemical shifts using artificial neural networks. *Journal of Biomolecular NMR*, *56*(3), 227–241. https://doi.org/10.1007/s10858-013-9741-y.

Shen, Y., Lange, O., Delaglio, F., Rossi, P., Aramini, J. M., Liu, G., et al. (2008). Consistent blind protein structure generation from NMR chemical shift data. *Proceedings of the National Academy of Sciences of the United States of America*, *105*(12), 4685–4690. https://doi.org/10.1073/pnas.0800256105.

Simons, K. T., Kooperberg, C., Huang, E., & Baker, D. (1997). Assembly of protein tertiary structures from fragments with similar local sequences using simulated annealing and bayesian scoring functions. *Journal of Molecular Biology*, *268*(1), 209–225. https://doi.org/10.1006/jmbi.1997.0959.

Söding, J. (2005). Protein homology detection by HMM-HMM comparison. *Bioinformatics (Oxford, England)*, *21*(7), 951–960. https://doi.org/10.1093/bioinformatics/bti125.

Söding, J., Biegert, A., & Lupas, A. N. (2005). The HHpred interactive server for protein homology detection and structure prediction. *Nucleic Acids Research*, *33*, W244–W248. [Web Server issue]. https://doi.org/10.1093/nar/gki408.

Sułkowska, J. I., Morcos, F., Weigt, M., Hwa, T., & Onuchic, J. N. (2012). Genomics-aided structure prediction. *Proceedings of the National Academy of Sciences of the United States of America*, *109*(26), 10340–10345. https://doi.org/10.1073/pnas.1207864109.

Tamaki, H., Egawa, A., Kido, K., Kameda, T., Kamiya, M., Kikukawa, T., et al. (2016). Structure determination of uniformly (13)C, (15)N labeled protein using qualitative distance restraints from MAS solid-state (13)C-NMR observed paramagnetic relaxation enhancement. *Journal of Biomolecular NMR*, *64*(1), 87–101. https://doi.org/10.1007/s10858-015-0010-0.

Tang, Y., Huang, Y. J., Hopf, T. A., Sander, C., Marks, D. S., & Montelione, G. T. (2015). Protein structure determination by combining sparse NMR data with evolutionary couplings. *Nature Methods*, *12*(8), 751–754. https://doi.org/10.1038/nmeth.3455.

Thompson, J., & Baker, D. (2011). Incorporation of evolutionary information into Rosetta comparative modeling. *Proteins*, *79*(8), 2380–2388. https://doi.org/10.1002/prot.23046.

Thompson, J. M., Sgourakis, N. G., Liu, G., Rossi, P., Tang, Y., Mills, J. L., et al. (2012). Accurate protein structure modeling using sparse NMR data and homologous structure information. *Proceedings of the National Academy of Sciences of the United States of America*, *109*(25), 9875–9880. https://doi.org/10.1073/pnas.1202485109.

Tyka, M. D., Keedy, D. A., André, I., Dimaio, F., Song, Y., Richardson, D. C., et al. (2011). Alternate states of proteins revealed by detailed energy landscape mapping. *Journal of Molecular Biology*, *405*(2), 607–618. https://doi.org/10.1016/j.jmb.2010.11.008.

Ulrich, E. L., Akutsu, H., Doreleijers, J. F., Harano, Y., Ioannidis, Y. E., Lin, J., et al. (2008). BioMagResBank. *Nucleic Acids Research*, *36*, D402–D408. https://doi.org/10.1093/nar/gkm957.

Wang, C., Bradley, P., & Baker, D. (2007). Protein-protein docking with backbone flexibility. *Journal of Molecular Biology*, *373*(2), 503–519. https://doi.org/10.1016/j.jmb.2007.07.050.

Wunderlich, Z., Acton, T. B., Liu, J., Kornhaber, G., Everett, J., Carter, P., et al. (2004). The protein target list of the northeast structural genomics consortium. *Proteins: Structure, Function, and Bioinformatics*, *56*(2), 181–187. https://doi.org/10.1002/prot.20091.

Würz, J. M., Kazemi, S., Schmidt, E., Bagaria, A., & Güntert, P. (2017). NMR-based automated protein structure determination. *Archives of Biochemistry and Biophysics*, *628*, 24–32. https://doi.org/10.1016/j.abb.2017.02.011.

Yagi, H., Pilla, K. B., Maleckis, A., Graham, B., Huber, T., & Otting, G. (2013). Three-dimensional protein fold determination from backbone amide pseudocontact shifts generated by lanthanide tags at multiple sites. *Structure*, *21*(6), 883–890. https://doi.org/10.1016/j.str.2013.04.001.

Zhang, Z., Porter, J., Tripsianes, K., & Lange, O. F. (2014). Robust and highly accurate automatic NOESY assignment and structure determination with Rosetta. *Journal of Biomolecular NMR*, *59*(3), 135–145. https://doi.org/10.1007%2Fs10858-014-9832-4.

CHAPTER THIRTEEN

Combining Evolutionary Covariance and NMR Data for Protein Structure Determination

Yuanpeng Janet Huang*,†, Kelly P. Brock‡, Yojiro Ishida*,§, Gurla V.T. Swapna*,†, Masayori Inouye*,§, Debora S. Marks‡, Chris Sander¶, Gaetano T. Montelione*,†,§,1

*Center for Advanced Biotechnology and Medicine, Rutgers, The State University of New Jersey, Piscataway, NJ, United States

†Department of Molecular Biology and Biochemistry, Rutgers, The State University of New Jersey, Piscataway, NJ, United States

‡Department of Systems Biology, Harvard Medical School, Boston, MA, United States

§Department of Biochemistry and Molecular Biology, Robert Wood Johnson Medical School, Rutgers, The State University of New Jersey, Piscataway, NJ, United States

¶Department of Cell Biology, Harvard Medical School and cBio Center, Dana-Farber Cancer Institute, Boston, MA, United States

[1]Corresponding author: e-mail address: gtm@rutgers.edu

Contents

Methods in Enzymology, Volume 614
ISSN 0076-6879
https://doi.org/10.1016/bs.mie.2018.11.004

Abstract

Accurate protein structure determination by solution-state NMR is challenging for proteins greater than about 20kDa, for which extensive perdeuteration is generally required, providing experimental data that are incomplete (sparse) and ambiguous. However, the massive increase in evolutionary sequence information coupled with advances in methods for sequence covariance analysis can provide reliable residue–residue contact information for a protein from sequence data alone. These "evolutionary couplings (ECs)" can be combined with sparse NMR data to determine accurate 3D protein structures. This hybrid "EC–NMR" method has been developed using NMR data for several soluble proteins and validated by comparison with corresponding reference structures determined by X-ray crystallography and/or conventional NMR methods. For small proteins, only backbone resonance assignments are utilized, while for larger proteins both backbone and some sidechain methyl resonance assignments are generally required. ECs can be combined with sparse NMR data obtained on deuterated, selectively protonated protein samples to provide structures that are more accurate and complete than those obtained using such sparse NMR data alone. EC–NMR also has significant potential for analysis of protein structures from solid-state NMR data and for studies of integral membrane proteins. The requirement that ECs are consistent with NMR data recorded on a specific member of a protein family, under specific conditions, also allows identification of ECs that reflect alternative allosteric or excited states of the protein structure.

1. INTRODUCTION

One key goal of structural bioinformatics over the past few decades has been the reliable prediction of tertiary structure contacts from protein sequence data (Gobel, Sander, Schneider, & Valencia, 1994; Neher, 1994; Shindyalov, Kolchanov, & Sander, 1994; Taylor & Hatrick, 1994; Thomas, Casari, & Sander, 1996). Residues that contact one another may covary in the course of evolution; e.g., when a large, buried hydrophobic residue which is packed against small hydrophobic residues in the protein core is mutated to a small hydrophobic residue, the space created may be compensated by mutation of the contacting residue(s) to larger hydrophobic residues. Detection of such sequence correlations from multiple sequence alignment (MSA) data can therefore potentially provide information about residue contacts, which in turn can be used to guide protein structure prediction and molecular modeling. Contact prediction has been pursued by many groups and has been a central activity of the Critical Assessment of Protein Structure Prediction experiment (Schaarschmidt, Monastyrskyy, Kryshtafovych, & Bonvin, 2018).

In interpreting contacts from sequence covariance, a challenge is created by *transitive correlations*, or relay effects; i.e., the challenge of distinguishing A–B covariation due to direct A → B interactions from A–B covariation due to relayed A → C and C → B interactions. Because multiple relayed pathways are possible, covariation due to such nondirect interactions may be more pronounced than covariance due to direct residue–residue interactions. Unless properly weighted, these relay effects can result in hot spots of covariance due to local networks in the protein structure (Morcos et al., 2011). These networks of correlations confound the extraction of directly coupled residue pairs useful for contact prediction and for providing distance restraints for molecular modeling.

Motivated by the potential of using sequence-based residue covariance restraints for molecular modeling and protein structure prediction, several groups have developed maximum entropy global statistical models, maximum likelihood parameter inference methods, and machine learning (e.g., neural nets) methods that can reliably distinguish direct evolutionary couplings (ECs) from transitive correlations, allowing analysis of direct residue–residue couplings from multiple alignments of structurally related protein sequences (Ekeberg, Lovkvist, Lan, Weigt, & Aurell, 2013; Jones, Buchan, Cozzetto, & Pontil, 2012; Kamisetty, Ovchinnikov, & Baker, 2013; Lapedes, Giraud, & Jarzynski, 2002; Marks et al., 2011; Morcos et al., 2011; Sulkowska, Morcos, Weigt, Hwa, & Onuchic, 2012). These ECs can provide accurate information about residue pair contacts (RPCs) present in the 3D structures of proteins (Hopf et al., 2012; Jones et al., 2012; Kamisetty et al., 2013; Marks et al., 2011; Marks, Hopf, & Sander, 2012; Morcos et al., 2011; Sulkowska et al., 2012). These predicted contacts can then be used, together with traditional spatial-restraint modeling methods, restrained molecular dynamics, knowledge-based, and/or energy minimization methods to model the native structure of the protein, often providing correct identification of the protein fold (Hopf et al., 2012; Marks et al., 2011; Ovchinnikov et al., 2016, 2015, 2017; Sheridan et al., 2015; Sulkowska et al., 2012). ECs have also been successfully used to identify RPCs at protein–protein interfaces (dos Santos, Morcos, Jana, Andricopulo, & Onuchic, 2015; Hopf et al., 2014), to predict the structures of protein–protein complexes (Hopf et al., 2014; Ovchinnikov, Kamisetty, & Baker, 2014), and to model tertiary structures of RNA molecules (Weinreb et al., 2016). The ability to identify direct covariance couplings from MSAs is a significant advance, which is having a huge impact on the accuracy of template-free protein structure predictions, as assessed by the Critical Assessment of Protein Structure Prediction

(CASP) (Abriata, Tamo, Monastyrskyy, Kryshtafovych, & Dal Peraro, 2018; Ovchinnikov et al., 2015; Schaarschmidt et al., 2018). For these reasons, contact prediction has evolved into a very important area of structural bioinformatics research (Abriata et al., 2018; Adhikari, Hou, & Cheng, 2018; Buchan & Jones, 2018; Kassem, Christoffersen, Cavalli, & Lindorff-Larsen, 2018; Schaarschmidt et al., 2018; Simkovic, Ovchinnikov, Baker, & Rigden, 2017; Wang, Sun, & Xu, 2018).

In addition to their role in modeling the native structures of proteins, protein complexes, and RNA, it is important to also recognize that high-confidence ECs may also reflect transient protein–protein interactions (Toth-Petroczy et al., 2016), alternative conformational or allosteric states (Morcos, Jana, Hwa, & Onuchic, 2013; Toth-Petroczy et al., 2016), and/or more subtle features of protein structural dynamics (Tang et al., 2015). Covariance pairs that are "false positives" with respect to the lowest energy conformational state of the protein structure may alternatively provide critical information about protein dynamics and/or function.

As outlined in our recent monograph (Huang, Brock, Sander, Marks, & Montelione, 2018), while a breakthrough in the field of protein structure prediction, the modeling of 3D structures from EC data alone has a number of limitations in generating accurate protein models. ECs are derived from MSAs, providing information about contacts that are conserved across a protein or protein domain family. These contacts may not all be present in the particular member of the protein family under investigation. Hence, "false positive" contact predictions may result from "structural drift" through the course of evolution. Some predicted contacts which do indeed occur in homologs may be inconsistent with the native structure of the subject protein, and if satisfied will result in an erroneous model. In addition, even highly ranked ECs generated from extensive MSA data may not reflect direct structural contacts in the predominantly populated native protein structure, as sequence covariance may result from contacts in alternative conformational states adopted by the target protein, allosteric networks, homooligomerization, and/or indirect residue interactions that are mediated through substrates or binding partners. They may also result from simple false positives (FPs) in the parameter inference computation, especially when insufficiently diverse sequences are available across the entire protein structure of interest, or when the sequence coverage is only sufficient for EC prediction over a segment of the target protein sequence. As a result, EC-derived models of proteins derived exclusively from EC data may distorted, or differ in crucial details, from the predominant native structure.

Solution-state NMR can generally provide accurate three-dimensional (3D) structures of small (MW $< \sim 20$ kDa) proteins (Mao, Guan, & Montelione, 2011; Mao, Tejero, Baker, & Montelione, 2014). However, for larger proteins efficient ^{1}H—^{1}H transverse spin relaxation results in broad NMR linewidths, preventing collection of sufficient data to allow structural analysis. Perdeuteration and selective reprotonation (i.e., replacement of most ^{1}H atoms with ^{2}H) decrease transverse relaxation rates of the remaining ^{1}H, ^{15}N, and ^{13}C nuclei, increasing the sensitivity and feasibility of NMR for larger proteins (Gardner, Rosen, & Kay, 1997). However, perdeuteration also reduces the number of ^{1}H's providing ^{1}H—^{1}H NOEs, providing much fewer structural restraints. This incompleteness of NOE data can be compensated to some degree using dihedral restraints based on chemical shift, and orientation restraints from residual dipolar coupling (RDC) data. Protein structure models based on such "sparse NMR data" can also be improved using advanced knowledge-based molecular modeling methods (Lange et al., 2012; Raman et al., 2010; Sgourakis et al., 2014). However, the protein structures derived from perdeuterated samples with limited sidechain resonance assignments are generally less accurate and precise than those obtained for smaller, fully protonated proteins with complete sidechain resonance assignments.

In recent work (Tang et al., 2015), we have demonstrated that spatial contact information derived from sparse NMR data and from ECs can provide highly complementary information. This creates the opportunity to combine NMR and EC data for more reliable structure determination than can be achieved using either data type alone. This hybrid approach is referred to as "EC–NMR." The method is particularly useful for proteins which cannot be modeled from a homologous template structure, and for which only sparse NMR data can be obtained.

Sparse NMR contact information is incomplete and often ambiguous in its assignment to specific ^{1}H—^{1}H interactions. Nonetheless, all (or most) of the NOE, chemical shift, and RDC data should be consistent with the 3D structure model(s), averaged over the dynamic states accessible to the protein structure at the temperature of study. Predicted contacts, if reliable, can complement this spectroscopic information to provide missing contact information and more accurate models. However, the ECs potentially predict interactions that are not consistent with the predominant structure of the subject protein under the conditions that the NMR data are acquired. The requirement that the overall structure be consistent with the experimental NMR data, however, provides "hard" constraints on the interpretation of ECs, allowing identification, and removal of proposed RPCs that are inconsistent with the dominant structure present in the solution conditions under investigation (Tang et al., 2015).

2. THE EC–NMR ALGORITHM AND IMPLEMENTATION

2.1 General Approach

The general EC–NMR approach as proposed by Tang et al. (2015) is outlined in Fig. 1. The overall process can be divided into three stages. Stage 1 involves generation of ECs from MSAs using maximum entropy or pseudo-likelihood models of the protein sequence (Ekeberg et al., 2013; Jones et al., 2012; Kamisetty et al., 2013; Marks et al., 2011; Morcos et al., 2011). In Stage 2, "sparse NMR data" is collected using uniformly ^{13}C,^{15}N-enriched and/or ^{2}H,^{13}C,^{15}N-enriched protein samples prepared with ^{1}H—^{13}C labeling of sidechain methyl groups (Gardner et al., 1997; Rosen et al., 1996; Tugarinov, Kanelis, & Kay, 2006). These data are used to determine backbone ^{1}H^{N}, ^{13}C, and ^{15}N, as well as sidechain amide ^{1}H^{N}—^{15}N and some methyl ^{13}CH$_3$ resonance assignments. Backbone dihedral angle restraints may also generated from the ^{13}C$^{\alpha}$ and ^{13}C$^{\beta}$ chemical shift data using programs like TALOS-N (Shen & Bax, 2015). Unassigned NOESY peak lists are then compiled from simultaneous 3D ^{15}N,^{13}C-NOESY spectra. In some cases, ^{15}N—^{1}H RDC data are also measured using one or more RDC alignment media. In Stage 3, automated NOESY analysis software (Huang, Tejero, Powers, & Montelione, 2006) is used to identify and iteratively refine a list of distance restraints using both EC and NMR data simultaneously. These formatted distance restraints are then used to generate a small ensemble of 3D structures. In this process, chemical shift, NOESY peak list, EC, and RDC data are interpreted *together* to determine NOESY cross-peak assignments, rule out ECs that are inconsistent with the NMR data and/or intermediate 3D structures, and to generate initial 3D models of the protein.

A key component of this process is the generation of accurate MSAs for large numbers of protein sequences. In generating these MSAs, it is important to choose an appropriate range of evolutionary neighbors: not too distant, so as to optimize specificity of structural constraints to the target of interest, and not too restricted, so as to retrieve as many sequences as possible at maximum sequence diversity (Huang et al., 2018; Marks et al., 2012). It is also critical to suppress transitive (indirect) correlations using methods such as global entropy maximization, pseudo-likelihood scoring,

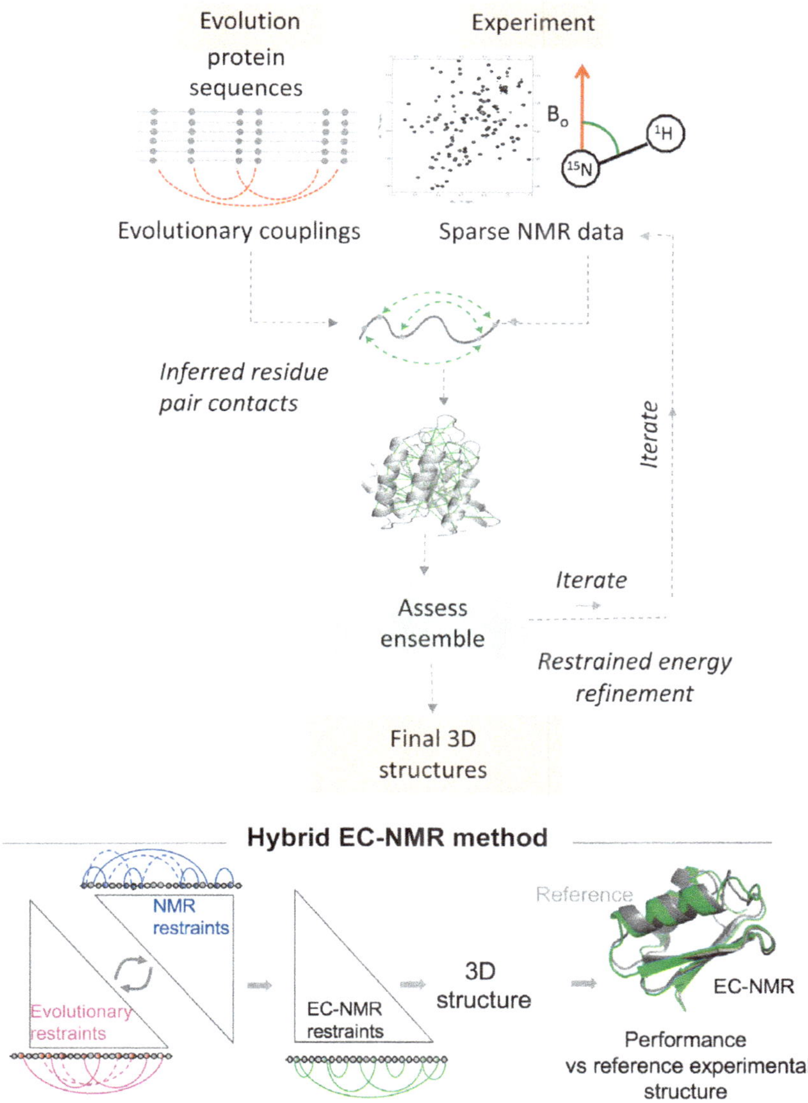

Fig. 1 3D structure determination by the hybrid EC–NMR method. The hybrid EC–NMR strategy combines evolutionary coupling (EC) information from multiple sequence alignments with sparse experimental nuclear magnetic resonance (NMR) data. *Reproduced from Tang, Y., Huang, Y. J., Hopf, T. A., Sander, C., Marks, D. S., & Montelione, G. T. (2015). Protein structure determination by combining sparse NMR data with evolutionary couplings.* Nature Methods, 12, *751–754. doi: 10.1038/nmeth.3455.*

or machine learning (Ekeberg et al., 2013; Jones et al., 2012; Kamisetty et al., 2013; Lapedes et al., 2002; Marks et al., 2011; Morcos et al., 2011; Sulkowska et al., 2012). In our published implementation of EC–NMR (Huang et al., 2018; Tang et al., 2015), evolutionary residue–residue couplings were computed using a pseudo-likelihood maximization algorithm (Ekeberg et al., 2013) implemented in the computer program *plmc*, part of the *EVcouplings* software suite (https://github.com/debbiemarkslab/plmc). However, other methods of contact prediction may also be used to generate ECs (Adhikari et al., 2018; Buchan & Jones, 2018; Jones et al., 2012; Kamisetty et al., 2013; Kassem et al., 2018; Marks et al., 2011; Morcos et al., 2011; Sulkowska et al., 2012; Wang et al., 2018).

Although alternative implementations may be anticipated, the automated combined analysis of NMR and EC data has been implemented in the NOESY assignment program *ASDP* (formally AutoStructure) (Huang, Mao, Xu, & Montelione, 2015; Huang et al., 2006; Tang et al., 2015). NMR and EC data are combined to generate distance restraints. Intermediate 3D structures are then generated from these distance restraints provided by *ASDP* using the program *CYANA* (Herrmann, Guntert, & Wuthrich, 2002), although other programs (e.g., NIH-XPLOR (Schwieters, Kuszewski, & Clore, 2006), CHARMM (Brooks et al., 2009), CNS (Adams et al., 2013), Rosetta (Lange et al., 2012; Lange & Baker, 2012), or ARIA (Nilges, 1995)) could be used to generate 3D structures. The resulting RPCs, derived by the combined analysis of EC and NMR data, are then deconvoluted into atom-specific distance restraints, which are then used to refine the protein structure using restrained energy minimization. In our published implementation (Tang et al., 2015), the refinement step used a specific restrained energy minimization and knowledge-based modeling protocol with the program *Rosetta*, as described by Mao et al. (2014), but alternative energy refinement protocols could also be used.

2.2 Perdeuterated Protein Sample Preparation and NMR Data Collection

Sparse NMR data (i.e., involving only backbone and some sidechain resonances) can generally be obtained for perdeuterated proteins with molecular weights as large as 40–70 kDa (Hiller et al., 2008; Lange et al., 2012; Raman et al., 2010) and have been used to determine chain folds for proteins as large as 82 kDa (Grishaev, Tugarinov, Kay, Trewhella, & Bax, 2008; Tugarinov, Choy, Orekhov, & Kay, 2005). In our laboratory, isotope-enriched proteins, and particularly perdeuterated $^{13}C,^{15}N$-enriched proteins, are generally produced using the condensed-phase *E. coli* single protein production (SPP) system. SPP provides high yields of perdeuterated soluble and membrane proteins with Ile(δ1), Leu, Val, and Ala methyl labeling in small volumes of defined media at reduced cost (Schneider, Inouye, Montelione, & Roth, 2009; Schneider et al., 2010; Vaiphei, Tang, Montelione, & Inouye, 2011). Backbone resonance assignments are determined using standard triple-resonance method with deuterium decoupling. NOESY data collection includes standard 3D N-NOESY, 3D C-NOESY (or 3D simultaneous N,C-NOESY), and for larger proteins HNCH and HCCH pseudo-4D NOESY (Diercks, Coles, & Kessler, 1999; Lange et al., 2012) data.

2.3 Calculation of ECs and Confidence Scores

In our published work, ECs were calculated using the *EVfold-plm* pipeline available at evfold.org, as described previously (Hopf et al., 2012). For structure modeling using ECs alone, EC pairs that are inconsistent with secondary structure prediction were removed (Hopf et al., 2012).

To assess the confidence of EC pairs, Hopf et al. (2014) have defined a score Q that measures how much the corresponding EC score is an outlier from the approximately Gaussian distribution expected from non-informative background coupling between the majority of positions. This measure depends on the shape of the EC scores distribution, which can be approximated by a mixed Gaussian-lognormal model, and has been shown to be a strong predictor of the accuracy of ECs (Hopf et al., 2014). Toth-Petroczy et al. (2016) have defined a related metric for defining "significant" ECs, the posterior probability P of membership in lognormal component of a Gaussian-lognormal mixed model fit to the distribution. In this measure, ECs with $P \geq 0.9$ (i.e., $\geq 90\%$ probability to be in the lognormal tail of the distribution) are considered to be significant. In our published work (Tang et al., 2015), we defined *high-confidence ECs* as all pairs with $Q(i,j) > 2$, i.e., couplings that exceed the background noise by a factor of at least 2; while for the contact maps presented here we plot ECs with $P \geq 0.9$.

2.4 Implementation in the Automated NOESY Assignment Program *ASDP*

The EC-NMR method is implemented within the automated NOESY cross-peak assignment program *ASDP* (Huang et al., 2006) (version 2.0), available from http://ec-nmr.nesg.org/. The following steps of the iterative EC–NMR analysis process (Tang et al., 2015) are outlined in Fig. 2. Additional details are provided in the original paper.

2.4.1 Step 1. Generate De Novo Initial NOE-Based Distance Constraints and EC-Based Ambiguous Residue Pair Contact (RPC) Distance Restraints

Here, initial NOE-based distance restraints are generated from NOESY peak list and chemical shift data using algorithms encoded in the *ASDP*

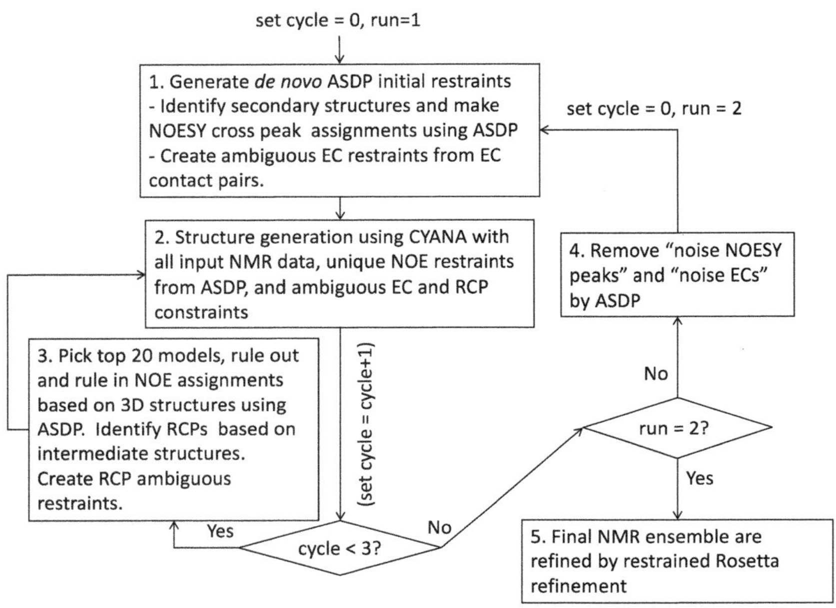

Fig. 2 Flowchart of iterative EC–NMR analysis process. The EC–NMR analysis process has been fully automated by integration into the automated NOESY NMR data assignment program *ASDP* (Huang et al., 2006; Huang, Powers, & Montelione, 2005). NOESY cross peak, $^{1}H—^{15}N$ resonance assignments, and $^{1}H—^{13}C$ methyl resonance assignments (where available) are combined with high-scoring EC contacts predicted from sequence covariance analysis. Distance restraints are identified by *ASDP* from these data and used together with RDC data (where available) to generate initial structural models with the program *CYANA* (Herrmann et al., 2002). Initial structures are then used to assess false-positive EC contact predictions and false-NOESY peaks (i.e., noise peaks), and to identify consistent residue pair contacts (RPCs). *ASDP* then uses the intermediate structured to rule in/rule out possible NOESY cross peak assignments and predicted EC residue–residue contacts. Once the analysis converges, these specific interatomic distance restraints are used for loop rebuilding and core repacking with the program *Rosetta* (Mao et al., 2014).

program (Huang et al., 2006). Secondary structures, including beta-strand alignments, are identified using previously described algorithms (Huang et al., 2006) that utilize chemical shift data (Wishart & Sykes, 1994) together with characteristic NOE contact patterns within and between secondary structure elements (Wuthrich, 1986). Additional NOE assignments are then ruled in and ruled out by *ASDP* based on uniqueness relative to the chemical shift list, NOESY cross-peak symmetry patterns, and the extensive network anchoring algorithms of *ASDP* (Huang et al., 2006). For perdeuterated proteins, a deuterium correction to the ^{13}C chemical shifts (Maltsev, Ying, & Bax, 2012) is applied in *ASDP* automatically, and longer distance cutoffs (up to 6 Å) are used for the NOEs since such interactions are often observable for longer distances in such perdeuterated proteins. For larger (>20 kDa) perdeuterated proteins, the NMR data also include 3D H^N-Me and Me-Me NOESY peak list data, which provide NOEs involving Val $^{13}C^{\gamma}H_3$, Leu $^{13}C^{\delta}H_3$, and Ile $^{13}C^{\delta 1}H_3$ methyl groups.

EC-based RPC ambiguous restraints are also generated in Step 1 as follows: Ambiguous distance restraints ($\leq$5 Å) are generated between every two carbon atoms [C_i, C_j] for each residue pair [i,j] in the EC list. ECs are ranked based on EC reliability scores (Sheridan et al., 2015). Only ECs with reliability score $\geq$0.5 were included in these calculations, with weights (w) initially set equal to their reliability scores. These parameters may be optimized differently for specific EC datasets.

2.4.2 Step 2. Initial 3D Structures Are Generated From These Initial RPC Restraints

One hundred decoy models (initial structures) are generated with distance restraints output by *ASDP* using the program *CYANA* (Herrmann et al., 2002). Specifically, input to CYANA (or an alternative structure generation program) includes $^1H—^{15}N$ RDC data (if available), dihedral angle restraints generated from backbone chemical shift data (Shen, Delaglio, Cornilescu, & Bax, 2009), together with unique NOE-based and EC-based ambiguous RPC distance restraints, identified by *ASDP* in Step 1.

The standard protocol of *Talos* + or *Talos-N* (Shen et al., 2009) is used to generate backbone dihedral angle restraints based on $^{13}C^{\alpha}$ and $^{13}C^{\beta}$ chemical shifts, and residues with "good" *Talos* scores (e.g., *Talos* + reliability score of 10) are restrained to phi and psi ranges of $\pm$20 degree. A deuterium correction to ^{13}C chemical shifts is also applied in *Talos* analysis for perdeuterated proteins (Shen et al., 2009).

2.4.3 Step 3. Initial 3D Structures Are Used to Rule In/Rule Out NOESY Peak Assignments, Identify Inconsistent EC Contacts, and to Define Additional RCPs

The "top" 20 models generated by *CYANA* are identified using a combined score comprised of the *NMR RPF* Recall (Huang et al., 2005) score and *CYANA* target function. The *NMR RPF* Recall score measures the fraction of NOESY cross peaks that can be explained by the decoy model structure. These "best scoring" 3D decoy structures are then used by *ASDP* to rule in and rule out potential NOE and EC assignments.

Using these decoy models, standard rules of *ASDP* are then used to make new NOESY cross-peak assignments, which are added as unique constraints to the distance restraint list. Structurally inconsistent NOE assignments from Step 1 are excluded using several network anchoring algorithms describe in original *AutoStructure* paper (Huang et al., 2006). Structurally inconsistent RPCs (SI-RPCs) are also identified when ambiguous RPC distance constraints are violated by >0.5 Å in more than 60% (e.g., 12 of 20) conformers. These SI-RPCs are excluded from the next cycle of structure generation calculations. Ambiguous RPC constraints that are satisfied by all 20 decoy conformers (i.e., no violation >0.5 Å) are reassigned weight $w=1.0$. No changes are made for the remaining RPCs, which have small violations among the 20 conformers. All RPCs are then again defined as ambiguous distance restraints, between all C atoms of residue i and all C atoms of residue j.

In this process, the *ASDP* software also identifies new RPCs, which are defined as long-range residue pairs (i.e., $|i-j|\geq 5$) that have at least one interatom (i.e., any H, N, or C atom with a resonance assignment) distance $\leq$5 Å apart in all of 20 conformers. These RPCs are added to the EC–NMR constraint list as ambiguous distance restraints between all C atoms of residue i and all C atoms of residue j, with weight $w=1.0$. These additional RPCs based on intermediate structures often, but not always, correspond to EC contact predictions with lower ranking scores in covariance analysis.

Steps 2 and 3 are then repeated two more cycles, resulting in an ensemble of 20 protein structure models.

2.4.4 Step 4. The Protein Structure Models From Cycle 3 Are Used to Identify NOE Peaks and RPCs That Are Inconsistent With These Intermediate Structures

RPCs and NOE peaks for which any possible assignment are structurally inconsistent in well-defined regions of these intermediate decoy models are then removed from the input data, and Steps 1–3 are then repeated again. Using intermediate structures to clean up the de novo initial distance restraints helps to regenerate better conformers for subsequent restrained energy optimization. "Noise NOESY cross peaks" are defined as all NOESY cross peaks with initial NOE assignments from Step 1 for which the corresponding constraint is violated by >10 Å in all 20 conformers from *Cycle* 3. "Noise ECs" are initial ECs from Step 1 for which the corresponding ambiguous restraint is violated with distance >10 Å in all 20 conformers from Cycle 3. These "noise" NOESY cross peaks and ECs are removed from the EC–NMR restraint list. It should be noted that "noise" NOESY peaks and RPCs identified by structural consistency are FPs with respect to the ground state of the protein sample under investigation, by may encode important information about allosteric states and other dynamic aspects of the protein structure. Variables Cycle and Run (Fig. 2) are used to control the repeated analyses of Steps 1–3. When the process begins, Cycle is set to 0 and Run is set to 1. After Steps 2 and 3 are repeated for three cycles, Step 4 is executed. If any "noise NOEs" and/or "noise ECs" are identified, Steps 1–3 are repeated again. The iterative process ends with Run = 2. No further runs are then executed to avoid potential overfitting.

2.5 Refinement of EC–NMR Structures With Rosetta

Each of the resulting NMR structure models are then further energy refined using the restrained Rosetta refinement protocol outlined in Mao et al. (2014). Detailed protocols for restrained Rosetta refinement are available online at http://www.nmr2.buffalo.edu/nesg.wiki/Rosetta_High_Resolution_Protein_Structure_Refinement_Protocol.

Atom–atom distance restraints are used by this Rosetta refinement protocol. The script *getCC.pl* from the *ASDP-2.0* package is used to generate specific atom–atom Rosetta distance restraints for each heavy atom (N, C, O, S) pair in RPCs which have interatomic distance ≤5 Å in all 20 models. These atom pairs are then assigned upper-bound restraints of 7 Å. NOE-based upper-bound distance restraints are loosened by 30%, in order to avoid over constraining these interatomic distances, before converting to the Rosetta restraint format. Restrained Rosetta refinement (Mao et al., 2014) is then done using either a stand-alone version of *Rosetta*, or using our Restrained Rosetta Refinement Server available online at: http://psvs-1_4-dev.nesg.org/consRosetta.html.

2.6 Access to ASDP Software and Tutorial for EC–NMR Calculations

Access to the *ASDP* software, together with a short tutorial for its use, is available online at: http://www-nmr.cabm.rutgers.edu/NMR software/asdp/Quick_Starts.html. Additional instructions for using *ASDP* are available at: http://www.nmr2.buffalo.edu/nesg.wiki/AutoStructure_ Structure_Determination_Program. A web-based tutorial for running EC–NMR calculations with *ASDP* is available online at: http://ec-nmr. nesg.org/tutorial.html. The online tutorial includes sample input and output data files (see online methods of Tang et al., 2015).

3. QUALITY CONTROL AND STRUCTURE VALIDATION

3.1 Assessment of Overall Structure Accuracy

In testing and developing the EC–NMR protocol, the primary assessment of structure accuracy was done by comparison with the corresponding X-ray crystal structure or high-quality conventional NMR structure determined using essentially complete backbone and sidechain resonance assignments. When X-ray crystal structures are used as the reference structure, hydrogens are added using the *Reduce* program of the *Molprobity* software package (Word, Lovell, Richardson, & Richardson, 1999). Backbone (defined as N, C^{α}, C', and O atoms) and all-heavy-atom (N, C, O, S) root mean square deviations (RMSDs) are computed using the *fit* command as implemented in the *PyMol* software (DeLano, 2002), or using the molecular structure superimposition tools of the *PDBStat* software package (Tejero, Snyder, Mao, Aramini, & Montelione, 2013). Note that these two software packages use somewhat different criteria for defining the subset of atoms to use in superimpositions, and hence generally provide slightly different RMSD values for the same coordinate pairs.

Structure quality evaluation also includes a model vs data assessment. NMR DP scores (Huang et al., 2005; Huang, Rosato, Singh, & Montelione, 2012), which are crystallographic *R*-factor-like metrics for NMR structures, provide a global measure of how well a protein structure model fits with the NMR NOE and chemical shift data. Several calibration studies using decoys distorted from the correct structure, or inaccurate NMR structures compared with reference structures, have demonstrated that accurate structure models generally have DP scores >0.70 (Huang et al., 2005, 2012; Rosato et al., 2012). NMR DP scores are computed by the *ASDP* program, or can be assessed independently using the *RPF-DP* server available online at: http://nmr.cabm.rutgers.edu/rpf/. The *RPF-DP* program can also be downloaded to run on local machines. Reliable EC–NMR structures also have normalized structure quality scores (Bhattacharya, Tejero, & Montelione, 2007) $Z > -2$ for *Procheck*(backbone), *Procheck*(all dihedral), *Verify3D*, *MolProbity*, and *Prosa II* knowledge-based structure quality assessment metrics. Structure quality Z scores are computed using the online *Protein Structure Validation Software Suite Server (PSVS)* accessible at: http://psvs-1_5-dev.nesg.org/. Detailed instructions on using the PSVS server are also available at: http://www.nmr2.buffalo.edu/nesg.wiki/PSVS.

3.2 Assessment of Core Sidechain Accuracy

In assessing NMR-derived model accuracy relative to reference structures, it is also useful to compare sidechain conformations. One useful metric is the comparison of χ_1 rotamer states. In this analysis, χ_1 rotamers for all residues in each reference X-ray crystal structure are assigned to the nearest g^+, t, or g^- conformational state. Generally, we limit this analysis to "buried" side chains, with solvent accessible surface area (SASA) less than 40 $Å^2$ in the reference X-ray crystal structure (calculated using the *Molmol* program, Koradi, Billeter, & Wuthrich, 1996), as surface sidechains are generally not limited to a single conformation. In considering NMR structure ensembles (e.g., the EC–NMR structure or a NMR reference structure obtained from the PDB), side chains whose χ_1 dihedral angle values have standard deviations of <30 degrees are considered as "converged side chains." To assess sidechain conformation agreement between model and reference structures, rotamer states for residues with both buried and converged side chains are compared between the reference X-ray crystal (or the "representative" NMR conformer) and each member of the ensemble of NMR structures. The representative model is defined as the medoid conformer in the NMR ensemble; that is the one conformer most like all of the others (Montelione et al., 2013; Tejero et al., 2013). This representative conformer of the ensemble can be selected, and sidechain rotamer statistics calculated. Using the program *PDBStat* (Tejero et al., 2013).

3.3 Assessment of FP Rates in Contact Predictions

The Precision (P), Recall (R), and Performance (F) statistics for contact prediction (ECs or RPCs) are computed relative to reference X-ray or NMR structures as:

$$\text{Precision (P)} = \text{TP}/(\text{TP} + \text{FP}) \tag{1}$$

$$\text{Recall (R)} = \text{TP}/(\text{TP} + \text{FN}) \tag{2}$$

$$F\_\text{measure (F)} = (2 \times \text{Recall} \times \text{Precision})/(\text{Recall} + \text{Precision}) \tag{3}$$

In this analysis, a TP contact is defined for residue pair (i,j) if any atom of residue i is ≤ 5 Å apart from any atom of residue j in the reference structure. An EC (or RPC) for which a contact is not indicated in the reference structure is a FP. A contact in the reference structure which is not included in the EC (or RPC) list is a FN.

4. REPRESENTATIVE EC–NMR RESULTS

4.1 Assessment on Eight Reference Structures

The EC–NMR method was initially tested using experimental NMR (chemical shift, NOESY, and RDC data) for eight proteins ranging in size from 6 to 41 kDa (Tang et al., 2015). The resulting EC–NMR structures were compared with "reference structures," determined either by X-ray crystallography or by NMR using essentially complete backbone and sidechain resonance assignments. The EC–NMR structures of smaller (<20 kDa) proteins were observed to have accurate backbone and all-heavy-atom positions; i.e., <2 Å backbone atom positional RMSDs and <3 Å all-heavy atom RMSDs relative to the reference structure. Larger proteins, including human p21 H-Ras (21 kDa) and maltose binding protein (MBP) (41 kDa) had no or limited RDC data, respectively, but were nevertheless reasonably accurate; both protein structures had backbone RMSDs <2.8 Å and all-heavy-atom RMSDs <3.6 Å relative to the corresponding X-ray crystal structures (Tang et al., 2015). The percentages of χ_1 states for buried and converged sidechains that are consistent between the representative NMR structure selected from the ensemble (defined as the medoid conformer in the NMR ensemble; Montelione et al., 2013; Tejero et al., 2013) and the reference X-ray crystal were also assessed. Averaged over all eight EC–NMR structures, ~80% of buried, well-defined side chains have χ_1 rotamers that match the corresponding reference structures (Tang et al., 2015). For the three largest proteins studied, 85%, 81%, and 65%

of the buried, well-defined side chains have χ_1 rotamers that match the corresponding X-ray crystal structures. Generally speaking, the EC–NMR structures determined using HN only, or HN and Me, NOEs are quite accurate, based on comparisons with the corresponding reference structures (Tang et al., 2015).

4.2 Impact of Updated Database of Protein Sequences

Recently, five of the EC–NMR structures originally described by Tang et al. (2015) were redetermined using the same archived NMR data, but an updated database of protein sequences (Huang et al., 2018). For the four smaller protein targets, with molecular weights of 6–15 kDa, the NMR data include only H^N—H^N NOE data, along with restraints on backbone dihedral angles computed from C^α/C^β chemical shifts using Talos-N (Shen & Bax, 2015). For two of these four proteins, ^{15}N—^{1}H RDCs were available in two different molecular alignment conditions, for a third ^{15}N—^{1}H RDCs were available only one alignment condition, and for the fourth no RDC data are available. EC's were generated using the *EVfold* software suite. Contact maps for the most confident ECs, with posterior probability $P>0.9$, are shown in Fig. 3 together with the contact maps of the corresponding reference structures. EC–NMR models were generated with the *ASDP/Cyana/Rosetta* pipeline described in Sections 2.4 and 2.5, and compared with the reference NMR structures determined with complete sidechain proton assignments and much more extensive NOESY data (Fig. 4). They were assessed based on (i) accuracy of atomic positions and (ii) accuracy of sidechain χ_1 rotamer states for well-defined (i.e., converged), buried (i.e., not on the protein surface) sidechains. In each case, the representative structure from the NMR ensemble (either the EC–NMR ensemble or the reference NMR structure ensemble) was selected as the medoid conformer of the ensembles, using *PDBStat* (Tejero et al., 2013). The backbone RMSD's between these EC–NMR structures and the corresponding reference structures ranges from 1.5 to 1.8 Å, while the RMSD's for all C, N, O, and S atoms (both backbone and sidechain) range from 2.4 to 2.9 Å (Huang et al., 2018). The χ_1 values of well-defined buried sidechains, compared for all conformers in the EC–NMR ensemble with all conformers in the reference ensemble, also agree in 73%–85% of pair-wise comparisons. These results are similar to those reported for corresponding earlier EC–NMR structures of these same proteins reported by Tang et al. (2015). In both studies, the EC–NMR structures are significantly more

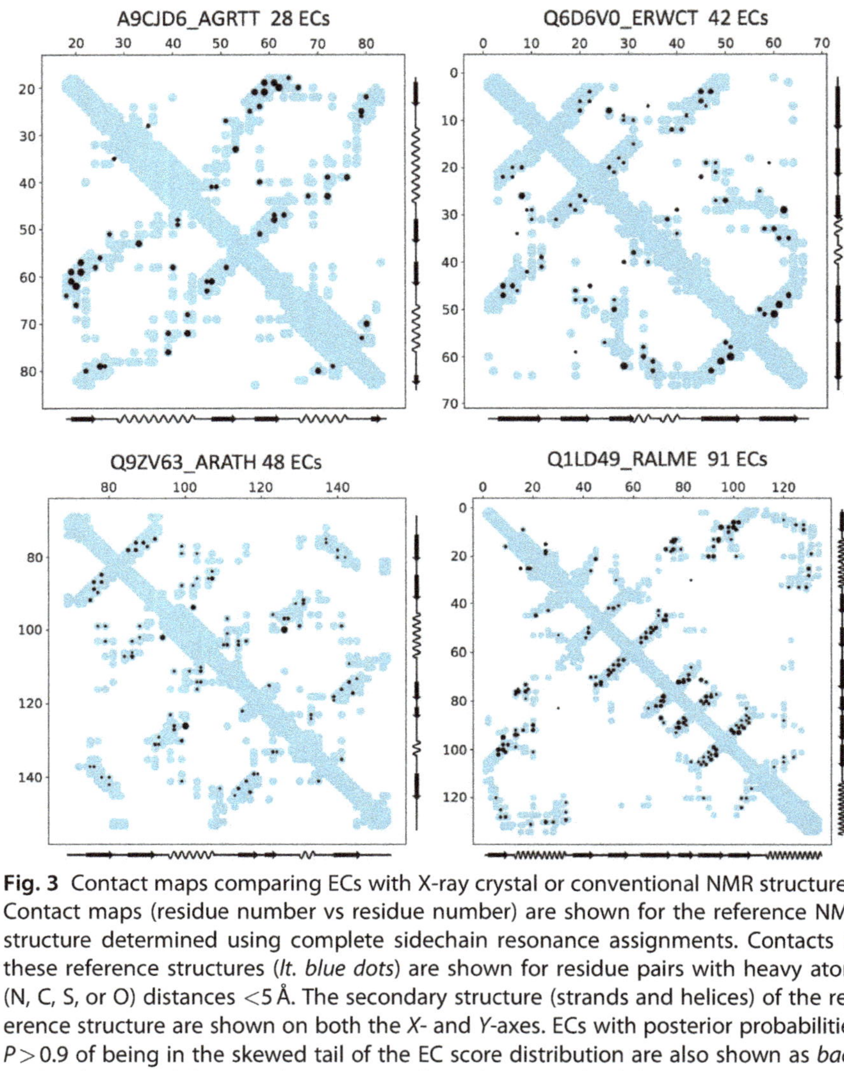

Fig. 3 Contact maps comparing ECs with X-ray crystal or conventional NMR structures. Contact maps (residue number vs residue number) are shown for the reference NMR structure determined using complete sidechain resonance assignments. Contacts in these reference structures (*lt. blue dots*) are shown for residue pairs with heavy atom (N, C, S, or O) distances <5 Å. The secondary structure (strands and helices) of the reference structure are shown on both the *X*- and *Y*-axes. ECs with posterior probabilities $P > 0.9$ of being in the skewed tail of the EC score distribution are also shown as *back circles*; the size of these circles corresponds to the strength of the EC.

accurate than models generated using either the EC or sparse NMR data alone. These EC–NMR structures, determined using only H^N—H^N NOE data together with ECs (and no methyl resonance assignments), have accuracies that compare with high-quality NMR structures determined with complete backbone and sidechain resonance assignments, *demonstrating*

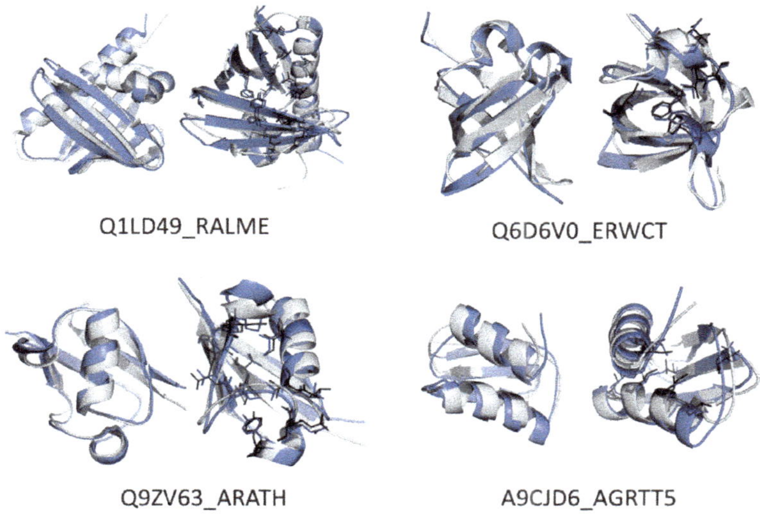

Fig. 4 EC–NMR structures determined using only H^N—H^N NOESY data, and ^1H—^{15}N RDC data, superimposed on reference conventional NMR structures. The representative (medoid) structure from the ensemble of conformers generated by the EC–NMR method (*blue*) is superimposed on a representative (medoid) structure from reference NMR structure ensemble. For each protein, the left image is a superimposition of backbone atoms, and the right image a superimposition of backbone and some well-defined core sidechain atoms.

that when good quality ECs are available for small (< 15 kDa) proteins, it may only be necessary to complete the majority of backbone resonance assignments in order to determine a high-quality solution NMR structure.

4.3 EC–NMR Structure of 41 kDa MBP

Tang et al. (2015) and Huang et al. (2018) also generated EC–NMR structures of the 41 kDa *E. coli* MBP bound to beta-cyclodextrin. The experimental NMR data for MBP include H^N—H^N NOE data, as well as Ile(δ1), Leu, and Val methyl ^{13}C and ^1H assignments, providing also Me-Me and HN-Me NOEs, along with restraints on backbone dihedral angles computed from C^α/C^β chemical shifts using Talos-N (Shen & Bax, 2015) and H^N—^{15}N RDC data in a single alignment media (Lange et al., 2012). These results (Fig. 5) demonstrate high-quality EC–NMR structures of MBP are produced, with backbone RMSD's to the corresponding X-ray crystal structure of 2.5 Å for backbone atoms, and 3.2 Å for all C, N, O, and S atoms (both backbone and sidechain). MBP

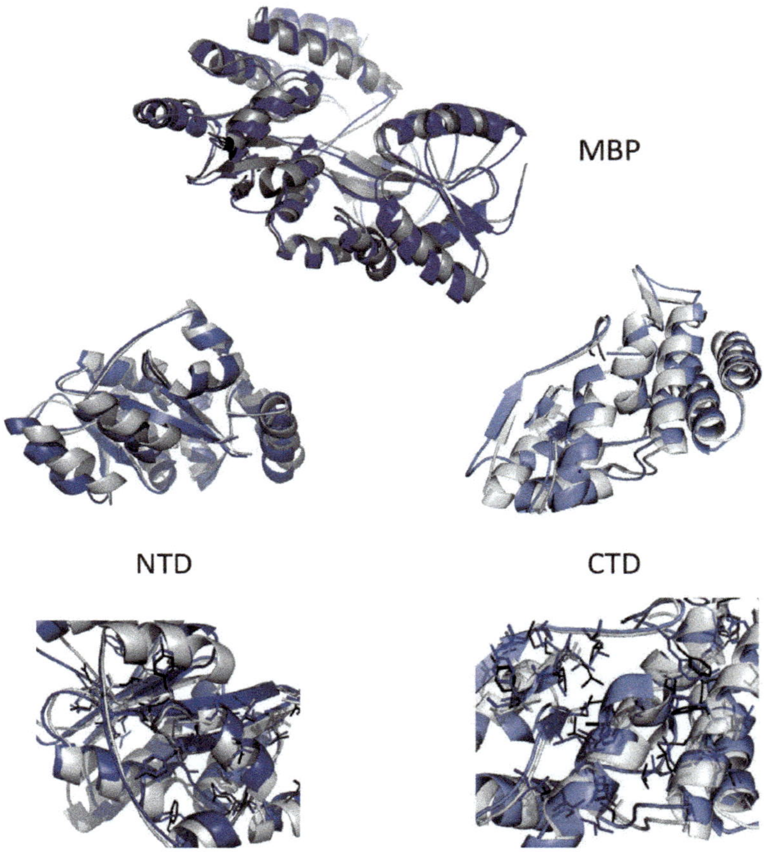

Fig. 5 EC–NMR structure of *E. coli* maltose binding protein superimposed on reference X-ray crystal structure. Superimposed backbone and core sidechain structures are for full-length MBP, and for the individual N-terminal domain (NTD) and C-terminal domain (CTD) in the full-length EC–NMR structure. *Blue ribbon structures*, final EC–NMR structure of MBP; *gray ribbons*, reference X-ray crystal structure.

is a two-domain protein, and the relative orientation of domains depends on which sugars are bound; the "open form" being preferred when bound to beta-cyclodextrin (Evenas et al., 2001). Considered separately, the two individual domains of MBP in the EC–NMR structure of the two-domain protein are even more accurate when compared to the reference X-ray crystal structure (N-terminal domain/C-terminal domain backbone RMSD 1.8 Å/1.7 Å, all-heavy-atom RMSD 2.7 Å/2.6 Å) than is apparent from rigid body superimposition for the entire protein.

We also compared the accuracy of the EC–NMR structure of MBP relative to several previously published NMR structures determined with more extensive sidechain assignments (Tang et al., 2015). The core sidechains of the EC–NMR structure are significantly more accurate than PDB ID 1EZP, determined using similar sparse NMR data together with five kinds of RDC data (Mueller et al., 2000). The core sidechain accuracy of the EC–NMR structure is similar to that of the solution NMR structure PDB ID 2D21, which was determined using extensive sidechain resonance assignments provided by the sophisticated and expensive stereoarrayed isotope labeling (SAIL) method (Kainosho et al., 2006). Based on RMSD relative to the X-ray crystal structure of beta-cyclodextrin-bound MBP, the overall structure of the EC–NMR models is more accurate than any previously published NMR structures. Hence, we conclude that the EC–NMR method of Tang et al. (2015) can deliver structures with accurate backbone and core sidechain atomic positions for larger (~40 kDa, or larger) proteins, with accuracy comparable or better than models obtained with sophisticated sidechain labeling methods.

4.4 Sensitivity to Numbers of Sequence Homologs

A prerequisite for the EC–NMR approach is extensive, diverse sequence data, required to obtain accurate ECs between the residues (Hopf et al., 2012; Kamisetty et al., 2013; Marks et al., 2011). Recent experience suggests that more than $2 \times L$ nonredundant sequences (N_{eff}) are generally required for confident predictions of overall protein fold from EC's alone, where L is the length of the target sequence (Hopf et al., 2014; Kamisetty et al., 2013; Marks et al., 2012; Michel et al., 2014; Ovchinnikov et al., 2014; Ovchinnikov et al., 2017). For a target protein that is 200 residues long, this typically requires on the order of 5000 sequences, before removal of redundancy.

For EC–NMR, the goal is more ambitious than simply determining a correct fold. Rather, we aim to obtain models with accuracies comparable to high-quality NMR structures; i.e., backbone positional RMSD's relative to reference structures <2.5 Å, and accurate core sidechain packing. In order to test the robustness of structure accuracy to the numbers of sequences used in the MSA, Tang et al. (2015) analyzed a series of MSAs, each with numbers of sequences ranging from $N_{eff}/L \sim 150$ down to $N_{eff}/L < 0.1$. This study is reproduced from the original publication in Fig. 6. In this analysis, using

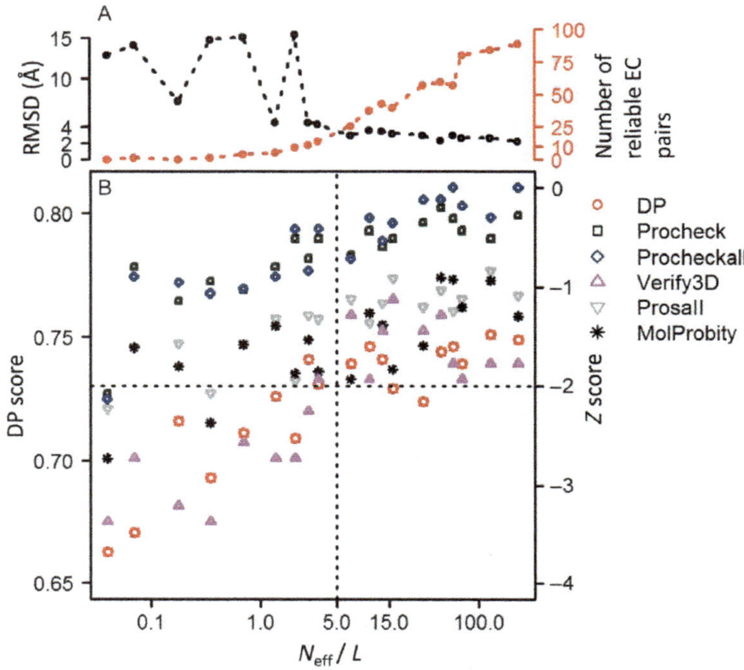

Fig. 6 Evaluation of structure quality statistics for 20 ensembles of EC–NMR structures. The 20 ensembles of EC–NMR structures generated in the sensitivity analysis outlined in the text were assessed using structural quality metrics of the Protein Structure Validation Software Suite (PSVS) (Bhattacharya et al., 2007), including NMR DP scores (Huang et al., 2005). The top panel plots backbone RMSD to reference structures (*black data points*) and the number of high-confidence ECs (*red data points*), as defined in the main text, for each of the 20 EC–NMR analyses carried out using different amounts of sequence data (N_{eff}/L). The bottom panel plots DP (*left axis*) or knowledge-based *Z* score (*right axis*) computed with the *PSVS* software. The metrics include the NMR DP score (Huang et al., 2005) (*red circles*), *ProCheck* (Laskowski, Moss, & Thornton, 1993) analysis of backbone dihedral angle distributions (Procheck, *green boxes*) and distributions for all backbone and sidechain dihedral angels (Procheck_all, *blue diamonds*), *Veify3D* (Luthy, Bowie, & Eisenberg, 1992) (*magenta triangles*), *Prosa II* (Sippl, 1993) (*gray triangles*), and *Molprobity* atomic clash scores (Lovell et al., 2003) (*black asterisk*). *Z* scores are normalized against a collection of high-resolution crystal structures (Bhattacharya et al., 2007). "Reliable" structures (backbone RMSD <3.5 Å from X-ray crystal structure) obtained using multiple sequence alignments of $N_{eff}/L > 5$, all have DP score $> \sim 0.73$ and knowledge-based *Z* scores > -2 (upper right quadrant), while less accurate structures all have DP $< \sim 0.73$ and *Z* scores for Verify3D < -2. *Reproduced from Tang, Y., Huang, Y. J., Hopf, T. A., Sander, C., Marks, D. S., & Montelione, G. T. (2015). Protein structure determination by combining sparse NMR data with evolutionary couplings.* Nature Methods, 12, *751–754. doi: 10.1038/nmeth.3455.*

good quality NMR data for a perdeuterated, Ile(δ1), Leu, and Val $^{13}CH_3$ methyl-labeled protein, the cutoff point for accurate modeling (<2.5 Å backbone RMSD) was estimated to be $N_{eff}/L \sim 5$, with little improvement in structural accuracy for higher values of N_{eff}/L.

More recently we have explored other real EC–NMR datasets with N_{eff}/L ranging from 6 to 0.1 nonredundant sequences per residue. Using our current implementation of the EC–NMR software pipeline, we confirmed the production of accurate models for data with $N_{eff}/L \sim 6$, but failure of the method for $N_{eff}/L < 1$. Although we anticipate continued growth of sequence data bases, and the quality of results will also depend on the quality and extent of the sparse NMR data, it remains an important benchmark for the EC–NMR to provide accurate models for targets with low numbers (e.g., $N_{eff}/L < 5$) of sequence homologs.

4.5 Structural Restraints Can Guide Identification of FP Contact Predictions

EC-based contacts can complement sparse NMR data to provide more contact information, and more accurate models. However, EC lists will generally include predicted interactions that are FP contacts for the predominant low-energy structure of the target protein. A key feature of the EC-NMR method is to use the constraints of structural consistency to properly assign NOE-based restraints and to identify FP ECs. The requirement that the structure model be consistent with all (or most) of the experimental NMR data provides "hard" constraints on the interpretation of ECs, allowing identification and removal of structurally inconsistent FP residue pair contacts (SI-RPCs) (Tang et al., 2015). In all of the calculations we carried out, the EC–NMR protocol increased the number of ECs that could be used reliably in structure calculations and reduced the FP contact rates (Huang et al., 2018; Tang et al., 2015). For example, for MBP, the number of FPs was reduced from 127 to 82 and the Precision of the ECs (relative to the reference X-ray crystal structure) was increased from 0.66 to 0.73 (see Tang et al., 2015, Supplementary Table 4B). The remaining FP EC contacts remaining at the end of the EC–NMR protocol are nearby in the 3D structures to true-positive interresidue contacts, and have little effect on the accuracy of the final structure. Similarly, for the other EC–NMR structures that were calculated, all of the FPs that are significantly inconsistent with the reference structures were eliminated automatically by the ASDP analysis process. Hence structural constraints that evolve in course of EC–NMR protocol are very effective in removing structurally inconstent FPs from the EC lists. Similar results have been recently reported using a different algorithm for imposing structural consistency in identifying FPs in EC analysis, based on the automated NOESY peak assignment program YAPP (Kassem et al., 2018).

5. FUTURE PROSPECTS

Contact predictions and sparse NMR data, used together with knowledge-based modeling, are highly complementary for protein structure determination. The EC–NMR approach improves the accuracy of models generated by EC data alone, by requiring that predicted EC-based contacts are consistent with internal structural constraints and experimental NMR data collected for one member of the protein family under specific conditions. This requirement eliminates potentially important, but confounding, EC-derived contact restraints that may arise from structural drift across the protein family, or from transient oligomeric interactions, allosteric networks, and/or alternative states which may also be detected as evolutionary sequence covariation. In this way, complementarity EC and NMR data provide much more complete and accurate residue contact information than can be obtained from either method alone.

The EC–NMR method requires large MSAs, which are currently available only for a fraction of proteins. However, as the sequence databases continue to grow, more proteins will be amenable to this approach. Fortunately, combining ECs together with sparse NMR data reduces the requirements for the amount and diversity of sequence information.

The EC–NMR method is largely automated and provides high-quality 3D structures with accurate backbone and core sidechain conformations (Huang et al., 2018; Tang et al., 2015). For small proteins and domains up to 150 residues ($<\sim$15 kDa) with extensive sequence information, EC–NMR is a new, powerful, and efficient approach for protein structure determination using only backbone NMR data. For larger proteins, up to 400–500 residues (40–60 kDa, or larger), for which extensive sidechain resonance assignment is challenging if not impossible with current methods, ECs can be combined with sparse NMR data obtained on perdeuterated protein samples to provide structures that are more accurate and complete than those obtained using such NMR data alone. In the method outlined here, ECs were combined with solution NMR data to determine both small and larger soluble protein structures, but the same approach should be applicable to membrane proteins (Hopf et al., 2012; Ovchinnikov et al., 2017) and for RNA structure determination (Weinreb et al., 2016). The EC–NMR method should be particularly powerful for modeling structures of biomolecules using solid-state NMR data.

In our work to date, we have used a simple restrained energy minimization protocol of *Rosetta* in the final refinement step (Mao et al., 2014). This protocol improves both backbone and sidechain structure accuracy. It is advantageous because is it relatively fast and can be implemented with limited computer resources. However, more sophisticated spatial-restrained modeling methods, such as the Resolution Adapted Structural RECombination protocol (RASREC) developed by Lange and Baker (Lange & Baker, 2012; Raman et al., 2010) has significant advantages for generating accurate structures of proteins from sparse NMR data (Lange et al., 2012; Raman et al., 2010). The RASREC protocol has also been used successfully for modeling protein structures with EC data alone (Braun, Koehler Leman, & Lange, 2015). While it is much more computationally demanding, currently limiting its broad application, the RASREC protocol has the potential to provide more accurate EC–NMR structures with less complete and/or more noisy EC and sparse NMR data.

The EC–NMR method also allows identification of ECs which are not consistent with the NMR data collected for the target protein under specific conditions. While these are "false positives" relative to the modeling the structure of this particular state of the protein, ECs with strong signals and high reliability that are not consistent with this particular state of the protein structure can provide information on alternative conformations that are accessible to the protein, excited states, and potentially provide information on allosteric networks, as residue contacts in these alternative states may also covary in the course of evolution. Further investigations of the combined use of ECs and NMR data to characterize the multiple conformational states of proteins and their energy landscapes is an exciting emerging area which can be explored using these powerful hybrid methods.

ACKNOWLEDGMENTS

This work was supported by National Institutes of Health grants 1R01-GM120574 (to G.T.M.) and 1R01-GM106303 (C.S. and D.M.). We thank all of the members of the Northeast Structural Genomics Consortium who generated and archived NMR data used in this work, particularly scientists in the laboratories of C. Arrowsmith, M. Kennedy, G. T. Montelione, T. Szyperski, and J. Prestegard.

REFERENCES

Abriata, L. A., Tamo, G. E., Monastyrskyy, B., Kryshtafovych, A., & Dal Peraro, M. (2018). Assessment of hard target modeling in CASP12 reveals an emerging role of alignment-based contact prediction methods. *Proteins*, *86*(Suppl. 1), 97–112. https://doi.org/10.1002/prot.25423.

Adams, P. D., Baker, D., Brunger, A. T., Das, R., DiMaio, F., Read, R. J., et al. (2013). Advances, interactions, and future developments in the CNS, Phenix, and Rosetta structural biology software systems. *Annual Review of Biophysics*, *42*, 265–287. https://doi.org/10.1146/annurev-biophys-083012-130253.

Adhikari, B., Hou, J., & Cheng, J. (2018). Protein contact prediction by integrating deep multiple sequence alignments, coevolution and machine learning. *Proteins*, *86*(Suppl. 1), 84–96. https://doi.org/10.1002/prot.25405.

Bhattacharya, A., Tejero, R., & Montelione, G. T. (2007). Evaluating protein structures determined by structural genomics consortia. *Proteins*, *66*(4), 778–795. https://doi.org/10.1002/prot.21165.

Braun, T., Koehler Leman, J., & Lange, O. F. (2015). Combining evolutionary information and an iterative sampling strategy for accurate protein structure prediction. *PLoS Computational Biology*, *11*(12), e1004661. https://doi.org/10.1371/journal.pcbi.1004661.

Brooks, B. R., Brooks, C. L., Mackerell, A. D., Nilsson, L., Petrella, R. J., Roux, B., et al. (2009). CHARMM: The biomolecular simulation program. *Journal of Computational Chemistry*, *30*, 1545–1614.

Buchan, D. W. A., & Jones, D. T. (2018). Improved protein contact predictions with the MetaPSICOV2 server in CASP12. *Proteins*, *86*(Suppl. 1), 78–83. https://doi.org/10.1002/prot.25379.

DeLano, W. L. (2002). *The PyMOL molecular graphics system, version 1.5.0.4.* Schrödinger, LLC.

Diercks, T., Coles, M., & Kessler, H. (1999). An efficient strategy for assignment of cross-peaks in 3D heteronuclear NOESY experiments. *Journal of Biomolecular NMR*, *15*(2), 177–180. https://doi.org/10.1023/A:1008367912535.

dos Santos, R. N., Morcos, F., Jana, B., Andricopulo, A. D., & Onuchic, J. N. (2015). Dimeric interactions and complex formation using direct coevolutionary couplings. *Scientific Reports*, *5*, 13652. https://doi.org/10.1038/srep13652.

Ekeberg, M., Lovkvist, C., Lan, Y., Weigt, M., & Aurell, E. (2013). Improved contact prediction in proteins: Using pseudolikelihoods to infer Potts models. *Physical Review. E, Statistical, Nonlinear, and Soft Matter Physics*, *87*, 012707.

Evenas, J., Tugarinov, V., Skrynnikov, N. R., Goto, N. K., Muhandiram, R., & Kay, L. E. (2001). Ligand-induced structural changes to maltodextrin-binding protein as studied by solution NMR spectroscopy. *Journal of Molecular Biology*, *309*, 961–974. https://doi.org/10.1006/jmbi.2001.4695.

Gardner, K. H., Rosen, M. K., & Kay, L. E. (1997). Global folds of highly deuterated, methyl-protonated proteins by multidimensional NMR. *Biochemistry*, *36*, 1389–1401.

Gobel, U., Sander, C., Schneider, R., & Valencia, A. (1994). Correlated mutations and residue contacts in proteins. *Proteins*, *18*, 309–317. https://doi.org/10.1002/prot.340180402.

Grishaev, A., Tugarinov, V., Kay, L. E., Trewhella, J., & Bax, A. (2008). Refined solution structure of the 82-kDa enzyme malate synthase G from joint NMR and synchrotron SAXS restraints. *Journal of Biomolecular NMR*, *40*(2), 95–106. https://doi.org/10.1007/s10858-007-9211-5.

Herrmann, T., Guntert, P., & Wuthrich, K. (2002). Protein NMR structure determination with automated NOE assignment using the new software CANDID and the torsion angle dynamics algorithm DYANA. *Journal of Molecular Biology*, *319*, 209–227.

Hiller, S., Garces, R. G., Malia, T. J., Orekhov, V. Y., Colombini, M., & Wagner, G. (2008). Solution structure of the integral human membrane protein VDAC-1 in detergent micelles. *Science*, *321*(5893), 1206–1210. https://doi.org/10.1126/science.1161302.

Hopf, T. A., Colwell, L. J., Sheridan, R., Rost, B., Sander, C., & Marks, D. S. (2012). Three-dimensional structures of membrane proteins from genomic sequencing. *Cell*, *149*, 1607–1621. https://doi.org/10.1016/j.cell.2012.04.012.

Hopf, T. A., Scharfe, C. P., Rodrigues, J. P., Green, A. G., Sander, C., Bonvin, A. M., et al. (2014). Sequence co-evolution gives 3D contacts and structures of protein complexes. *eLife 3*, e03430. https://doi.org/10.7554/eLife.03430.

Huang, Y. J., Brock, K., Sander, C., Marks, D. S., & Montelione, G. T. (2018). A hybrid approach for protein structure determination combining sparse NMR with evolutionary coupling sequence data. In J. L. Markley, S. Burley, H. Nakamura, & G. Kleywegt (Eds.), Integrative structural biology with hybrid methods. Singapore: Springer.

Huang, Y. J., Mao, B., Xu, F., & Montelione, G. T. (2015). Guiding automated NMR structure determination using a global optimization metric, the NMR DP score. *Journal of Biomolecular NMR, 62*, 439–451. https://doi.org/10.1007/s10858-015-9955-2.

Huang, Y. J., Powers, R., & Montelione, G. T. (2005). Protein NMR recall, precision, and F-measure scores (RPF scores): Structure quality assessment measures based on information retrieval statistics. *Journal of the American Chemical Society, 127*, 1665–1674. https://doi.org/10.1021/ja047109h.

Huang, Y. J., Rosato, A., Singh, G., & Montelione, G. T. (2012). RPF: A quality assessment tool for protein NMR structures. *Nucleic Acids Research, 40*, W542–W546. https://doi.org/10.1093/nar/gks373.

Huang, Y. J., Tejero, R., Powers, R., & Montelione, G. T. (2006). A topology-constrained distance network algorithm for protein structure determination from NOESY data. *Proteins, 62*, 587–603. https://doi.org/10.1002/prot.20820.

Jones, D. T., Buchan, D. W., Cozzetto, D., & Pontil, M. (2012). PSICOV: Precise structural contact prediction using sparse inverse covariance estimation on large multiple sequence alignments. *Bioinformatics, 28*, 184–190. https://doi.org/10.1093/bioinformatics/btr638.

Kainosho, M., Torizawa, T., Iwashita, Y., Terauchi, T., Mei Ono, A., & Guntert, P. (2006). Optimal isotope labelling for NMR protein structure determinations. *Nature, 440*, 52–57. https://doi.org/10.1038/nature04525.

Kamisetty, H., Ovchinnikov, S., & Baker, D. (2013). Assessing the utility of coevolution-based residue–residue contact predictions in a sequence- and structure-rich era. *Proceedings of the National Academy of Sciences of the United States of America, 110*, 15674–15679. https://doi.org/10.1073/pnas.1314045110.

Kassem, M. M., Christoffersen, L. B., Cavalli, A., & Lindorff-Larsen, K. (2018). Enhancing coevolution-based contact prediction by imposing structural self-consistency of the contacts. *Scientific Reports, 8*, 11112. https://doi.org/10.1038/s41598-018-29357-y.

Koradi, R., Billeter, M., & Wuthrich, K. (1996). MOLMOL: A program for display and analysis of macromolecular structures. *Journal of Molecular Graphics, 14*, 51–55. 29–32.

Lange, O. F., & Baker, D. (2012). Resolution-adapted recombination of structural features significantly improves sampling in restraint-guided structure calculation. *Proteins, 80*, 884–895.

Lange, O. F., Rossi, P., Sgourakis, N. G., Song, Y., Lee, H. W., Aramini, J. M., et al. (2012). Determination of solution structures of proteins up to 40 kDa using CS-Rosetta with sparse NMR data from deuterated samples. *Proceedings of the National Academy of Sciences of the United States of America, 109*, 10873–10878. https://doi.org/10.1073/pnas.1203013109.

Lapedes, A., Giraud, B., & Jarzynski, C. (2002). *Using sequence alignments to predict protein structure and stability with high accuracy*. arXiv. 1207.2484.

Laskowski, R. A., Moss, D. S., & Thornton, J. M. (1993). Main-chain bond lengths and bond angles in protein structures. *Journal of Molecular Biology, 231*, 1049–1067. https://doi.org/10.1006/jmbi.1993.1351.

Lovell, S. C., Davis, I. W., Arendall, W. B., 3rd, de Bakker, P. I., Word, J. M., Prisant, M. G., et al. (2003). Structure validation by Calpha geometry: Phi,psi and Cbeta deviation. *Proteins*, *50*, 437–450. https://doi.org/10.1002/prot.10286.

Luthy, R., Bowie, J. U., & Eisenberg, D. (1992). Assessment of protein models with three-dimensional profiles. *Nature*, *356*, 83–85. https://doi.org/10.1038/356083a0.

Maltsev, A. S., Ying, J., & Bax, A. (2012). Deuterium isotope shifts for backbone (1)H, (1)(5) N and (1)(3)C nuclei in intrinsically disordered protein alpha-synuclein. *Journal of Biomolecular NMR*, *54*, 181–191. https://doi.org/10.1007/s10858-012-9666-x.

Mao, B., Guan, R., & Montelione, G. T. (2011). Improved technologies now routinely provide protein NMR structures useful for molecular replacement. *Structure*, *19*, 757–766. https://doi.org/10.1016/j.str.2011.04.005.

Mao, B., Tejero, R., Baker, D., & Montelione, G. T. (2014). Protein NMR structures refined with Rosetta have higher accuracy relative to corresponding X-ray crystal structures. *Journal of the American Chemical Society*, *136*, 1893–1906. https://doi.org/10.1021/ja409845w.

Marks, D. S., Colwell, L. J., Sheridan, R., Hopf, T. A., Pagnani, A., Zecchina, R., et al. (2011). Protein 3D structure computed from evolutionary sequence variation. *PLoS One, 6*, e28766. https://doi.org/10.1371/journal.pone.0028766.

Marks, D. S., Hopf, T. A., & Sander, C. (2012). Protein structure prediction from sequence variation. *Nature Biotechnology*, *30*, 1072–1080. https://doi.org/10.1038/nbt.2419.

Michel, M., Hayat, S., Skwark, M. J., Sander, C., Marks, D. S., & Elofsson, A. (2014). PconsFold: Improved contact predictions improve protein models. *Bioinformatics*, *30*, i482–i488. https://doi.org/10.1093/bioinformatics/btu458.

Montelione, G. T., Nilges, M., Bax, A., Guntert, P., Herrmann, T., Richardson, J. S., et al. (2013). Recommendations of the wwPDB NMR validation task force. *Structure*, *21*, 1563–1570. https://doi.org/10.1016/j.str.2013.07.021.

Morcos, F., Jana, B., Hwa, T., & Onuchic, J. N. (2013). Coevolutionary signals across protein lineages help capture multiple protein conformations. *Proceedings of the National Academy of Sciences of the United States of America*, *110*, 20533–20538. https://doi.org/10.1073/pnas.1315625110.

Morcos, F., Pagnani, A., Lunt, B., Bertolino, A., Marks, D. S., Sander, C., et al. (2011). Direct-coupling analysis of residue coevolution captures native contacts across many protein families. *Proceedings of the National Academy of Sciences of the United States of America*, *108*, E1293–E1301. https://doi.org/10.1073/pnas.1111471108.

Mueller, G. A., Choy, W. Y., Yang, D., Forman-Kay, J. D., Venters, R. A., & Kay, L. E. (2000). Global folds of proteins with low densities of NOEs using residual dipolar couplings: Application to the 370-residue maltodextrin-binding protein. *Journal of Molecular Biology*, *300*, 197–212. https://doi.org/10.1006/jmbi.2000.3842.

Neher, E. (1994). How frequent are correlated changes in families of protein sequences? *Proceedings of the National Academy of Sciences of the United States of America*, *91*, 98–102.

Nilges, M. (1995). Calculation of protein structures with ambiguous distance restraints. Automated assignment of ambiguous NOE crosspeaks and disulphide connectivities. *Journal of Molecular Biology*, *245*, 645–660.

Ovchinnikov, S., Kamisetty, H., & Baker, D. (2014). Robust and accurate prediction of residue–residue interactions across protein interfaces using evolutionary information. *eLife, 3*, e02030. https://doi.org/10.7554/eLife.02030.

Ovchinnikov, S., Kim, D. E., Wang, R. Y., Liu, Y., DiMaio, F., & Baker, D. (2016). Improved de novo structure prediction in CASP11 by incorporating coevolution information into Rosetta. *Proteins*, *84*, 67–75. https://doi.org/10.1002/prot.24974.

Ovchinnikov, S., Kinch, L., Park, H., Liao, Y., Pei, J., Kim, D. E., et al. (2015). Large-scale determination of previously unsolved protein structures using evolutionary information. *eLife 4*, e09248. https://doi.org/10.7554/eLife.09248.

Ovchinnikov, S., Park, H., Varghese, N., Huang, P. S., Pavlopoulos, G. A., Kim, D. E., et al. (2017). Protein structure determination using metagenome sequence data. *Science*, *355*, 294–298. https://doi.org/10.1126/science.aah4043.

Raman, S., Lange, O. F., Rossi, P., Tyka, M., Wang, X., Aramini, J., et al. (2010). NMR structure determination for larger proteins using backbone-only data. *Science*, *327*, 1014–1018. https://doi.org/10.1126/science.1183649.

Rosato, A., Aramini, J. M., Arrowsmith, C., Bagaria, A., Baker, D., Cavalli, A., et al. (2012). Blind testing of routine, fully automated determination of protein structures from NMR data. *Structure*, *20*, 227–236. https://doi.org/10.1016/j.str.2012.01.002.

Rosen, M. K., Gardner, K. H., Willis, R. C., Parris, W. E., Pawson, T., & Kay, L. E. (1996). Selective methyl group protonation of perdeuterated proteins. *Journal of Molecular Biology*, *263*, 627–636.

Schaarschmidt, J., Monastyrskyy, B., Kryshtafovych, A., & Bonvin, A. (2018). Assessment of contact predictions in CASP12: Co-evolution and deep learning coming of age. *Proteins*, *86*(Suppl. 1), 51–66. https://doi.org/10.1002/prot.25407.

Schneider, W. M., Inouye, M., Montelione, G. T., & Roth, M. J. (2009). Independently inducible system of gene expression for condensed single protein production (cSPP) suitable for high efficiency isotope enrichment. *Journal of Structural and Functional Genomics*, *10*, 219–225. PMC4898478.

Schneider, W. M., Tang, Y., Vaiphei, S. T., Mao, L., Maglaqui, M., Inouye, M., et al. (2010). Efficient condensed-phase production of perdeuterated soluble and membrane proteins. *Journal of Structural and Functional Genomics*, *11*, 143–154. PMC4119428.

Schwieters, C. D., Kuszewski, J. J., & Clore, G. M. (2006). Using Xplor-NIH for NMR molecular structure determination. *Progress in Nuclear Magnetic Resonance Spectroscopy*, *48*, 47–62. https://doi.org/10.1016/J.Pnmrs.2005.10.001.

Sgourakis, N. G., Natarajan, K., Ying, J., Vogeli, B., Boyd, L. F., Margulies, D. H., et al. (2014). The structure of mouse cytomegalovirus m04 protein obtained from sparse NMR data reveals a conserved fold of the m02-m06 firal immune modulator family. *Structure*, *22*, 1263–1273. https://doi.org/10.1016/j.str.2014.05.018.

Shen, Y., & Bax, A. (2015). Protein structural information derived from NMR chemical shift with the neural network program TALOS-N. *Methods in Molecular Biology*, *1260*, 17–32. https://doi.org/10.1007/978-1-4939-2239-0_2.

Shen, Y., Delaglio, F., Cornilescu, G., & Bax, A. (2009). TALOS+: A hybrid method for predicting protein backbone torsion angles from NMR chemical shifts. *Journal of Biomolecular NMR*, *44*, 213–223. https://doi.org/10.1007/s10858-009-9333-z.

Sheridan, R., Fieldhouse, R. J., Hayat, S., Sun, Y., Antipin, Y., Yang, L., et al. (2015). EVfold.org: Evolutionary couplings and protein 3D structure prediction. *bioRxiv*, 021022. https://doi.org/10.1101/021022.

Shindyalov, I. N., Kolchanov, N. A., & Sander, C. (1994). Can three-dimensional contacts in protein structures be predicted by analysis of correlated mutations? *Protein Engineering*, 7, 349–358.

Simkovic, F., Ovchinnikov, S., Baker, D., & Rigden, D. J. (2017). Applications of contact predictions to structural biology. *IUCrJ*, *4*(Pt. 3), 291–300. https://doi.org/10.1107/S2052252517005115.

Sippl, M. J. (1993). Recognition of errors in three-dimensional structures of proteins. *Proteins*, *17*, 355–362. https://doi.org/10.1002/prot.340170404.

Sulkowska, J. I., Morcos, F., Weigt, M., Hwa, T., & Onuchic, J. N. (2012). Genomics-aided structure prediction. *Proceedings of the National Academy of Sciences of the United States of America*, *109*, 10340–10345. https://doi.org/10.1073/pnas.1207864109.

Tang, Y., Huang, Y. J., Hopf, T. A., Sander, C., Marks, D. S., & Montelione, G. T. (2015). Protein structure determination by combining sparse NMR data with evolutionary couplings. *Nature Methods*, *12*, 751–754. https://doi.org/10.1038/nmeth.3455.

Taylor, W. R., & Hatrick, K. (1994). Compensating changes in protein multiple sequence alignments. *Protein Engineering*, 7, 341–348.

Tejero, R., Snyder, D., Mao, B., Aramini, J. M., & Montelione, G. T. (2013). PDBStat: A universal restraint converter and restraint analysis software package for protein NMR. *Journal of Biomolecular NMR*, *56*, 337–351. https://doi.org/10.1007/s10858-013-9753-7.

Thomas, D. J., Casari, G., & Sander, C. (1996). The prediction of protein contacts from multiple sequence alignments. *Protein Engineering*, *9*(11), 941–948.

Toth-Petroczy, A., Palmedo, P., Ingraham, J., Hopf, T. A., Berger, B., Sander, C., et al. (2016). Structured states of disordered proteins from genomic sequences. *Cell*, *167*(158–170), e112. https://doi.org/10.1016/j.cell.2016.09.010.

Tugarinov, V., Choy, W. Y., Orekhov, V. Y., & Kay, L. E. (2005). Solution NMR-derived global fold of a monomeric 82-kDa enzyme. *Proceedings of the National Academy of Sciences of the United States of America*, *102*(3), 622–627. https://doi.org/10.1073/pnas.0407792102.

Tugarinov, V., Kanelis, V., & Kay, L. E. (2006). Isotope labeling strategies for the study of high-molecular-weight proteins by solution NMR spectroscopy. *Nature Protocols*, *1*, 749–754. nprot.2006.101 [pii]https://doi.org/10.1038/nprot.2006.101.

Vaiphei, S. T., Tang, Y., Montelione, G. T., & Inouye, M. (2011). The use of the condensed single protein production (cSPP) system for isotope- labeled outer membrane proteins, OmpA and OmpX in *E. coli*. *Molecular Biotechnology*, *47*, 205–210. PMC4190416.

Wang, S., Sun, S., & Xu, J. (2018). Analysis of deep learning methods for blind protein contact prediction in CASP12. *Proteins*, *86*(Suppl. 1), 67–77. https://doi.org/10.1002/prot.25377.

Weinreb, C., Riesselman, A. J., Ingraham, J. B., Gross, T., Sander, C., & Marks, D. S. (2016). 3D RNA and functional interactions from evolutionary couplings. *Cell*, *165*, 963–975. https://doi.org/10.1016/j.cell.2016.03.030.

Wishart, D. S., & Sykes, B. D. (1994). The 13C chemical-shift index: A simple method for the identification of protein secondary structure using 13C chemical-shift data. *Journal of Biomolecular NMR*, *4*, 171–180.

Word, J. M., Lovell, S. C., Richardson, J. S., & Richardson, D. C. (1999). Asparagine and glutamine: Using hydrogen atom contacts in the choice of side-chain amide orientation. *Journal of Molecular Biology*, *285*(4), 1735–1747. https://doi.org/10.1006/jmbi.1998.2401.

Wuthrich, K. (1986). *NMR of proteins and nucleic acids*. New York: Wiley.

CHAPTER FOURTEEN

Combining NMR Spectroscopy and Molecular Dynamic Simulations to Solve and Analyze the Structure of Protein–RNA Complexes

Sebastien Campagne*[,1], Miroslav Krepl[†,‡,1], Jiri Sponer[†,‡,1], Frederic H-T. Allain*[,1]

*Department of Biology, ETH Zürich, Institute of Molecular Biology and Biophysics, Zürich, Switzerland
[†]Institute of Biophysics of the Czech Academy of Sciences, Brno, Czech Republic
[‡]Department of Physical Chemistry, Faculty of Science, Regional Centre of Advanced Technologies and Materials, Palacky University Olomouc, Olomouc, Czech Republic
[1]Corresponding authors: e-mail address: campagne@mol.biol.ethz; krepl@seznam.cz; sponer@ncbr.muni.cz; allain@mol.biol.ethz.ch

Contents

Methods in Enzymology, Volume 614
ISSN 0076-6879
https://doi.org/10.1016/bs.mie.2018.09.002

Abstract

Understanding the RNA binding specificity of protein is of primary interest to decipher their function in the cell. Here, we review the methodology used to solve the structures of protein–RNA complexes using solution-state NMR spectroscopy: from sample preparation to structure calculation procedures. We also describe how molecular dynamics simulations can help providing additional information on the role of key amino acid side chains and of water molecules in protein–RNA recognition.

1. INTRODUCTION

Protein–RNA interactions control gene expression from the birth of the messenger RNA to his death, i.e., transcription (Gomez et al., 2013), RNA processing (Licatalosi & Darnell, 2010), translation (Duss et al., 2014), and decay (Karousis, Nasif, & Muhlemann, 2016). Eukaryotic genomes encode for a plethora of signaling pathways that can be adjusted at any time in response to internal or environmental cues, and protein–RNA interactions are critically involved in these adjustments. Failure or misfunction of protein–RNA interactions has been associated with numerous genetic diseases and cancers (Corbett, 2018). Consequently, understanding the atomic details of protein–RNA interactions is of fundamental interest to decipher their functions and to rationally develop innovative therapeutic strategies. RNA binding proteins (RBPs) are generally composed of small RNA binding domains (RBDs) as the RNA recognition motif (RRM) (Maris, Dominguez, & Allain, 2005), the double-strand RNA binding domain (dsRBD) (Masliah, Barraud, & Allain, 2013), zinc finger domains (Berg, 2003), the hnRNP K homology domain or KH domain (Nicastro, Taylor, & Ramos, 2015), and many others. Small RBDs can bind sequence specifically fairly short RNA sequences (2–8 nucleotides) and are generally combined in order to increase sequence specificity. Understanding the molecular basis of the specific RNA recognition of an RBP requires solving its three-dimensional structure in complex with its RNA target. The small size of the RNA–RBD complexes and the intrinsic flexibility of RNA molecules are two critical features that make nuclear magnetic resonance (NMR) spectroscopy particularly suitable to study the RNA binding process and solving the structure of the complex. Indeed, among 311 protein–RNA complexes structures lower than 40 kDa into the Protein Data Bank, 116 have been solved by solution-state NMR spectroscopy (http://www.rcsb.org/). In this review, we describe the conventional strategy used to solve solution structures of protein–RNA complexes by means of solution-state NMR, from RNA sequence optimization to structure analysis.

2. SAMPLE PREPARATION

2.1 Protein Preparation

NMR spectroscopy is not a sensitive method since milligram quantities of isotopically enriched protein are required for solution studies. Procedures allowing efficient preparation of isotopically labeled proteins (^{15}N, ^{13}C, ^{2}H) have been previously described (Gardner & Kay, 1998). The easiest approach consists in producing the protein of interest in *Escherichia coli* that leads to high yield; however, in the case of eukaryotic RBPs, the presence of rare codons could be an important limiting factor. To overcome this problem, codon-optimized gene synthesis and/or the choice of the appropriated bacterial strain are two important factors to increase insufficient yields. Since the RBP will be assembled with its RNA target, it is essential to remove any contamination from ribonuclease (RNAse) activity. RNAse-free protein solutions are generally obtained by setting up a three-step native purification that generally combined an affinity chromatography (that depends on the fusion tag used), followed by an ion-exchange chromatography and finally a size exclusion purification. In most cases, this purification protocol leads to a sufficiently pure protein solution for NMR, devoid of RNAse and nucleic acid contamination.

2.2 RNA Preparation

RBD targets generally short sequences of RNA (6–10 nucleotides) that can be prepared by chemical synthesis (Scaringe, Kitchen, Kaiser, & Marshall, 2004), since enzymatic synthesis of short RNA is not very efficient. This method is convenient but results in short RNA fragments that are not isotopically labeled. When the RNA sequence is longer (over 18 nucleotides), T7 RNA polymerase-based enzymatic production is the most common way to produce RNA. The RNA can be ^{13}C and ^{15}N labeled using ^{13}C, ^{15}N-labeled NTPs (Dieckmann & Feigon, 1997). When the RNA sequence is between 18 and 30 nucleotides, small DNA oligonucleotides can be used as transcription templates. However, when the RNA sequence is longer, we generally use a template embedded in a plasmid that is purified from bacteria and subsequently linearized. This leads to purer products for the transcription reactions. A key point of this approach is the RNA purification since the RNA polymerase is prone to generate a complex mixture composed of several products including aborts. To reduce the complexity of the mixture, 5′ and/or 3′ ribozymes are of great utility to ensure homogeneous RNA boundaries since uniform ends are resulting from catalytic cleavage (Price, Ito, Oubridge, Avis, & Nagai, 1995). In our laboratory, the transcription mixtures are submitted to denaturing anion exchange chromatography on an HPLC system (8 *M* urea, 80°C) to isolate the RNA of interest (Anderson, Scaringe, Earp, & Frederick, 1996; Duss, Maris, von Schroetter, & Allain, 2010). Other laboratories have shown that the RNA product can be purified by reverse phase (Feigon, Butcher, Finger, & Hud, 2001) or on denaturing polyacrylamide gels (Feigon et al., 2001). When the RNA is even longer (over 50 nucleotides), recently developed segmental labeling approaches have shown that a single segment of the RNA can be isotopically labeled (Duss, Diarra Dit Konte, & Allain, 2015; Duss, Lukavsky, & Allain, 2012); however, getting NMR compatible amounts of segmentally labeled RNA is quite tedious and time consuming.

2.3 Monitoring Protein–RNA Complex Formation Using NMR Spectroscopy

Solution-state NMR spectroscopy offers an easy way to monitor complex formation using ^{15}N-labeled protein and unlabeled RNA. This method is commonly referred as NMR titration. It consists in following the chemical shifts of the amide signals of the protein using ^{1}H–^{15}N heteronuclear single quantum correlations (HSQC) as a function of the RNA concentration. This experiment leads to the determination of the strength of the interaction (Markin & Spyracopoulos, 2012) and of the exchange regime. In general, if the interaction is strong (K_D below 0.1 n*M*), the exchange regime is slow, the protein–RNA complex dissociates slower than the NMR timescale and both free and bound resonances of the protein can be observed during the titration until saturation (Fig. 1). When the protein binds the RNA with a moderate affinity ($0.25\,\mathrm{n}M < K_D < 5\,\mu M$), the exchange regime is, in most cases, a mixture between intermediate and fast and an excess of one partner is required to saturate the resonances of its counterpart. In this case, the chemical shift is the population-weighted average of the bound and free states.

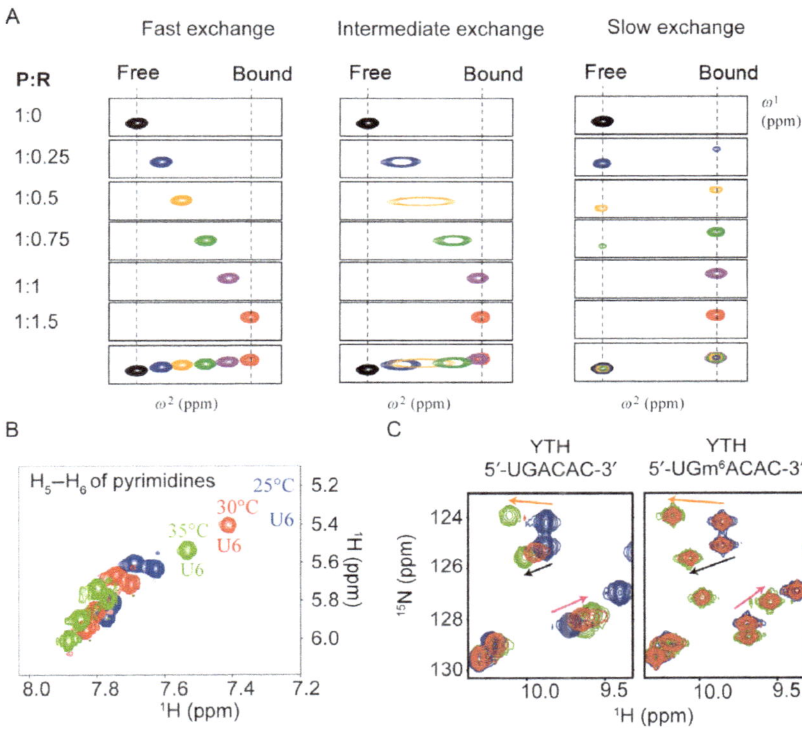

Fig. 1 Optimization of the conditions. (A) Scheme that illustrates the behaviors of one amide group from the protein upon addition of the RNA in function of the different exchange regimes. On the left, the protein:RNA ratio is indicated for each titration step. The *left panel* illustrates the fast exchange, the *middle one* corresponds to the intermediate exchange, and on the *right*, the slow exchange regime is depicted. For fast and intermediate exchanges, the chemical shift corresponds to the population-weighted average between the free and bound states. In the case of slow exchange, the correlations are observed at the chemical shifts of the free and bound forms and their intensities are weighted with respect to the different populations. (B) Portion of the 2D ^{1}H–^{1}H TOCSY spectrum of the 5′-ACCUUUG-3′ RNA oligonucleotide when in complex with the RRM2 domain of RBM39. On this panel, spectra recorded at three different temperatures are shown (25, 30, and 35°C). One can clearly see that the optimal temperature is 35°C. (C) Portion of the 2D ^{15}N–^{1}H HSQC spectra of the YTH domain titrated by the RNA target 5′-UGACAC-3′ with (*right panel*) or without (*left panel*) methylation in position 6 of the adenine 3. For each NMR titration, the NMR spectra are colored according to the molecular ratio protein:RNA. Ratios 1:0, 1:0.5, and 1:1 are colored in *blue*, *red*, and *green*, respectively. Changes of three different amide groups of the protein are illustrated by the *arrows*. The presence of the m6A epigenetic mark strengthens the protein–RNA interaction as illustrated by the change in exchange regime that switches from fast/intermediate to slow exchange upon methylation of the adenine 3.

One main advantage of solution-state NMR spectroscopy is that it also enables structural studies even with weak interactions.

It is also of primary interest to monitor the resonances of the RNA signals upon formation of the complex. In general, we use the homonuclear 2D ^{1}H–^{1}H total correlation spectroscopy (TOCSY) to follow the resonance of the pyrimidine signal (H_5–H_6) that fall into a spectral window where protein resonances are mostly absent (Fig. 1B). When the RNA does not contain any pyrimidine base, it also possible to monitor the RNA resonances using 2D ^{1}H–^{13}C HSQC (using ^{13}C-labeled RNA or natural abundance). Optimization of the conditions (salt concentration, pH, and temperature) is crucial to ensure good signal dispersion and homogeneous intensity of the RNA signals. As illustrated in Fig. 1B, by varying the temperature of the protein–RNA complex formed by the RRM2 of RBM39 and its RNA target, optimal conditions for structure determination were identified.

2.4 Finding the Optimal Conditions

Several molecular biology approaches have been developed to identify the RNA sequence targeted by an RBD. In vitro, the systematic evolution of ligands by exponential enrichment is one of the most popular method to identify the binding preferences of an RBP (Joyce, 1994). Genome-wide in vivo approaches like cross-linking immunoprecipitation (CLIP) and its various derivatives are the methods of choice to identify the natural RNA targets of an RBP (Lee & Ule, 2018). When no data on the RNA binding specificity of an RBP is available, a strategy based on NMR spectroscopy consists in determining the length of the RNA sequence and the specificity of the different positions. To reduce the cost due to RNA synthesis, RNA can be replaced sometimes by single-strand DNA oligonucleotides for the sequence optimization. The protein is first tested by NMR titration against polyA, polyC, polyG, and polyT oligonucleotides. The preferred base is placed in the middle of the motif and the optimal length of the oligonucleotide has to be determined by playing on boundaries. Then, for each position, the four bases are tested and the best one is selected to optimize the next position. How one can determine the optimal base for each position? In general, the chemical shift perturbations of the protein amide groups are determined, the magnitude and the directionally of the shifts are of major importance for selecting the optimal base. This iterative sequence optimization method is referred as scaffold-independent analysis (Beuth, Garcia-Mayoral, Taylor, & Ramos, 2007). An important parameter that has to be carefully monitored is the exchange regime of the interaction. For example, optimization of the RNA target for the YTH domain was performed using the SIA approach and despite optimization, the exchange regime remained fast to intermediate. A major breakthrough came when it was discovered that the YTH domain was in fact a reader of the epigenetic *N*6-methyladenine modification (Dominissini et al., 2012). Upon methylation of the central adenine of the motif, the interaction became stronger, the exchange regime switched from fast to slow (Theler, Dominguez, Blatter, Boudet, & Allain, 2014) and offered ideal conditions for the structure determination of the protein–RNA complex using NMR (Fig. 1C). As for most other methods of structural biology, optimization of the conditions is a crucial step that facilitates data analysis and structure determination using NMR spectroscopy.

3. DATA ANALYSIS AND STRUCTURE CALCULATIONS

3.1 The Exchange Regime Determines the Strategy

The exchange regime determines which strategy to adopt in order to solve the structure of the protein–RNA complex in solution. If the interaction is strong and the exchange regime slow, the protein–RNA complex dissociates slower than the NMR timescale and both, protein and RNA resonances, can be observed in a single sample (Fig. 2A). When the protein binds the RNA with a moderate affinity, the exchange regime is generally a mixture between intermediate and fast, and a slight excess of the partner is required to saturate the resonances of its counterpart (Fig. 2B). In this case, the chemical shifts and the NOE effects are the population-weighted average of the bound and free states. Therefore, two samples are required, one to observe the saturated resonances of the protein (P sample) and one for the RNA resonances (R sample). With the P and R samples, the intramolecular distances for the protein and the RNA bound states are extracted, respectively. In addition, a sample close to the 1:1 ratio is used to measure the edited filtered experiments in order to extract the intermolecular distances. This last sample is the most important, several protein–RNA ratios have to be tested in order to extract as many intermolecular distances as possible (Fig. 2C). Although finding the optimal ratio is tedious and time consuming, these data will define the RNA binding interface and are therefore crucial for the structure calculation of the complex.

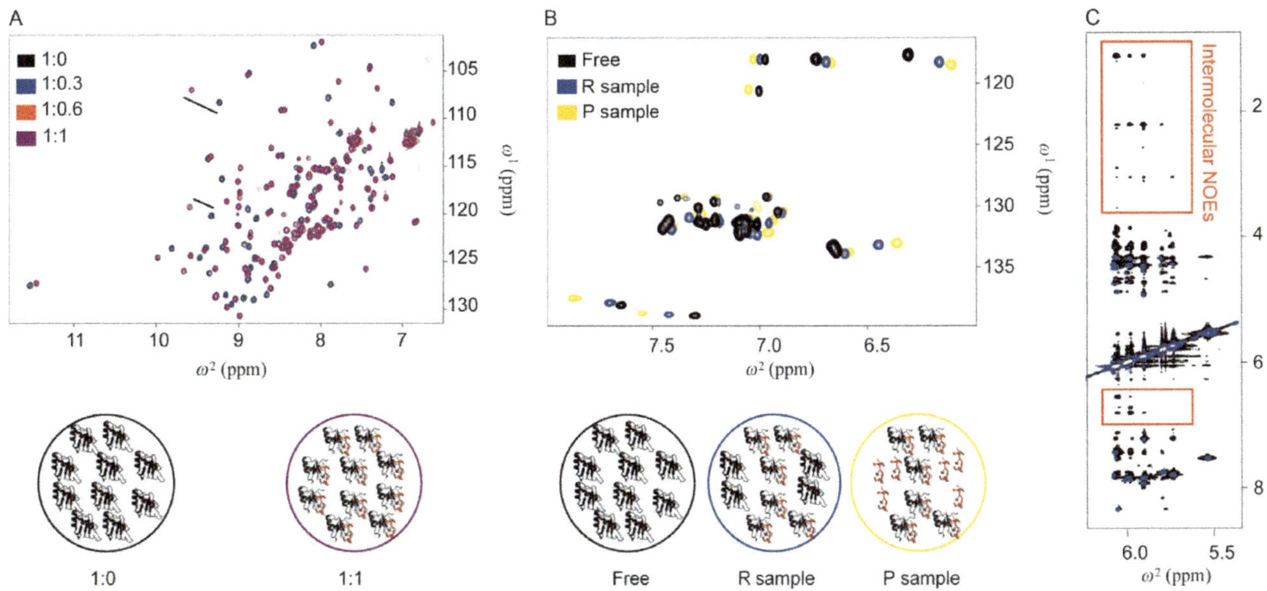

Fig. 2 Two scenarios: slow or fast exchange. (A) NMR titration of the RNA binding domain of human Fox-1 with 5′-UGCAUGU-3′. The titration was monitored by 2D ^{1}H–^{15}N HSQC spectra recorded at different ratios protein:RNA (1:0; 1:0.3; 1:0.6; 1:1). Below the spectra, a *cartoon* illustrates the sample composition at ratio 1:0 and 1:1. This panel illustrated the slow exchange regime, at equimolar amount of protein and RNA, all the resonances are saturated. (B) Overlay of the 2D ^{1}H–^{13}Caro HSQC spectra of RBM39 RRM2 in its free form (*black*), in the R sample (*blue*), and in the P sample (*yellow*). Below the spectra, a *cartoon* illustrates the sample composition for the free protein, the R sample, and the P sample. (C) Overlay of the 2D F1f F2f NOESY (*blue*) and 2D F2f NOESY (*black*) centered on the H1′ region. The signals that are unique to the *black* spectra correspond to intermolecular NOEs. This experiment was recorded with the R sample in which there is an excess of protein (ratio protein:RNA = 1.5:1).

3.2 The NMR Experiments to Extract NOE-Derived Distances

Assignment of the protein resonances in the protein–RNA complex is based on the classical approach developed for protein assignment (Sattler, Schleucher, & Griesinger, 1999). NMR spectra of RNA in the protein–RNA complexes are more difficult to assign than the protein due to the lack of unambiguous correlation-based assignment methodology, poor chemical dispersion, and unfavorable relaxation properties. Assignment of the short and unlabeled RNA relied heavily on 2D homonuclear experiments (Wuthrich, 1986) and on the assignment of the free RNA. When the RNA becomes longer than 10 nucleotides, isotope labeling of the RNA is preferred to overcome resonance overlap and more advanced three-dimensional pulse programs are used to assign it (Dominguez, Schubert, Duss, Ravindranathan, & Allain, 2011; Krahenbuhl, Lukavsky, & Wider, 2014). In order to calculate the structure of the protein–RNA complexes, the accuracy and the completeness of the resonance assignment are essential.

On the other hand, the structure calculation protocol requires experimental nuclear overhauser enhancement and exchange spectroscopy (NOESY) spectra recorded at high-field spectrometer with appropriate mixing time (60–100 ms). In order to identify intramolecular nuclear overhauser enhancement effects (NOE) for the protein, a complete set of NOESY experiments is recorded with the sample P and contains a 3D ^{1}H–^{15}N HSQC NOESY, a 3D ^{1}H–^{13}Cali HSQC NOESY, and a 3D ^{1}H–^{13}Caro HSQC NOESY. Those two last experiments can be combined into a single 3D ^{1}H–^{13}C HMQC NOESY thanks to the recent development of broadband heteronuclear decoupling scheme (Schilling et al., 2014) and are recorded in 10% D_2O as well as in 100% D_2O. On the RNA side, using the sample R, one should record a 2D ^{13}C F1-filtered F2-filtered NOESY in 100% D_2O to get access to intramolecular NOEs of the RNA. In addition, using a sample containing ^{15}N-labeled protein and unlabeled RNA in 10% D_2O, recording a 2D ^{1}H–^{1}H NOESY at low temperature helps to identify protected imino protons and intermolecular NOEs between the exchangeable protons and the protein. Finally, to unambiguously identify intermolecular NOEs, samples with opposite labeling of the two components are used (either the RNA is unlabeled and the protein ^{15}N/^{13}C labeled or vice versa) together with 2D and 3D NOESY spectra that select, for example, protons attached to ^{13}C in one dimension and to ^{12}C in another dimension. Selecting protons covalently attached to NMR active nuclei (e.g., ^{1}H[^{13}C] or ^{1}H[^{15}N]) is called editing, whereas suppressing such resonances and thus indirectly selecting $^{1}H[^{12}C/^{14}N]$ is called filtering. In the case of small unlabeled RNA in complex with $^{15}N/^{13}C$ protein, we generally combine the 2D ^{13}C F1-filtered F2-filtered NOESY (to observe NOEs which are intramolecular of the RNA) with the 2D ^{13}C F2-filtered NOESY (one can observe intramolecular RNA and intermolecular ones). Both experiments are recorded in 100% D_2O and if all the amide protons are not exchanged by deuterium, it is also important to add ^{15}N filtering. In case of strong resonance overlap on the protein side, there is also the possibility to record the same type of experiment by adding an editing step; this experiment is called 3D ^{13}C F1-edited F3-filtered HMQC-NOESY and is slightly less sensitive than the 2D version due to its higher dimensionality. The NOESY experiments commonly used to determine the solution structure of protein–RNA complexes are summarized in Table 1 (for a detailed descriptions of these NMR experiments, see Dominguez et al., 2011).

3.3 Initial Structure Calculations

Structure determination using solution-state NMR spectroscopy primarily consists in the interpretation of the NOESY cross-peaks into proton–proton distances. Thus, it requires several NOESY spectra and the assigned chemical shifts of the complex. The NOESY spectra are picked and integrated; this defines the positions of the cross-peaks and their intensities. The cross-peaks are associated with specific ^{1}H–^{1}H interactions on the basis of the established resonance assignments, and their intensity is interpreted as inversely proportional ($1/d^6$) to the distance d that separates both protons. Peak peaking of the NOESY spectra is performed "semiautomatically." The program is peaking the correlations automatically; however, the peak lists have to be carefully checked manually prior to structure calculations. Accordingly, ATNOS/CANDID (Herrmann, Guntert, & Wuthrich, 2002), CYPICK (Wurz & Guntert, 2017), CCP-NMR (Skinner et al., 2016), and CV-Peak Picker (Klukowski, Walczak, Gonczarek, Boudet, & Wider, 2015) are adapted programs for peak picking. The next steps consist in the assignment of the NOEs, their calibration, and the evaluation of the NOE network (referred as "network anchoring"). These three steps are combined in two structure calculations packages: CYANA "noeassign" (Guntert, 2004) or ARIA (Linge, Habeck, Rieping, & Nilges, 2003). These programs generate a list of intramolecular protein distance restraints (based

Table 1 NOESY Experiments for Protein–RNA Complexes

Protein	RNA	Solvent	Experiment	Purpose
^{15}N	Nonlabeled	H_2O	2D NOESY	Extracting intermolecular NOEs between imino protons and the protein
			3D ^{15}N-ed NOESY	Extracting protein–protein and protein–RNA NOEs
^{15}N	Nonlabeled	D_2O	2D NOESY	Extracting NOEs for whole complex
$^{13}C/^{15}N$	Nonlabeled	H_2O	3D ^{13}Cali-ed NOESY	Extracting protein–protein and protein–RNA NOEs
			3D ^{13}Caro-ed NOESY	Extracting protein–protein and protein–RNA NOEs
		D_2O	3D ^{13}Cali-ed NOESY	Extracting protein–protein and protein–RNA NOEs
			3D ^{13}Caro-ed NOESY	Extracting protein–protein and protein–RNA NOEs
			2D F1fF2f NOESY	Extracting RNA intramolecular NOEs
			2D F2f NOESY	Extracting protein–RNA intermolecular NOEs
^{15}N	$^{13}C/^{15}N$	D_2O	3D ^{13}Cali-ed NOESY	Extracting RNA–RNA and protein–RNA NOEs
			3D ^{13}Cali-ed NOESY	Extracting RNA–RNA and protein–RNA NOEs
			3D F1eF3f NOESY	Extracting protein–RNA intermolecular NOEs

on the interpretation of the NOESY spectra) that is used for the structure calculation of the protein–RNA complex. Concerning the intramolecular RNA–RNA and the intermolecular RNA–protein NOEs, these programs are less suited, mainly because of the low density of protons present in RNAs resulting in small network of NOEs. For these NOEs, the assignment and the calibration are therefore generally performed manually. Structure calculations in our laboratory are performed by torsion angle dynamics (TAD) simulated annealing (Guntert, Mumenthaler, & Wuthrich, 1997). TAD

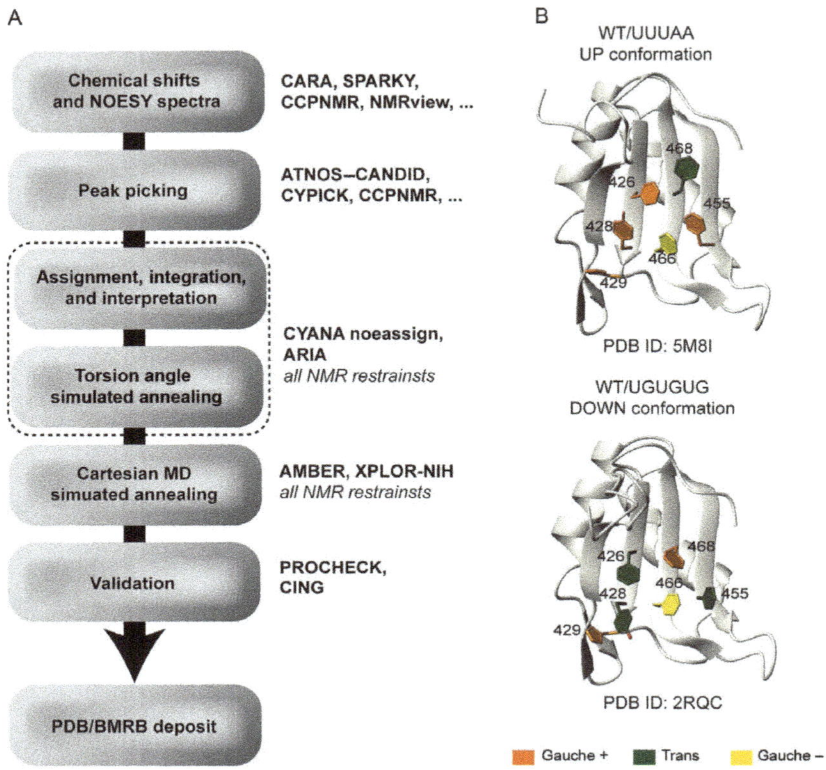

Fig. 3 Structure calculation procedure and aromatic side chain conformational switch. (A) Diagram showing the structure calculation procedure used to solve protein–RNA complex structures using solution NMR. (B) *Ribbon* representation of the structure of CUG-BP RRM3 bound to AU-rich sequence (*top*) or to GU-rich sequence (*bottom*). Conserved aromatic side chains are illustrated as *stick* models and colored according to the conformation of the side chains.

utilizes internal coordinates instead of Cartesian coordinates and the only degrees of freedom are torsion angles, while bond lengths, bond angles, chirality, and planarity of the peptide bond are kept fixed at their optimal values. This allows the number of degrees of freedom to be decreased by approximately 10-fold resulting in a simpler potential energy function and therefore faster calculations. Structure calculations are performed iteratively by several cycles of NOE assignment, calibration, and structure calculation that are repeated to increase the agreement between the experimental data and the resulting NMR structures (Fig. 3A). It should be noted that, in the case of nucleic acids, CYANA parameter sets do not contain a bond between the

C4′ and the O4′ atoms in order to allow the sugar ring to remain flexible. Therefore, artificial distance restraints (lower bound and upper bound) between these two atoms for each sugar ring of the nucleic acid molecule need to be added.

3.4 Refinement and Validation

Output structures from simulated annealing performed using TAD are generally not in agreement with the ideal geometry and need to be refined in the Cartesian space. Therefore, the output of the initial structure calculation (NMR structure ensemble and NMR restraints) is used to perform a simulated annealing procedure in Cartesian space using appropriate force fields (AMBER or CHARMM; Mackerell & Nilsson, 2008) that have been developed for both proteins and nucleic acids and therefore, suitable for the refinement of protein–RNA structures. Refinements of the structures are performed using the SANDER (simulated annealing with NMR-derived energy restraints) module of the molecular dynamics simulation package AMBER (Case et al., 2005) or using XPLOR-NIH that include parameter sets derived from AMBER and CHARMM force fields. Both the SANDER module of AMBER and XPLOR can incorporate most NMR-derived restraints. Additionally, both packages include different solvent models for refinement. After refinement, the agreement between the experimental data and the NMR structure is evaluated (Doreleijers et al., 2012) together with its agreement with ideal geometry (Laskowski, Rullmannn, MacArthur, Kaptein, & Thornton, 1996).

3.5 Convergence and Restraint Density

In order to illustrate the number of NMR restraints, the violations of those NMR restraints, the agreement with the expected geometry, and the convergence of the final calculation, structural statistics of a bundle of 10 NMR structures of protein–RNA complexes solved by following the described procedure are summarized in Table 2. In addition, the final NMR ensemble obtained for the solution structure determination of the RRM of SRSF2 bound to GGAG containing RNA is shown in Fig. 4 (Daubner, Clery, Jayne, Stevenin, & Allain, 2012). This superimposition gives a rough idea of the expected convergence of the NMR structure of protein–RNA complex solved by following the approach described along this chapter. In addition, a stereo view centered on the first G of the GGAC motif illustrates

Table 2 Structural and Refinement Statistics of 10 Protein–RNA Solution Structures Solved Using the Method Described in This Chapter

PDB Code	5MPG	5MPL	2MTV	2M8D	2LEB	2LEC	2KXN	2KFY	2KG0	2KG2	Mean
Protein length	97	101	155	90	101	101	95	102	102	104	
RNA length	7	6	6	8	6	6	6	6	6	6	
NOE restraints											
Intramolecular protein	2356	2083	4497	1400	1933	2098	1107	1312	2180	982	19.03[a]
Intramolecular RNA	87	59	49	49	73	83	112	82	101	140	13.25[b]
Intermolecular	117	111	214	38	107	83	93	56	63	60	14.95[b]
Dihedrals restraints											
Protein	0	0	234	0	0	0	0	0	0	0	0.22[a]
RNA	4	4	7	8	6	6	6	6	6	6	0.93[b]
Max. violations (mean)											
Protein–RNA (Å)	0.25	0.27	0.32	0.43	0.32	0.30	0.45	0.37	0.37	0.35	0.34
Geometry (mean)											
Bond length (Å)	0.0037	0.0035	0.0036	0.0041	0.0041	0.0042	0.0040	0.0041	0.0041	0.0044	0.0039
Bond angle (degrees)	0.51	0.49	1.7	1.74	1.67	1.77	1.67	1.82	1.59	1.88	1.48
R.m.s.d. (mean)											

Protein											
Backbone (Å)	0.40	0.45	0.21	0.46	0.63	0.78	1.31	0.83	0.36	1.14	0.65
Heavy atoms (Å)	0.83	0.78	0.51	0.92	1.08	1.43	1.95	1.36	0.72	1.49	1.10
RNA											
Heavy atoms (Å)	0.62	0.78	0.64	0.60	1.11	1.3	0.92	1.90	1.30	0.75	0.99
Protein–RNA											
Heavy atoms (Å)			0.54	0.91	1.11	1.47	1.87	1.54	0.90	1.42	1.22

[a]These values are mean per residues of protein.
[b]These values are mean per ribonucleotide.

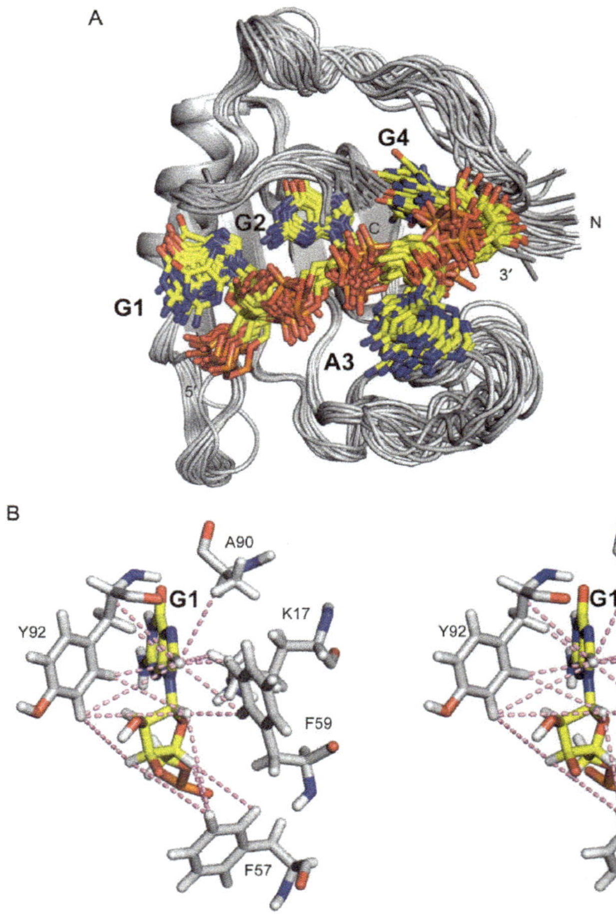

Fig. 4 Graphical overview of the solution structure of the RRM of SRSF2 bound to GGAG-containing single-stranded RNA. (A) Superimposition of the 20 NMR structures of the RRM of SRSF2 bound to GGAG-containing RNA motif (PDB ID 2LEC). The ensemble of NMR structures overlay with a RMSD of 1.47 Å (all heavy atoms). The protein and the RNA are represented as *ribbon* and *stick* models, respectively. The protein and the RNA are colored in *gray* and *yellow*, respectively. (B) Stereo view of the intermolecular NOEs experimentally identified for the unambiguous positioning of the guanine 1 with respect to amino acid side chains of the protein. The intermolecular NOEs are shown as *pink dashed lines*. In order to simplify the figure, intermolecular NOEs involving equivalent atoms (i.e., methyl group) are marked only on one atom.

graphically the 17 intermolecular NOE that unambiguously position the guanine 1 in its binding pocket. According to the statistics presented in Table 2, this network of intermolecular NOEs is as dense as the average network since ~15 intermolecular NOEs are generally observed per nucleotide.

3.6 Other Types of NMR Restraints

Besides NOE-derived distances, other kinds of structural information can be used to drive structure calculations. First, by adding liquid crystalline media in the NMR sample, partial anisotropy is reintroduced and brakes the averaging of the dipolar interaction (Tolman, Flanagan, Kennedy, & Prestegard, 1995). Consequently, partial dipolar couplings can be extracted. The residual dipolar coupling (RDC) amplitude is dependent on the orientation of the atomic bond with respect to the molecular frame. RDC has to be measured at least in two independent liquid crystalline media to avoid symmetric solutions of the alignment tensors. In most of the cases, these orientation restraints are included during the structure refinement stage to increase the accuracy of the solution structure calculated on the basis of NOE restraints. They are notably useful to restrict the conformational space of the stem part of an RNA stem–loop when the protein binds the loop part of such RNA (Dorn et al., 2017). Another powerful application of RDC restraints consists in using them to orient independent domains with respect to each other (Dosset, Hus, Marion, & Blackledge, 2001). Notably, this approach was used for the structure determination of both N-terminal dsRBDs of TRBP in complex with a functionally asymmetric siRNA. In this example, RNA recognition of the dsRNA is not sequence specific but rather depends on the RNA shape. Thus, two pseudosymmetrical complexes are observed in the same NMR sample and RDC was very important to deconvolute the dense network of intermolecular NOEs (Masliah et al., 2018). Another kind of NMR restraint can be derived from the paramagnetic spin labeling of the RNA or the protein. The addition of a chemical group containing an unpaired electron gives rise to paramagnetic relaxation enhancement (PRE). NMR signals located in close proximity with the paramagnetic center are broadened, and the PRE effect is proportional to the distance between the observed spin and the paramagnetic electron (Clore, 2011). This method is preferentially used to verify the orientation of the RNA when bound to a protein (Beusch, Barraud, Moursy, Clery, & Allain, 2017; Mackereth et al., 2011) since precise intermolecular distance calculations are very tedious and require a lot of measurement time. To extract precise long-range distances using paramagnetic labels, we generally prefer to use two spin labels and take advantage of EPR (Duss et al., 2014; Masliah et al., 2018). Apart from these NMR restraints, any kind of restraints that can be translated into interatomic distances can be combined with the NOEs-derived distances during the structure calculation procedure.

4. STRUCTURAL ENSEMBLE AND FUNCTIONAL VALIDATION

4.1 Improving the Precision of the Final Structural Ensemble

Defining the final structural ensemble is not a simple and easy task as the procedure described above very rarely results in a satisfying structural ensemble, i.e., precise and accurate. The protein–RNA interface is often more dynamic than the rest of the complex resulting in severe line broadening of the protein side chain and RNA resonances. Very frequently, we observe conformational exchange line broadening caused by slow ring flip of aromatic side chains when stacking with RNA bases or by the slow motion of arginine and lysine side chains contacting multiple sites in the RNA bases or backbone. Solution to the line-broadening problem can be to increase the temperature of the sample in order to sharpen the linewidth as the tumbling of the complex and the aromatic ring flip will be faster. Protein–RNA complex is often fairly stable and key intermolecular NOEs can sometimes only be observed above 50°C (Allain, Bouvet, Dieckmann, & Feigon, 2000). Measurement at different field strengths can also have an impact as this can shift the exchange regime. Decreasing the temperature can also be critical to observe key resonances. Quite classically, the observation of RNA imino resonances at the protein–RNA interface is often improved at lower temperature as the exchange rate with the solvent is reduced (Clery et al., 2011, 2013). Lower temperature can also be critical to prevent the exchange between multiple binding registers as we saw when studying the RRM of hnRNP C (Cienikova, Damberger, Hall, Allain, & Maris, 2014). Overall, taking NMR spectra at different temperature to determine the structure of a protein–RNA complex is very frequently done in order to maximize the number of structural constraints.

During the refinement process, it is often frequent that several conformations fit equally well with the input constraints which will of course increase the precision of the ensemble. Often, these multiple conformations result from a lack of constraints but not always. Sometimes, the structure calculation can result into several discrete conformations of low energy which are not always accurate. In such false conformation, two protons can be very close in space although no NOE was observed between them. In such cases, after careful inspection of the NOESY spectra a lower bound restraint could be added between the two protons in order to avoid "falling" into this conformation space. Although one should be very cautious in adding such "repulsive" restraints, using such restraints help defining the final ensemble and improving the precision and accuracy of the structure. Such restraints are not always needed in the final calculation and not always deposited but they certainly help during the refinement process.

Another fairly frequent case found when solving NMR structures of protein–RNA complex is the presence of NOEs originating from minor conformations. These can originate from multiple registers in double-stranded RNA or single-stranded RNA which are in fast exchange. Even if a minor conformation is only present 10% of the time, a strong NOE can be visible. Often, such NOEs are incompatible with the main conformation and need to be ignored. Yet, it is important to report them in papers (Loughlin et al., 2011; Stefl et al., 2010) to indicate that such minor conformations exist in solution.

Using such tips and tricks, precise structural ensemble of protein–RNA complexes can be obtained using the approach described earlier. Often an RMSD below 1 Å at the RNA–protein interface can be achieved when enough intermolecular NOEs can be detected and is usually sufficient to decipher how RNA binding recognition is achieved.

4.2 Accuracy and Functional Validation of the Solution Structure

Often, since the interpretation of the solution structure is not always straightforward due to some imprecisions in the position of the RNA and the protein side chains at the interface. It is important to independently validate the structural findings especially when new modes of binding or unusual contact are found. One immediate way to validate the final ensemble is to see if the structure of the complex explains well the chemical shift changes observed upon RNA binding. Large chemical shift changes can originate from hydrogen bonding (amide or carbonyl) or from the proximity of an aromatic ring (RNA bases). Large chemical shift can also occur when RNA binding triggers a conformational change in the protein. The structure of the complex should explain all if not the great majority of these changes.

The second way to check the accuracy of the structure is to perform mutagenesis studies in the protein or in the RNA. One expects that mutating the protein side chains involved in RNA recognition should result in a decrease of the affinity of the complex. Similarly, if an RNA base is sequence specifically recognized, one expect that mutating this nucleotide by any of the three other nucleotides should also decrease the affinity. In the vast majority of the protein–RNA complexes solved in our laboratory and others, mutations of key residues (amino acids or nucleotides) found to interact in the structures result in a decrease of affinity or even to no binding. Yet, in a few exceptions, such mutations resulted in an affinity increase which then triggered very interesting subsequent studies. Two examples are discussed below.

The first example is when we solved the structure of the RRM of SRSF2 bound to RNA (Daubner et al., 2012). We initially solved the solution structure of the RRM bound to a CCAG-containing RNA. The structure revealed the sequence-specific recognition of both cytosines and the guanine, yet when the two cytosines were mutated to a guanine the affinity slightly increased and indeed SRSF2 RRM binds GGAG with higher affinity than CCAG. These findings made us solve subsequently a second structure with a GGAG-containing RNA that could explain this unexpected slight increase in affinity. Indeed, although the two most 5′ guanines are bound in the same binding pockets as the two cytosines in the CCAG–RRM complex, these two guanines adopt a *syn*-conformation while the two cytosines adopt an *anti*-conformation. This "adaptation" of the RNA to the protein surface could then explain these surprising affinity measurements.

The second example is more recent and involves amino acid mutations on the surface of the RRM of CUG-BP2. The RRM is one of the most abundant protein domains in eukaryotes. While the structure of this domain is well characterized by the packing of two alpha-helices on a four-stranded beta-sheet, it generally recognizes a small sequence of single-stranded RNA. RRMs often expand two stretches of conserved residues called RNP1 and RNP2 that contains conserved aromatic residues on the β-sheet surface of the domain (Maris et al., 2005). The conserved aromatic residues are involved in RNA recognition and stack against nucleobases. In a recent publication, we solved the solution structure of the CUG-BP2 RRM3 bound to an AU-rich sequence with NMR spectroscopy (Diarra Dit Konte et al., 2017). When validating the role of the aromatic residues by mutagenesis, we unexpectedly observed an increase in affinity with some of the mutants while expecting no binding. We soon realized that the conformation of the aromatic residues in the complex is very similar to their conformation in the structure of the free form (UP conformation, Fig. 3B). Yet when some aromatic residues are mutated to alanine, a change of the aromatic conformation takes place (DOWN conformation, Fig. 3B) that is even more compatible with RNA binding and found in CUG-BP1

RRM bound to high-affinity targets (Tsuda et al., 2009). Analysis of the conformation of the aromatic residues in the different forms of the protein could be performed using NMR spectroscopy (Diarra Dit Konte et al., 2017). In support of our findings, molecular dynamics simulations of the free protein, of the complexes, and of the mutants allowed us to visualize how the aromatic side chains can move between a conformation that will bind weak affinity targets to a conformation that will bind strong affinity targets. This was one example of how MD simulation can complement NMR solution structure of protein–RNA complexes. Below, more examples of synergies between the two methods are described.

5. MOLECULAR DYNAMIC SIMULATIONS PROVIDE ADDITIONAL INSIGHT TO THE SOLUTION STRUCTURES

Classical molecular dynamics (MD) simulations can provide specific structural information about protein–RNA complexes that is difficult to obtain experimentally and thus complement the NMR structural studies. The current status of the field of protein–RNA complexes has been recently reviewed (Šponer et al., 2018, 2017).

The greatest advantage of MD simulation technique is that it provides essentially unlimited temporal and spatial resolution of the studied systems, including positions of all the hydrogen atoms and solvent molecules. However, the outcome of MD studies is predetermined by the properties of the starting structures (including all their errors and uncertainties) and influenced by two major limitations, namely, quality of the simulation force field (empirical potential) and the affordable simulation timescale (Šponer et al., 2018). The latter needs to be compared with real timescale of the structural dynamics. With increasing power of the computers, the force field is gradually becoming a more crucial limitation of the MD simulations, provided the starting structure is of sufficient quality (Šponer et al., 2018). MD simulation studies should always be accompanied by careful discussion of the limitations and the declared results should be documented in detail. Unfortunately, without these steps, it is easy to make wide-reaching biochemical speculations based even on totally corrupted MD simulation trajectories. Due to easy access to fast hardware and the contemporary push for "hot-topic" papers, many MD publications show an astonishing reluctance to report force-field problems and to discuss limitations of the modeling, which is especially notorious in self-confident communication-style papers, often published in presumably high-quality journals. Another obstacle related to the growing power of computers is that it is becoming increasingly difficult to convert the overwhelming amounts of raw simulation data into humanly comprehensible information (Šponer et al., 2018).

The gold MD standard is performing unbiased (standard, plain) MD simulations. This is essentially a single-molecule technique that models the thermal structural dynamics of a given system starting from a given starting structure on the given simulation timescale. The standard MD should be relatively easily understandable for nonspecialists, as one should try to imagine what a real molecule would do on the simulation timescale starting from the used starting structure. For a detailed overview of basic principles, advantages and also limitations of so-called enhanced-sampling MD simulation methods applicable to protein–RNA complexes see this recent review (Šponer et al., 2018).

Simulations of protein–RNA complexes must be carefully prepared and executed, with the appropriate selection of the starting structure. There is typically more than one model of the structure available in a refined NMR ensemble. Fortunately, in high-quality NMR structures with large amounts of experimental constraints, the differences between the individual models are usually small and it is safe to select a single frame as the starting structure for MD simulations (for further details of the structure selection, see Krepl et al., 2018). Another important step is the choice of the biomolecular force field used to describe the complex. A combination of state-of-the-art AMBER force fields, such as ff12SB or ff14SB for proteins (Maier et al., 2015) and bsc0χ_{OL3} for RNA (Zgarbova et al., 2011), was shown to perform satisfactorily for protein–RNA complexes (Krepl, Cléry, Blatter, Allain, & Sponer, 2016; Krepl et al., 2015). It also should be noted that the force field performance for protein–RNA complexes is not uniform, i.e., the capability of the force field to reliably describe a given protein–RNA system depends on its complexity and type of molecular interactions. It appears that interfaces with complex networks of H-bonds and exposed to competition with solvent molecules can be difficult to balance (Krepl et al., 2015).

MD is unique in its capability to capture the interplay between structural dynamics of the solute and of the enveloping solvent. By their design, MD simulations allow studying the movement of every individual water molecule present in the system. Further, the timescales relevant to the water dynamics are well accessible using the current computers and the performance of even relatively simple water models is solid enough to provide information about the specific hydration sites present in the protein/RNA complexes (Šponer et al., 2018). The explicit solvent model must always be used to properly address the solute–solvent interactions as these can significantly contribute to the complex stability. For example, the simulations have shown that specific hydration sites can contribute to the RNA binding affinity of the Fox-1 RRM (Krepl et al., 2017). This contribution can be facilitated by the water molecule directly by forming simultaneous interaction with protein and RNA atoms—a water bridge. Alternatively, the binding of the water molecule can stabilize specific protein or RNA conformation which is conducive to formation and stability of the protein–RNA complex—an allosteric effect (Fig. 5A).

In our MD simulations of the NMR structures, in addition to standard equilibration protocol (Krepl et al., 2015), we always apply the experimentally determined NMR restraints in the initial stages of the production simulations (Krepl et al., 2016). This procedure can prevent abrupt departures from the geometry of the experimental structure and allows the system more time to settle down which significantly reduces instability in later, unrestrained parts of the simulations. In essence, temporary use of the NMR restraints allows the structure to smoothly cross from the world of NMR data to the world of MD simulations. However, note that sufficiently long fully unrestrained simulations must always be carried out following the initial use of the restraints. In fact, one of the aspects of protein–RNA studies where the simulations can complement the NMR experiments is the free solution dynamics of the protein–RNA interface (in addition, infinitely long MD would converge to equilibrium population of states due to the ergodic principle). The signals (from which the NMR restraints are derived) collected during the NMR measurement correspond to all molecules in the sample collected over time. The obtained information is thus ensemble averaged and time averaged and certain aspects of structural dynamics are missing. This is also the reason why we do not recommend imposing the ensemble-averaged NMR distance restraints across the full course of the production simulations, since they would distort the shape of the ensemble by entirely eliminating substates in which the distances are longer than the upper bound of restraints. This problem is magnified by the highly nonlinear relation between the NOE data and the associated interatomic distances. The MD simulations can be used to uncover the structure averaging and obtain the actual, real-time conformers of the structure. This complementarity between NMR and MD is exemplified by the study of the CUG-BP2 RRM3 (see above and Fig. 3). While NMR experiments have provided

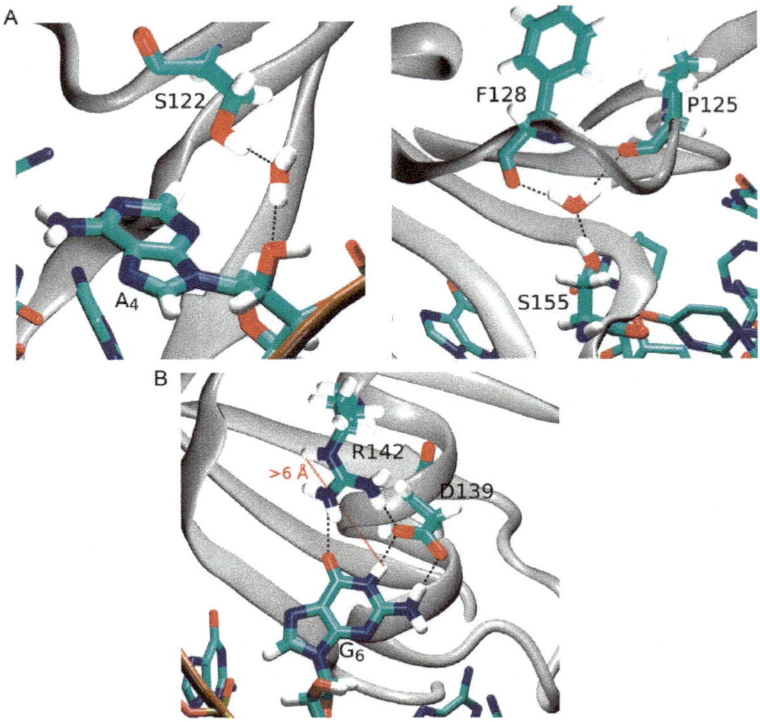

Fig. 5 Illustration of the synergy between NMR structure and MD of protein–RNA complexes. (A) MD simulations have revealed existence of two hydration sites in the Fox-1 RRM which influence the protein/RNA affinity. The water molecule near residue S122 (*left*) directly bridges the protein and RNA atoms, while the water molecule near S155 (*right*) stabilizes the protein's structure and increases its affinity to RNA via allosteric effect. (B) Simulations have shown that the interaction between R142 and G_6 is an integral part of the G_6 recognition pattern by the SRSF1 RRM2. The R142/G_6 interaction was not identified in the original NMR experiments because the intermolecular distance (indicated by *red dashed line*) between the two nearest observable protons was above 6 Å. Hydrogen bonds are indicated by *black dashed lines*.

accurate information about the relative populations of the aromatic side chain conformations in the CUG-BP2 RRM3, the simulations were used to observe and define the exact transitional pathway between the UP and DOWN conformations of the aromatic side chains (Diarra Dit Konte et al., 2017). The work shows that MD could be exceptionally useful in studies of protein–RNA complexes where the recognition cannot be fully described by one static structure.

The results of the MD simulations must be carefully analyzed and comparison with the available structural and biochemical data such as the NMR

NOE distances or mutation studies should always be made. Also, potential force-field deficiencies should be considered. One of the greatest pitfalls in MD simulation studies is that sometimes the simulation instabilities resulting from various force-field problems are being misinterpreted and assigned a biological significance. However, well-executed and critically analyzed MD simulations can have great predictive power (Šponer et al., 2018) and can supply information about protein/RNA interactions that cannot be observed in the NMR experiments due to, e.g., lack of observable proton/proton distances. This was the case in the study of the SRSF1 RRM2, where the involvement of the R142 residue in RNA binding was first identified by MD simulations and subsequently confirmed by ITC measurements of the R142A mutants (Krepl et al., 2016). This protein–RNA interaction could not be observed in the original NMR experiment (Cléry et al., 2013) because, despite the direct R142/G_6 binding, the closest observable hydrogens G_6(H1) and R142(HE) are distanced more than 6 Å (Fig. 5B).

An attractive option would be to integrate the primary experimental data directly into the MD simulations, in order to obtain MD structural ensembles satisfying the measured data. However, it is far from being trivial, since diverse ensembles can satisfy the primary data when the system is underdetermined. These approaches, despite lot of recent efforts, are still in their infancy (Bonomi, Heller, Camilloni, & Vendruscolo, 2017), and may be quite challenging for protein–RNA complexes (Borkar et al., 2016; Šponer et al., 2018).

6. CONCLUSION

We have reviewed here our current approaches for solving medium-size (less than 20kDa) protein–RNA complexes using solution-state NMR spectroscopy. Our approach have proven to be robust and accurate since when confronted with crystal structures of identical or related complexes, our solution structure was found to be very similar and with the same RNA recognition mode (Allain et al., 1996; Beusch et al., 2017; Daubner et al., 2014; Loughlin et al., 2011; Oberstrass et al., 2006; Theler et al., 2014). More recently, the use of MD simulation to complement our structural work has proven to be very powerful to further interpret the solution structure of various protein–RNA complexes. MD provides crucial information to understand the role of water molecule at the protein–RNA interface of Fox RRM (Krepl et al., 2017), on the

RNA recognition of SRSF1 RRM2 (Krepl et al., 2016) and on the role of the aromatic side chains of CUG-BP2 in RNA recognition (Diarra Dit Konte et al., 2017). More are to come.

When confronted by larger protein–RNA complexes (above 25 kDa), the sole use of NMR spectroscopy to solve atomic structures becomes very challenging due to increase resonance linewidth and the increased resonance overlap. In such case, combining NMR with other methodologies such as mass spectrometry (Dorn et al., 2017), EPR (Duss et al., 2014; Masliah et al., 2018), FRET (Masliah et al., 2018), SAXS/SANS (Lapinaite et al., 2013), solid-state NMR (Marchanka, Simon, Althoff-Ospelt, & Carlomagno, 2015), and even electron microscopy is the way to go at the moment as shown by us and others. Yet, there are still a lot to find in small systems. In particular, very little is known about the dynamic of protein–RNA interface although such interface offers certainly new grounds for drug discovery considering the numerous diseases originating from defect in protein–RNA recognition.

ACKNOWLEDGMENTS

S.C. and F.A. acknowledge support from the NCCR RNA and Disease funded by the Swiss National Science Foundation. M.K. and J.S. acknowledge support from the Czech Science Foundation [P305/12/G034]; the Ministry of Education, Youth and Sports of the Czech Republic [LO1305] and the Praemium Academiae.

REFERENCES

Allain, F. H., Bouvet, P., Dieckmann, T., & Feigon, J. (2000). Molecular basis of sequence-specific recognition of pre-ribosomal RNA by nucleolin. *The EMBO Journal*, *19*(24), 6870–6881.

Allain, F. H., Gubser, C. C., Howe, P. W., Nagai, K., Neuhaus, D., & Varani, G. (1996). Specificity of ribonucleoprotein interaction determined by RNA folding during complex formulation. *Nature*, *380*(6575), 646–650.

Anderson, A. C., Scaringe, S. A., Earp, B. E., & Frederick, C. A. (1996). HPLC purification of RNA for crystallography and NMR. *RNA*, *2*(2), 110–117.

Berg, J. M. (2003). Fingering nucleic acids: The RNA did it. *Nature Structural Biology*, *10*(12), 986–987.

Beusch, I., Barraud, P., Moursy, A., Clery, A., & Allain, F. H. (2017). Tandem hnRNP A1 RNA recognition motifs act in concert to repress the splicing of survival motor neuron exon 7. *eLife*, *6*.

Beuth, B., Garcia-Mayoral, M. F., Taylor, I. A., & Ramos, A. (2007). Scaffold-independent analysis of RNA-protein interactions: The Nova-1 KH3-RNA complex. *Journal of the American Chemical Society*, *129*(33), 10205–10210.

Bonomi, M., Heller, G. T., Camilloni, C., & Vendruscolo, M. (2017). Principles of protein structural ensemble determination. *Current Opinion in Structural Biology*, *42*, 106–116.

Borkar, A. N., Bardaro, M. F., Camilloni, C., Aprile, F. A., Varani, G., & Vendruscolo, M. (2016). Structure of a low-population binding intermediate in protein-RNA recognition. *Proceedings of the National Academy of Sciences of the United States of America, 113*(26), 7171–7176.

Case, D. A., Cheatham, T. E., 3rd, Darden, T., Gohlke, H., Luo, R., Merz, K. M., Jr., et al. (2005). The amber biomolecular simulation programs. *Journal of Computational Chemistry, 26*(16), 1668–1688.

Cienikova, Z., Damberger, F. F., Hall, J., Allain, F. H., & Maris, C. (2014). Structural and mechanistic insights into poly(uridine) tract recognition by the hnRNP C RNA recognition motif. *Journal of the American Chemical Society, 136*(41), 14536–14544.

Clery, A., Jayne, S., Benderska, N., Dominguez, C., Stamm, S., & Allain, F. H. (2011). Molecular basis of purine-rich RNA recognition by the human SR-like protein Tra2-beta1. *Nature Structural & Molecular Biology, 18*(4), 443–450.

Cléry, A., Sinha, R., Anczuków, O., Corrionero, A., Moursy, A., Daubner, G. M., et al. (2013). Isolated pseudo–RNA-recognition motifs of SR proteins can regulate splicing using a noncanonical mode of RNA recognition. *Proceedings of the National Academy of Sciences of the United States of America, 110*(30), E2802–E2811.

Clore, G. M. (2011). Exploring translocation of proteins on DNA by NMR. *Journal of Biomolecular NMR, 51*(3), 209–219.

Corbett, A. H. (2018). Post-transcriptional regulation of gene expression and human disease. *Current Opinion in Cell Biology, 52*, 96–104.

Daubner, G. M., Brummer, A., Tocchini, C., Gerhardy, S., Ciosk, R., Zavolan, M., et al. (2014). Structural and functional implications of the QUA2 domain on RNA recognition by GLD-1. *Nucleic Acids Research, 42*(12), 8092–8105.

Daubner, G. M., Clery, A., Jayne, S., Stevenin, J., & Allain, F. H. (2012). A syn-anti conformational difference allows SRSF2 to recognize guanines and cytosines equally well. *The EMBO Journal, 31*(1), 162–174.

Diarra Dit Konte, N., Krepl, M., Damberger, F. F., Ripin, N., Duss, O., Sponer, J., et al. (2017). Aromatic side-chain conformational switch on the surface of the RNA recognition Motif enables RNA discrimination. *Nature Communications, 8*(1), 654.

Dieckmann, T., & Feigon, J. (1997). Assignment methodology for larger RNA oligonucleotides: Application to an ATP-binding RNA aptamer. *Journal of Biomolecular NMR, 9*(3), 259–272.

Dominguez, C., Schubert, M., Duss, O., Ravindranathan, S., & Allain, F. H. (2011). Structure determination and dynamics of protein-RNA complexes by NMR spectroscopy. *Progress in Nuclear Magnetic Resonance Spectroscopy, 58*(1–2), 1–61.

Dominissini, D., Moshitch-Moshkovitz, S., Schwartz, S., Salmon-Divon, M., Ungar, L., Osenberg, S., et al. (2012). Topology of the human and mouse m6A RNA methylomes revealed by m6A-seq. *Nature, 485*(7397), 201–206.

Doreleijers, J. F., Sousa da Silva, A. W., Krieger, E., Nabuurs, S. B., Spronk, C. A., Stevens, T. J., et al. (2012). CING: An integrated residue-based structure validation program suite. *Journal of Biomolecular NMR, 54*(3), 267–283.

Dorn, G., Leitner, A., Boudet, J., Campagne, S., von Schroetter, C., Moursy, A., et al. (2017). Structural modeling of protein-RNA complexes using crosslinking of segmentally isotope-labeled RNA and MS/MS. *Nature Methods, 14*(5), 487–490.

Dosset, P., Hus, J. C., Marion, D., & Blackledge, M. (2001). A novel interactive tool for rigid-body modeling of multi-domain macromolecules using residual dipolar couplings. *Journal of Biomolecular NMR, 20*(3), 223–231.

Duss, O., Diarra Dit Konte, N., & Allain, F. H. (2015). Cut and paste RNA for nuclear magnetic resonance, paramagnetic resonance enhancement, and electron paramagnetic resonance structural studies. *Methods in Enzymology, 565*, 537–562.

Duss, O., Lukavsky, P. J., & Allain, F. H. (2012). Isotope labeling and segmental labeling of larger RNAs for NMR structural studies. *Advances in Experimental Medicine and Biology, 992*, 121–144.

Duss, O., Maris, C., von Schroetter, C., & Allain, F. H. (2010). A fast, efficient and sequence-independent method for flexible multiple segmental isotope labeling of RNA using ribozyme and RNase H cleavage. *Nucleic Acids Research*, *38*(20), e188.

Duss, O., Michel, E., Yulikov, M., Schubert, M., Jeschke, G., & Allain, F. H. (2014). Structural basis of the non-coding RNA RsmZ acting as a protein sponge. *Nature*, *509*(7502), 588–592.

Feigon, J., Butcher, S. E., Finger, L. D., & Hud, N. V. (2001). Solution nuclear magnetic resonance probing of cation binding sites on nucleic acids. *Methods in Enzymology*, *338*, 400–420.

Gardner, K. H., & Kay, L. E. (1998). The use of 2H, 13C, 15N multidimensional NMR to study the structure and dynamics of proteins. *Annual Review of Biophysics and Biomolecular Structure*, *27*, 357–406.

Gomez, J. A., Wapinski, O. L., Yang, Y. W., Bureau, J. F., Gopinath, S., Monack, D. M., et al. (2013). The NeST long ncRNA controls microbial susceptibility and epigenetic activation of the interferon-gamma locus. *Cell*, *152*(4), 743–754.

Guntert, P. (2004). Automated NMR structure calculation with CYANA. *Methods in Molecular Biology*, *278*, 353–378.

Guntert, P., Mumenthaler, C., & Wuthrich, K. (1997). Torsion angle dynamics for NMR structure calculation with the new program DYANA. *Journal of Molecular Biology*, *273*(1), 283–298.

Herrmann, T., Guntert, P., & Wuthrich, K. (2002). Protein NMR structure determination with automated NOE-identification in the NOESY spectra using the new software ATNOS. *Journal of Biomolecular NMR*, *24*(3), 171–189.

Joyce, G. F. (1994). In vitro evolution of nucleic acids. *Current Opinion in Structural Biology*, *4*, 331–336.

Karousis, E. D., Nasif, S., & Muhlemann, O. (2016). Nonsense-mediated mRNA decay: Novel mechanistic insights and biological impact. *Wiley Interdisciplinary Reviews RNA*, 7(5), 661–682.

Klukowski, P., Walczak, M. J., Gonczarek, A., Boudet, J., & Wider, G. (2015). Computer vision-based automated peak picking applied to protein NMR spectra. *Bioinformatics*, *31*(18), 2981–2988.

Krahenbuhl, B., Lukavsky, P., & Wider, G. (2014). Strategy for automated NMR resonance assignment of RNA: Application to 48-nucleotide K10. *Journal of Biomolecular NMR*, *59*(4), 231–240.

Krepl, M., Blatter, M., Cléry, A., Damberger, F. F., Allain, F. H. T., & Sponer, J. (2017). Structural study of the Fox-1 RRM protein hydration reveals a role for key water molecules in RRM-RNA recognition. *Nucleic Acids Research*, *45*(13), 8046–8063.

Krepl, M., Cléry, A., Blatter, M., Allain, F. H. T., & Sponer, J. (2016). Synergy between NMR measurements and MD simulations of protein/RNA complexes: Application to the RRMs, the most common RNA recognition motifs. *Nucleic Acids Research*, *44*(13), 6452–6470.

Krepl, M., Havrila, M., Stadlbauer, P., Banas, P., Otyepka, M., Pasulka, J., et al. (2015). Can we execute stable microsecond-scale atomistic simulations of protein-RNA complexes? *Journal of Chemical Theory and Computation*, *11*(3), 1220–1243.

Krepl, M., Vögele, J., Kruse, H., Duchardt-Ferner, E., Wöhnert, J., & Sponer, J. (2018). An intricate balance of hydrogen bonding, ion atmosphere and dynamics facilitates a seamless uracil to cytosine substitution in the U-turn of the neomycin-sensing riboswitch. *Nucleic Acids Research*, *46*(13), 6528–6543. gky490-gky490.

Lapinaite, A., Simon, B., Skjaerven, L., Rakwalska-Bange, M., Gabel, F., & Carlomagno, T. (2013). The structure of the box C/D enzyme reveals regulation of RNA methylation. *Nature*, *502*(7472), 519–523.

Laskowski, R. A., Rullmannn, J. A., MacArthur, M. W., Kaptein, R., & Thornton, J. M. (1996). AQUA and PROCHECK-NMR: Programs for checking the quality of protein structures solved by NMR. *Journal of Biomolecular NMR*, *8*(4), 477–486.

Lee, F. C. Y., & Ule, J. (2018). Advances in CLIP technologies for studies of protein-RNA interactions. *Molecular Cell*, *69*(3), 354–369.

Licatalosi, D. D., & Darnell, R. B. (2010). RNA processing and its regulation: Global insights into biological networks. *Nature Reviews. Genetics*, *11*(1), 75–87.

Linge, J. P., Habeck, M., Rieping, W., & Nilges, M. (2003). ARIA: Automated NOE assignment and NMR structure calculation. *Bioinformatics*, *19*(2), 315–316.

Loughlin, F. E., Gebert, L. F., Towbin, H., Brunschweiger, A., Hall, J., & Allain, F. H. (2011). Structural basis of pre-let-7 miRNA recognition by the zinc knuckles of pluripotency factor Lin28. *Nature Structural & Molecular Biology*, *19*(1), 84–89.

Mackerell, A. D., Jr., & Nilsson, L. (2008). Molecular dynamics simulations of nucleic acid-protein complexes. *Current Opinion in Structural Biology*, *18*(2), 194–199.

Mackereth, C. D., Madl, T., Bonnal, S., Simon, B., Zanier, K., Gasch, A., et al. (2011). Multi-domain conformational selection underlies pre-mRNA splicing regulation by U2AF. *Nature*, *475*(7356), 408–411.

Maier, J. A., Martinez, C., Kasavajhala, K., Wickstrom, L., Hauser, K., & Simmerling, C. (2015). ff14SB: Improving the accuracy of protein side chain and backbone parameters from ff99SB. *Journal of Chemical Theory and Computation*, *11*(8), 3696–3713.

Marchanka, A., Simon, B., Althoff-Ospelt, G., & Carlomagno, T. (2015). RNA structure determination by solid-state NMR spectroscopy. *Nature Communications*, *6*, 7024.

Maris, C., Dominguez, C., & Allain, F. H. (2005). The RNA recognition motif, a plastic RNA-binding platform to regulate post-transcriptional gene expression. *The FEBS Journal*, *272*(9), 2118–2131.

Markin, C. J., & Spyracopoulos, L. (2012). Increased precision for analysis of protein-ligand dissociation constants determined from chemical shift titrations. *Journal of Biomolecular NMR*, *53*(2), 125–138.

Masliah, G., Barraud, P., & Allain, F. H. (2013). RNA recognition by double-stranded RNA binding domains: A matter of shape and sequence. *Cellular and Molecular Life Sciences*, *70*(11), 1875–1895.

Masliah, G., Maris, C., Konig, S. L., Yulikov, M., Aeschimann, F., Malinowska, A. L., et al. (2018). Structural basis of siRNA recognition by TRBP double-stranded RNA binding domains. *The EMBO Journal*, *37*(6), e97089.

Nicastro, G., Taylor, I. A., & Ramos, A. (2015). KH-RNA interactions: Back in the groove. *Current Opinion in Structural Biology*, *30*, 63–70.

Oberstrass, F. C., Lee, A., Stefl, R., Janis, M., Chanfreau, G., & Allain, F. H. (2006). Shape-specific recognition in the structure of the Vts1p SAM domain with RNA. *Nature Structural & Molecular Biology*, *13*(2), 160–167.

Price, S. R., Ito, N., Oubridge, C., Avis, J. M., & Nagai, K. (1995). Crystallization of RNA-protein complexes. I. Methods for the large-scale preparation of RNA suitable for crystallographic studies. *Journal of Molecular Biology*, *249*(2), 398–408.

Sattler, M., Schleucher, J., & Griesinger, C. (1999). Heteronuclear multidimensional NMR experiments for the structure determination of proteins in solution employing pulsed field gradients. *Progress in Nuclear Magnetic Resonance Spectroscopy*, *34*, 93–158.

Scaringe, S. A., Kitchen, D., Kaiser, R. J., & Marshall, W. S. (2004). Preparation of 5'-silyl-2'-orthoester ribonucleosides for use in oligoribonucleotide synthesis. *Current Protocols in Nucleic Acid Chemistry*. Chapter 2, Unit 2.10.

Schilling, F., Warner, L. R., Gershenzon, N. I., Skinner, T. E., Sattler, M., & Glaser, S. J. (2014). Next-generation heteronuclear decoupling for high-field biomolecular NMR spectroscopy. *Angewandte Chemie (International Ed. in English)*, *53*(17), 4475–4479.

Skinner, S. P., Fogh, R. H., Boucher, W., Ragan, T. J., Mureddu, L. G., & Vuister, G. W. (2016). CcpNmr AnalysisAssign: A flexible platform for integrated NMR analysis. *Journal of Biomolecular NMR*, *66*(2), 111–124.

Šponer, J., Bussi, G., Krepl, M., Banáš, P., Bottaro, S., Cunha, R. A., et al. (2018). RNA structural dynamics as captured by molecular simulations: A comprehensive overview. *Chemical Reviews*, *118*(8), 4177–4338.

Šponer, J., Krepl, M., Banáš, P., Kührová, P., Zgarbová, M., Jurečka, P., et al. (2017). How to understand atomistic molecular dynamics simulations of RNA and protein–RNA complexes? *Wiley Interdisciplinary Reviews. RNA*, *8*(3), e1405.

Stefl, R., Oberstrass, F. C., Hood, J. L., Jourdan, M., Zimmermann, M., Skrisovska, L., et al. (2010). The solution structure of the ADAR2 dsRBM-RNA complex reveals a sequence-specific readout of the minor groove. *Cell*, *143*(2), 225–237.

Theler, D., Dominguez, C., Blatter, M., Boudet, J., & Allain, F. H. (2014). Solution structure of the YTH domain in complex with N6-methyladenosine RNA: A reader of methylated RNA. *Nucleic Acids Research*, *42*(22), 13911–13919.

Tolman, J. R., Flanagan, J. M., Kennedy, M. A., & Prestegard, J. H. (1995). Nuclear magnetic dipole interactions in field-oriented proteins: Information for structure determination in solution. *Proceedings of the National Academy of Sciences of the United States of America*, *92*(20), 9279–9283.

Tsuda, K., Kuwasako, K., Takahashi, M., Someya, T., Inoue, M., Terada, T., et al. (2009). Structural basis for the sequence-specific RNA-recognition mechanism of human CUG-BP1 RRM3. *Nucleic Acids Research*, *37*(15), 5151–5166.

Wurz, J. M., & Guntert, P. (2017). Peak picking multidimensional NMR spectra with the contour geometry based algorithm CYPICK. *Journal of Biomolecular NMR*, *67*(1), 63–76.

Wuthrich, K. (1986). *NMR of proteins and nucleic acids*. New York: John Wiley.

Zgarbova, M., Otyepka, M., Sponer, J., Mladek, A., Banas, P., Cheatham, T. E., et al. (2011). Refinement of the Cornell et al. nucleic acids force field based on reference quantum chemical calculations of glycosidic torsion profiles. *Journal of Chemical Theory and Computation*, 7(9), 2886–2902.